Supplement A
The chemistry of
double-bonded functional groups
Part 1

THE CHEMISTRY OF FUNCTIONAL GROUPS

A series of advanced treatises under the general editorship of
Professor Saul Patai

The chemistry of alkenes (published in 2 volumes)
The chemistry of the carbonyl group (published in 2 volumes)
The chemistry of the ether linkage (published)
The chemistry of the amino group (published)
The chemistry of the nitro and nitroso groups (published in 2 parts)
The chemistry of carboxylic acids and esters (published)
The chemistry of the carbon–nitrogen double bond (published)
The chemistry of amides (published)
The chemistry of the cyano group (published)
The chemistry of the hydroxyl group (published in 2 parts)
The chemistry of the azido group (published)
The chemistry of acyl halides (published)
The chemistry of the carbon–halogen bond (published in 2 parts)
The chemistry of the quinonoid compounds (published in 2 parts)
The chemistry of the thiol group (published in 2 parts)
The chemistry of the hydrazo, azo and azoxy groups (published in 2 parts)
The chemistry of amidines and imidates (published)
Supplement A: The chemistry of double-bonded functional groups
(published in two parts)

$$\diagup C=C \diagdown \qquad \diagup C=O \qquad \diagup C=N{-} \qquad {-}N=N{-}$$

Supplement A
The chemistry of
double-bonded
functional groups
Part 1

Edited by

SAUL PATAI

The Hebrew University, Jerusalem

1977

JOHN WILEY & SONS

LONDON—NEW YORK—SYDNEY—TORONTO

An Interscience ® Publication

Library of Congress Catalog Card No. 75-6913.

ISBN 0 471 99464 2 (Pt 1).
ISBN 0 471 99465 0 (Pt 2).
ISBN 0 471 66940 7 (Set).

Produced in Great Britain by Technical Filmsetters Europe Ltd,
76 Great Bridgewater Street, Manchester M1 5JY.

Printed in Great Britain by Unwin Brothers Limited, The
Gresham Press, Old Woking, Surrey.

To
my grand-daughter **Tal** and the
double-bond between us

Contributing Authors

N. R. Barot   Chemistry Department, University of Nairobi, Kenya

K. B. Becker   Institute of Organic Chemistry, University of Basel, Switzerland

G. Bianchi   Institute of Organic Chemistry, University of Pavia, Italy

N. Calderon   The Goodyear Tire & Rubber Co., Akron, Ohio, U.S.A.

A. F. Cockerill   Lilly Research Centre Ltd, Erl Wood Manor, Windlesham, Surrey, U.K.

C. De Micheli   Institute of Organic Chemistry, University of Pavia, Italy

J. A. Elvidge   Chemistry Department, University of Surrey, U.K.

O. Exner   Institute of Organic Chemistry and Biochemistry, Czechoslovak Academy of Science, Prague, Czechoslovakia

A. J. Fry   Wesleyan University, Middletown, Connecticut, U.S.A.

R. Gandolfi   Institute of Organic Chemistry, University of Pavia, Italy

D. G. Garratt   Department of Chemistry, University of Toronto, Toronto, Ontario, Canada

C. A. Grob   Institute of Organic Chemistry, University of Basel, Switzerland

R. G. Harrison   Lilly Research Centre Ltd, Erl Wood Manor, Windlesham, Surrey, U.K.

P. M Henry   Department of Chemistry, University of Guelph, Guelph, Ontario, Canada

D. E. James   University of Iowa, Iowa City, Iowa, U.S.A.

G. L. Lange   Department of Chemistry, University of Guelph, Guelph, Ontario, Canada

A. P. Marchand   Department of Chemistry, University of Oklahoma, Norman, Oklahoma, U.S.A.

R. G. Reed   The Mary Imogene Bassett Hospital, Cooperstown, New York, U.S.A.

G. H. Schmid   Department of Chemistry, University of Toronto, Toronto, Ontario, Canada

R. Shaw   Physical Sciences Division, Stanford Research
                         Institute, Menlo Park, California, U.S.A.
J. K. Stille   University of Iowa, Iowa City, Iowa, U.S.A.
J. P. Van Meter   Eastman Kodak Company, Rochester, New York,
                         U.S.A.

Foreword

Most of the originally planned volumes of the series *The Chemistry of the Functional Groups* have appeared already or are in the press. The first two books of the series, *The Chemistry of Alkenes* (1964) and *The Chemistry of the Carbonyl Group* (1966) each had a second volume published in 1970, with chapters not included in the plans of the original volumes and others which were planned but failed to materialize.

This book is the first of a set of supplementary volumes which should include material on more than a single functional group. For these volumes a division into five categories is envisaged, and supplementary volumes in each of these categories will be published as the need arises. These volumes should include 'missing chapters' as well as chapters which give a unified and comparative treatment of several related functional groups together.

The planned division is as follows:

Supplement A: The Chemistry of Double-Bonded Functional Groups ($C=C$; $C=O$; $C=N$; $N=N$ etc.).

Supplement B: The Chemistry of Acid Derivatives ($COOH$; $COOR$; $CONH_2$ etc.).

Supplement C: The Chemistry of Triple-Bonded Functional Groups ($C\equiv C$; $C\equiv N$; $-\overset{+}{N}\equiv N$ etc.).

Supplement D: The Chemistry of Halides and Pseudohalides ($-F$; $-Cl$; $-Br$; $-I$; $-N_3$; $-OCN$; $-NCO$ etc.).

Supplement E: Will include material on groups which do not fit any of the previous four categories ($-NH_2$; $-OH$; $-SH$; $-NO_2$ etc.).

In the present volume, as usual, the authors have been asked to write chapters in the nature of essay-reviews not necessarily giving extensive or encyclopaedic coverage of the material. Once more, not all planned chapters materialized, but we hope that additional volumes of Supplement A will appear, when these gaps can be filled together with coverage of new developments in the various fields treated.

Jerusalem, March 1976   SAUL PATAI

The Chemistry of Functional Groups
Preface to the series

The series 'The Chemistry of Functional Groups' is planned to cover in each volume all aspects of the chemistry of one of the important functional groups in organic chemistry. The emphasis is laid on the functional group treated and on the effects which it exerts on the chemical and physical properties, primarily in the immediate vicinity of the group in question, and secondarily on the behaviour of the whole molecule. For instance, the volume *The Chemistry of the Ether Linkage* deals with reactions in which the C—O—C group is involved, as well as with the effects of the C—O—C group on the reactions of alkyl or aryl groups connected to the ether oxygen. It is the purpose of the volume to give a complete coverage of all properties and reactions of ethers in as far as these depend on the presence of the ether group but the primary subject matter is not the whole molecule, but the C—O—C functional group.

A further restriction in the treatment of the various functional groups in these volumes is that material included in easily and generally available secondary or tertiary sources, such as Chemical Reviews, Quarterly Reviews, Organic Reactions, various 'Advances' and 'Progress' series as well as textbooks (i.e. in books which are usually found in the chemical libraries of universities and research institutes) should not, as a rule, be repeated in detail, unless it is necessary for the balanced treatment of the subject. Therefore each of the authors is asked *not* to give an encyclopaedic coverage of his subject, but to concentrate on the most important recent developments and mainly on material that has not been adequately covered by reviews or other secondary sources by the time of writing of the chapter, and to address himself to a reader who is assumed to be at a fairly advanced post-graduate level.

With these restrictions, it is realized that no plan can be devised for a volume that would give a *complete* coverage of the subject with *no* overlap between chapters, while at the same time preserving the readability of the text. The Editor set himself the goal of attaining *reasonable* coverage with *moderate* overlap, with a minimum of cross-references between the chapters of each volume. In this manner, sufficient freedom is given to each author to produce readable quasi-monographic chapters.

The general plan of each volume includes the following main sections:

(a) An introductory chapter dealing with the general and theoretical aspects of the group.

(b) One or more chapters dealing with the formation of the functional group in question, either from groups present in the molecule, or by introducing the new group directly or indirectly.

(c) Chapters describing the characterization and characteristics of the functional groups, i.e. a chapter dealing with qualitative and quantitative methods of determination including chemical and physical methods, ultraviolet, infrared, nuclear magnetic resonance and mass spectra: a chapter dealing with activating and directive effects exerted by the group and/or a chapter on the basicity, acidity or complex-forming ability of the group (if applicable).

(d) Chapters on the reactions, transformations and rearrangements which the functional group can undergo, either alone or in conjunction with other reagents.

(e) Special topics which do not fit any of the above sections, such as photochemistry, radiation chemistry, biochemical formations and reactions. Depending on the nature of each functional group treated, these special topics may include short monographs on related functional groups on which no separate volume is planned (e.g. a chapter on 'Thioketones' is included in the volume *The Chemistry of the Carbonyl Group*, and a chapter on 'Ketenes' is included in the volume *The Chemistry of Alkenes*). In other cases, certain compounds, though containing only the functional group of the title, may have special features so as to be best treated in a separate chapter, as e.g. 'Polyethers' in *The Chemistry of the Ether Linkage*, or 'Tetraaminoethylenes' in *The Chemistry of the Amino Group*.

This plan entails that the breadth, depth and thought-provoking nature of each chapter will differ with the views and inclinations of the author and the presentation will necessarily be somewhat uneven. Moreover, a serious problem is caused by authors who deliver their manuscript late or not at all. In order to overcome this problem at least to some extent, it was decided to publish certain volumes in several parts, without giving consideration to the originally planned logical order of the chapters. If after the appearance of the originally planned parts of a volume it is found that either owing to non-delivery of chapters, or to new developments in the subject, sufficient material has accumulated for publication of a supplementary volume, containing material on related functional groups, this will be done as soon as possible.

The overall plan of the volumes in the series 'The Chemistry of Functional Groups' includes the titles listed below:

The Chemistry of Alkenes (published in two volumes)
The Chemistry of the Carbonyl Group (published in two volumes)
The Chemistry of the Ether Linkage (published)
The Chemistry of the Amino Group (published)
The Chemistry of the Nitro and the Nitroso Group (published in two parts)
The Chemistry of Carboxylic Acids and Esters (published)
The Chemistry of the Carbon–Nitrogen Double Bond (published)
The Chemistry of the Cyano Group (published)
The Chemistry of Amides (published)
The Chemistry of the Hydroxyl Group (published in two parts)
The Chemistry of the Azido Group (published)
The Chemistry of Acyl Halides (published)
The Chemistry of the Carbon–Halogen Bond (published in two parts)
The Chemistry of the Quinonoid Compounds (published in two parts)
The Chemistry of the Thiol Group (published in two parts)
The Chemistry of the Carbon–Carbon Triple Bond (in preparation)
The Chemistry of Amidines and Imidates (published)
The Chemistry of the Hydrazo, Azo and Azoxy Groups (published)
The Chemistry of the Cyanates and their Thio-derivatives (in press)
The Chemistry of the Diazonium and Diazo Groups (in press)
The Chemistry of Cumulenes and Heterocumulenes
Supplement A: The Chemistry of Double-Bonded Functional Groups (published in two parts)
Supplement B: The Chemistry of Acid Derivatives (in preparation)
Supplement C: The Chemistry of Triple-Bonded Functional Groups
Supplement D: The Chemistry of Halides and Pseudo-halides
Supplement E: Other Functional Groups

Advice or criticism regarding the plan and execution of this series will be welcomed by the Editor.

The publication of this series would never have started, let alone continued, without the support of many persons. First and foremost among these is Dr. Arnold Weissberger, whose reassurance and trust encouraged me to tackle this task, and who continues to help and advise me. The efficient and patient cooperation of several staff-members of the Publisher also rendered me invaluable aid (but unfortunately their code of ethics does not allow me to thank them by name). Many of my friends and colleagues in Israel and overseas helped me in the solution of various major and minor matters, and my thanks are due to all of them, especially to Professor Z. Rappoport. Carrying out such a long-range project would be quite impossible without the non-professional but none the less essential participation and partnership of my wife.

The Hebrew University,   SAUL PATAI
Jerusalem, ISRAEL

Contents

CHAPTER **1**

Dipole moments, configurations and conformations of molecules containing X═Y groups

OTTO EXNER

Institute of Organic Chemistry and Biochemistry,
Czechoslovak Academy of Sciences, Prague, Czechoslovakia

I. INTRODUCTION

In this chapter results of stereochemical studies are reviewed which have been obtained essentially by a single approach, namely measurements of dipole moments, almost always in solution. This delimitation includes the rare studies based on the Kerr constant since determination of this constant is always accompanied by a dipole moment measurement whose value is necessary for the calculation. Most frequently the Kerr constant is used to supplement and/or refine the results from the dipole moment approach; information obtained from the two methods is essentially of the same kind. On the other hand the results from the n.m.r. spectroscopy are different in character and often complementary. This method is being used most widely for stereochemical problems; the various applications have been reviewed recently in general[1] as well as with particular regard to the configuration on C=C and C=N bonds[2]. These results will be referred to only as far as necessary in the context; the same applies to the other methods such as X-ray, electron diffraction, ultrasonic relaxation, microwave and i.r. spectroscopy etc.

The stereochemistry of the X=Y bonds has not been reviewed very frequently. Apart from a review on the C=N bond in Czech[3], short items on the C=C bond[4] and C=N bond[5] appeared in the previous volumes of this Series; the N=N bond is treated more thoroughly[6]. Some of the material included in this chapter, concerning the stereochemistry of partial double bonds, belongs strictly to other volumes of this series, in particular into the chemistry of carboxylic acids and esters[7], amides[8] and imidates[9]. It is, however, useful to treat all these compounds in one place in a comparative manner. This is justified since the stereochemical aspects were treated in the companion volumes[7,8] only very briefly if at all. Recent reviews on particular classes of compounds with a partial double bond are available, viz. on esters[10] and amides[11].

II. STEREOCHEMISTRY OF MOLECULES CONTAINING X=Y GROUPS

A. Classical and Partial Double Bonds

The presence of a double bond generally has profound consequences for the stereochemistry; it gives the molecule more rigidity and restricts the number of possible forms. The main feature is the coplanarity of all atoms adjacent to a double bond as well as of the double-bonded atoms

themselves. In this way *E,Z*-isomerism appears in the case of molecules
substituted unsymmetrically on each end of the double bond; this possibility exists for C=C, C=N and N=N bonds, and exceptionally also
for the C=S bond.

In addition the presence of a double bond also affects the rotational
barriers of the adjacent single bonds and the conformation on them.
This effect is most pronounced when these formally single bonds are
directly conjugated and thus acquire a partial double-bond character.
In this way two possibilities were presented as to the systemization of this
Chapter. These partial double bonds could either be included into the
term 'double bond' and treated separately in a particular section, or the
conformation on such bonds may be considered in the stereochemistry
of the adjacent classical double bond, e.g. the conformation of enol ethers

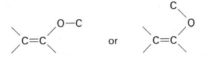

may be treated either in a special section devoted to the partial double
bond C⋯O or included in a treatment of the C=C double bond under a
subheading 'conformation on the adjacent bonds'. We have chosen the
first possibility and will deal with configurations on the double bonds
C=C, C=N and N=N, then with conformations on partial double
bonds C⋯C, C⋯N, C⋯O etc. The distinction between the two types is
not quite sharp and the concept of a partial double bond will neither be
explained in detail nor criticized. In a purely formal manner the simplest
mesomeric formula without charge separation will be decisive for classification purposes, e.g. in amides

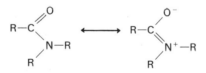

the C=O bond is double, while the C—N bond is partially double;
in imidates the opposite is true.

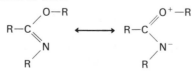

The above classification fits—with few exceptions[12]—the statement that
the double bond gives rise to stable geometrical isomers, isolable under

suitable conditions, while the partial double bond is manifested only by restricted rotation and more or less populated conformers. Of course the distinction between configuration and conformation is also indistinct. In fact a continuous scale of rotational barriers exists, decreasing from those typical for the double bonds down to the values typical for single bonds. In this chapter even bonds of a slight double bond character are included insofar as these are clearly reflected in the stereochemistry.

B. Concerning the Concept of Conformation

Conformational studies carried out by different experimental approaches yield results which differ in character; in addition even the terminology differs somewhat in studies with prevailing physicochemical or organic orientation. This may cause, and has already caused, some confusion.

Let us consider the simplest case when the conformation is determined by a single dihedral angle τ. A continuous range of rotamers exists, with τ from 0° to 360°, of which all are present in a real sample under specified conditions. However, their relative abundance may be very different since it is controlled by their free enthalpy ΔG. The exact physicochemical solution of the problem involves determining ΔG as a function of τ, like the upper curve in Figure 1. The abundance of individual forms then follows simply from the Boltzmann law and is represented by the lower curve in Figure 1. Such a solution has been achieved in some cases and is always approximate or even tentative. Mostly, the lower parts of the curve and/or the heights of the peaks are determined by experiment. In calculations ΔG is often approximated by ΔH or even by the potential energy E_p.

In organic chemistry two approximations are commonly used. Firstly, the less abundant forms, representing together a small percentage of the total, are simply neglected; i.e. the upper part of the curve is cut off (in Figure 1 at the value of 2·5 kcal relative to the lowest energy value). Secondly, the conformations with similar values of τ are counted together as a single conformation, e.g. in Figure 1 the conformations with $\tau = -15°$ to 15° are denoted as E, with $\tau = 156°$ to 194° as Z. Their abundance, x, is generally given by the Boltzmann distribution:

$$x = \int_{\tau_1}^{\tau_2} \exp\left(-\Delta G/RT\right) d\tau \Big/ \int_0^{2\pi} \exp\left(-\Delta G/RT\right) d\tau$$

In the case under consideration the E and Z forms as defined represent 9% and 89%, respectively, of the molecules present†.

---

† When dealing with partial double bonds, we prefer the symbols E and Z. In another system of symbols[13], sp, ap, sc, ac, the range of possible values of τ has been explicitly defined, e.g. synperiplanar (sp) means $\tau = -30°$ to 30°.

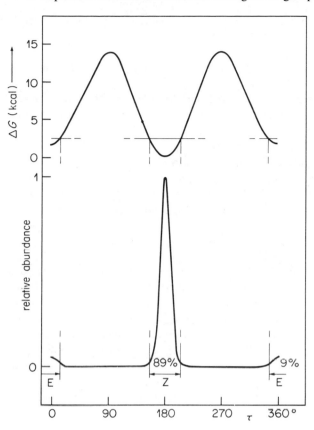

FIGURE 1. The free energy function and population of rotamers (corresponds approximately to a secondary amide).

The statements that a compound exists 'only' or 'prevailingly' in a certain conformation must be understood always in the sense just described, otherwise they would be inacceptable from the physicochemical point of view. Since the dipole moment method is of relatively low accuracy, it yields only results of this type. However, the approximations involved may sometimes become apparent even with inaccurate methods. In particular the mean effective value of a physical quantity for a certain range of near conformations need not correspond exactly to the value for the middle of the interval. If a quantity y is additive in character† and is

---

†Additive is e.g. the molar Kerr constant $_mK$; in the case of dipole moment it is not directly μ but its square, μ^2.

a function of the dihedral angle τ, its effective value $\bar{y}$ in the interval between τ_1 and τ_2 is given by

$$\bar{y} = \int_{\tau_1}^{\tau_2} y \exp(-\Delta G/RT) \, d\tau \bigg/ \int_{\tau_1}^{\tau_2} \exp(-\Delta G/RT) \, d\tau$$

and is generally not identical with y corresponding to the middle of the interval $(\tau_1 + \tau_2)/2$. It is particularly striking when y is at minimum at this point, see e.g. Section V.B.1.

In some cases the curve $\Delta G = f(\tau)$ may be very flat, or, alternatively it may exhibit several minima of almost the same depth, situated at regular intervals. In such cases it may be of value to calculate the effective quantity $\bar{y}$ under the assumption of 'free rotation', i.e. independency of ΔG of τ. Such a mean value is relatively easily obtained as

$$\bar{y} = \int_0^{2\pi} y \, d\tau/2\pi$$

In the case of dipole moment, relatively simple formulae are deduced by substituting μ^2 for y, which will not be reproduced here (see Reference 14).

The values calculated for 'free rotation' often agree quite well with experiment. To avoid any misunderstanding one must always bear in mind that 'free rotation' actually does not exist; the term denotes the kind of calculation rather than any physical reality. Reasonable agreement with experimental values may be achieved even with considerable rotational barriers. The probable interpretation of such cases is that several conformations are almost equivalent (e.g., all staggered conformations).

III. THE DIPOLE MOMENT APPROACH

A. Some Specific Features of the Method

The dipole moment of a molecule† (μ) characterizes the gross charge separation by a vector, oriented from the centre of gravity of the positive charges to the centre of gravity of the negative charges. By most experimental methods, in particular by common solution measurements, only the absolute value of the vector is obtained (still commonly expressed in the units $1 D = 10^{-18}$ e.s.u.), while its direction remains more or less hypothetical. However, the vector character of this quantity must be never forgotten when calculating the total dipole moment from its components, comparing experimental and predicted values, etc. Only when

† See Refs. 14 and 15 for the theory, experimental technique and application of dipole moments in organic chemistry.

determined from the Stark effect in the microwave spectra, is the direction of the dipole moment obtained directly. However, these data are of less importance for stereochemical problems since microwave spectroscopy yields, as a rule, all the necessary information immediately.

The structure problems are in principle solved by comparing experimental dipole moments with those calculated, or qualitatively predicted for individual structures. When compared with e.g. spectroscopic methods, the amount of information involved in one number is very small. In addition, this one experimental value is not very accurate; the inaccuracy is due not only to the experimental error but also to the inherent approximate character of the theory[14,15]. Even so, the total combined error is usually smaller than the discrepancy between calculated and found values, and will not be analysed here. Only the unique error distribution is worth mentioning: The smallest values of μ are determined with the largest absolute error, so that they often cannot be distinguished from zero. In practice this unfavourable situation may often be avoided, e.g. the dipole moment is enhanced by suitable substitution.

Summarizing the characteristic features of the dipole moment, we can deduce that this quantity is most suitable for merely qualitative problems, i.e. decision between two or, say, four possibilities. This is just in the case of configurations on the double bond. When the number of forms involved increases, or when their abundance is to be determined quantitatively, the small amount of information and limited accuracy of the method make themselves felt. Complex problems cannot be solved partially or stepwise, since all the structural features must be determined at once. These inherent restrictions of the method are counterbalanced by several advantages. First of all it is a simple concept which permits the solving of many problems without theoretical considerations and without model compounds; only the structural formulae and some universal empirical constants are necessary. Another advantage is work in solution in non-polar solvents (as compared to X-ray and m.w. spectroscopy) and the possibility of studying stable and unstable forms by the same method.

B. Application to Stereochemical Problems

The given problem, to which the dipole moment is applied, may acquire various forms according to the stability of isomers, their abundance, the possibility of isolation, etc. In particular the following tasks may be encountered:

(a) Assignment of configuration to two stable, isolated isomers, e.g. E and Z dichloroethylenes (1)

(b) Assignment of configuration to one stable isomer when the isolation of the second failed, e.g. 1,2-dinitro-1,2-diphenylethylene (**2**).

(c) Estimation of population of two or more isomers in an equilibrium solution, e.g. conformations of dimethyl fumarate (**12**).

(d) Determination of the prevailing conformation with one variable angle, e.g. the dihedral angle $C-C-C-C$ in diacetyl (**89c**).

(e) More complex patterns, as two variable angles or configuration combined with conformation, e.g. in imidoates (**28**) configuration on $C=N$ and conformation on $C-O$.

Although the tasks are successively more difficult, the approach itself is not principally different. Calculation of predicted dipole moments and comparison with the experimental value may be used in each case. In addition, there are simpler procedures, based on symmetry and other qualitative considerations, which are applicable only to simpler problems. In more complex problems these qualitative considerations are applied at the start, to obtain approximate solution, eliminate some structures, etc. The available methods and their combinations have been discussed in some detail[14]; summarizing, the following possibilities may be distinguished:

(a) Symmetry considerations. A non-zero dipole moment may occur only in symmetry groups, C_1, C_s, C_p, and C_{pv}, i.e. among symmetry elements an inversion centre, or a rotation–reflexion axis, or more than one rotation axis, or, finally, a rotation axis perpendicular to a mirror plane all imply zero moment. In particular the presence of an inversion centre is frequent and conspicuous and may itself serve to solve many problems (e.g., see formulae **1a, b**).

(b) Qualitative predictions. When two isomers are available, it is sometimes possible to predict which of them has the larger dipole moment without calculating the values. This may happen when strongly polar groups with large moments are present (e.g., **6a, b**).

(c) Model compounds. Compounds with known sterical arrangement, which is mostly secured by ring closure, may serve as models of certain conformations, e.g. phenanthridine (**25**) as model of benzalaniline in the Z-configuration (**23b**).

(d) Vector addition of bond moments. The dipole moment for a given conformation or configuration is calculated as the vector sum of bond moments situated along the individual bonds. These are assumed to be roughly constant, or, in a more sophisticated approach, several refinements are introduced. In spite of many inherent approximations this approach is the most productive one, see e.g. **8a, b**. Several tabulations of bond moments are available[14,15].

(e) Quantum chemical calculations. An accuracy comparable with the preceding case may be achieved when calculating the dipole moment by the CNDO or INDO methods. These calculations are, however, seldom carried out for this purpose only, see e.g. 3-fluoroacetophenone (**156**).

(f) Substitution. The amount of information is significantly enhanced when a substituent is introduced into the molecule. It must have a known dipole moment and must not introduce any additional possibility of rotation. Most popular is the substitution at the *para* position of a benzene nucleus, see e.g. (**23**) and (**24**).

(g) Graphical methods. In complex problems the comparison of various calculated and experimental values is greatly facilitated by a graphical representation; in particular the reliability of results may be better estimated (see Figure 2, p. 24 as an example).

The reliability in all cases increases with the number of compounds investigated and decreases with the number of possible forms. The dipole moments yield suitable examples for application of probability calculus to estimate the reliability of results in a quantitative manner. For instance, when a configuration has been assigned, one can ask with which proba bility the result is right, or, before the experiment has been carried out, one can give the probability that the assignment will succeed. Hitherto, the approach was successful insofar as various problems may be quantitatively compared, but the probability can be hardly given an absolute value[16] (see e.g. Table 14).

C. Combination with other Physical Methods

The relatively small amount of information, obtained from the dipole moments, may be often enhanced by other physical methods. The second method may either yield essentially the same kind of information as the dipole moments, or quite another one, not obtainable by this approach.

Into the first category belongs particularly the molar Kerr constant†, $_mK$, which expresses quantitatively the birefringence caused by an electric field and is related to the anisotropic polarizability of the molecule similarly as the dipole moment is related to the asymmetric charge distribution. The merit of the Kerr constant over the dipole moment is the larger extent of possible values and greater experimental accuracy, e.g. many symmetrical molecules with zero dipole moments differ markedly in the Kerr constant, the position of alkyl groups is clearly reflected, etc. On the other hand, the experimental work and the calculations are much

---

† See Refs. 17 and 18 for the theoretical background and application of the Kerr constant in organic chemistry.

more difficult. In current applications the expected dipole moment μ is computed first for the assumed molecular geometry as outlined in the preceding section, then the value of $_mK$ from μ and from bond polariza- bility tensors, which are tabulated[17]. The two quantities, μ and $_mK$, are compared with experimental values and reasonable agreement in both must be obtained. The comparison may be facilitated by graphical representation[19].

Complementary data on the anisotropic polarizability may be obtained from light-scattering but this approach is difficult and seldom used in organic stereochemistry (see Reference 17). The third possibility is measur- ing of the Cotton–Mouton effect, i.e. the birefringence raised by the magnetic field instead of the electric one. The application to stereochemical problems is still at the exploratory stage[20].

The inherent weakness of the dipole moment method, as well as of other methods just mentioned, is their insufficient accuracy when less abundant forms are to be detected, or when the ratio of two forms is to be determined exactly. In such cases n.m.r. spectroscopy is usually the most efficient. However, while n.m.r. spectroscopy is able to solve many stereochemical problems, it may fail particularly in the case of functional groups not containing hydrogen atoms, or it may give only the ratio of forms but not the assignment. Also, an assignment based only on chemical shifts may need either a direct comparison with model compounds, or many em- pirical rules. When n.m.r. spectroscopy and dipole moments are applied together, it is logical to determine first the number of forms present and their abundance from n.m.r. spectra, then to assign the conformation on the basis of dipole moments finally the chemical shifts may be discussed in terms of conformation. In this way for example, the configuration of nitronic acid esters (**59**) was solved.

When the rotamers interconvert rapidly, their mixture may behave either as two compounds or as a single compound in the n.m.r. spectro- scopy, depending on the rate of interconversion and on the character- istics of spectra and apparatus, including temperature. From this point of view, dipole moments are always a 'slow' method; whether the isomers interconvert or not, only the weighted average value is obtained. When the conversion is fast, both methods behave in the same manner and the n.m.r. spectrum is of no assistance.

Another available method is i.r. spectroscopy. However, neither the heights of peaks nor the integrated intensities allow direct calculation of the ratio of the pertinent forms. Sometimes even care must be taken in interpreting doubled maxima as due to presence of rotamers, see e.g. the problem of chloroformates (**99**) or urethanes (**103**). An indisputable

advantage of i.r. spectroscopy is that it is always a 'fast' method, i.e. the
two forms give separate signals, however fast their interconversion may be.

IV. CONFIGURATION ON THE X=Y BONDS

A. C=C Bonds

The configuration on the C=C bond is a classical problem. Although the
dipole moment method was applied at its early stage of development,
most of the configurations had already been solved at that time from the
chemical reactivity and the dipole moments confirmed only the configurations already known. Nevertheless, in the classical examples this
confirmation is so convincing and so simple that it is of fundamental
importance. Let us consider one of the first historical examples[21], E-
and Z-1,2-dichloroethylenes (1).

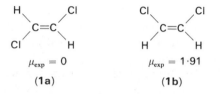

$$\mu_{exp} = 0 \qquad \mu_{exp} = 1.91$$

(1a)   (1b)

No theory, nor any comparison with other molecules is necessary:
the knowledge of structural formulae is sufficient by itself. The zero
moment of the E form follows from its C_{2h} symmetry, the calculation of
the supposed moment of the Z form is not necessary. These dipole moments
thus represented an early direct proof of the reality of structural formulae.

The example of E- and Z-1,2-difluoroethylenes is quite similar, the
values of μ are 0 and 2.42 D respectively[22]. It is characteristic that again
in this case the dipole moment was indispensable to determine configuration. Since it was obtained from microwave spectroscopy, the geometry of the molecule investigated was determined simultaneously in detail.

The assignment of configuration in the preceding examples was particularly easy since one isomer had an inversion centre and hence a zero
moment. Only one difficulty may arise, since as mentioned low dipole
moments are determined with low accuracy and sometimes may become
indistinguishable from zero. This is, of course, irrelevant when the second
isomer is available and its moment is relatively high.

Alternatively the problem may be solved by calculating the expected
dipole moment for the second unknown isomer. Dipole moments of
1,2-dinitro-1,2-diphenylethylene (2) and 1,2-bis-(4-nitrophenyl)-ethylene
(3) are approximately zero[23] and the second isomer was not available in

either case. Even so, the assignment of E configuration is quite safe, when the supposed dipole moment of the Z configuration is estimated ($c.$ 7 D in both cases).

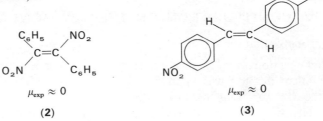

$\mu_{exp} \approx 0$

(2)

$\mu_{exp} \approx 0$

(3)

The molecule of E-4,β-dinitrostyrene (**4a**) has no other symmetry element than a mirror plane which does not exclude a dipole moment. However, the two nitro groups, which contribute most to the total moment, are arranged in practically opposite directions, giving the molecule a certain symmetrical character. Hence one may expect a very low dipole moment. Comparison of the experimental and calculated values allows the configuration to be assigned[24]. Note the difference as compared with the preceding examples: The zero dipole moments of the E forms of (**1**), (**2**) and (**3**) follow strictly from symmetry conditions, the approximate zero moment of (**4a**) is deduced from several approximations (the same moment of the two different nitro groups, no induction in the benzene nucleus, etc.).

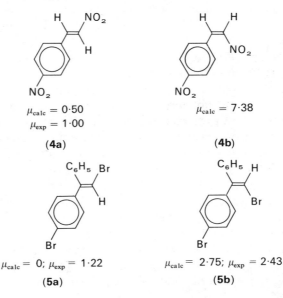

$\mu_{calc} = 0.50$
$\mu_{exp} = 1.00$

(**4a**)

$\mu_{calc} = 7.38$

(**4b**)

$\mu_{calc} = 0; \mu_{exp} = 1.22$

(**5a**)

$\mu_{calc} = 2.75; \mu_{exp} = 2.43$

(**5b**)

In 1-bromo-2-(4-bromophenyl)-2-phenylethylene (**5**) the assignment of configuration is less certain although both isomers were available. On the basis of the experimental moments the assignment was made by qualitative comparison[25]. By simple vector addition of bond moments we can calculate 0 and 2·75 D for the *E* and *Z* forms, respectively. Hence, it is very probable that the former isomer is *E*, although its dipole moment is somewhat high†.

The molecules of *E*- and *Z*-4-bromo-4′-nitrostilbene (**6**) bear two different substituents and none of the isomers possesses a centre of symmetry. However, one can safely assume that the partial moments of both substituents are oriented equally, with the negative end away from the benzene nucleus. Hence the isomer with the larger dipole moment[26] may be assigned the *Z* configuration. A more sophisticated approach is possible either by computing the expected dipole moments with proper allowance for the conjugation of both groups, or by simple comparison with the dipole moments of *para* and *ortho* bromonitrobenzenes (**7a, b**). The discrepancies may be explained in terms of induction or mesomerism in the prolonged π-bond system, steric hindrance, etc. At any rate the assignment is safe.

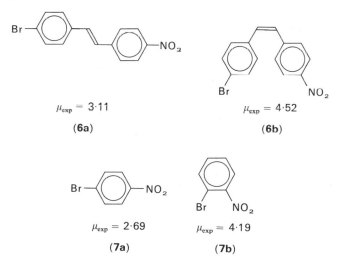

$\mu_{exp} = 3·11$

(**6a**)

$\mu_{exp} = 4·52$

(**6b**)

$\mu_{exp} = 2·69$

(**7a**)

$\mu_{exp} = 4·19$

(**7b**)

The configurations of 1-chloro-2-iodoethylenes (**8a, b**) have an interesting history. The simple plausible assumption that C—Cl and C—I

---

† The enhanced value may be partly due to the old technique[25] of extrapolation to infinite dilution, compare References 14 and 15.

moments have the same direction (actually with the negative end towards the halogen) leads to the conclusion that the Z isomer has a higher dipole moment. However, the result of the first measurement[26a] was at variance with the configurations determined from reactivities. Some rather forced theories were used to explain that the dipole moment of the E isomer could be higher and finally the configuration was questioned[27]. Since the reliability of the result was estimated[16] to the probability of 0·998, it was concluded that it can hardly be disproved by arguments based on reactivity. The controversy has been recently explained[28] by the mistaken identity of samples during the first measurement. Hence the true assignment is as follows:

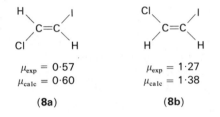

$$\mu_{exp} = 0·57 \qquad\qquad \mu_{exp} = 1·27$$
$$\mu_{calc} = 0·60 \qquad\qquad \mu_{calc} = 1·38$$

(8a)   (8b)

Another example when the substituents do not differ sharply by their polarity is represented by E- and Z-1-fluoro-1,2-dibromoethylenes (9). The original reasoning[29] was that in (9a) the two C—Br bonds cancel, resulting in a dipole moment equal to that of vinyl fluoride (1·5 D); in (9b) the C—F bond is slightly more polar than the opposite C—Br and the resulting dipole moment is slightly lower than that of vinyl bromide (1·41 D). The comparison with experimental values is not convincing. More straightforward reasoning would be that if the dipole moment of vinyl fluoride is higher than of vinyl bromide, then the moment of (9a) is higher than of (9b). The reliability of the assignment would depend only on the assumption that the difference of experimental values is real.

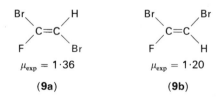

$$\mu_{exp} = 1·36 \qquad\qquad \mu_{exp} = 1·20$$

(9a)   (9b)

In all the preceding examples only substituents were involved the dipole moments of which have an unambiguous position in space, co-linear with their bond to the rest of molecule. They may be either simple linear groups (halogens, CN), or their group moments may coincide with the axis of their rotation (NO_2, CF_3). Sometimes these are denoted as 'regular' substituents. When the molecule contains also other ('angular') substituents, the problem of determining the configuration is more complex. Either the conformations on all the possible axes of rotation must be known from another source, or they must be included as additional unknowns to be determined. The number of possible forms thus increases and it is essential for the dipole moment approach that the whole structure must be determined at once. Quite often the assumption of 'free rotation' is included in the place of one possible conformation, although this concept is only hypothetical (see Section II.B). The neglecting of vector direction and possible conformations of these substituents has been criticized[30].

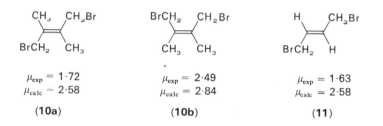

$$\mu_{exp} = 1 \cdot 72$$
$$\mu_{calc} = 2 \cdot 58$$

(10a)

$$\mu_{exp} = 2 \cdot 49$$
$$\mu_{calc} = 2 \cdot 84$$

(10b)

$$\mu_{exp} = 1 \cdot 63$$
$$\mu_{calc} = 2 \cdot 58$$

(11)

When dealing with configurations of the E- and Z-1,4-dibromo-2,3-dimethyl-2-butenes[31] (10), it would be such an error to estimate the dipole moment of the E-isomer (10a) to be zero. Actually the molecule may exist in various conformations the dipole moments of which differ strongly. Since the barrier at the C_{sp^2}—C_{sp^3} bonds is as a rule very low, we may try to apply the concept of free rotation. The calculated value is distinctly higher than the experimental one, showing that the rotation is not entirely free and conformers with nearly syn-parallel C—Br bonds are suppressed. Steric hindrance by the methyl groups may also play a role. In the original literature[31] the assignment was based simply on the belief that the Z isomer must have a higher dipole moment, however, this is not unconditionally true since the rotation could be hindered to a different extent in the two isomers. For the only isolated stereoisomer of 1,4-dibromo-2-butene (11) the dipole moment of 1·63 D was found[31]. The assignment of the E configuration should rely merely on the comparison with (10a) than on the calculation which would yield a similar discrepancy as above.

The example of dimethyl esters of fumaric (**12**) and maleic (**13**) acids is still more complex since rotation around the C—C as well as C—O bonds should be considered *a priori*. However, as shown later (Section V.B), the conformation of the ester group is remarkably uniform and rigid and the group moment has approximately the direction of the C=O bond. As to the conformation on the C—C bonds, their partial double bond character suggests the preference of planar forms. We thus have the following conformations of the fumarate:

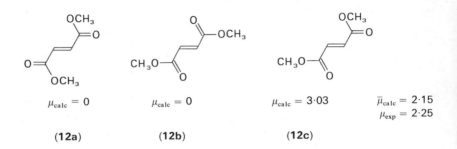

$\mu_{calc} = 0$ $\mu_{calc} = 0$ $\mu_{calc} = 3\cdot03$ $\bar{\mu}_{calc} = 2\cdot15$
                                                                                                          $\mu_{exp} = 2\cdot25$

(**12a**)   (**12b**)   (**12c**)

and of the maleate:

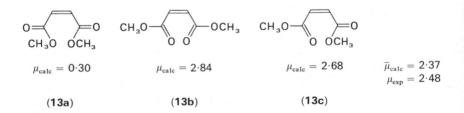

$\mu_{calc} = 0\cdot30$ $\mu_{calc} = 2\cdot84$ $\mu_{calc} = 2\cdot68$ $\bar{\mu}_{calc} = 2\cdot37$
                                                                                                          $\mu_{exp} = 2\cdot48$

(**13a**)   (**13b**)   (**13c**)

It lies in the essence of the approach that all the features of the structure must be solved simultaneously. The conformation on the partial double bonds C—C will be dealt with later (Section V.A), here only the consequences for determination of configuration are examined. When we assume that the three conformations **a**, **b**, **c** are populated in the statistical ratio 1:1:2, we get the mean values $\bar{\mu}$. The experimental values[32] are slightly higher but their ratio is reproduced quite exactly. The assumption made is in fact quite reasonable since steric hindrance in the three forms (**13a–c**) is nearly equal, while in (**12a–c**) it is absent; only an electrostatic repulsion could destabilize the form **13b**.

Exactly the same values as above would be calculated with the assumption of free rotation instead of the statistical population of planar forms. This coincidence is general in such simple cases and explains the preference for the free rotation hypothesis. Generally, the configurations may be assigned in similar symmetrically-substituted molecules as far as the assumptions of free rotation or of statistical distribution hold; under such conditions the Z isomer has always a higher dipole moment. This statement applies even to more complex molecules with several rotational axes, although in these the difference between isomers diminishes. In E- and Z-2-butene-1,4-diol (14) there are four axes in the C—C and C—O bonds, and the difference is already within the experimental error[33].

CH$_2$OH

HOCH$_2$

$\mu_{exp} = 2\cdot45$
$\mu_{calc} = 2\cdot54$

(14a)

HOCH$_2$ CH$_2$OH

$\mu_{exp} = 2\cdot48$
$\mu_{calc} = 2\cdot62$

(14b)

Unsymmetrically substituted molecules sometimes require a still more detailed analysis, when the determination of configuration is possible at all. The case of the isomeric α,β-dibromocinnamates (15) is relatively advantageous since there is only one angular substituent.

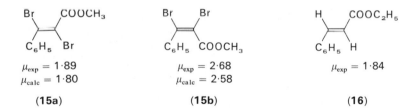

Br   COOCH$_3$

C$_6$H$_5$ Br

$\mu_{exp} = 1\cdot89$
$\mu_{calc} = 1\cdot80$

(15a)

Br   Br

C$_6$H$_5$ COOCH$_3$

$\mu_{exp} = 2\cdot68$
$\mu_{calc} = 2\cdot58$

(15b)

H   COOC$_2$H$_5$

C$_6$H$_5$ H

$\mu_{exp} = 1\cdot84$

(16)

The original assignment[23] was based on the simple assumption that in the E isomer (15a) the two C—Br moments cancel, the resulting dipole moment is essentially equal to that of an alkyl cinnamate (16). The moment of the Z isomer (15b) was assumed to be larger. The latter statement need not be a priori true and depends on the values of partial moments of C—Br and COOR. Nevertheless, our calculations agree well with the experiment.

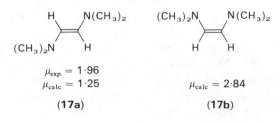

$$\mu_{exp} = 1\cdot96$$
$$\mu_{calc} = 1\cdot25 \qquad\qquad\qquad \mu_{calc} = 2\cdot84$$

(17a)   (17b)

When both isomers are not available, the problem is more delicate. The dipole moment[34] of E-bis-dimethylaminoethylene (17a) is understandable since the dimethylamino group is as a rule not planar. However, the confirmation of configuration does not seem to be possible from this one value. Recently, the group moment of 1·58 D at the angle of 34° to the C—N bond was attributed to the $N(CH_3)_2$ group bonded to the aromatic nucleus[35]. With this adopted value, we calculated the dipole moments, given at the structures, for which however there exists no convincing proof. The second isomer or a model compound would clearly be desirable.

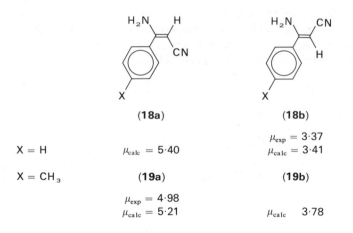

(18a)   (18b)

$$\mu_{exp} = 3\cdot37$$
X = H $\mu_{calc} = 5\cdot40$ $\mu_{calc} = 3\cdot41$

X = CH$_3$ (19a)   (19b)

$$\mu_{exp} = 4\cdot98$$
$$\mu_{calc} = 5\cdot21 \qquad\qquad\qquad \mu_{calc} \quad 3\cdot78$$

According to the hybridization on N, the geometry of the amino group may gradually change and finally become almost planar. In this way an angular substituent is turned into a 'regular' one. When computing expected dipole moments for the enaminonitriles (18) and (19), the amino group was taken as planar with respect to its conjugation with the nitrile group[36]. Remarkably enough, the configuration of the single isolated isomer is not the same in each case but depends on the substitution at a remote place. Again, the isolation of the lacking stereoisomers would be here of great importance.

Examples where the configuration could not be established by the dipole moment approach include mostly molecules with non-polar substituents or with two substituents of comparable polarity. Thus the dipole moments of ethyl cinnamate and allocinnamate are practically equal[23], 1·84 and 1·77 D, respectively. The difference, if real, is due more to different intensity of conjugation and/or induction, than to different polarities of hydrogen and phenyl. Such differences cannot at present be accounted for by a theoretical approach. On the other hand, an experimental approach is often helpful in similar cases, and is achieved by the introduction of a substituent. For evident reasons the 'regular' substituents are preferred, the most convenient site being the *para* position of the benzene nucleus. The dipole moment of benzalacetophenone[23] (**20**) does not allow the determination of its configuration, which combined with the conformation on the C—C bond demands that four planar forms should be considered:

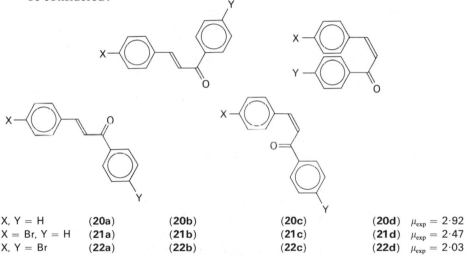

X, Y = H	(**20a**)	(**20b**)	(**20c**)	(**20d**)	$\mu_{exp} = 2\cdot92$
X = Br, Y = H	(**21a**)	(**21b**)	(**21c**)	(**21d**)	$\mu_{exp} = 2\cdot47$
X, Y = Br	(**22a**)	(**22b**)	(**22c**)	(**22d**)	$\mu_{exp} = 2\cdot03$

When the last form is excluded as sterically impossible, decision is possible by introducing a 4′-bromine substituent (**21**). For the *E* isomer, either in the form (**20a**) or (**20b**) one would expect a moderate lowering, and for the form (**20c**) a marked rise of the dipole moment. The assignment is further confirmed by the dibromo derivative (**22**). In the case of form (**c**) the dipole moment of (**22c**) would be essentially unchanged with respect to (**20c**), and in the case of form (**d**) even raised. We may conclude that the *E*

configuration has been proved while the decision between the conformations (20a) and (20b) is not possible on the basis of dipole moment
data. No quantitative calculations have been made; in the original
literature[23] the problem is presented in a rather simplified form. Conformation of similar conjugated systems will be dealt with later (Section
V.A).

Another class of compounds which brings about difficulties in the
dipole moment approach are some evidently symmetrical molecules
whose experimental moment apparently differs from zero[14]. An example
with the C=C bond is tetracyanoethylene[37] with $\mu = 0.82$ D in benzene.
The explanation is not easy and several hypotheses have been advanced,
each of which may be true in different cases. One possibility is an exceptionally high atomic polarization[38] for which the usual correction used
in solution measurements is insufficient. This means in fact that the
molecule is deformed in the electric field, and another explanation[39]
is thus approached that some unsymmetrical conformations are present
even in the absence of an electric field. Since most of these molecules
possess several electron-attracting groups and may act as strong acceptors,
a possibility of charge-transfer complexes with solvent occurs[40] when the
dipole moment would be oriented perpendicular to the molecular plane,
from the associated molecule of the solvent. This explanation is probable
for tetracyanoethylene since the dipole moment is enhanced in dioxan[37]
but in other cases no solvent dependence was observed[38]. Finally the
experimental error may contribute appreciably to the dipole moments
lower than 1 D. It can be seen that the problem is rather complex; in
the case of volatile compounds it will likely be solved by microwave
spectroscopy. The practical consequence is a certain caution needed
when dealing with molecules of the acceptor type.

The foregoing examples revealed that dipole moments have been of
fundamental importance for simple problems of configuration on the
C=C bond, while they are relatively seldom used in more complex
cases; most frequently they are combined with other methods. Nevertheless, the possibilities are not exhausted, and even some of the older
problems seem worth a reinvestigation, in particular a more detailed
calculation of the expected dipole moments.

B. C=N Bonds

In an older review it was suggested[41] that configurations on the C=N
bond are less suitable for the dipole moment approach owing to the

intrinsic C=N moment. Nevertheless, this approach has been used lately quite often. One of the reasons may be that the absence of hydrogen atoms may make the application of n.m.r. spectroscopy less efficient†. In addition the two 'substituents' on nitrogen, one of which is usually the lone electron pair, differ sufficiently from each other in their dipole moments. Of course, the value of the C=N bond moment, which may partly account also for the lone pair moment, is of primary importance. Studies confined to this problem[44,45] revealed that the value of 1·8 D is reasonably reliable and constant in various molecules. Surprisingly, it is not necessary to give any separate moment to the lone electron pair, only that part collinear with the C=N bond is implicitly involved in the formal bond moment.

1. Imines and imidates

The configuration of the only known stereoisomer of benzalaniline (23) was solved in a classical manner[46], showing nicely the importance of proper substitution. The dipole moment of the 4,4'-dichloro derivative (24) is exactly equal to that of the mother substance; this is compatible only with the E configuration (23a, 24) having antiperiplanar C—Cl bonds. This configuration is nowadays well established from X-ray data[47]. The second stereoisomer Z (23b) has remained unknown; the calculation based on bond moments suggests that its dipole moment would not be appreciably different from that of 23a. This is substantiated by comparison[48] with phenanthridine (25) which may be considered as a rigid model of (23b).

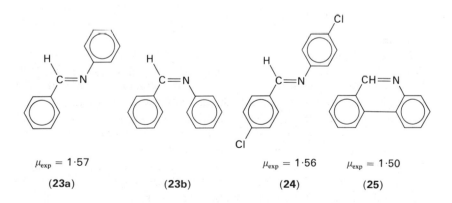

$\mu_{exp} = 1·57$ $\mu_{exp} = 1·56$ $\mu_{exp} = 1·50$

(23a)   (23b)   (24)   (25)

† Quite recently, determination of configurations on C=N were advanced based on ^{13}C shifts[42] and on [13]C–[15]N coupling constants[43].

More recent dipole moment studies of substituted benzalanilines are based on the determined configuration and deal with interaction of substituents through the conjugated chain[49], or with hydrogen bonds in 2-hydroxy and 2'-hydroxy derivatives[50,51]. The hydrogen-bonded derivatives have also been used to investigate the finer stereochemical details, like the position of the aniline ring which is not coplanar with the rest of molecule. From the dipole moment of the model compound (26) it was inferred[50] that the distortion is 30–60°.

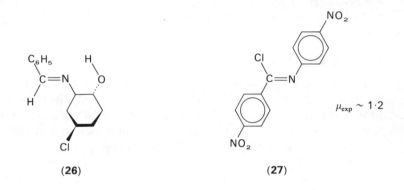

(26)   (27)

Both stereoisomers, more or less stable, are known in the case of certain N-alkylimines, and their equilibrium and interconversion has been studied. Recent studies make use mostly of n.m.r. and i.r. spectroscopy even for assignment of configuration (e.g. Reference 52) without referring to dipole moments. On the other hand, the dipole moments were interpreted in terms of substituent interaction[53], the configuration being assumed as known.

The Z-configuration (i.e., aryl or alkyl groups *trans*) of imidoyl chlorides was established using the same approach as in the preceding paragraph. The dipole moment of 4-nitro-(4'-nitrophenyl)benzimidoyl chloride (27), although not determined with a high accuracy, yields convincing proof[54] since the expected moment for the E configuration would be as high as 6 D. Interestingly, nitro substitution is not suitable in the case of benzalaniline since its 4,4'-dinitro derivative has as high dipole moment[55] as 3·59 D. This suggests that a moderate substituent, like halogen, is the best choice; too strong electron-attacking substituents may appreciably shift the electron distribution by the mesomeric mechanism.

Alternatively the Z configuration of imidoyl chlorides may be deduced from dipole moments of several derivatives with a single substituent[45].

The pertinent data in Table 1, show the importance of various derivatives and also throw light on the problem of choosing the proper substituent.

TABLE 1. Experimental and calculated dipole moments of imidoyl chlorides

Compound 4—XC₆H₄CCl=NR		μ_{exp}	μ_{calc}	μ_{calc}	$P(Z)$
X	R	(cf.[45])	Z	E	
H	C₆H₅	1·16	1·19	1·95	0·88
Cl	C₆H₅	0·25	0·41	0·62	0·59
NO₂	C₆H₅	3·34	3·31	2·68	0·89
H	C₄H₉	0·54	0·89	2·18	0·9996
Cl	C₄H₉	0·96	0·70	0·92	0·45
NO₂	C₄H₉	3·74	3·61	2·60	0·986
				Total probability	0·9999998

Table 1 as a whole reveals immediately that configuration Z is correct, but inspection of individual compounds separately yields much less convincing proof. In order to get quantitative insight, the probabilities of the Z configuration were calculated[16] assuming the normal distribution of error with the standard deviation 0·4 D. (It is the inaccuracy in calculating the expected moment which is decisive for this value, not the experimental uncertainty.) The probabilities listed in the last column of Table 1 vary strongly; note that the fourth derivative is more important than all the remaining ones. Assuming that all these compounds must possess the same configuration, either E or Z, we get an astonishing probability of 0·9999998 that it is Z. This calculation illustrates the importance of measurements on several compounds.

The configuration of imidoates (28) represents a more complex problem since the conformation on the C—O bond must be considered simultaneously, as in the case of esters (see Section V.B). Hence the four planar forms 28a–d are to be considered as most probable:

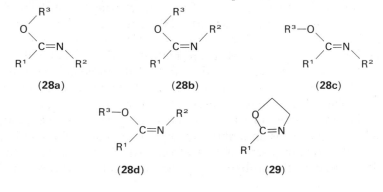

(28a)   (28b)   (28c)

(28d)   (29)

Otto Exner

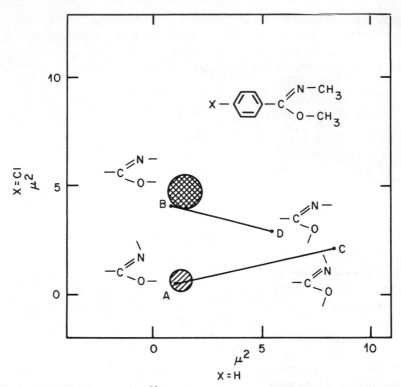

FIGURE 2. Graphical method[58] for determining conformation by comparing the calculated dipole moments (A–D) with the experiment (hatched point for the compound **28**, screened for **29**). Reproduced, with permission, from Exner and Schindler, *Helv. Chim. Acta*, **55**, 1921 (1972).

The problem was attacked by calculation of the expected dipole moments from bond moments[56], by introducing substituents in the *para* position and by graphical comparison[57], and finally by referring to the closed ring derivative[57] (**29**) as model for the form (**28b**). The results of the graphical comparison are reproduced in Figure 2 in order to show how this approach[58] works. The values of μ^2 are plotted on the x axis for the unsubstituted and on the y axis for the substituted compound. Then each form is represented by a single point. The connecting lines of two points may represent either various mixtures or transitory conformations arising by rotation around the single bond C—O. (For this reason μ^2 was plotted instead of μ, since μ^2 is an additive quantity.) The comparison with the experimental point may serve not only to choose the most probable form but also to estimate the probability of other possibilities. Figure 2 reveals in a convincing manner that (**28a**) is either the prevailing or even the only present form; the model compounds (**29**) correspond fairly to the form (**28b**). The main drawback of this approach is a little sensitivity to the presence of minority forms. The n.m.r. studies[59,60] revealed later the presence of form (**28b**) in some derivatives (not in those dealt with in Figure 2). Imidothioates have not been studied by dipole moments, and the results of n.m.r. spectroscopy[61] are questionable (compare Ref. 59).

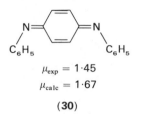

$$\mu_{exp} = 1.45$$
$$\mu_{calc} = 1.67$$

(**30**)

Benzoquinone dianil represents a simple molecule with two C=N bonds. The non-zero dipole moment[62] is compatible only with the Z configuration (**30**); the calculated moment for the centrosymmetrical E form would be of course zero.

The thiazetidine derivative (**31**), has two non-equivalent C=N bonds. The configuration of the aryl group was assigned easily[63], when a nitro group was introduced into the *para* position. The configuration of the cyclohexyl group is less certain due to the low polarity of the C—N bond but may be still considered as reasonably certain, although the difference between the two calculated values did not exceed 1 D.

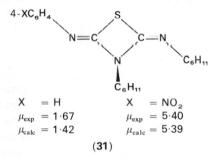

X	= H	X	= NO$_2$
μ_{exp}	= 1.67	μ_{exp}	= 5.40
μ_{calc}	= 1.42	μ_{calc}	= 5.39

(**31**)

The configurational problem of N-chloroimines (**32a, b**) was not solved in such a convincing manner, since the difference between the isomers is rather small[64]. One may assume that the polarity of the N—Cl bond is indeed small but this does not comply with the high dipole moment of the unsubstituted benzophenone-N-chloroimine[64] (2.96 D). The problem would be worth reinvestigating on a broader experimental basis.

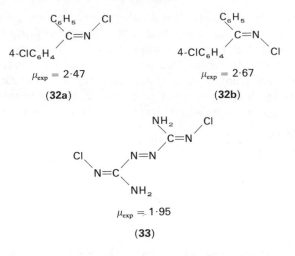

$\mu_{exp} = 2 \cdot 47$

(32a)

$\mu_{exp} = 2 \cdot 67$

(32b)

$\mu_{exp} = 1 \cdot 95$

(33)

Another chloroimine (**33**) may serve as an example of overestimating the dipole moment method[65]. Even when the NH_2 group is assumed to be planar, there are still configurations on three double bonds to be determined in addition to conformations on two C—N bonds. For this purpose, the dipole moment value of a single compound is clearly insufficient. Even the modest conclusion[65] that the configuration on N=N is *trans* must be denoted as unproven.

2. Hydroxylamine derivatives

Among hydroxylamine derivatives many examples of stereoisomerism are encountered which have been extensively studied. The configuration of simple oximes, object of much controversy in the past, is now established beyond any doubt by X-ray diffraction[66]. The application of dipole moments meets with two difficulties[67]. Firstly, oximes associate in non-polar solvents, but the extrapolation to infinite dilution seems to be generally feasible. This inconvenience may be also removed by studying e.g. the *O*-methyl derivatives instead of free oximes; these possibilities have not been sufficiently exploited. The second problem lies in the conformation on the N—O bond, which must be solved simultaneously. Owing to the appreciable conjugation involving the oxygen lone pair, the planar conformations **34a** or **34b** may be considered more probable. In fact the practically planar *E* form **34a** with H and C *trans* was found uniformly in recent X-ray[66,68] and microwave[69] studies as well as by dipole moment measurements on many simple oximes[67,70,71]. Exceptions have been found only for some more complex derivatives, where they are explained by hydrogen bridges (see formulae **37b**, **43b**).

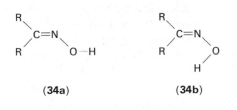

(**34a**)   (**34b**)

Even with the knowledge of the shape of the =NOH group the results were seldom such as to allow determining the configuration or confirming that one which was already known. This is understandable in the case of unsubstituted oximes, e.g. benzaldoximes[70] (**35**), but even with 4-chlorobenzaldoximes[70] (**36**) there is little difference between stereoisomers and the agreement between calculated and experimental values is at best fair; this discrepancy may be partly due to the use of different solvents for the two isomers.

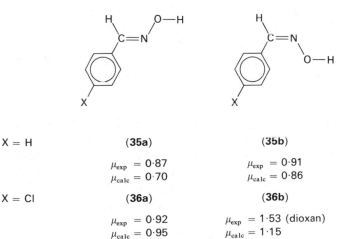

X = H   (**35a**)   (**35b**)

μ_{exp} = 0·87 μ_{exp} = 0·91
μ_{calc} = 0·70 μ_{calc} = 0·86

X = Cl   (**36a**)   (**36b**)

μ_{exp} = 0·92 μ_{exp} = 1·53 (dioxan)
μ_{calc} = 0·95 μ_{calc} = 1·15

Generally the Z aldoximes possess somewhat higher dipole moments but the difference between isomers is very low as a rule. This is caused by the unfavourable position of the group moment of the C=NOH group; its vector is nearly co-linear with the C=N bond and substitution in the two positions has almost the same effect. This is shown most clearly on the dipole moments of ketoximes, e.g. for *E*- and *Z*-4-chloro-benzophenone oximes the values do not differ significantly (2·32 and 2·38 D, respectively[72]). More significant differences could be obtained with substituents not symmetrical with respect to the C=N bond which may take different positions in the two isomers. However, these positions enter the calculations as additional unknowns, making the assignment difficult. The case of furfuraldoximes (**37**) is relatively favourable since the rotation is clearly restricted in the *Z* isomer by the hydrogen bond[67], distorting the conformation of the =NOH group (**37b**). The unequal conformation of stereoisomers and in addition the contribution from the hydrogen bond itself raise the difference between the stereoisomers.

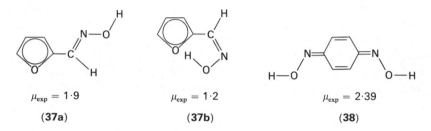

$\mu_{exp} = 1\cdot9$ $\mu_{exp} = 1\cdot2$ $\mu_{exp} = 2\cdot39$

(**37a**)   (**37b**)   (**38**)

In the stereoisomeric piperonaldoximes[73] the unsymmetrical substitution makes the difference still perceptible (1·55 and 1·75 D, respectively) but at the same time the interpretation and assignment of configuration becomes difficult. In the original literature[73] no conclusion concerning stereochemistry was drawn. Like benzoquinone dianil (**30**), benzoquinone dioxime (**38**) seems to possess the *Z* configuration. The dipole moment is clearly different from zero[74].

First steps only have been taken to exploit the Kerr constant in the stereochemistry of oximes[75]. The most efficient method at present seems to be n.m.r. spectroscopy. In addition to ¹H chemical shifts, suitable mainly for aldoximes[76], even the ^{13}C shifts[42], ¹H–¹H and ^{13}C–^{15}N coupling constants[43,77] may serve to assign configuration. The possibilities may be broadened by applying shift reagents[78] or aromatic solvents[79].

The oxime *O*-alkyl ethers could be somewhat more suitable for dipole moment studies than free oximes since association is excluded. However, they have been little studied. A detailed investigation of aliphatic derivatives[80] did not deal with the problem of configuration. The early work on 4-nitrobenzophenone oxime *O*-methyl ether revealed a sufficient difference between stereoisomers[81] (4·26 and 3·75 D for *E* and *Z*, respectively) but calculations from bond moments have not been carried out.

The *E*- and *Z*-benzaldoxime *O*-tritylethers (**39**) differ slightly in their dipole moments[82]. While their structure was deduced from the spectral properties, the configuration assignment was based simply on the relative stability of isomers and on preparation of **39a** from *E*-benzaldoxime.

$\mu_{exp} = 0.84$

(**39a**)

$\mu_{exp} = 1.23$

(**39b**)

The problem of 1,2-dioximes (glyoximes) is very complex. Three stereoisomers are possible in the case of symmetrical derivatives, four in the case of unsymmetrical ones; in addition the conformation on the central C—C bond must be dealt with and possible hydrogen bonds represent a further complicating factor. Although many dipole moments have been measured[83,84], there is no single case where the configuration has been deduced from them, or where all possible stereoisomers were clearly characterized and distinguished by dipole moments. In the old literature[83] values for the three benzil dioximes (denoted α, β, γ) are given, but they contain an error in computation[84]. A recent study[84] deals only with selected compounds and focuses attention on the conformation on C—C (probably antiperiplanar). The configurations formerly deduced mainly from chemical behaviour are nowadays correlated with n.m.r. shifts[85]. The importance of hydrogen bonds is shown by the case of 1,2-quinone-dioxime, assigned[84] the *E,Z* configuration ('amphi'). The dipole moments calculated for the chelated (**40a**) and non-chelated (**40b**) forms differ sharply.

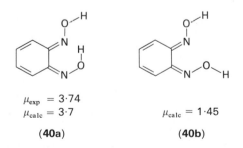

$\mu_{exp} = 3.74$
$\mu_{calc} = 3.7$

(**40a**)

$\mu_{calc} = 1.45$

(**40b**)

With the monoximes of 1,2-dicarbonyl compounds the number of stereoisomers is reduced but the problems of conformation and hydrogen

bonds persist. The configuration of benzilmonoximes (**41**) is firmly
established by an X-ray study of the *O*-(4-bromobenzoyl)derivative[86].
The dipole moments of the stereoisomers are practically equal and do not
allow safe conclusions. The antiperiplanar conformations were preferred[84] in spite of the rather poor agreement with the calculated value
(see **41a**); however, the hydrogen bonded conformation comes strongly
into consideration for the *Z* isomer (**41b**).

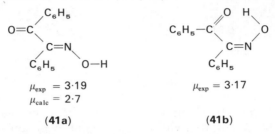

$$\mu_{exp} = 3 \cdot 19$$
$$\mu_{calc} = 2 \cdot 7$$

(**41a**)

$$\mu_{exp} = 3 \cdot 17$$

(**41b**)

In a study of 2-oximinocarboxylates[87] the configuration was already
known and the dipole moments were discussed in terms of the N—O and
C—C conformations and hydrogen bonds.

From oximes with larger substituents in the functional group, the
O-acyl derivatives received particular attention[88]. Eight planar forms are
to be considered altogether since we have to determine the configuration
on C=N (**42a** or **42c**), the conformation on N—O (**42a** ⇌ **42b**) and the
conformation on C—O (**42a** ⇌ **42d**); all possible combinations of these

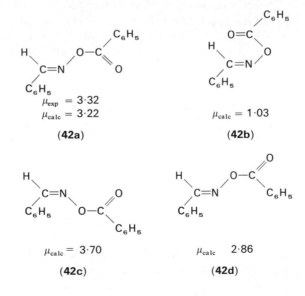

$$\mu_{exp} = 3 \cdot 32$$
$$\mu_{calc} = 3 \cdot 22$$

(**42a**)

$$\mu_{calc} = 1 \cdot 03$$

(**42b**)

$$\mu_{calc} = 3 \cdot 70$$

(**42c**)

$$\mu_{calc} \quad 2 \cdot 86$$

(**42d**)

factors yield the remaining four forms. The decision in favour of **42a** was made possible by introducing substituents into either benzene nucleus and by graphical comparison. The result is supported by the conformation of some derivatives in the crystalline state[86,89] and the partial conformations of the molecular segments agree with those of the simple model compounds: the C—O conformation with that of esters (Section V.B), the N—O conformation with oximes (see e.g. structures **35**). Such cases indicate the principal means of predicting the conformation of a complex molecule in parts (see also structure **45**).

Oximes substituted on the double bonded C atom are classified as hydroximic acid derivatives. Their stereochemistry had been obscure for a long time but was attacked successfully by several methods, in particular by dipole moments. O-Alkylhydroximic acids, e.g. (**43**), have been long known in two stereoisomeric forms but Werner's original assignment[90], based on a kind of Beckmann rearrangement, was wrong. The correct configurations were determined from the dipole moments of (**43a, b**) and their 4-nitro derivatives[91,92] and confirmed by X-ray analysis[93]. The whole story thus repeated exactly the well known history of the oxime configuration (see References 3, 94).

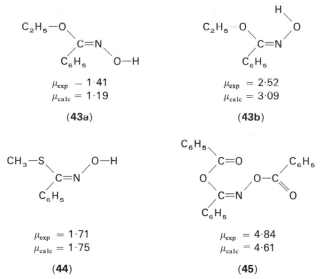

μ_{exp} — 1·41
μ_{calc} = 1·19

(**43a**)

μ_{exp} = 2·52
μ_{calc} = 3·09

(**43b**)

μ_{exp} = 1·71
μ_{calc} = 1·75

(**44**)

μ_{exp} = 4·84
μ_{calc} = 4·61

(**45**)

The solution of the problem was rather difficult since the conformations on the N—O and C—O bonds are to be solved simultaneously; the details of the latter are in fact still obscure[94]. On the other hand the fact that both stereoisomers are available was favourable. This is not the case with their isopropyl esters[95] nor with their thio analogues. For instance the compound **44** was isolated only in one form whose Z conformation was established[96] by the same approach as in the case of **43a, b**. The result agrees with the independent study of similar aliphatic derivatives[97], whose configurations were determined by the Beckmann rearrangement; the Z isomers were more stable.

The diacylderivative (**45**) called in the old literature α-tribenzoyl-hydroxylamine is of interest from the methodological point of view. It possibly represents the most complex problem[98] solved with some probability on the basis of a single dipole moment value. In addition to the configuration on C=N, conformations on the N—O bond and two C—O bonds are to be determined. The decision was possible only due to the lucky chance that the experimental value[99] agreed approximately with the highest of all computed ones. Note that the whole grouping $C=NOCOC_6H_5$ corresponds exactly in conformation to **42a**, and both COO groups are in Z conformation like in esters (Section V.B). The principle of retaining partial conformations in more complex molecules[100] is met once more.

Configurations of the compounds **43a, b** and **45** were also deduced from the n.m.r. ^{1}H shifts[101]. The assignment needed several model compounds of known configuration, hence in this case the n.m.r. approach was less conclusive than the dipole moments.

The amidoximes are also known in one stereoisomer only. In this case one can assume[92] that the two stereoisomers **46a** and **46b** cannot be isolated since they are interconverted through the tautomeric form (**47**)

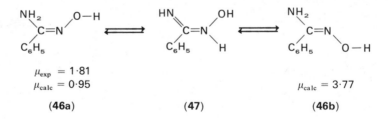

μ_{exp} = 1·81
μ_{calc} = 0·95 μ_{calc} = 3·77

(**46a**)   (**47**)   (**46b**)

The prevailing or solely present Z form (**46a**) was established[92] using the 4-nitro derivative and the graphical comparison; the result complies with the X-ray[102] and i.r.[103] investigations of formamidoxime. On the other hand the Z configuration of N,N-diethylbenzamidoxime, determined simultaneously[92], has been questioned recently, since N,N-dimethyl-acetamidoxime has the E configuration in the crystalline state[104]. Hence the configuration is either sensitive to substitution, or influenced by state of aggregation and solvent. With compounds **43–46** the alternative conformation on the N—O bond was taken into consideration, too. The experimental results agreed always with the E (antiperiplanar) conformation, as in the case of simple oximes (**34a**). The only exception in **43a**, not quite firmly established[92,94], may be ascribed to a weak hydrogen bond. In N-phenylbenzamidoximes (**48**) the problem of configuration is still combined with the conformation on C—N. While the latter is not firmly fixed in all derivatives, the former is uniformly Z as in the preceding derivatives[105].

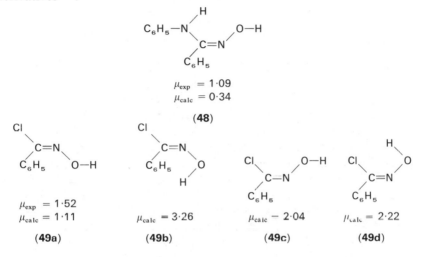

$\mu_{exp} = 1.09$
$\mu_{calc} = 0.34$

(**48**)

$\mu_{exp} = 1.52$
$\mu_{calc} = 1.11$

(**49a**)

$\mu_{calc} = 3.26$

(**49b**)

$\mu_{calc} = 2.04$

(**49c**)

$\mu_{calc} = 2.22$

(**49d**)

The configuration of hydroxamoyl chlorides, e.g. **49**, represents a problem apparently more difficult than those mentioned hitherto. The two stereoisomers have never been isolated, one case described by Werner[106] being very doubtful. As in the preceding cases the conformation on the N—O bond must be accounted for when investigating the configuration by dipole moments (**49a–d**). In this particular case, however, the former affects the resulting moments more than the latter; the situation is not much changed by introducing substituents. By chance it is only the form **49b**, *a priori* sterically improbable, which can be excluded with certainty on the basis of dipole moments. Nevertheless, the E configuration (**49a**) was deduced[92] as the most probable from the comparison with the 4-nitro derivative. However, another calculation, with slightly differing bond moments, was in favour of the Z configuration[84] (**49c**), which was also supported by n.m.r. spectroscopy. A reinvestigation[107] based on the 4-chloro derivative and several *ortho* derivatives confirmed the E configuration but as yet no agreement between the two groups of workers was reached[108]. An X-ray investigation has not yet been undertaken and the possibilities of the dipole moment approach seem to be exhausted. Recently the methyl ether of compound (49) has been prepared and the isolation of the two isomers was successful[109]. Their configurations were determined without difficulty, by the same approach as above, using the 4-nitro derivative. The relative position of vectors involved is more favourable in the methyl ether and in addition the existence of the second stereoisomer is of much assistance. The E isomer is thermodynamically stable[109], hence one may also infer that the known isomer of 49 is E.

3. Hydrazine derivatives

Regarding the stereoisomerism of hydrazones the isolation of isomers has been made difficult by their rapid interconversion and the simultaneous occurrence of polymorphic and phototropic forms; assignment of configuration may be complicated by the possible tautomeric equilibrium with the azo form. Progress in this direction has mostly been achieved by spectroscopic methods, and a recent review[110] does not even mention dipole moments among the methods for determining configuration. It is clear that such a determination would be reliable, or even possible, only in the presence of polar groups in the aldehyde (ketone) moiety. The two stereoisomers of phenylglyoxal phenylhydrazone (50) differ sufficiently in their dipole moments[111] (α-isomer 1·70 D; β-isomer 2·72 D) but no assignment has been attempted. Similarly as with the oximes (41a, b) the calculation is prevented by the unknown conformation on the C—C and N—N bonds and a possible hydrogen bond in 50b.

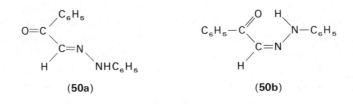

(50a)   (50b)

The stereoisomeric acetaldehyde phenylhydrazones (51) have practically equal dipole moments[111,112], and the assignment advanced[112] cannot be considered dependable. Only one isomer of benzaldehyde phenylhydrazone has been isolated and the dipole moment recorded[111,112].

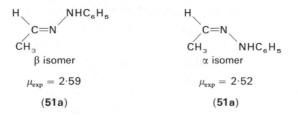

β isomer   α isomer

$\mu_{exp} = 2\cdot59$ $\mu_{exp} = 2\cdot52$

(51a)   (51a)

Understandably, studies of more complex compounds have met with no success. For glyoxal osazones (52) three possible configurations (*EE*, *ZZ*, and *EZ*), two conformations on the C—C bond (sp, ap), and three possibilities as to the conformation on N—N [(i) the four R groups on the same side of the N—C—C—N plane, (ii) two R groups above and two below this plane, (iii) an equimolecular mixture of the two preceding forms] were taken into consideration[113]. The comparison of 18 computed values with the single experimental value is evidently hopeless. Finally, the *EE* configuration was inferred from n.m.r. spectra and the ap conformation (52) preferred but not proven, from dipole moments[113]. A study of formazanes[114] had still less prospect of success so that the assignment of configurations 53a and 53b is purely tentative; details of the conformation have not been considered at all.

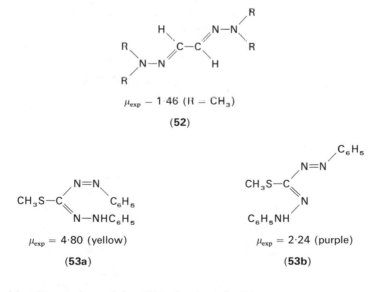

$\mu_{exp} - 1\cdot46$ (R = CH_3)

(52)

$\mu_{exp} = 4\cdot80$ (yellow) $\mu_{exp} = 2\cdot24$ (purple)

(53a)   (53b)

Evidently, much work is still to be done in this area.

4. Compounds with two substituents on nitrogen

This group of compounds includes imine N-oxides (nitrones), nitrone imines, and nitronic esters. The presence of the semi-polar bond N—O or N—N gives these molecules a strongly polar character and makes them a suitable object for dipole moment studies. Most of the work has been devoted to nitrones (see Reference 115). The assignment of configuration to the stereoisomeric N-methyl-4-nitrobenzophenone imine-N-oxides (**54**) is the classical example and one of the first achievements of the dipole moment method[116]. This example is particularly advantageous: In the presence of two strongly polar groups N^+—O^- and C—NO_2, the remaining contributions to the dipole moment value may simply be neglected and the assignment made by qualitative comparison of the two isomers. The difference between their moments is certainly convincing. In the case of nitrones (**55**) the same reasoning was followed. In this early work the systematic substitution was already exploited, the substituents 2-Cl, 3-Cl, 4-Cl, 2-CH_3, 3-CH_3, 4-CH_3 being introduced into the N-phenyl nucleus[117]. Since the graphical methods were not known, the results were evaluated only qualitatively; the dipoles C≡N and N^+—O^- are in fact so large that the qualitative pattern is not changed even by chloro substituents. Note the difference between the significance of these substituents in compounds **55** and the nitro substituent in **54**: The former were introduced to enlarge the amount of information, while the nitro group in **54** is an inherent structural feature, without which there would be no configurational problem.

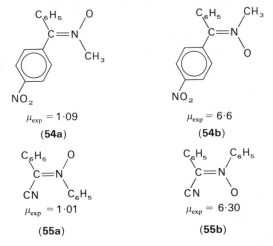

Nitrones derived from aldehydes have been known only in the Z configuration (e.g. **56**) until recently. The assignment is based on an

X-ray study of N-methyl-4-chlorobenzaldimine N-oxide[66] and on other arguments[115]; among these even the dipole moment of **56** is completely convincing[118] although the second isomer is not known. In a recent paper the dipole moments of substituted aldonitrones are discussed in terms of substituent interactions[119], the configuration being considered certain. One pair of the recently discovered[120] stereoisomers has also been investigated by the dipole moment technique[121]. The experimental difference between **57a** and **57b** is remarkably high in the absence of any polar substituent. It may be accounted for in terms of induced dipoles but it is not easily explained in full.

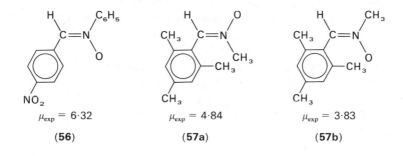

$$\mu_{exp} = 6{\cdot}32$$

(56)

$$\mu_{exp} = 4{\cdot}84$$

(57a)

$$\mu_{exp} = 3{\cdot}83$$

(57b)

The structure of aliphatic nitrones has been in some cases disproved[122] and is being reinvestigated in others[123]. From this point of view also the relatively low reported dipole moment[124] of N-benzyl-butyraldimine N-oxide (2·96 D) is understandable; the assumed structure is probably wrong.

Of the relatively new class of compounds, nitrone imines, the dipole moment of **58** was reported[125] and referred to in a discussion of the structure, while there is, of course, no problem of configuration.

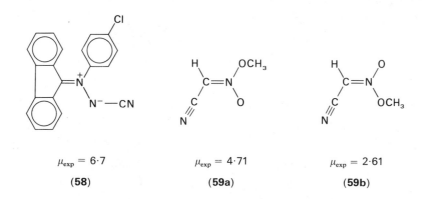

$$\mu_{exp} = 6{\cdot}7$$

(58)

$$\mu_{exp} = 4{\cdot}71$$

(59a)

$$\mu_{exp} = 2{\cdot}61$$

(59b)

In spite of their instability the isolation of isomeric nitronic esters was successful in some instances[126]. While **59a** was measured as the pure substance, the dipole moment of **59b** was calculated from measurement on an enriched mixture, the abundance of isomers being known from the n.m.r. spectra[127]. The assignment is quite safe since the unknown conformation on N—O is of little importance.

C. N=N Bonds

The stereochemistry of the N=N double bond presents many problems for the dipole moment approach, and as yet all the configurational problems cannot be viewed as definitely solved. The configuration of E and Z azobenzenes (**60**) represents a classical problem[128]. While the former possesses the C_{2h} symmetry, the latter may have a mirror plane at best, i.e. at the proper position of the benzene rings. Its dipole moment is oriented with the negative end towards the nitrogen atoms[129], as follows e.g. from the comparison with E azoxybenzene (formula **69b**) in which the total moment is the sum of the partial moments of the C—N=N—C and the N—O groups which are of the same direction. The relatively high dipole moment of **60b** was attributed simply to the lone electron pairs[129]; however, a mesomeric interaction should be also taken into account, since the effect of lone pairs is not great enough to be manifested, e.g. in the C=N compounds. Substituted azobenzenes have not been studied except for 2-, 3-, and 4-azopyridines in connection with the conformation of the pyridine ring[129]. The non-zero apparent moments found for E-azobenzene[129] and E-4,4'-dichloroazobenzene[130] are to be attributed to the atomic polarization which was not sufficiently accounted for, or even to the presence of the Z isomer. The same reason may cause the apparent dipole moment[131] of the aliphatic derivative **61**; its E configuration is supported by comparison with the model compound **62**, which has a fixed Z configuration.

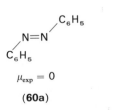

$\mu_{exp} = 0$

(60a)

$\mu_{exp} = 3 \cdot 0$

(60b)

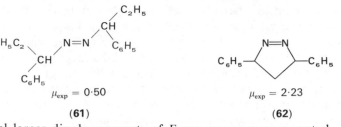

$\mu_{exp} = 0.50$ $\mu_{exp} = 2.23$

(61)   (62)

Several larger dipole moments of E azo arenes were reported, up to 1·07 for 2,2′-azonaphthalene[132], but they were explained[133] by an erroneous value of the electronic polarization. In general the assignment of configuration of azo compounds is nowadays firmly established, especially by X-ray and electron diffraction[6]. A recent dipole moment study of hydroxy azo compounds deals mainly with tautomeric equilibria[134], the E configuration being assumed in all cases.

The molecule of E-ethyl azodicarboxylate represents a more complex case. It would be formally symmetrical only if the conformations were neglected, hence the actual dipole moment of 2·58 D is understandable[32]. Neither the dipole moment value nor the Kerr constant were sufficient to prove the configuration and conformation simultaneously[32]. The free rotation hypothesis seems plausible in such cases; if it is adopted, the experimental result is compatible with the E configuration. Azodibenzoyl presents similar problems while 2,2,2-trichloroethyl azodicarboxylate[135] has two rotational axes more. These compounds were used as models to demonstrate the relatively independent conformation of moderately remote groups[135]; the E configuration was assumed as almost self-evident.

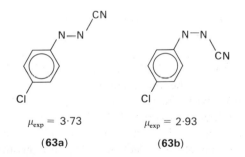

$\mu_{exp} = 3.73$ $\mu_{exp} = 2.93$

(63a)   (63b)

Unsymmetrical azo derivatives also exist in two stereoisomeric forms. For the assignment of configuration the dipole moments were decisive, problems were connected mainly with the isolation of the unstable isomer and/or with the rapid interconversion. The configuration of diazocyanides was first solved[136] for 4-substituted derivatives, e.g. **63**.

Besides its moment the substituent is important in that it makes the stereoisomers crystalline and isolable. The expected dipole moments were not calculated in that early work, the assignment being based on rather complex qualitative reasoning. In a simplified manner we may say that the dipole moment of **63a** is near to the *algebraic* difference of the $C\equiv N$ and $C-Cl$ moments, while in **63b** these two components are subtracted *vectorially* from the moment of Z-azobenzene. The configuration of 1- and 2-naphthalenazocyanides[137] and 4-biphenylazocyanides[138] was solved in a similar way, in all the compounds the *E* isomer has a greater dipole moment. From this regularity it was finally deduced that the oily specimen of benzenediazocyanide consists mainly of the *E* form[137]. The latter result is, however, not quite conclusive since there are unexplained differences up to 1·3 D among the derivatives of various hydrocarbons with corresponding configurations.

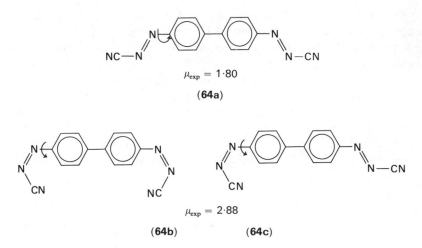

$\mu_{exp} = 1\cdot80$

(**64a**)

$\mu_{exp} = 2\cdot88$

(**64b**)   (**64c**)

The configuration of 4,4'-biphenylbisdiazocyanide (**64**) has not been completely solved[139]. The more stable isomer was assigned the *EE* configuration (**64a**) with the lowest expected moment, while the labile isomer could be either *ZZ* (**64b**) or *EZ* (**64c**). In addition to the unknown conformation on the $C-N$ bond, there is still another complicating factor: The presence of two electron attracting groups may give an apparent non-zero dipole moment, (due to the atomic polarization and/or solvation, see Section IV.A), even for symmetrical molecules. For this reason the moment of **64a** is as high as 1·80 D, although it should be exactly zero in the centrosymmetrical C_{2h} conformation and almost zero in the C_{2v} conformation (shown in **64a**). The same effect is also involved in the dipole moment of the labile isomer, in addition to the actual asymmetry; both effects can hardly be separated.

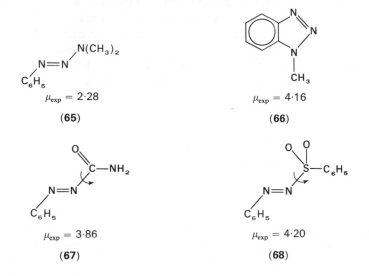

$\mu_{exp} = 2\cdot28$

(65)

$\mu_{exp} = 4\cdot16$

(66)

$\mu_{exp} = 3\cdot86$

(67)

$\mu_{exp} = 4\cdot20$

(68)

In all the cases mentioned, the stable configuration on the N=N bond was E. Hence the same configuration could be anticipated even for compounds isolated only in one form, and it was indeed found for phenyl-triazenes[140] (65), benzenediazocarboxamide[141] (67) and benzenediazo-sulphones[142] (68). However, the proof from dipole moments was more or less convincing in individual cases and had to be supported by arguments from u.v. spectroscopy. The result on phenyltriazenes is most dependable since several substituted derivatives were also studied[140]. A 4-chloro substituent enhances the dipole moment of 1-phenyl-3,3-dimethyl-triazene (65) by 1·73 D, i.e. almost by the value of the C—Cl moment, indicating that the moment of 65 is oriented approximately from N to $C_{(1)}$. This is compatible with the E configuration even when the con-figuration on $N_{(3)}$ is not planar. The result is confirmed by the comparison with 1-methylbenzotriazole (66) as a model for the Z configuration.

The E configuration of benzenediazocarboxamide (67) was simply deduced from the comparison with benzamide ($\mu = 3\cdot77$ D); for the Z isomer an approximate value of 5 D was expected, assuming free rotation around the N—C bond[141]. A study of substitution was prevented by the low solubility. The configuration of benzenediazosulphone (68) was also assigned merely tentatively[142] from the comparison of dipole moments of 68 and its 4-chloro derivative (4·45 D) with diphenylsulphone (5·09 D) and 4-chlorodiphenylsulphone (4·42 D), respectively. Calculations of the expected dipole moments, e.g. for the free rotation around the N—S bond, have not been carried out; additional arguments were taken from the u.v. spectra. The assignments in all the cases mentioned seem to be correct, but the interpretation of dipole moments would be worth a re-investigation using more detailed calculations, graphical comparison, model compounds etc. Configuration of some azo compounds was also mentioned in Section IV.B, see formulae **33** and **53a, b**.

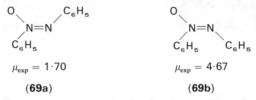

$\mu_{exp} = 1·70$ $\mu_{exp} = 4·67$

(**69a**)   (**69b**)

In the field of azoxy compounds the fundamental assignment is also based on dipole moments[143]. It is in accord with other evidence[6], in particular with the chemical correlation of Z and E azoxybenzenes (**69a, b**) with the corresponding stereoisomers of azobenzene (**60a, b**). The differences of dipole moments between stereoisomers are also similar in these two cases. The values[143] for the azoxybenzenes (**69a, b**) themselves are relatively low for compounds containing a semipolar bond; this is particularly striking for **69a** where the observed dipole moment should virtually equal the moment of this bond. The value calculated for a complete charge separation and the bond length N—O of 1·25 Å is $4·80 \times 1·25 = 6·0$ D. The much smaller experimental value may be accounted for in terms of mesomeric formulation, placing a partial negative charge on nitrogen.

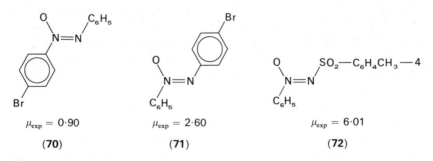

$\mu_{exp} = 0·90$ $\mu_{exp} = 2·60$ $\mu_{exp} = 6·01$

(**70**)   (**71**)   (**72**)

A recent study of azoxybenzenes[144] assumes the configuration being known and deals rather with stereochemical details; a non-exact parallel-ism of the two N—C bonds was claimed. Dipole moments were also able to distinguish between the structural isomers **70** and **71**, and to assign to both the *Z* configuration[143]. The azoxy derivative **72** represents a more complex problem. Like in many preceding examples, the expected dipole moments were not calculated; the *Z* configuration inferred from the u.v. spectra was not at variance with the dipole moment[145].

The chemistry of azodioxy compounds (diimine dioxides) is rather complex[146,147] and many misunderstandings arose in the past. These compounds—when symmetrically substituted as usual—represent the dimeric form of nitroso compounds and are in equilibrium with the monomeric form:

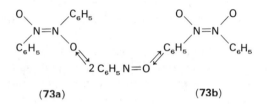

(**73a**)   (**73b**)

The interconversion of stereoisomers **73a** ⇌ **73b** is realized through the monomeric form, or even also directly[148]. In the case of aliphatic derivatives, which are more stable as dimers, an irreversible isomerization to oximes takes place in addition. The configuration of the dimeric forms have been established by a variety of methods. Generally the *E* form is more stable. Both the *E* and *Z* forms of the aliphatic derivatives have often been isolated as two different solids, while the aromatic derivatives are either *E* or monomeric in the crystalline state. The complete equilibrium in solution is less known and may depend sensitively on subtle structural features[146,147].

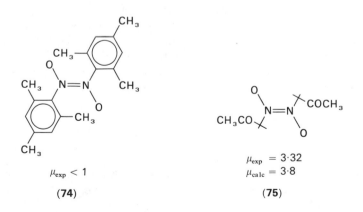

$\mu_{exp} < 1$

(**74**)

$\mu_{exp} = 3.32$
$\mu_{calc} = 3.8$

(**75**)

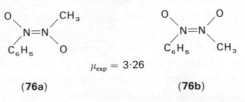

$$\mu_{exp} = 3 \cdot 26$$

(76a)   (76b)

With nitrosomesitylene the equilibrium is shifted more in favour of the dimer (74) due to the steric inhibition of conjugation in the monomer. Hence an attempt was made to determine the dipole moment of 74 from measurements on moderately dilute solutions and from the estimated polarization (or dipole moment) of the monomer[149]. However, this estimation was wrong[150]. The value claimed[149] (1·63 D) is thus certainly too high but a recalculation was not made; the E configuration (74) seems possible[150].

3-Nitroso-3-methyl-2-butanone exists as the dimer (75) even in solution. Its dipole moment was interpreted[151] as a mixture of the E form (75) and the Z form, assuming free rotation around the C—N and C—C bonds. However, the assumption and calculations are far from reliable. Hence the virtually zero dipole moment of E azomethane dioxide[150] is the only one known with certainty; the Z isomer was too unstable to be measured.

From unsymmetrical derivatives benzeneazomethane dioxide (76) and its derivatives were investigated[152]. The high dipole moment is certainly incompatible with the E configuration (76a) and was interpreted in terms of some rotational freedom around the N=N bond[152]. This explanation should be rather replaced by assuming the equilibrium 76a ⇄ 76b, as far as the structure is correct at all.

D. Other Types of Bonds

The C=S double bond may give rise to configurational problems in some rare classes of compounds. The assignment was straightforward in the case of S-oxythiobenzoyl chlorides[153] (77) since both isomers were known. The only necessary assumptions are that the C—Cl dipole is directed to Cl and the S=O dipole to O, both being self-evident. Hence the Z isomer (77b) has the larger dipole moment.

$$\mu_{exp} = 2 \cdot 63$$ $$\mu_{exp} = 3 \cdot 97$$

(77a)   (77b)

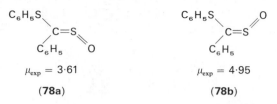

$\mu_{exp} = 3.61$ $\mu_{exp} = 4.95$

(78a)   (78b)

The second example is more complex since in **78a** and **b** the conformation on the C—S bond should be solved simultaneously. Although the solution would certainly be possible with the aid of a *para* substituted compound and after deriving the S=O bond moment from **77**, another route was preferred[154]: The C-phenyl group was replaced by mesityl and configurations were determined by n.m.r. spectroscopy from the shifts of the *ortho* methyl protons. Dipole moments served only to assign the corresponding phenyl and mesityl derivatives to each other.

The S=N bond is involved in thionylamines (**79**). Although their dipole moments[155–157] and even the Kerr constant[157] have been studied rather extensively, the only conclusion was that the N=S—O group is not linear. This follows from the non-zero moment of benzene-1,4-bisthionylamine (**80**) which is assumed to exist as a 1:1 mixture of the conformations **80a** and **80b**. Interpretations as to the configuration on the S=N bond (**79a** or **79b**) differed, and it was either inferred that an equilibrium exists[156] or that the E-form was *a priori* preferred[157].

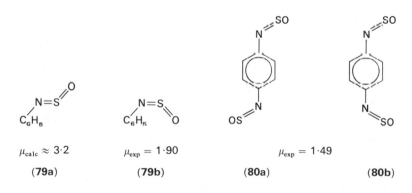

$\mu_{calc} \approx 3.2$ $\mu_{exp} = 1.90$ $\mu_{exp} = 1.49$

(79a)   (79b)   (80a)   (80b)

Recently the low dipole moment of the sterically hindered compound (**81**) has revealed that the form (**81c**) is not present[158], hence there is no equilibrium of configurations. A more detailed analysis of the component vectors (C—N, N=S, S—O and lone pairs on N and S) then suggested the Z configuration (**79b**) for all derivatives. The n.m.r. spectra supported this view by additional, less conclusive arguments[158]; there is, however, a striking disagreement both with CNDO calculations[159] and with simple estimation of the steric hindrance[157].

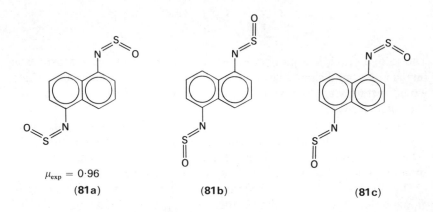

$\mu_{\text{exp}} = 0.96$

(81a)   (81b)   (81c)

The remarkable compound called originally 'azothiobenzene' (*N,N'*-diphenylsulphurdiimide, **82**), includes two N=S bonds. The dipole moments of this compound and its derivatives[130] were interpreted in terms of the assumed structure **83** and have thus not helped in determining the configuration. This was shown to be *E, Z* by an X-ray investigation[160].

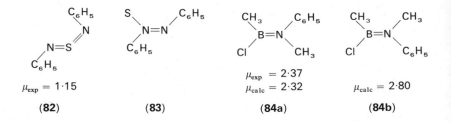

$\mu_{\text{exp}} = 1.15$ $\mu_{\text{exp}} = 2.37$
                                                              $\mu_{\text{calc}} = 2.32$ $\mu_{\text{calc}} = 2.80$

(82)   (83)   (84a)   (84b)

A rare configurational problem on the B=N double bond was encountered[161] with compounds **84a** and **b**. It was solved by calculating the expected dipole moments from contributions derived on similar symmetrical derivatives. A conventional computation from average bond moments would be less precise since the differences between *N*-alkyl and *N*-aryl bonds are rather small and uncertain. Even so, the decision seems not to be completely dependable.

V. CONFORMATION ON PARTIAL DOUBLE BONDS

A. C—C Bonds in Conjugated Systems

In this section only such C—C bonds are included which may be considered to have at least some degree of double bond character; no attempt will be made to cover all cases of a hindered rotation. More particularly, single bonds come into consideration which are situated between two real double bonds. A single bond adjacent to a triple bond cannot give rise to conformational problems; bonds adjacent to an aromatic nucleus are dealt with separately in Section V.E.

Conjugated hydrocarbons are not a suitable object for dipole moment studies since they are either non-polar or have almost negligible dipole moments. Introducing a suitable substituent is not always feasible and in many cases does not help much, owing to unfavourable symmetry conditions. The dipole moment of perchlorobutadiene is very small[162] and probably not real. If it were real, it would exclude rigorously the ap conformation† **85a**, which is exactly non-polar (symmetry C_{2h}). For the sp conformation **85b** the zero dipole moment is predicted within the bond moment scheme by the following consideration: Tetrachloroethylene is non-polar, hence the moment of one C—Cl bond equals the moment of the remainder $-CCl=CCl_2$. If one Cl atom in tetrachloroethylene is substituted by this remainder, the resulting molecule is non-polar, irrespective of the conformations. Hence the form **85b** and any other rotamer are non-polar, as far as the bond moment scheme holds exactly. If the moments of different C—Cl bonds are not exactly equal, a small dipole moment may arise. The main difficulty lies, therefore, in determining such small moments.

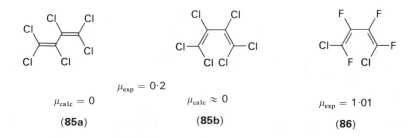

$\mu_{exp} = 0.2$

$\mu_{calc} = 0$

$\mu_{calc} \approx 0$

$\mu_{exp} = 1.01$

(85a)   (85b)   (86)

† For partial double bonds we retain the symbols E and Z in such cases where the double bond character is strongly expressed. Otherwise the terms antiperiplanar (ap) and synperiplanar (sp) are preferred[13].

The rather high moment of 1,2,3,4-tetrafluoro-1,4-dichlorobutadiene[162] (**86**) is best explained by the E, Z configuration and sp conformation as indicated; however, the result is clearly not very dependable.

In α,β-unsaturated aldehydes, ketones, or esters, the functional group accounts for a sufficient dipole moment. Even so, the second moiety is non-polar and introducing a suitable substituent may be difficult. Since the problems of conformation on C—C and configuration on C=C are always connected, the matter was already dealt with in Section IV.A. The example of benzalacetophenone (**20**) demonstrated clearly the difficulties encountered. While the configuration was essentially determined, the conformation remained unsolved. A quite similar molecule is E benzalacetone (**87**) whose dipole moment does not allow a decision between conformations (**87a**) and (**87b**); in the framework of the bond moment system the predicted dipole moments are equal. A *para* substituent does not help since it lies at the same angle to the C=O bond in both forms; an α or β substituent could itself influence the conformation. However, in this case the problem was solved[163] by referring to the model compound **88**, which must have the sp conformation for steric reasons, and has a smaller dipole moment. Hence the ap conformation (**87a**) is deduced for benzalacetone. A theoretical explanation of the different values for **87a** and **88** would require to apply the concept of mesomeric dipole moments[14,15] and to correlate the degree of conjugation with the conformation of the conjugated chain.

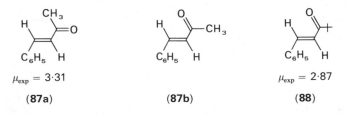

$\mu_{exp} = 3\cdot31$ $\mu_{exp} = 2\cdot87$

     (**87a**)   (**87b**)   (**88**)

The steric arrangement of dibenzalacetone involves configuration on two C=C bonds and conformation on two C—C bonds. It was solved rather tentatively from dipole moments alone, although several substituted derivatives were available[164]. Measurement of the Kerr constant[165] favoured the ap conformation on C—C and E configurations on C=C. The conformations of dialkyl fumarates (**12**) and maleates (**13**) were already mentioned in Section IV.A.

The conformation on the central bond of 1,2-dicarbonyl compounds has been the subject of much discussion. The definitely non-zero dipole moments of biacetyl, benzil, oxalyl chloride, ethyl oxalate, and many other similar compounds can in principle be accounted for by: (i) mixture of the sp and ap forms; (ii) one non-planar form; (iii) free rotation; (iv) an anomalous atomic polarization (if the observed dipole moment is not too large). In the older literature the least probable hypothesis of the free rotation was usually applied (e.g. Reference 166), although the dipole moment calculated for (i)–(iii) can be exactly the same. Combining the dipole moment with the Kerr constant seems to have decided in favour of a single conformation[167]. The dipole moments of various twisted conformations of biacetyl (**89c**) are intermediate between those for the ap (**89a**) and sp (**89b**) form. On the other hand the Kerr constant is negative in a certain range of the dihedral angles τ (dipole moment almost perpendicular to the main polarizability axis), while it is very large and positive in **89b** (dipole moment and main polarizability axis colinear) and is small and positive in **89a** (no dipole moment). The negative experimental Kerr constant favours the form **89c** with $\tau \approx 160°$ against any possible mixture of **89a** and **89b**. (In estimating τ a reasonable agreement must be achieved for μ and $_m K$ simultaneously.)

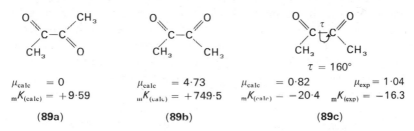

$\mu_{calc} = 0$

$_m K_{(calc)} = +9.59$

(**89a**)

$\mu_{calc} = 4.73$

$_m K_{(calc)} = +749.5$

(**89b**)

$\tau = 160°$

$\mu_{calc} = 0.82$ $\mu_{exp} = 1.04$

$_m K_{(calc)} - -20.4$ $_m K_{(exp)} = -16.3$

(**89c**)

Although this result[167] seems to be conclusive, it is opposed by the electron diffraction of biacetyl[168] and by the m.w. spectroscopy of glyoxal[169]; in both cases the pure ap form was found in agreement with theoretical calculations[170]. For oxalyl chloride a mixture of ap and ac ($\tau = 125°$) forms was claimed[171].

The problem thus seems unresolved even for biacetyl and the same uncertainty is encountered with similar classes of compounds. Diethyl oxalate is believed to exist as an almost equimolar mixture of ap and sp forms; the conclusion reached on the basis of dipole moments and Kerr constant[172] was supported by i.r. spectroscopy[173]. However, a recent reinvestigation[174] by the same approach prefers different conformations within the ester grouping† (compare Section V.B.). For aromatic

† There is a mistake in this paper[174] since the values for μ were averaged for various conformations instead of μ^2 (see Refs. 14 and 15).

α-keto-esters one non-polar form, for the aliphatic ones the mixture of rotamers was preferred[175]; the distinction was based on the i.r. spectra. The conformation of the free pyruvic acid is unique[176] and is determined by the hydrogen bond (see Section V.B., structure **98**). In conclusion, the results are not uniform within this class of compounds, and there is no agreement[168,176] even as to the double bond character of the C—C bond.

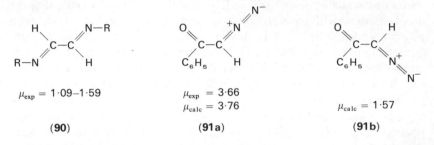

$\mu_{exp} = 1 \cdot 09\text{--}1 \cdot 59$ $\mu_{exp} = 3 \cdot 66$

                             $\mu_{calc} = 3 \cdot 76$ $\mu_{calc} = 1 \cdot 57$

       (90)   **(91a)**   **(91b)**

The conformation of 1,2-diimines was studied[177] relying on the already known E configuration on the C=N bond. A non-planar conformation was preferred, although the possibility of a mixture cannot be rejected on the basis of dipole moments alone. In the case of 1,2-dioximes[84] and osazones[113] (**52**) the C=N configuration had to be solved simultaneously; hence the problems were mentioned in Section IV.B. The same applies to monoximes of 1,2-diketones[84] (**41**) and to 2-oximinocarboxylic esters[87]. A remarkable example with a C—C bond between one C=O and one C=N bond is represented by α-diazoketones[178]. The unsubstituted derivative of acetophenone as well as most substituted ones are clearly in the sp conformation (**91a**), only the dipole moment of the 4-nitro derivative seems to correspond better to **91b**.

B. Esters and Related Compounds

1. Simple esters and carboxylic acids

The conformation of esters has been studied by a variety of methods with essentially concordant results. They are summarized in an excellent recent review[10] and will not be repeated here; attention will instead be focused on dipole moments which are given little space in that review.

The planar Z conformation (**92a**) of simple esters is quite general and remarkably stable. It has been found by X-ray analysis of crystals of many esters[179], while in the gas phase the most important results were obtained by electron diffraction and m.w. spectroscopy (see Reference 10). As to the conformation in solution the simplest and one of the most conclusive methods is based on the dipole moments.

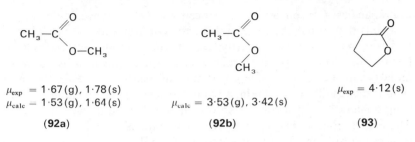

$\mu_{exp} = 1\cdot67\,(g),\ 1\cdot78\,(s)$
$\mu_{calc} = 1\cdot53\,(g),\ 1\cdot64\,(s)$

(92a)

$\mu_{calc} = 3\cdot53\,(g),\ 3\cdot42\,(s)$

(92b)

$\mu_{exp} = 4\cdot12\,(s)$

(93)

The early comparison of esters with butyrolactone[180] (**93**) as a model compound for the E conformation is quite convincing. So also is the approach using *para* substitution in the aromatic derivatives[181]. Dipole moment studies which reached accordant results are numerous, e.g. References 182–186. A recent study of pyridine derivatives is already based on the Z conformation as certain, and deals mainly with the conformation of pyridine rings[187].

No further proof of the general and prevailing Z conformation seems necessary, current interest being focused on more detailed problems: (i) possible exceptions with particular structures; (ii) possible slight deviations from planarity; (iii) detection of the E form in minute amounts; (iv) reasons for the stability of the Z form.

Claims that some esters exist prevailingly or solely in the E conformation have usually been disproved. The claim concerning liquid methyl formate[188] was contradicted by the dipole moment in solution[185]. It is true, however, that certain formates contain the E conformation in a low abundance[189–191] (see below). Further cases, chloroformates and carbamates are dealt with in Section V.B.2. No ester-like compound is thus known at present, which is stable in the E conformation although the generality of the phenomenon has been tested widely (Sections V.B.2–4).

The question arises in connection with the exact planarity of form (**92a**) at which value of the dihedral angle τ does the energy minimum lie. The answer may be quite difficult when this minimum is flat, and it will always depend on the precision of available methods. While the more exact physical methods did not reveal any deviation from planarity (except one older electron diffraction study[192] which allows some distortion), it was recognized[180,195] that the calculated dipole moments are always somewhat low. Of course the calculations use either different model compounds or mean bond moments, which are given different values by various authors (see in particular[193]); in addition a mesomeric moment[194,195] expressing the contribution from structure **94** may or may not be included. Hence not much importance can be attached to small discrepancies. Certain esters have been claimed to have higher dipole moments than usual and hence a distorted conformation[195,196]. There is, however, no structural similarity among these anomalous esters apart from a somewhat larger molecule. Electrostatic induction in the larger moiety may be also considered, if the differences are real. The most significant results were obtained from the Kerr constant, indicating for simple esters[197] a distortion from planarity by the dihedral angle of 30°. The calculation of the expected Kerr constant is complex and may be subject to criticism as is the calculation of dipole moments; nevertheless, the final agreement with experiments is good. When evaluating these results, one must consider that in a real compound a mixture of conformations is always present according to the Boltzmann distribution. The mean effective value of a measured quantity may differ significantly from its value at the energy minimum. Different methods may even yield different results. In conclusion, certain librations around the planar position must be taken into consideration in the conformation (92a), however, a value of 30° for the dihedral angle[197] seems definitely too high.

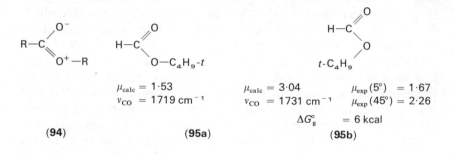

$$\mu_{calc} = 1\cdot53$$
$$\nu_{CO} = 1719 \text{ cm}^{-1}$$

(94)

$$\mu_{calc} = 3\cdot04$$
$$\nu_{CO} = 1731 \text{ cm}^{-1}$$

(95a)

$$\mu_{exp}(5°) = 1\cdot67$$
$$\mu_{exp}(45°) = 2\cdot26$$
$$\Delta G_g^\circ = 6 \text{ kcal}$$

(95b)

The detection of the minor form (92b) is a question of sensitivity of the method on the one hand, and of the free energy difference $\Delta G°$ between the two forms 92a and 92b, on the other. With respect to some methods (n.m.r.) even the activation energy $\Delta G^{\neq}$ (or $\Delta H^{\neq}$—the rotational barrier) plays some role. The latter being of the order of 10 kcal (see Reference 190), the equilibrium is fast enough; the second form can be observed only if $\Delta G°$ is not too high. The recorded values[10] of $\Delta G°$ or $\Delta H°$ differ considerably between methods and even within one method, covering the range of c. 0·5–7 kcal. Hence two forms could be observable but it seems that t-butyl formate (95) is the only ester where they have been proved beyond any doubt. The high value of $\Delta G°$ is lowered in solvents so that a double C=O band[189,191] in i.r., doubled formyl proton signal[189,190] in n.m.r. at −100°C and even a steep temperature dependence of dipole moments[189] were registered. (Note that dipole moments of common esters are remarkably independent of temperature[180].) Similar but less detailed observations are described for mesityl formate and some other formates[189,191]. With respect to the problem of chloroformates and carbamates (Section V.B.2), proofs from the i.r. spectroscopy only must be accepted critically. In the i.r. spectra of acetates and propionates[198], or in the n.m.r. spectra of longer chain esters at variable temperatures[199], no indications of the equilibrium were obtained. Considerable reserve is necessary as to the ultrasonic relaxation measurements[10,200] which claimed equilibria even for acetates with very low ΔH values[201]. The dependence of dipole moments on temperature is itself also hardly reliable owing to the low precision of the method and to the temperature dependent solvent effects. A study of nitro-substituted esters, whose conformational change would affect the dipole moment particularly markedly, did not discover any significant proof[202], nor did a study of the dielectric loss reveal any change of conformation[186].

To account for the *a priori* unexpected uniform and highly stable conformation of esters, several theories have been advanced[10,203]: (i) mesomerism within the functional group[180,190] expressed by structure **94**; (ii) electrostatic repulsion[204,205] of nearly parallel dipoles in the E form **92b**; (iii) repulsion of lone electron pairs[206,207]; (iv) purely steric hindrance in the same form[189,203]; (v) a kind of hydrogen bond[208] stabilizing the Z form **92a**. The last hypothesis can be easily disproved[10].

When discussing these individual theories, one must first differentiate which factor affects ΔG° and which $\Delta G^{\neq}$. It is commonly accepted that mesomerism within the ester group exists and is an important factor for the conformation. The existence of the contributing structure (**94**) and its approximate weight may be judged from the lengthening of the C=O and shortening of the C(O)—O bonds, which are quite general (e.g., Ref. 192, 209, 210). However, in the dipole moment values the contribution of **94** is not manifested clearly by a dipole oriented from O^+ to O^-; the result of various computations[194,203] differ according to the bond moments introduced. The importance of mesomerism was proved by comparing the dipole moments of lactones with increasing size of the ring[180,211]. If the electrostatic factors were deciding, one would expect gradual changes of conformation and a steady decrease of dipole moments along the series. Actually the dipole moments drop rather suddenly between the 8- and 9-membered cycle. This means that the Z conformation (**92a**) is always preferred; if this is not possible, the also-planar E conformation (**92b**) is still more advantageous than any non-planar one.

Hence mesomerism is the deciding factor for the $\Delta G^{\neq}$ value, since (in organic chemical terminology) it prevents the rotation around the C—O bond. To explain the value of $\Delta G°$, i.e. the preference of the Z against the E form, either one of the remaining theories must be adopted in addition, or the assumption[203] is to be made that the mesomerism energy is not equal in the two forms **92a** and **b**. The explanation by electrostatic forces has essentially the same meaning whether the attraction (in **92a**) and repulsion (in **92b**) of dipoles is considered[204], or similar interactions between the partial charges[205] on =O and (O)R. If this theory were correct, all similar compounds would exist in conformations of the minimum dipole moment. This is definitely not true for sulphonates[181], cyanoformates[209] (structure **102** in Section V.B.2) and N-cyano-amides[212] (structure **132** in Section V.C).

More promising is the idea of repulsion between the lone electron pairs. This factor was claimed to be the only one responsible for the conformation, mesomerism being unimportant[207]; the sp^3 hybridization on the ether oxygen and tetrahedral arrangement of its lone pairs was assumed. This view is clearly contradicted by the above-mentioned experiments with lactones[180,211], by the comparison of sulphonic and carboxylic esters[181] as well as by the values of bond lengths. In a more correct form[206] the hybridization on the ether oxygen is assumed to be sp^2, the lone pair lying in the O—C—O plane. The interaction of the electron clouds on the two oxygen atoms in the conformation **92b** is responsible for its destabilization compared to **92a**, mesomerism being still the main reason for the planar arrangement. In this form the theory is able to account for the conformation of carboxylic esters, nitrites[213] (**146**), vinyl ethers[206] (**112**) and may be accommodated to explain the difference showed by sulphinates[214] or thiosulphinates[215]. Nevertheless, the behaviour of some more complex compounds does not comply with this principle: The conformation of each moiety is retained as it is in simple derivatives[100] regardless of the interaction of the close electron pairs; cf. anhydrides (structure **124a**) and carbonates (structure **119a**) in Section V.B.4. The purely steric effects on the conformation have seldom been considered[189,203] although they are certainly of importance. In all esters the interaction between the two alkyls prefers the Z form **92a**, the only exception are formates[189] where the interaction between =O and R in **95a** is stronger than between H and R in **95b**. The effect is the more pronounced the larger the group R; hence t-butyl formate (**95**) is the most favourable case for observing the presence of the E form.

We conclude that the reasons for the stability of conformation **92a** are not understood in detail. In particular the possibility should be still examined that the mesomeric interaction itself depends on the conformation, i.e. on the position of the lone electron pairs; hence e.g. the double bond character of the C=O bond could be different in **92a** and **92b**.

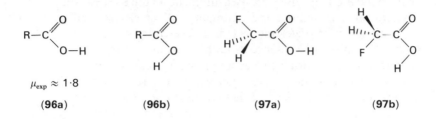

$\mu_{exp} \approx 1{\cdot}8$

(96a)   (96b)   (97a)   (97b)

The whole discussion concerning esters may be adapted to carboxylic acids with but few exceptions. The Z conformation (**96a**) was proved by a variety of methods, in solution also by dipole moments[58,216–218]. The association in non-polar solvents brings up certain problems which are usually circumvented by measuring in dioxan[58,216–218]; even in benzene the extrapolation to infinite dilution is usually possible.

The main difference against esters lies in the possibility that the hydrogen atom is engaged in a hydrogen bond, and the latter may shift the conformational equilibrium. Thus, fluoroacetic acid exists in two conformations (**97a, b**) at equilibrium[219] while in pyruvic acid only the form **98** was found[176]. Hydrogen-bonded forms were already considered[220] for several α-keto and α-alkoxy acids.

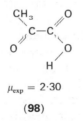

$\mu_{exp} = 2{\cdot}30$

(98)

2. α-Substituted esters

Some compounds of this group are important since they may represent possible exceptions from the general preference of the Z conformation for esters. Hence, they were repeatedly studied but the results often disagreed. In particular, α-halogenoformates have been the object of much confusion (see Reference 10). The dipole moments of alkyl chloroformates lead either to preference[221,222] of the form **99b** or to a slightly distorted[193] form of **99a**. The evidence was not conclusive since the calculated values do not differ sufficiently, owing to the relatively close values of the bond moments $C=O$ and $C-Cl$. In addition, the results of the calculations depend greatly on the bond moments used[193]. Measurements of the Kerr constants[193,197] were not able to remove the ambiguity. Older electron diffraction investigations[192,223] preferring **99a**, do not seem to be very reliable. An i.r. study[224] claimed the presence of both forms in equilibrium but the only argument was the doubled carbonyl bond. Only recently was a convincing proof presented, when microwave spectroscopy revealed **99a** as the only detectable form[225] of methyl chloroformate.

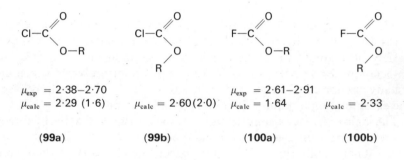

$$\mu_{exp} = 2\cdot38-2\cdot70$$
$$\mu_{calc} = 2\cdot29\ (1\cdot6)$$

$$\mu_{calc} = 2\cdot60(2\cdot0)$$

$$\mu_{exp} = 2\cdot61-2\cdot91$$
$$\mu_{calc} = 1\cdot64$$

$$\mu_{calc} = 2\cdot33$$

(99a)   **(99b)**   **(100a)**   **(100b)**

This result threw doubts on the calculation of dipole moments within this class of compounds[10]. However, deviations from the bond moment scheme occur quite often when two strongly polar bonds are linked to the same atom[14,15]; the case under consideration is particularly unfavourable since the difference between the two calculated values is small. Hence the calculation needs a more sophisticated procedure using methyl formate as the model compound. The computed values[193] for **99a** and **b** differ markedly from those obtained by a more conventional procedure[222] (in parentheses). The situation is still worse with fluoroformates for which the form **100b** was preferred on the basis of simple calculations of dipole moments[226]; this result has been questioned[10] too. The E form (**100a**) is preferred by the microwave data[209] which are, however, not quite conclusive due to the similarity of atomic masses of oxygen and fluorine.

In two points some uncertainty remains. The first one is the strong temperature dependence of dipole moments of chloroformates[221], not observed with other esters. The second point concerns aromatic esters[227], where the introduction of substituents into the *para* position enhances the difference of dipole moments between stereoisomers. The conclusions of the original paper[227], as well as our calculations by the simple bond moment scheme seem to prefer the Z form, **101b**. It is still questionable whether the calculations could be so much modified as to reverse the assignment, or whether the aromatic derivatives possess the opposite conformation.

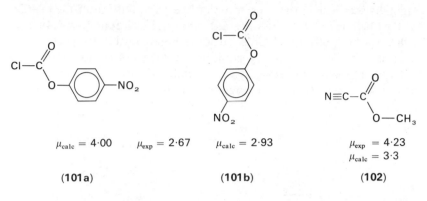

$\mu_{calc} = 4\cdot00$ $\mu_{exp} = 2\cdot67$	$\mu_{calc} = 2\cdot93$	$\mu_{exp} = 4\cdot23$ $\mu_{calc} = 3\cdot3$
(101a)	**(101b)**	**(102)**

We conclude that α-halogenoesters most probably have the same conformation as other esters, i.e. alkyl *cis* to the oxygen. It is, however, possible that the free energy difference between the two forms is lowered. If they had the reversed conformation or at least a shifted equilibrium, the α-halogenoesters would give a support to the electrostatic theories of conformation (see the preceding section). From this point of view methyl cyanoformate **(102)** is important as its conformation[209] proves the irrelevance of electrostatic effects. Note the disagreement of experimental and calculated dipole moments, as with chloroformates.

Another class of compounds about which some confusion arose are carbamates. When they are unsymmetrically substituted on nitrogen a (restricted) rotation around the C—N bond has to be considered in addition to that around the C—O bond. Hence four planar forms are possible:

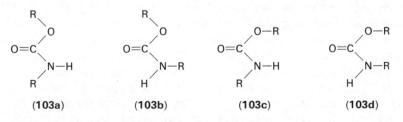

(103a)	**(103b)**	**(103c)**	**(103d)**

The conformation **103a** corresponds to that of esters in one moiety and to monosubstituted amides (see Section V.C) in the second. It may be considered as normal, while **103b** and **103c** are anomalous within the amide or ester grouping, respectively. When presence of two forms is observed, it may be difficult to decide which conformations are involved. Thus double n.m.r. signals in certain more complex carbamates were originally ascribed[228] to the ester moiety, i.e. to equilibrium **103a** ⇄ **103c**, but they belong in fact[229] to the amide moiety, i.e. to equilibrium **103a** ⇄ **103b**. More recently the double carbonyl bonds in the i.r. spectra were interpreted[230] in terms of the equilibrium **103a** ⇄ **103c** for derivatives substituted either symmetrically or unsymmetrically on nitrogen, while the system **103a** ⇄ **103b** is preferred for thiourethanes[231].

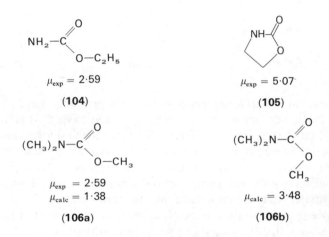

Considering the conformation of the ester grouping, the derivatives unsubstituted or symmetrically substituted on nitrogen are most instructive. The dipole moment of ethyl carbamate was explained[232] by the Z conformation (**104**), referring to the model compound **105**. The conformation **104** corresponds to that of simple esters and agrees with an X-ray investigation of more complex carbamates[233]. However, the very similar value for methyl N,N-dimethylcarbamate is believed[230] to agree with the equilibrium **106a** ⇄ **106b** in a ratio of almost 1:1. This interpretation was based mainly on the i.r. spectra and seems rather suspicious. It is also clearly at variance with the dipole moments of aromatic derivatives[234], e.g. **107**. In these compounds the presence of a strongly polar substituent made the results of the calculations much more convincing. Similar results were obtained for N-monosubstituted carbamates[234]. Also the dipole moments of dithiocarbamates comply with a conformation[235] analogous to **106a**.

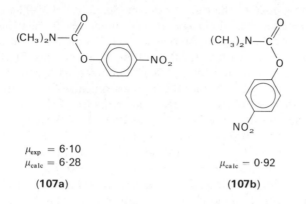

$\mu_{exp} = 6\cdot10$
$\mu_{calc} = 6\cdot28$

(107a)

$\mu_{calc} - 0\cdot92$

(107b)

It follows that the conformation of the ester group of carbamates is the same as that of other esters and their amide moiety corresponds[234,236] to that of monosubstituted amides (compare Section V.C).

3. Ester analogues

In this section compounds are included in which either the oxygen atoms of the COO group are replaced by other elements or the whole COO group is bonded to atoms other than carbon. The results in this class are less controversial; there are very few exceptions in comparison with the behaviour of esters.

The Z conformation (108, 109) has been deduced uniformly from dipole moments of aliphatic or aromatic thiol esters[182,194,203], thione esters[237,238], dithioesters[194,203], aromatic selenoesters[239] and telluroesters[240]. The substitution of oxygen by sulphur has no influence and the dipole moments of thiocarboxylic acids[218,241] revealed the same conformation as that of carboxylic acids. For thiochloroformates the Z conformation (110) was claimed[242] as for chloroformates but this

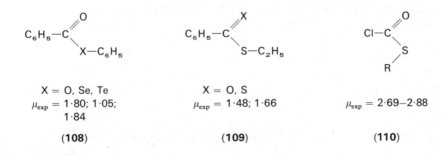

X = O, Se, Te
$\mu_{exp} = 1\cdot80; 1\cdot05;$
    $1\cdot84$

(108)

X = O, S
$\mu_{exp} = 1\cdot48; 1\cdot66$

(109)

$\mu_{exp} = 2\cdot69 - 2\cdot88$

(110)

assignment is open to the same criticism (Section V.B.2). The analogy of dithiocarbamates[235] and carbamates has already been mentioned.

If the carbonyl oxygen in esters is replaced by an $=NH$ or an $=NOH$ group, compounds with a $C=N$ double bond are produced and the problem of conformation on $C-O$ is combined with configuration on $C=N$. The former affects more the dipole moments of imidoates[56,57] and was established with certainty. The conformation of the whole molecule is represented by structure **28a** (Section IV.B.1). With hydroximic esters, on the other hand, the $C=N$ configuration has been safely assigned to the two stereoisomers[91,93] (structure **43** in Section IV.B.2) but the conformation on $C-O$ is less certain and probably less rigid. One may judge that the resonance structure **111** is less represented and the $C-O$ bond acquires less double bond character than in esters. On the contrary the mesomeric dipole moments in the functional groups $-C(=X)OR$ were claimed[194] to increase in the order $X = O, S, NH$. At any rate the behaviour of these compounds does not violate the general preference for the Z form in esters.

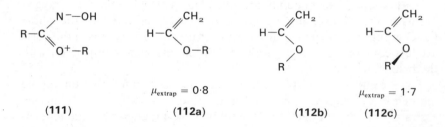

(**111**)   (**112a**)   (**112b**)   (**112c**)

$\mu_{extrap} = 0.8$ (under 112a) $\mu_{extrap} = 1.7$ (under 112c)

Finally, the stereochemistry of vinyl ethers may be mentioned in this connection since these are formally derived by replacing the $=O$ atom by the $=CH_2$ or $=CR_2$ groups. Two forms (**112a, b**) have been believed to exist in equilibrium[206,243]. In methyl vinyl ether the Z form (**112a**) slightly predominates ($\Delta G° = 1.2$ kcal), however, the second form is now claimed to be not E (**112b**) but a non-planar form[244] (**112c**). In t-butyl vinyl ether the steric hindrance destabilizes the form **112a**, the actual conformation is considered[243,245] either to be (**112b**) or (**112c**). The different conformation of vinyl ethers with various alkyl groups is reflected in the variability of their dipole moments[246] (0.96–1.84 D), while their temperature dependence[247] proves the coexistence of rotamers. From the data for ethyl vinyl ether the supposed extrapolated dipole moments were calculated[247] for the forms **112a** and **c** and ΔH was estimated to be 0·8 kcal. Hence the overall pattern for vinyl ethers does not fundamentally contradict that for esters; the smaller energy difference between rotamers is explained[206] by the absence of lone electron pairs at =CH$_2$ as compared with =O. The conformation of vinyl alkyl sulphides yields probably a similar picture but has not been studied so thoroughly. Dipole moments were reported for several unsaturated sulphides, sulphoxides and sulphones[248], for more complex derivatives[249], and also for vinyl selenides[250]. Sometimes the problem of conformation was combined with the configuration on C=C.

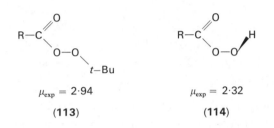

$$\mu_{exp} = 2·94 \qquad\qquad \mu_{exp} = 2·32$$

$$\textbf{(113)} \qquad\qquad\qquad \textbf{(114)}$$

The COO group not bonded to carbon is involved in peroxy acids and esters, hydroxylamine derivatives and some derivatives of polyvalent iodine; for the latter see Section V.B.4, formula **128**. Peroxy esters have been little studied owing to their instability; dipole moments of several t-butyl peroxy esters have been explained[251] by a conformation near to **113** with certain deviations from planarity on the bonds C—O and/or O—O.

Dipole moments of peroxy acids[252] were interpreted in terms of a non-planar conformation **114**, however, the calculation did not account for a possible contribution of the hydrogen bond to the dipole moment. At any rate the heavy-atom skeleton is planar and the conformation on C—O is Z; the rotation around the O—O bond seems not to be very strongly hindered. While the i.r. spectra[253] indicate the presence of a hydrogen bond to the carbonyl oxygen, in the crystalline state H and CO are anticlinal in the molecule of peroxypelargonic acid[254] and engaged in intermolecular hydrogen bonds.

On the whole, the conformation of the COO moiety in peroxy compounds is similar to that observed on esters. The same is true of hydroxylamine derivatives. The only exceptions seem to be the O-acylhydroxylamine derivatives **115** and **116** in the crystalline state[255,256]; it is, however, not known whether these forms persist in solution.

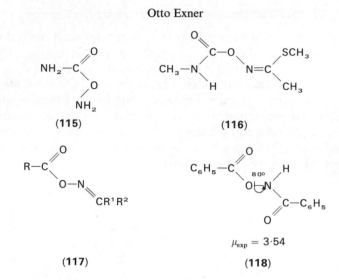

(115)   (116)

(117)   (118)

$\mu_{exp} = 3.54$

Simple O-acyl derivatives of hydroxylamine, $RCOONH_2$, have not been investigated by dipole moments owing to their instability. On the other hand a dipole moment study of acylated oximes[88] permitted the determination of the conformations of the ester-like moiety and of the oxime moiety at the same time, and in addition the configuration on C=N was also confirmed, see structure **42a** in Section IV.B.2. The conformation on the C—O bond (**117**) corresponds to that of esters. The conformation of N,O-diacylhydroxylamines[92] (**118**) confirms the principle of relative independence of conformations on individual bonds (see Reference 100, 179). The ester-like moiety repeats exactly the conformations of esters, the amide-like moiety that of mono substituted amides (Section V.C.). The non-planar arrangement of the four inner atoms C—O—N—C is that found in H_2O_2 and diacyl peroxides or disulphides (structure **126**). The conformation **118** was established quite reliably since it was possible to introduce substituents into both *para* positions[92], and the results were subsequently confirmed by n.m.r.[257] and X-ray[258] investigations. Even the values of the dihedral angle C—N—O—C estimated from dipole moments[92] (*c.* 70°) and found in the crystalline state[258] (81°) agree sufficiently.

4. Derivatives with two axes of rotation

Many esters of dibasic acids[174,183,184,197] or of polyhydric alcohols[259] have been investigated by the dipole moment technique but the results revealed nothing new in comparison with simple esters. This section will be devoted to compounds where two COO groups are bonded more

closely or even overlap so that the two rotational axes are either immediately adjacent or separated by one or two other bonds. A study[100],
based mainly on dipole moment data, was devoted to conformation on
two equivalent bonds in a molecule of the general formula $R-X-Y-$
$X-R$. The question is, whether the conformation on the two $X-Y$ bonds
can be predicted on the basis of a simple model compound $R-X-Y-R$.
The results are promising: the conformation of the monofunctional
derivative is retained in some 90% of the cases, the exceptions being
caused by steric hindrance in the more crowded bifunctional compound.

Taking an ester molecule as the monofunctional compound, two types
of bifunctional derivatives, dialkyl carbonates and carboxylic anhydrides,
may be derived. Three planar forms (**119a–c**) are possible for carbonates,
among which the *ZZ* conformation (**119a**) would correspond to simple
esters in both moieties. This actually agrees with the low dipole moments
found in the gas phase[221] as well as in solution[260,261]; also instructive
is the comparison with ethylene carbonate[261] (**120**). An infrared study
pointed to the same form for dimethyl carbonate although with certain
deviations from planarity[262], while the molecule of bistrichloromethyl
carbonate in the crystal is quite exactly planar[263]. At variance with these
results, the slightly distorted form **119b** has recently been claimed to be
the most abundant one[264]. This claim was based on the doubled carbonyl
bond and broadening of n.m.r. signals; the most important proofs were
obtained with methyl *t*-butyl carbonate, including also the decrease of its
dipole moment with temperature. However, the conformation of this
overcrowded compound may be an exception, since dipole moments of
other alkyl carbonates show a reversed temperature dependence[260].
Recent i.r. spectroscopic results allow at most 2% of the form **119b** for
dimethyl carbonate and even for methyl *t*-butyl carbonate this form is
questioned[265]. Note that conclusions based on the doubled carbonyl
frequency[224,230] were at variance with other results even for other
classes of compounds, such as chloroformates and carbamates (Section
V.B.2).

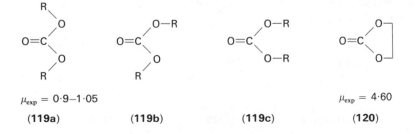

$\mu_{exp} = 0.9-1.05$ $\mu_{exp} = 4.60$

(**119a**)   (**119b**)   (**119c**)   (**120**)

For aromatic carbonates the ZZ form is again clearly preferred[266]. The introduced *para* nitro group enhances the dipole moment differences between some rotamers as shown by structures **121a–c**. No indication of the presence of any other form was detected, neither any signs of deviation from planarity, which could be recognized rather sensitively. Theoretical calculations[267] also preferred the ZZ form (**119a**, R = H) although the energy difference with respect to **119b** was only 1 kcal.

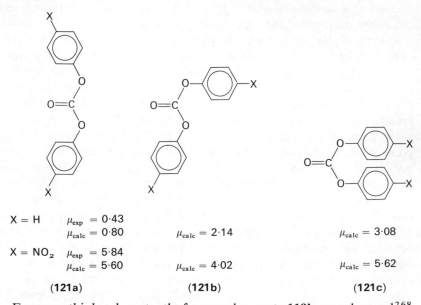

X = H	μ_{exp} = 0·43		
	μ_{calc} = 0·80	μ_{calc} = 2·14	μ_{calc} = 3·08
X = NO$_2$	μ_{exp} = 5·84		
	μ_{calc} = 5·60	μ_{calc} = 4·02	μ_{calc} = 5·62
	(121a)	**(121b)**	**(121c)**

For monothiol carbonates the form analogous to **119b** was advanced[268], the two possibilities (**122a, b**) remained undecided. The arguments, based on i.r. and n.m.r. spectra and dipole moments, are similar as in the case of carbonates[264]; they are also open to the same criticism. For trithio-carbonates the ZZ form, analogous to **119a** was claimed as the most abundant[269], in addition to EZ (like **119b**). The most important proof was obtained from the i.r. spectra, while the dipole moments would be worth a reinvestigation. In particular the HMO calculations of dipole moments are not dependable.

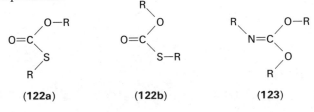

| **(122a)** | **(122b)** | **(123)** |

The only real exceptions where the Z conformation of esters is not retained seem to be the imidocarbonates[270] (**123**). Here the reversal of conformation in one moiety is easily understood by the steric hindrance of the unsymmetrical =NR group.

The conformation of carboxylic anhydrides and diacyl sulphides seems to have been solved definitely since all recent investigations accord. Dipole moment studies on acetic anhydride[271] and diacetyl sulphide[272] did not meet with success and led to the erroneous assignment in favour of the conformation **124b**. Investigation of aromatic derivatives[100,273] enabled to introduce *para* substituents and to apply the graphical method[58]. Hence it was possible to distinguish between the three planar (**124a–c**) and various non-planar forms. The graphical method is illustrated in Figure 3 for dibenzoyl sulphide[100] (x-axis) and its 4,4'-dichloro

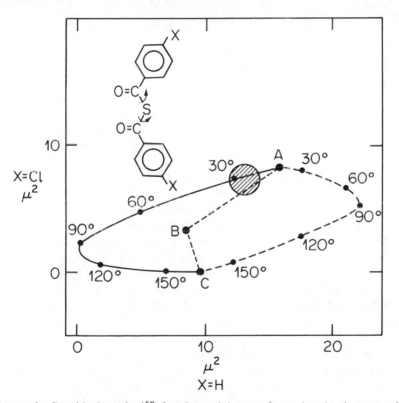

FIGURE 3. Graphical method[58] for determining conformation in the case of a complex motion. Reproduced, with permission, from Exner and co-workers, *J. Chem. Soc. Perkin II*, 1870 (1973).

derivative (y-axis) in order to demonstrate application to a more complex case (compare Figure 2 for a simple case). The three planar conformations (**124a–c**) are represented by the points A–C, respectively. The conversion of the *ZZ* conformation (**124a**) into *EE* (**124c**) may be accomplished by several paths: (a) by rotating simultaneously the two acyl groups in the same direction so that one carbonyl oxygen is always above the C—O—C plane and the other one below this plane, the transitory conformations have the C_2 symmetry (full curve); (b) by rotating the two acyl groups in opposite directions so that both carbonyl oxygens are situated on the same side, the transitory conformations have the C_s symmetry (broken curve); (c) by turning first one and then the other acyl group (dotted lines). Comparison with the experimental point reveals the C_2 conformation corresponding to the slightly distorted form **124a**. The dihedral angles $\tau = \sphericalangle$ O—C—O—C may be estimated to be 20–30° for both dibenzoyl sulphide[100] and benzoic anhydride[273].

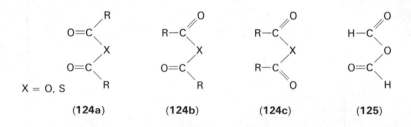

(**124a**)   (**124b**)   (**124c**)   (**125**)

Essentially the same conformation was found for acetic anhydride and for diacetyl sulphide on the basis of the Kerr constants[274] and dipole moments ($\tau = 47°$ and 30–40°, respectively), for acetic anhydride by electron diffraction[275] ($\tau = 48°$), and for monochloroacetic anhydride[276] in the crystal ($\tau \approx 23°$). The exceptional conformation of formic anhydride[277] (**125**) is understandable in terms of hydrogen bonding rather than of steric factors. Dipole moments of bisthiocarbamoyl sulphides[235] were also discussed in favour of a conformation analogous to **124a**.

In the aforementioned classes of compounds the conformation of esters was retained—or almost retained—in spite of some steric hindrance and even though the cross-conjugation could weaken the resonance interaction. There is no such hindrance in diacyl peroxides and diacyl disulphides and the ester-like conformation is repeated exactly. From dipole moments of dibenzoyl peroxide[278,279] and disulphide[279,280] not only the conformation on the C—X bonds but also on the central X—X bond was deduced (**126**). In the former the 4,4-dichloro substitution is relevant[278], the dipole moment is only slightly changed. The estimated dihedral angle[278] for peroxides (c. 100°) compares favourably with that for H_2O_2 (111°). Once more the conformation of the simplest compound is retained in quite complex derivatives. The dipole moments of bisthio-carbamoyl disulphides ('thiuramdisulphides') are in accord with an analogous conformation[235].

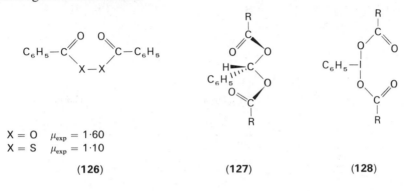

$$X = O \quad \mu_{exp} = 1 \cdot 60$$
$$X = S \quad \mu_{exp} = 1 \cdot 10$$

(126)　　　　　　　　(127)　　　　　　　　(128)

In acylals the two COO groups are separated by one atom and may be expected to behave independently. When discussing the conformation of benzylidene diacetate (127) the conformation of the COO groups was assumed to be Z-like in esters, and attention was focused on the rotation around the two adjacent C—O bonds[58]. Most probably the two acetyl groups are situated on the same side of the O—C—O plane, almost perpendicular to it. However, the result based on the dipole moments of 127 and its 4-nitro derivative[281] would be worth a confirmation.

In 1,1-diacetoxyiodobenzene (128) two rotational axes I—O are collinear (dsp³ hybridization on I). The dipole moments revealed[282] the planar form 128. This is remarkable since a polar bond (O—C) is placed between two electron pairs on I contrary to the Edward–Lemieux rule[283]; in addition there could be some electrostatic repulsion between the two carbonyl oxygens, and the dipole moment in the actual conformation is at maximum.

C. Amides and Related Compounds

The stereochemistry of amides has been much studied[8], particularly in connection with the peptide linkage. However, the dipole moment approach was less used and less efficient than in the case of esters, mainly because of the association in solution and insufficient solubility of some derivatives. For this reason the present section is shorter than the preceding one, also some arguments will not be repeated. A recent review[11] deals mainly with i.r. spectroscopy and mentions only seldom the dipole moments; another review[284] covers n.m.r. results including the stereochemical aspects.

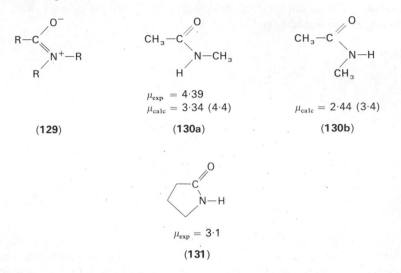

$\mu_{exp} = 4\cdot39$
$\mu_{calc} = 3\cdot34\ (4\cdot4)$

$\mu_{calc} = 2\cdot44\ (3\cdot4)$

(129)   (130a)   (130b)

$\mu_{exp} = 3\cdot1$

(131)

Because of the resonance within the amide group (structure **129**), which is more pronounced than in esters, planar configuration on nitrogen as well as the coplanarity of the C—O—N and C—N—R planes are expected and have been indeed found. In the case of secondary amides two forms are thus possible, e.g. (**130a, b**). Dipole moment studies[285–289] revealed the preference (or exclusive presence) for the Z form (**130a**); again, comparison with a cyclic model compound[261] (**131**) is decisive even though the difference between the two calculated values[287] is less and the agreement with experiment worse than usual. The latter depends on the method of computation, i.e. whether a particular resonance contribution[286,287,290] from structure **129** was applied or not. With such a correction (values in parentheses[287], structures **130a, b**) the agreement is very good. An additional argument for the assignment arises from the concentration dependence of dipole moments[261,285,287]. The lactam **131**—like the E form **130b**, if present—dimerizes to a cyclic centrosymmetrical dimer, hence the effective dipole moment drops. The Z form (**130a**) dimerizes or polymerizes linearly and its dipole moment increases with concentration.

The Z conformation (**130a**) has been also found, as the single present one, by electron diffraction[291], even for N-methylformamide[292], although the latter had been the object of some controversy[11]. The results of crystallographic studies, e.g. References 293, 294, are also concordant. Nevertheless, the conformational stability of amides is not so great as that of esters and can be diminished or even reversed by steric effects. Hence in N-t-butylformamide the E form is represented to a minor extent and in 2-methylformanilide it already predominates[11,236]. The latter compound was isolated in both conformers[295]. There are still other amides in which the E form is more abundant, e.g. N-alkyl-phenylacetamides[236]; the reasons are not clear. The relative abundance of conformers is known for numerous derivatives from the n.m.r.[284] and i.r.[236] data. A temperature dependence of dipole moments was not detected[286]. An exceptional, distinctly non-planar conformation was claimed to occur[289,296] with acetanilide and its derivatives; it had been even assumed that the configuration on nitrogen might be tetrahedral[58,296]. An authoritative combined study[297] of dipole moments and Kerr constants, favours planar forms like **130a**; it is only the phenyl ring itself which is rotated from the O—C—N plane. According to its i.r. spectrum[288] acetanilide is almost entirely Z, while in formanilide the two forms are represented in comparable amounts.

In physicochemical terms we may summarize the behaviour of secondary amides as follows: The rotational barrier $\Delta G^{\neq}$ is large[11] (c. 14 kcal), but the free energy difference $\Delta G°$ between rotamers is only of the order of a few kilocalories and in certain derivatives approaches zero or even changes sign. The conformation is influenced by steric and in some extent also by electronic[298] effects. Theoretical calculations[267] yielded $\Delta H° = 3$ kcal for N-methylformamide compared with 5·5 kcal for methyl formate.

The conformation of tertiary amides is less interesting and is not reflected in the dipole moment values, unless polar substituents are introduced. The preference of alkyl groups seems to be governed by steric factors: for numerical data see Reference 284.

As far as the finer details of conformation are concerned, the question of the exact planarity of the amide group has recently received attention[299] in connection with the peptide structure. Since the dipole moments and many other methods are insufficiently sensitive, the evidence is based mostly on X-ray results[300] and theoretical calculations[299]. There is some evidence that a non-planar configuration on N (both substituents on the same side of the O—C—N plane) is involved rather that a twisting of the O—C—N and C—N—C planes; the configuration on the carbonyl carbon remains exactly planar. A deviation arising merely by twisting around the C—N bond, was claimed for amides of particular structures[301]; e.g. N-t-butylacetamide (in addition to the prevailing Z form). This finding is based only on i.r. spectroscopy, and has been questioned[302].

Non-planar amide groups were studied on a series of specially prepared cyclic model compounds[303,304].

The preference for the planar conformation of amides, and among them the preference of Z before E, may be explained in terms of the same theories as in the case of esters (Section V.B.1), i.e. (i) mesomerism within the functional group (**129**), (ii) electrostatic repulsion of dipoles, (iii) repulsion of lone electron pairs, (iv) steric factors. The importance of resonance, giving the C—N bond a double bond character, is beyond any discussion[8]; it is reflected most significantly in the shortening of the C—N bond and lengthening of the C=O bond[291–294]; the two bond lengths are inversely proportional[8]. In the dipole moment values the resonance is manifested by a component[286,287,290,305] of 0.7–1.7 D oriented from N to O, which corresponds to the contribution of structure **129** (say 5–15%, as far as such figures are meaningful). Another proof of the isomerism was sought in studying cyclic lactams of variable ring sizes[306], as in the case of lactones[211]. However, the dipole moments are only little altered with the conformational change so that E conformation (5– 8– cycle) and Z conformation (10– 19–cycle) were distinguished only by different behaviour in the association.

Of course resonance explains only the rotational barrier, i.e. $\Delta G^{\neq}$, and not ΔG°, the preference for the E form before Z. When accounting for this difference, the electrostatic factors may be rejected still more conclusively than for esters: While the actual form of esters (**92a**) has the minimum dipole moment and minimum repulsive interaction, with secondary amides the actual form (**130a**) has the highest moment and maximum interaction since the bond moment N—H is greater than N—CH$_3$. Most conclusive is the comparison with N-cyano amides[212] (**132**). Although the electronic character of the CN and CH$_3$ groups is opposite, the preferred conformation is the same. Nor can the idea of lone pair repulsion[206] be applied to amides since the nitrogen atom is sp^2 hydridized. Steric effects—and maybe also some attractive non-bonded interactions—certainly play an important role, as can be seen from comparison of derivatives with substituents of variable size[236,284], in particular of tertiary amides. However, there must be an intrinsic important factor stabilizing the E form (**130a**) in addition to steric effects and sometimes against them. A possible dependence of conjugation on conformation could be expressed only in quantum chemical terms[267]; its existence may be suggested by the electronic effects influencing the abundance of rotamers[298].

N-Alkylcarbamates were already discussed in Section V.B.2 (structures **103**). Other analogues of amides yield an essentially unchanged pattern,

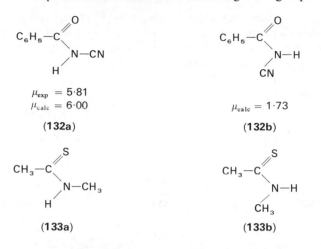

μ_{exp} = 5·81
μ_{calc} = 6·00

(132a)

μ_{calc} = 1·73

(132b)

(133a)

(133b)

although in details the steric hindrance may be greater or smaller. In thioamides the partial double bond character of the C—N bond is strengthened so that some rotamers, e.g. **133a, b**, were isolated as two distinct compounds[12]; the dipole moments are also quoted in connection with the assignment of the conformation (or, say, configuration) but the main arguments were from i.r. and n.m.r. spectroscopy. The dipole moments of thioamides are not well understood. It has been argued[307] that the *E* form **133b** should have a greater moment than **133a**, considering the position of the C=S and N—CH$_3$ vectors only. This conclusion seemed to be supported by the dipole moments of *N*-methylthioformamide (4·56 D) and *N*-*t*-butylthioformamide (5·17 D), which exist prevailingly as *Z* or *E* conformers, respectively[307]. However, in the argument the N—H bond moment has been neglected although it is certainly greater than N—CH$_3$. It seems that the mesomeric moment plays an important role, more so than in the case of amides, and that its magnitude could depend on conformation. Owing to the larger atomic radius of sulphur, the steric interaction in the *Z* conformer is enhanced; hence the proportion of the *E* form in thioamides generally exceeds that in corresponding amides[11,236,284]. For instance in thioacetanilide the *E* form already prevails and in thioformanilide it is present almost exclusively[236,288].

The dipole moment of a tertiary selenoamide was reported[308]. In amide analogues of the type RC(=X)NMe$_2$ the double bond character of the C—N bond increases[309] for X in the order NH, O, S, NH$_2^+$. The mesomeric moments increase in a similar order NH, O, S, Se, although all the values reported[308] seem to be too high. In *N*-aryl benzamidoximes (**48**) the problem of the C—N conformation arises together with the configuration and conformation of the =NOH group and was already mentioned in Section IV.B.2. The reversed E conformation found[105] is understandable in terms of the steric requirements of the =NOH group as compared to the carbonyl oxygen.

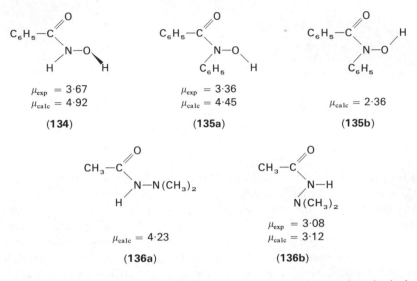

$\mu_{exp} = 3\cdot67$
$\mu_{calc} = 4\cdot92$

(134)

$\mu_{exp} = 3\cdot36$
$\mu_{calc} = 4\cdot45$

(135a)

$\mu_{calc} = 2\cdot36$

(135b)

$\mu_{calc} = 4\cdot23$

(136a)

$\mu_{exp} = 3\cdot08$
$\mu_{calc} = 3\cdot12$

(136b)

The compounds with the CONH group not bonded to carbon include, among others, hydroxamic acids, hydrazides and nitrosamines. The conformation of hydroxamic acids as deduced from dipole moments[92] seems to be not quite planar. Although the value for the dipole moment of benzohydroxamic acid[92] was in error[71], the actual conformation of this compound remains near to **134** with certain deviations from planarity, particularly on the N—O bond. The conformation of acetohydroxamic acid (hemihydrate) in the crystalline phase[310] differs from **134** only by the reversed position of the OH hydrogen, engaged in intermolecular hydrogen bonds. N-Phenylhydroxamic acids have been investigated by the dipole moment approach[311] using substituents on the N-aryl group; the probable form is between **135a** and **135b**. Dipole moments of several N,N-dimethylhydrazides, e.g. **136**, served to determine the conformation[305] which is reversed as compared with secondary amides. The pyramidal configuration on the second nitrogen and free rotation around the N—N bond were assumed. The abundance of the majority form is 93% in benzene and depends on solvent[312]. The assignment of conformation agrees with an independent n.m.r. study[313]. On the other hand for unsubstituted benzohydrazide the conformation (**137a**) stabilized by one or two hydrogen bonds, was deduced from dipole moments[314] while thiosemicarbazide exists in crystal as the *E* conformation[315] (**137b**). An n.m.r. study of dithiocarbazic acid derivatives (**138**) did not allow distinction between the four possible forms[316] (cf. carbamic acid derivatives (**103**)). Dipole moments were not applied in the two aforementioned cases.

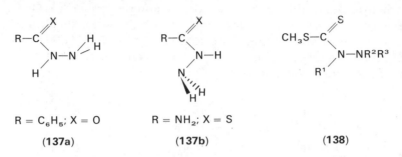

R = C₆H₅; X = O   R = NH₂; X = S

(137a)   (137b)   (138)

Rotation around a C—N bond is also involved in nitroso amides and acyl azides. The conformation of the former (**139**) was solved by a classical approach[317], comparing with the cyclic model compound (**140**) The difference of dipole moments is so great that the unknown conformation of the NO group can not invalidate the assignment. From the dipole moment of benzoyl azide no stereochemical conclusion was drawn[318] but the *Z* conformation (**141**) seems very probable with respect to the dipole moment of phenyl azide (1·44 D) and its orientation from phenyl to the functional group.

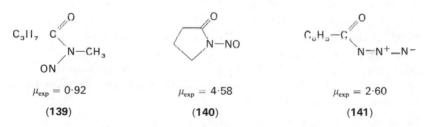

$\mu_{exp} = 0.92$ $\mu_{exp} = 4.58$ $\mu_{exp} = 2.60$

(139)   (140)   (141)

From compounds with two axes of rotation, substituted ureas yield a similar picture to carbonates (**119**) but they have been less studied. Infrared spectroscopy revealed only the *Z, Z* form (**142a**) for 1,3-dimethylurea, while larger substituents ·shift the equilibrium and the *E,Z* form (**142b**) is detected, too[236]. Similar results were obtained for thioureas with the *E,Z* form more populated[319]; the presence of one aryl group favours this form still more through a hydrogen bond to the aromatic π electrons. In the diphenyl derivative **143** the conformation with the two benzene rings parallel is preferred in crystal as well as in solution[320]. Dipole moments of polysubstituted ureas must be measured in rather polar solvents, in addition a mesomeric moment is of importance[321]. Hence no definite results have been obtained concerning their conformation[321]; a proposal made[314] must be considered to be purely speculative. Thio-carbohydrazide exists in the *E,Z* conformation in crystal[322]. One can conclude that the *Z,Z* form (142a) is the preferred one for simple dialkyl-ureas and their analogues, but it can be changed to *E,Z* (142b) easily by moderate steric effects; compare the lower conformational stability of secondary amides than of esters.

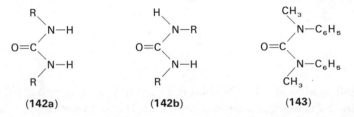

(142a)   (142b)   (143)

The conformation of *N*-hydroxythioureas represents a still more complex problem, since in addition the conformation on the N—O bond and numerous possible hydrogen bonds must be considered. The results obtained[323,324] from dipole moments and i.r. spectroscopy are in our opinion not quite conclusive.

Diacylamides (144) yield a similar picture to anhydrides (124) but more conformational freedom and more importance of steric effects is to be expected. Indeed the *Z,Z* form (144a) corresponding strictly to the conformation of secondary amides, has been found only in the solid state[325], while in solution the *E,Z* form (144b) is preferred, i.e. the con-formation of amides is retained in one moiety only. This follows from the dipole moments, e.g. of *N*-methyldiacetamide compared with the model compounds[326] like 145, as well as from a combined approach using dipole moments, i.r., n.m.r. and u.v. spectra[325].

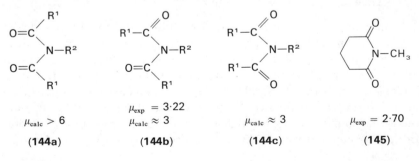

$\mu_{exp} = 3\cdot22$

$\mu_{calc} > 6$ $\mu_{calc} \approx 3$ $\mu_{calc} \approx 3$ $\mu_{exp} = 2\cdot70$

(144a)   (144b)   (144c)   (145)

The conformation is also sensitive to steric hindrance. While ditrimethyl-acetamide exists in the form **144a** ($R^1 = t\text{-}C_4H_9$, $R^2 = H$) even in solution[325], for its N-phenyl derivative the distorted form **144c** ($R^1 = t\text{-}C_4H_9$, $R^2 = C_6H_5$) was claimed[327] in addition to the prevailing **144b**. Triacyl-amides were studied by n.m.r. spectroscopy, and the C_{3h} form seems most probable[328]. Dipole moments of 1,2-dibenzoylhydrazine agree with a similar conformation[314] as that of diacyl peroxides or disulphides (**126**); the dihedral angle C—N—N—C was estimated to be 130°.

D. Other Types

In the preceding two sections virtually all compounds involving a partial double C—O or C—N bond, have been included. In this section some other types of bonds will be mentioned which are related to those already discussed. The N—O bond in oximes may be attributed some double bond character owing to the conjugation of the C=N bond with the lone electrons on oxygen. Actually this bond is somewhat shortened[66,68,93] with respect to non-conjugated hydroxylamine derivatives, and the conformation on it is planar[66-71]. Since this conformation is an inherent feature of the stereochemistry of oximes, it was dealt with already in Section IV.B.2. The E form (**34a**) is preferred in oximes (**35, 36**), their functional derivatives (**39, 42a**), and hydroximic acid derivatives (**43a, 44, 45, 46, 48, 49a**); several exceptions (**37b, 40a, 41b**, possibly **43b**) are caused by hydrogen bonds. The preference for the E form (**34a**) may be explained in terms of repulsion between the lone electron pairs[206] in the form **34b**. In hydroxamic acids (**134**) the double bond character of the N—O bond is doubtful; although it is shortened[310], the conformation on it is probably not planar[92,311]. In O-acylhydroxylamines, e.g. **115**, there is no shortening of the N—O bond[255].

$\mu_{calc} = 3\cdot05$ $\mu_{exp} = 1\cdot94$ $\mu_{calc} = 1\cdot1$
$_mK_{(calc)} = 233$ $_mK_{(exp)} = 66$ $_mK_{(calc)} = 14$

    (**146a**)   (**146b**)

Alkyl nitrites bear some resemblance to esters; on the basis of this analogy the planar forms (**146a, b**) could be considered, and from these **146b** is the preferred one. The n.m.r. spectroscopy revealed the presence of two rotamers[329] although there is no agreement as to the assignment[213].

Dipole moments[213,330] as well as the Kerr constant[330] were interpreted in terms of the equilibrium (146a) ⇄ (146b). In methyl nitrite both forms are present in comparable amounts which depend on state and solvent. With the increasing size of the alkyl group the E form (146a) becomes more populated up to t-butyl nitrite ($\mu = 2{\cdot}37$ D; 80–100% E). It was pointed out that the two forms need not be just the exactly planar ones[330]. The E form (146a) is relatively more stable than in esters and this difference is an important support of the theory of lone pair repulsion[206]: In esters there is a repulsive 1,3-interaction in the E form (92b), in nitrites the same interaction in (146a) is counterbalanced by a 1,2-interaction in (146b). In alkyl nitrates there is no problem of conformation; the double bond character of the N—O(R) bond is revealed by the planarity of the whole molecule[331].

The N—N bond in hydrazones (50, 51) may give rise to a similar picture to the N—O bond in oximes (34), but there is little evidence as to the actual conformation (Section IV.B.3). In acyl hydrazides (136, 137) hardly any double-bond character of the N—N bond can be expected, the hypothesis of free rotation was sufficient to account for the dipole moment data[305]. In nitrosamines the partially double N—N bond gives rise to two rotamers, revealed by n.m.r. spectroscopy[332]. Dipole moments of N-nitroso-N-methylaniline and its derivatives—which exist in the E form (147) according to the n.m.r. data[332]—were not interpreted in terms of conformation[333]. Nor was the extent of conjugation estimated, although the high dipole moment points to a significant contribution from the resonance structure 148. The dipole moments of N-nitrosoamides (139) were used[306] only to derive the conformation on the C—N bond, not on the N—N bond. In triazenes (65) an N—N bond of similar type is involved, however, the conformation was not investigated.

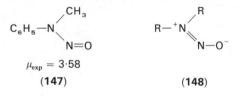

$\mu_{exp} = 3{\cdot}58$

(147)   (148)

E. Substituents on the Benzene Nucleus

In this section conformations of aromatic derivatives will be discussed and especially the questions of the coplanarity of the substituent and of the preference among the possible planar forms. The material is restricted to some recent dipole moment investigations because the whole subject exceeds the scope of this chapter.

The double-bond character of a particular bond may be judged in the area of dipole moments according to the so-called mesomeric moment, which is essentially the vector difference between moments of an aromatic and an aliphatic derivative[14,15]. The concept was subject to criticism since it was argued that it overestimated the degree of conjugation, e.g. for nitrobenzene[334]. However, in conformation studies the fact that conjugation is able to compel coplanarity is sufficient and this is fulfilled even in the case of nitrobenzene.

The benzene nucleus itself being non-polar, the conformation of monosubstituted benzenes is not immediately reflected in their dipole moments, but it is in the Kerr constant since the benzene nucleus is strongly anisotropic. This experimental approach served to prove the planarity of benzaldehyde[335], and acetophenone[335]; a slight, almost insignificant, deviation from planarity in phenol[336] and a larger one in anisole[337] (the dihedral angle C—C—O—C of c. 18°). The position of the benzene nucleus played also an important role in investigating the conformation of all aromatic derivatives, e.g. esters[197], anilides[297], and ketones[165] on the basis of the Kerr constant.

In one case the twisting of the substituent may be observable on the basis of the dipole moment alone: if this substituent is strongly conjugated, the steric repression of the conjugation may be detected by the reduction of the mesomeric dipole moment (i.e., either by reduction or increase of the total moment). This second-order effect is usually rather small and its interpretation is not quite unambiguous[35,334]. In addition it may be observed only if several derivatives with different dihedral angles are available for comparison; this is usually achieved by alkyl groups in the *ortho* positions[338,339]. Only rough estimation of the dihedral angle may be obtained at best.

More significant results are obtained—for conjugated or non-conjugated substituents—if the second substituent on the same nucleus enables determination of its position.

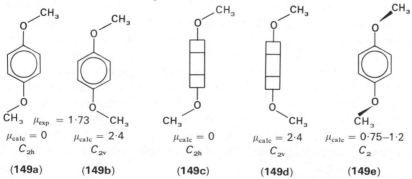

(149a) $\mu_{exp} = 1\cdot73$ (149b) $\mu_{calc} = 2\cdot4$ (149c) $\mu_{calc} = 0$ (149d) $\mu_{calc} = 2\cdot4$ (149e) $\mu_{calc} = 0\cdot75$–$1\cdot2$

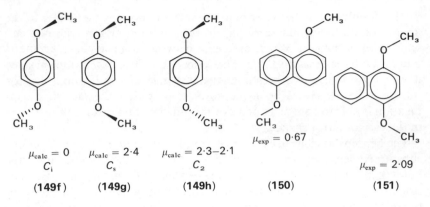

$\mu_{calc} = 0$ $\mu_{calc} = 2\cdot4$ $\mu_{calc} = 2\cdot3\text{–}2\cdot1$ $\mu_{exp} = 0\cdot67$ $\mu_{exp} = 2\cdot09$

C_i C_s C_2

(149f)   **(149g)**   **(149h)**   **(150)**   **(151)**

The dipole moments of 1,4-dimethoxybenzene[337,340,341] **(149)**, 1,4-diacetylbenzene[335] **(152)**, terephthalic acid and esters[183] and similar symmetrical 1,4-derivatives are particularly instructive. They all differ from zero, although they are consistently lower than those calculated for the pure C_{2h} (*trans*) forms **(149a)**. This experimental result is compatible with the following hypotheses: (a) mixture of the two planar forms with the C_{2h} and C_{2v} symmetry (**149a** and **149b**, respectively); (b) free rotation around the C_{ar}—O bond, i.e. all conformations equally populated; (c) mixture of the non-planar C_{2h} and C_{2v} forms (**149c** and **149d**, respectively); (d) various mixtures of non-planar, less-symmetrical forms, e.g. **149e–h**; (e) one particular non-planar unsymmetrical conformation. To distinguish between these possibilities an ingenious approach was advanced[342], referring to the naphthalene derivatives **150** and **151**. According to hypotheses (c) or (e) the dipole moments would be equal to that of **149**; according to (b) the free rotation should be only partly hindered and the difference between **150** and **151** less than it actually is. Hence an approximately 1:1 mixture of the planar forms **149a** and **149b** was accepted as the most probable result which, however, is capable of refinement in two directions. Firstly the ratio of these rotamers need not be exactly 1:1. Although the rotating groups are rather apart, their electrostatic interaction was considered as a factor able to destabilize the form **149b**. An estimation of the electrostatic energy[340] yielded a low value of 60 cal mol^{-1} which is in accord with the observed independence of the dipole moment of temperature[341]. On the other side it was inferred from the conformation of anisole that the methoxy groups in 1,4-dimethoxybenzene are not planar and a mixture of the four forms **149e–h** (with a dihedral angle of 18–30°) was advanced as the best description of the actual state[337]. This finding is supported particularly by the Kerr constant measurement. Similar conformations were preferred for 1,4-diacetylbenzene[335] **(152)**, tere-phthaldehyde[335] and 1,4-dibenzoylbenzene[343]; in the last case the dihedral angle is enhanced to 40°. On the other hand, non-planar struc-tures with the functional group in a perpendicular plane to the benzene nucleus **(153a, b)** were advanced for 1,4-bis-isothiocyanatobenzene[344]; the main evidence arose from the i.r. spectra. The practically zero dipole moment of this compound[345] is remarkable and is not explained by this hypothesis[344]. It seems certain that the group moment of the NCS group lies at a small angle to the C_{ar}—N bond; in addition there is the problem of safely determining small values of dipole moments. The moment of terephthaloyl chloride is also very low[346] so that the conformation **154** should prevail.

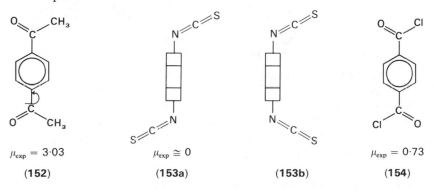

$\mu_{exp} = 3{\cdot}03$ $\mu_{exp} \cong 0$ $\mu_{exp} = 0{\cdot}73$

(152)   **(153a)**   **(153b)**   **(154)**

If the two substituents are in the *meta* position, it is sufficient that only one of them should be angular for a conformational problem to arise, e.g. in **155** or **156**. The experimental dipole moment lies as a rule between the two calculated values and it is interpreted[347] as a mixture of the two planar forms rather than a non-planar conformation, or a more or less free rotation. The inductive interaction between the two groups may be a complicating factor in computing the expected moments[348]; although it represents only a small correction, it may influence the calculated abund-ance of rotamers. In an extensive series of benzene *meta* derivatives, chlorine was chosen as a standard reference substituent and the position of the second, angular substituent was estimated from the experimental dipole moment[347]. The result was that the rotamer with the smaller dipole moment always predominated. The population was also predicted from the calculated electrostatic energy. Although the agreement was very good, see e.g. 3-chlorobenzaldehyde **(155)**, it seems that the accuracy of the approach has been overestimated. A study[349] of 3-fluoroacetophenone and 3-trifluoromethylacetophenone differed in that the expected dipole moments were calculated by the INDO method, which was also used to calculate the theoretical populations. The results are here expressed in the same form as for the preceding example see **156a, b**. Even here the accuracy attained is not sufficient to reveal a real difference in the stability of the rotamers and an estimate of equal populations would be quite acceptable. Often this statistical population was simply assumed *a priori*[348].

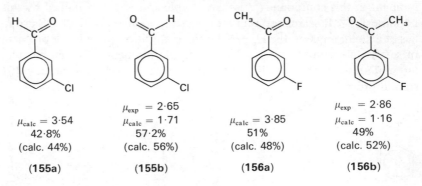

$\mu_{calc} = 3.54$	$\mu_{exp} = 2.65$ $\mu_{calc} = 1.71$	$\mu_{calc} = 3.85$	$\mu_{exp} = 2.86$ $\mu_{calc} = 1.16$
42.8%	57.2%	51%	49%
(calc. 44%)	(calc. 56%)	(calc. 48%)	(calc. 52%)
(155a)	**(155b)**	**(156a)**	**(156b)**

If two angular *meta*-substituents are present, there are three planar forms possible[183,340,346,350] whose purely statistical population is $1:1:2$. For isophthalaldehyde the population of rotamers **157a–c** was calculated either from the electrostatic energy[340] or by the semi-empirical π-SCFMO method[350] (in parentheses); the agreement is surprisingly good. Also the expected dipole moments calculated from bond moments[340] and semi-empirically with the π-moment contribution[351] (in parentheses) agree very well. However, when we calculate the expected effective moments from either theory, the agreement with the experiment is somewhat worse.

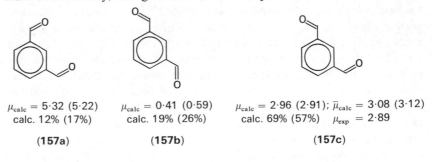

$\mu_{calc} = 5.32 (5.22)$ calc. 12% (17%)	$\mu_{calc} = 0.41 (0.59)$ calc. 19% (26%)	$\mu_{calc} = 2.96 (2.91); \bar{\mu}_{calc} = 3.08 (3.12)$ calc. 69% (57%) $\mu_{exp} = 2.89$
(157a)	**(157b)**	**(157c)**

In *ortho* derivatives the steric hindrance is of much importance and the rotamers cannot be assumed, even approximately, to be equally populated. The INDO calculation[349] of this population and of the final effective dipole moment for 2-fluoro-acetophenone gave still good results, while the electrostatic calculations[340] are much less satisfactory. The insufficiency of the theory, deducing conformations from electrostatic effects alone, has already been mentioned in Section V.B.2. On the other hand it is possible to calculate[340,351,352] the abundance of rotamers from the experimental dipole moments. In 5-halogeno-2-methylacetophenones the sterically less hindered form (158a) always predominates and its abundance depends only insignificantly on the 5-substitution[351]. *Ortho* substituted benzaldehydes exist virtually only in the non-hindered form[340], so do also many other *ortho* derivatives.

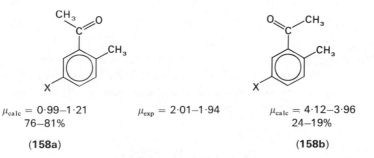

$\mu_{\text{calc}} = 0\cdot99-1\cdot21$ $\mu_{\text{exp}} = 2\cdot01-1\cdot94$ $\mu_{\text{calc}} = 4\cdot12-3\cdot96$
76–81%   24–19%

(158a)   (158b)

Much attention has been devoted[353-358] to conformation of compounds with two benzene nuclei, like diphenyl oxides, sulphides and sulphones, diphenyl disulphides, benzophenones, etc., bearing one, or more frequently two symmetrical substituents in various positions. The problem is rather difficult since in addition to a relatively weak conjugation strong steric effects are in play. Hence unsymmetrical conformations are possible and the two dihedral angles τ_1 and τ_2 must be considered as two independent unknowns. When only dipole moments are applied, the result may be visualized in the form of a contour map with coordinates τ_1 and τ_2; the contour lines connect points with equal calculated moments. By this approach the mean conformation 159 was found for 2,2'- dinitrodiphenyl sulphide[355] with $\tau_1 = 90°$, $\tau_2 = 216°$. Often the dipole moment results were confronted with those from n.m.r.[358] or u.v.[359] spectroscopy, Kerr constant measurement[354], or light scattering[357]. The last method applied to 3-chlorobenzophenone yielded the result that the equilibrium (160a) $\rightleftarrows$ (160b) is in the best accord both with the experimental molar optical anisotropy γ^2 and the dipole moment [357] ($\tau_1 = \tau_2 = 34°$).

Recently all these assignments have been criticized[360] since the rotational barrier is usually low in these compounds and many conformations can coexist. It was suggested that the theoretical conformational energies should always be calculated and depicted in the second contour map.

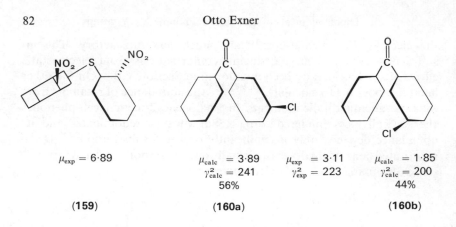

$\mu_{exp} = 6\cdot89$

$\mu_{calc} = 3\cdot89$ $\mu_{exp} = 3\cdot11$ $\mu_{calc} = 1\cdot85$
$\gamma^2_{calc} = 241$ $\gamma^2_{exp} = 223$ $\gamma^2_{calc} = 200$
56%   44%

(159)   **(160a)**   **(160b)**

Unless there are any pronounced energy minima, it is useless to speak about any actual or even 'effective' conformation. This point of view is certainly right, however the question remains, how much the theoretical calculations may be relied upon.

VI. REFERENCES

1. F. A. L. Anet and R. Anet in *Determination of Organic Structures by Physical Methods*, Vol. 3 (Ed. E. A. Braude and J. J. Zuckerman), 2nd ed., Academic Press, New York, 1971, p. 344.
2. G. J. Martin and M. L. Martin, *Progr. in NMR Spectroscopy*, **8**, 163, (1972).
3. O. Exner, *Chem. Listy*, **67**, 699 (1973).
4. *The Chemistry of Alkenes Parts I/II* (Ed. S. Patai, J. Zabicky), Interscience, New York, 1964, 1970.
5. C. G. McCarty in *The Chemistry of the Carbon–Nitrogen Double Bond* (Ed. S. Patai), Interscience, New York, 1970, p. 363.
6. Y. Shvo in *The Chemistry of Hydrazo, Azo and Azoxy Groups* (Ed. S. Patai), Interscience, New York, 1975.
7. M. Simonetta and S. Carrà in *The Chemistry of Carboxylic Acids and Esters* (Ed. S. Patai), Interscience, New York, 1969, p. 1.
8. M. B. Robin, F. A. Bovey, and H. Basch in *The Chemistry of Amides* (Ed. J. Zabicky), Interscience, New York, 1970, p. 1.
9. G. Fodor and B. A. Phillips, in *The Chemistry of Amidines and Imidates* (Ed. S. Patai), Interscience, New York 1975, p. 85.
10. G. I. L. Jones and N. L. Owen, *J. Mol. Structure*, **18**, 1 (1973).
11. H. E. Hallam and C. M. Jones, *J. Mol. Structure*, **5**, 1 (1970).
12. W. Walter and E. Schaumann, *Chem. Ber.*, **104**, 3361 (1971).
13. W. Klyne and V. Prelog, *Experientia*, **16**, 521 (1960).
14. O. Exner, *Dipole Moments in Organic Chemistry*, Thieme, Stuttgart, 1975.
15. V. I. Minkin, O. A. Osipov, and Yu. A. Zhdanov, *Dipole Moments in Organic Chemistry*, (English translation), Plenum Press, New York, 1970.
16. O. Exner, *Coll. Czech. Chem. Commun.*, **35**, 187 (1970).

17. R. J. W. Le Fèvre, *Adv. Phys. Org. Chem.*, **3**, 1 (1965).
18. R. J. W. Le Fèvre, *Rev. Pure Appl. Chem.*, **20**, 67 (1970).
19. A. N. Vereshchagin, O. N. Nuretdinova, L. Z. Nikonova, and B. A. Arbuzov, *Izv. Akad. Nauk S.S.S.R. Ser. Khim.*, 758 (1971).
20. C. L. Cheng, D. S. N. Murthy, and G. L. D. Ritchie, *Austral. J. Chem.*, **25**, 1301 (1972).
21. J. Errera, *J. Phys. Radium*, **6**, 390 (1925).
22. V. W. Laurie, *J. Chem. Phys.*, **34**, 291 (1961).
23. E. Bergmann, *J. Chem. Soc.*, 402 (1936).
24. V. N. Vasileva, V. V. Perekalin, and V. G. Vasilev, *Zh. Obshch. Khim.*, **31**, 2171 (1961).
25. E. Bergmann, L. Engel, and H. Meyer, *Chem. Ber.*, **B65**, 446 (1932).
26. K. B. Everard, L. Kumar, and L. E. Sutton, *J. Chem. Soc.*, 2807 (1951).
26a. J. Errera, *Physik. Z.*, **29**, 689 (1928).
27. E. L. Eliel, *Stereochemistry of Carbon Compounds*, McGraw-Hill, New York, 1962, p. 326.
28. G. Henderson and A. Gajjar, *J. Org. Chem.*, **36**, 3834 (1971).
29. A. Demiel, *J. Org. Chem.*, **27**, 3500 (1962).
30. W. N. Moulton, *J. Chem. Educ.*, **38**, 522 (1961).
31. O. J. Sweeting and J. R. Johnson, *J. Amer. Chem. Soc.*, **68**, 1057 (1946).
32. C. G. Le Fèvre, R. J. W. Le Fèvre, and W. T. Oh, *Austral. J. Chem.*, **10**, 218 (1957).
33. N. Q. Trinh, *C. R. Acad. Sci. Paris*, **227**, 393 (1948).
34. A. Halleux and H. G. Viehe, *J. Chem. Soc. (C)*, 1726 (1968).
35. J. D. Hepworth, J. A. Hudson, D. A. Ibbitson, and G. Hallas, *J. Chem. Soc. Perkin II*, 1905 (1972).
36. J. Kuthan, V. Jehlička, and E. Hakr, *Coll. Czech. Chem. Commun.*, **32**, 4309 (1967).
37. H. Huber and G. F. Wright, *Can. J. Chem.*, **42**, 1446 (1964).
38. I. E. Coop and L. E. Sutton, *J. Chem. Soc.*, 1269 (1938).
39. S. G. Vulfson, A. N. Vereshchagin, and B. A. Arbuzov, *Dokl. Akad. Nauk S.S.S.R.*, **209**, 885 (1973).
40. L. Krasnec and J. Veselovská, in press.
41. E. Bergmann and A. Weizmann, *Chem. Rev.*, **29**, 553 (1941).
42. G. E. Hawkes, K. Herwig, and J. D. Roberts, *J. Org. Chem.*, **39**, 1017 (1974).
43. R. L. Lichter, D. E. Dorman, and R. Wasylishen, *J. Amer. Chem. Soc.*, **96**, 930 (1974).
44. H. Lumbroso and G. Pifferi, *Bull. Soc. Chim. France*, 3401 (1969).
45. A. Dondoni and O. Exner, *J. Chem. Soc. Perkin II*, 1908 (1972).
46. V. de Gaouck and R. J. W. Le Fèvre, *J. Chem. Soc.*, 741 (1938).
47. J. Bernstein and G. M. J. Schmidt, *J. Chem. Soc. Perkin II*, 951 (1972).
48. V. de Gaouck and R. J. W. Le Fèvre, *J. Chem. Soc.*, 1392 (1939).
49. T. Prot, W. Waclawek, and V. I. Minkin, *Roczniki Chem.*, **47**, 2159 (1973).
50. V. I. Minkin, Yu. A. Zhdanov, A. D. Garnovskii, and I. D. Sadekov, *Zh. Fiz. Khim.*, **40**, 657 (1966).
51. V. I. Minkin, Yu. A. Zhdanov, A. D. Garnovskii, and I. D. Sadekov, *Dokl. Akad. Nauk S.S.S.R.*, **162**, 108 (1965).
52. J. Bjørgo, D. R. Boyd, C. G. Watson, and W. B. Jennings, *J. Chem. Soc. Perkin II*, 757 (1974).

53. D. Pitea, D. Grasso, and G. Favini, *J. Chem. Soc.* (*B*), 2290 (1971).
54. B. Greenberg and J. G. Aston, *J. Org. Chem.*, **25**, 1894 (1960).
55. K. A. Jensen and N. H. Bang, *Liebigs Ann. Chem.*, **548**, 106 (1941).
56. H. Lumbroso and D. M. Bertin, *Bull. Soc. Chim. France*, 1728 (1970).
57. O. Exner and O. Schindler, *Helv. Chim. Acta*, **55**, 1921 (1972).
58. O. Exner and V. Jehlička, *Coll. Czech. Chem. Commun.*, **30**, 639 (1965).
59. C. O. Meese, W. Walter, and M. Berger, *J. Amer. Chem. Soc.*, **96**, 2259 (1974).
60. M. Buděšinský, O. Schindler, and O. Exner, unpublished results.
61. E. L. Yeh, R. M. Moriarty, C. L. Yeh, and K. C. Ramey, *Tetrahedron Letters*, 2655 (1972).
62. F. Feichtmayr and F. Würstlin, *Ber. Bunsen Ges. Phys. Chem.*, **67**, 434 (1963).
63. O. Exner, V. Jehlička, and A. Dondoni, *Coll. Czech. Chem. Commun.*, in press.
64. W. Theilacker and K. Fauser, *Liebigs Ann. Chem.*, **539**, 103 (1939).
65. W. D. Kumler, *J. Amer. Chem. Soc.*, **75**, 3092 (1953).
66. K. Folting, W. N. Lipscomb, and B. Jerslev, *Acta Crystallogr.*, **17**, 1263 (1964).
67. K. E. Calderbank and R. J. W. Le Fèvre, *J. Chem. Soc.*, 1462 (1949).
68. F. Bachechi and L. Zambonelli, *Acta Crystallogr.*, **B28**, 2489 (1972).
69. R. S. Rogowski and R. H. Schwendeman, *J. Chem. Phys.*, **50**, 397 (1969).
70. S. Basu and C. W. N. Cumper, *J. Chem. Soc. Perkin II*, 325 (1974).
71. J. Barassin, J. Armand, and H. Lumbroso, *Bull. Soc. Chim. France*, 3409 (1969).
72. G. S. Parsons and C. W. Porter, *J. Amer. Chem. Soc.*, **55**, 4745 (1933).
73. R. J. W. Le Fèvre and J. Northcott, *J. Chem. Soc.*, 2235 (1949).
74. M. Milone and G. Tappi, *Atti X° Congr. Intern. Chim, Rome*, **2**, 352 (1938); *Chem. Abstr.*, **33**, 7633 (1939).
75. R. J. W. Le Fèvre, R. K. Pierens, and K. D. Steel, *Austral. J. Chem.*, **19**, 1769 (1966).
76. I. Pejković-Tadić, K. Hranisavljević-Jakovljević, S. Nešić, S. Pascual, and W. Simon, *Helv. Chim. Acta*, **48**, 1157 (1965).
77. F. J. Weigert, *J. Org. Chem.*, **37**, 1314 (1972).
78. K. D. Berlin and S. Rengaraju, *J. Org. Chem.*, **36**, 2912 (1971).
79. G. J. Karabatsos and R. A. Taller, *J. Amer. Chem. Soc.*, **75**, 3624 (1963).
80. A. Reiser, V. Jehlička, and K. Dvořák, *Coll. Czech. Chem. Commun.*, **16**, 13 (1951).
81. T. W. J. Taylor and L. E. Sutton, *J. Chem. Soc.*, 63 (1933).
82. E. J. Grubbs and J. A. Villarreal, *Tetrahedron Letters*, 1841 (1969).
83. M. Milone, *Gazz. Chim. Ital.*, **65**, 94 (1935).
84. C. Pigenet, J. Armand, and H. Lumbroso, *Bull. Chem. Soc. France*, 2124 (1970).
85. J. P. Guetté, J. Armand, and L. Lacombe, *C.R. Acad. Sci. Paris Ser. C.*, **264**, 1509 (1967).
86. K. A. Kerr, J. M. Robertson, and G. A. Sim, *J. Chem. Soc.* (*B*), 1305 (1967).
87. S. L. Spassov, G. Heublein, A. Jovtscheff, H. Reinheckel, and V. Jehlička, *Monatsh. Chem.*, **98**, 1682 (1967).
88. O. Exner, J. Hollerová, and V. Jehlička, *Coll. Czech. Chem. Commun.*, **32**, 2096 (1967).
89. E. Fischmann, C. H. McGillavry, and C. Romers, *Acta Crystallogr.*, **14**, 753 (1961).
90. A. Werner, *Chem. Ber.*, **26**, 1561 (1893).

91. O. Exner, V. Jehlička, and A. Reiser, *Coll. Czech. Chem. Commun.*, **24**, 3207 (1959).
92. O. Exner, *Coll. Czech. Chem. Commun.*, **30**, 652 (1965).
93. I. K. Larsen, *Acta Chem. Scand.*, **25**, 2409 (1971).
94. I. K. Larsen and O. Exner, *Chem. Commun.*, 254 (1970).
95. O. Exner and O. Schindler, *Helv. Chim. Acta*, **52**, 577 (1969).
96. O. Exner, M. H. Benn, and F. Willis, *Can. J. Chem.*, **46**, 1873 (1968).
97. J. H. Davies, R. H. Davis, and P. Kirby, *J. Chem. Soc.*, (*C*), 431 (1968).
98. O. Exner, *Coll. Czech. Chem. Commun.*, **27**, 2284 (1962).
99. E. M. Usova and E. M. Voroshin, *Dokl. Akad. Nauk S.S.S.R.*, **113**, 1306 (1957).
100. O. Exner, P. Dembech, G. Seconi, and P. Vivarelli, *J. Chem. Soc. Perkin II*, 1870 (1973).
101. D. Šnobl and O. Exner, *Coll. Czech. Chem. Commun.*, **34**, 3325 (1969).
102. D. Hall and F. J. Llewellyn, *Acta Cryst.*, **9**, 108 (1956).
103. W. J. Orville-Thomas and A. E. Parsons, *Trans. Faraday Soc.*, **54**, 460 (1958).
104. D. Bright, H. A. Plessius, and J. de Boer, *J. Chem. Soc. Perkin II*, 2106 (1973).
105. O. Exner, V. Jehlička, A. Dondoni, and A. C. Boicelli, *J. Chem. Soc. Perkin II*, 567 (1974).
106. A. Werner, *Chem. Ber.*, **25**, 27 (1892).
107. A. Battaglia, A. Dondoni, and O. Exner, *J. Chem. Soc. Perkin II*, 1911 (1972).
108. C. Pigenet, J. Armand, and H. Lumbroso, *C.R. Acad. Sci. Paris, Ser. C.*, **275**, 63 (1972).
109. J. E. Johnson, E. A. Nalley, and. C. Weidig, *J. Amer. Chem. Soc.*, **95**, 2051 (1973).
110. Yu. P. Kitaev, B. I. Buzykin, and T. V. Troepolskaya, *Uspekhi Khim.*, **39**, 961 (1970).
111. K. A. Jensen and B. Bak, *J. Prakt. Chem.*, **151**, 167 (1938).
112. Yu. P. Kitaev, S. A. Flegontov, and T. V. Troepolskaya, *Izv. Akad. Nauk S.S.S.R., Ser. Khim.*, **12**, 2086 (1966).
113. C. Pigenet, J. M. Kliegman, and H. Lumbroso, *C.R. Acad. Sci. Paris, Ser. C.*, **271**, 106 (1970).
114. R. G. Dubenko, P. S. Pelkis, and I. A. Sheka, *Ukr. Khim. Zh.*, **26**, 48 (1960).
115. J. Hamer and A. Macaluso, *Chem. Rev.*, **64**, 473 (1964).
116. L. E. Sutton and T. J. W. Taylor, *J. Chem. Soc.*, 2190 (1931).
117. F. Barrow and F. J. Thorneycroft, *J. Chem. Soc.*, 773 (1939).
118. T. Kubota, M. Yamakawa, and Y. Mori, *Bull. Chem. Soc. Japan*, **36**, 1552 (1963).
119. V. I. Minkin, E. A. Medyantseva, I. M. Andreeva, and G. V. Gorshkova, *Zh. Org. Khim.*, **9**, 148 (1973).
120. J. Bjørgo, D. R. Boyd, and D. C. Neill, *J. Chem. Soc. Chem. Commun.*, 478 (1974).
121. D. R. Boyd, personal communication (1974).
122. A. D. Baker, J. E. Baldwin, D. P. Kelly, and J. DeBernardis, *Chem. Commun.*, 344 (1969).
123. O. Exner and V. Hanuš, unpublished results.
124. F. Nerdel and I. Huldschinsky, *Chem. Ber.*, **86**, 1005 (1953).
125. R. Huisgen, *Angew. Chem.*, **75**, 742 (1963).
126. R. Grée and R. Carrié, *Tetrahedron Letters*, 4117 (1971).
127. R. Grée, personal communication, 1972.

128. G. S. Hartley and R. J. W. Le Fèvre, *J. Chem. Soc.*, 531 (1939).
129. D. J. W. Bullock, C. W. N. Cumper, and A. I. Vogel, *J. Chem. Soc.*, 5316 (1965).
130. L. Jannelli, O. Sciacovelli, and C. Carpanelli, *Ann. Chim. Rome*, **53**, 1541 (1963).
131. C. G. Overberger, J.-P. Anselme, and J. R. Hall, *J. Amer. Chem. Soc.*, **85**, 2752 (1963).
132. E. A. Shott-Lvova and Ya. K. Syrkin, *Izv. Akad. Nauk S.S.S.R., Otdel. Khim. Nauk*, 381 (1954).
133. R. S. Armstrong and R. J. W. Le Fèvre, *Aust. J. Chem.*, **19**, 29 (1966).
134. S. Millefiori, F. Zuccarello, A. Millefiori, and F. Guerrera, *Tetrahedron*, **30**, 735 (1974).
135. J. Firl and O. Exner, unpublished results.
136. R. J. W. Le Fèvre and H. Vine, *J. Chem. Soc.*, 431 (1938).
137. R. J. W. Le Fèvre and J. Northcott, *J. Chem. Soc.*, 333 (1949).
138. H. C. Freeman and R. J. W. Le Fèvre, *J. Chem. Soc.*, 3128 (1950).
139. R. J. W. Le Fèvre and H. Vine, *J. Chem. Soc.*, 1878 (1938).
140. R. J. W. Le Fèvre and T. H. Liddicoet, *J. Chem. Soc.*, 2743 (1951).
141. H. C. Freeman, R. J. W. Le Fèvre, and I. R. Wilson, *J. Chem. Soc.*, 1977 (1951).
142. H. C. Freeman, R. J. W. Le Fèvre, J. Northcott, and I. Youhotsky, *J. Chem. Soc.*, 3381 (1952).
143. K.-A. Gehrekens and E. Müller, *Liebigs Ann. Chem.*, **500**, 296 (1933).
144. V. Baliah and V. Chandrasekharan, *Indian J. Chem.*, **8**, 1096 (1970).
145. H. C. Freeman, R. J. W. Le Fèvre, J. Northcott, and C. V. Worth, *J. Chem. Soc.*, 3384 (1952).
146. B. G. Gowenlock and W. Lüttke, *Quart. Rev.*, **12**, 321 (1958).
147. J. H. Boyer in *The Chemistry of the Nitro and Nitroso Groups* (Ed. H. Feuer), Interscience, New York, 1969, p. 215.
148. M. Azoulay, T. Drakenberg, and G. Wettermark, *Tetrahedron Letters*, 2243 (1974).
149. D. L. Hammick, R. G. A. Nero, and R. B. Williams, *J. Chem. Soc.*, 29 (1934).
150. J. W. Smith, *J. Chem. Soc.*, 1124 (1957).
151. K. A. Jensen and A. Berg, *Liebigs Ann. Chem.*, **548**, 110 (1941).
152. M. V. George, R. W. Kierstead, and G. F. Wright, *Can. J. Chem.*, **37**, 679 (1959).
153. J. F. King and T. Durst, *J. Amer. Chem. Soc.*, **85**, 2676 (1963).
154. B. Zwanenburg, L. Thijs, and J. Strating, *Rec. Trav. Chim.*, **90**, 614 (1971).
155. G. Kresze and H. Smalla, *Chem. Ber.* **92**, 1042 (1959).
156. L. Jannelli, U. Lamanna, and H. Lumbroso, *Bull. Soc. Chim., France*, 3626 (1966).
157. J. S. Bonham, C. L. Chang, R. J. W. Le Fèvre, and G. L. D. Ritchie, *Aust. J. Chem.*, **26**, 421 (1973).
158. H. F. van Woerden and S. H. Bijl-Vlieger, *Rec. Trav. Chim.*, **93**, 85 (1974).
159. J. R. Grunwell and W. C. Danison, *Tetrahedron*, **27**, 5315 (1971).
160. G. Leandri, V. Busetti, B. Valle, and M. Mammi, *Chem. Commun.*, 413 (1970).
161. H. J. Becher and H. Diehl, *Chem. Ber.*, **98**, 526 (1965).
162. E. Rutner and S. H. Bauer, *J. Amer. Chem. Soc.*, **82**, 298 (1960).

163. M. E. Kronenberg and E. Havinga, *Rec. Trav. Chim.*, **84**, 979 (1965).
164. S. V. Tsukerman, V. D. Orlov, Yu. N. Surov, and V. F. Lavrushin, *Zh. Strukt. Khim.*, **9**, 67 (1968).
165. R. Bramley and R. J. W. Le Fèvre, *J. Chem. Soc.*, 56 (1962).
166. G. T. O. Martin and J. R. Partington, *J. Chem. Soc.*, 1178 (1936).
167. P. H. Cureton, C. G. Le Fèvre, and R. J. W. Le Fèvre, *J. Chem. Soc.*, 4447 (1961).
168. K. Hagen and K. Hedberg, *J. Amer. Chem. Soc.*, **95**, 8266 (1973).
169. D. R. Lide, *Trans. Amer. Cryst. Assoc.*, **2**, 106 (1966).
170. L. V. Vilkov, N. I. Sadova, and N. A. Tarasenko, *Zh. Strukt. Khim.*, **10**, 403 (1969).
171. K. Hagen and K. Hedberg, *J. Amer. Chem. Soc.*, **95**, 1003 (1973).
172. M. J. Aroney, D. Izsak, and R. J. W. Le Fèvre, *J. Chem. Soc.*, 3997 (1962).
173. R. A. Abramovitch, *Can. J. Chem.*, **37**, 1146 (1959).
174. P. Stäglich, K. Thimm, and J. Voss, *Liebigs. Ann. Chem.*, 671 (1974).
175. G. Oehme and A. Schellenberger, *Chem. Ber.*, **101**, 1499 (1968).
176. K.-M. Marstokk and H. Møllendal, *J. Mol. Struct.*, **20**, 257 (1974).
177. O. Exner and J. M. Kliegman, *J. Org. Chem.*, **36**, 2014 (1971).
178. S. Sorriso, G. Piazza, and A. Foffani, *J. Chem. Soc.* (*B*), 805 (1971).
179. R. W. Baker, G. P. Jones, and P. Pauling, *Nature*, **242**, 326 (1973).
180. R. J. B. Marsden and L. E. Sutton, *J. Chem. Soc.*, 1383 (1936).
181. O. Exner, Z. Fidlerová, and V. Jehlička, *Coll. Czech. Chem. Commun.*, **33**, 2019 (1968).
182. V. Baliah and K. Ganapathy, *Trans. Faraday Soc.*, **59**, 1784 (1963).
183. E. N. Guryanova and N. I. Grishko, *Zh. Strukt. Khim.*, **4**, 368 (1963).
184. E. N. Guryanova, G. N. Ignatova, and N. I. Grishko, *Zh. Strukt. Khim.*, **11**, 665 (1970).
185. E. Bock, *Can. J. Chem.*, **45**, 2761 (1967).
186. G. Krishna, B. Prakash, and S. V. Mahadane, *J. Phys. Chem.*, **73**, 3697 (1969).
187. S. Fasone, D. Grasso, and C. Gandolfo, *J. Mol. Struct.*, **20**, 449 (1974).
188. T. L. Pendred, A. A. Pritchard, and R. E. Richards, *J. Chem. Soc.* (*A*), 1009 (1966).
189. M. Ōki and H. Nakanishi, *Bull. Chem. Soc. Japan*, **43**, 2558 (1970).
190. T. Drakenberg and S. Forsén, *J. Phys. Chem.*, **76**, 3582 (1972).
191. J. S. Byrne, P. F. Jackson, K. J. Morgan, and N. Unwin, *J. Chem. Soc. Perkin II*, 845 (1973).
192. J. M. O'Gorman, W. Shand, and V. Schomaker, *J. Amer. Chem. Soc.*, **72**, 4222 (1950).
193. M. J. Aroney, R. J. W. Le Fèvre, R. K. Pierens, and H. L. K. The, *Aust. J. Chem.*, **22**, 1599 (1969).
194. H. Lumbroso and P. J. W. Schuijl, *C.R. Acad. Sci. Paris, Ser. C.*, **264**, 925 (1967).
195. B. Krishna, S. C. Srivastava, and S. V. Mahadane, *Tetrahedron*, **23**, 4801 (1967).
196. B. Krishna and R. K. Upadhyay, *J. Chem. Soc.* (*A*), 3144 (1970).
197. R. J. W. Le Fèvre and A. Sundaram, *J. Chem. Soc.*, 3904 (1962).
198. M. Ōki and H. Nakanishi, *Bull. Chem. Soc. Japan*, **44**, 3144 (1971).
199. O. Rosado-Lojo, C. K. Hancock, and A. Danti, *J. Org. Chem.*, **31**, 1899 (1966).

88 Otto Exner

200. J. Bailey, S. Walker, and A. M. North, *J. Mol. Struct.*, **6**, 53 (1970).
201. W. M. Slie and T. A. Litovitz, *J. Chem. Phys.*, **39**, 1538 (1963).
202. I. Cibulka, Thesis, Institute of Chemical Technology, Prague 1973.
203. O. Exner, V. Jehlička, and J. Firl, *Coll. Czech. Chem. Commun.*, **36**, 2936 (1971).
204. J. E. Piercy and S. V. Subrahmanyam, *J. Chem. Phys.*, **42**, 1475 (1965).
205. H. Wennerström, S. Forsén, and B. Ross, *J. Phys. Chem.*, **76**, 2430 (1972).
206. N. L. Owen and N. Sheppard, *Proc. Chem. Soc.* (*London*), 264 (1963).
207. D. Cook, *J. Amer. Chem. Soc.*, **80**, 49 (1958).
208. J. Karpovich, *J. Chem. Soc.*, **22**, 1767 (1954).
209. G. Williams, N. L. Owen, and J. Sheridan, *Trans. Faraday Soc.*, **67**, 922 (1971).
210. S. Merlino, *Acta Cryst.*, **27B**, 2491 (1971).
211. R. Huisgen and H. Ott, *Tetrahedron*, **6**, 253 (1959).
212. P. Janák and O. Exner, *Coll. Czech. Chem. Commun.*, **40**, 2502 (1975).
213. R. F. Grant, D. W. Davidson, and P. Gray, *J. Chem. Phys.*, **33**, 1713 (1960).
214. O. Exner, P. Dembech, and P. Vivarelli, *J. Chem. Soc.* (*B*), 278 (1970).
215. P. Dembech, P. Vivarelli, V. Jehlička, and O. Exner, *J. Chem. Soc. Perkin II*, 488 (1973).
216. C. S. Brooks and M. E. Hobbs, *J. Amer. Chem. Soc.*, **62**, 2851 (1940).
217. C. Béguin and T. Gäumann, *Helv. Chim. Acta*, **41**, 1376 (1958).
218. N. Saraswathi and S. Soundararajan, *J. Mol. Struct.*, **4**, 419 (1969).
219. B. P. van Eijck, G. van der Plaats, and P. H. van Roon, *J. Mol. Struct.*, **11**, 67 (1972).
220. M. Ōki and M. Hirota, *Bull. Chem. Soc. Japan*, **34**, 374, 378 (1960).
221. S. Mizushima and M. Kubo, *Bull. Chem. Soc. Japan*, **13**, 174 (1938).
222. E. Bock and D. Iwacha, *Can. J. Chem.*, **45**, 3177 (1967).
223. M. Kashima, *Bull. Chem. Soc. Japan*, **25**, 79 (1952).
224. M. Ōki and H. Nakanishi, *Bull. Chem. Soc. Japan*, **45**, 1552 (1972).
225. D. G. Lister and N. L. Owen, *J. Chem. Soc. Faraday II*, **69**, 1036 (1973).
226. E. Bock, D. Iwacha, H. Hutton, and A. Queen, *Can. J. Chem.*, **46**, 1645 (1968).
227. R. Bacaloglu, I. Bacaloglu, C. Csunderlik, and G. Ostrogovich, *J. Prakt. Chem.*, **313**, 331 (1971).
228. S. van der Werf, T. Olijnsma, and J. B. F. N. Engberts, *Tetrahedron Letters*, 689 (1967).
229. S. van der Werf and J. B. F. N. Engberts, *Rec. Trav. Chim.*, **89**, 423 (1970).
230. M. Ōki and H. Nakanishi, *Bull. Chem. Soc. Japan*, **44**, 3148 (1971).
231. R. A. Bauman, *Tetrahedron Letters*, 419 (1971).
232. C. M. Lee and W. D. Kumler, *J. Amer. Chem. Soc.*, **83**, 4596 (1961).
233. I. J. Tickle and J. B. F. N. Engberts, *J. Chem. Soc. Perkin II*, 2031 (1973).
234. O. Exner and K. Bláha, *Coll. Czech. Chem. Commun.*, in press.
235. E. N. Guryanova and L. S. Kuzina, *Zh. Fiz. Khim.*, **30**, 616 (1956).
236. C. N. R. Rao, K. G. Rao, A. Goel, and D. Balasubramanian, *J. Chem. Soc.* (*A*), 3077 (1971).
237. H. Lumbroso and P. Reynaud, *C.R. Acad. Sci. Paris, Ser. C.*, **262**, 1739 (1966).
238. O. Exner, V. Jehlička, and A. Ohno, *Coll. Czech. Chem. Commun.*, **36**, 2157 (1971).

239. K. Šindelář and O. Exner, *Coll. Czech. Chem. Commun.*, **37**, 2734 (1972).
240. V. Jehlička, J. L. Piette, and O. Exner, *Coll. Czech. Chem. Commun.*, **39**, 1577 (1974).
241. H. Lumbroso and C. Pigenet, *C.R. Acad. Sci. Paris, Ser. C.*, **266**, 735, 1653 (1968); **267C**, 631 (1968).
242. A. Queen, T. A. Nour, and E. Bock, *Can. J. Chem.*, **47**, 343 (1969).
243. E. Taskinen and P. Liukas, *Acta Chem. Scand.*, **B28**, 114 (1974).
244. N. L. Owen and H. M. Slip, *Chem. Phys. Letters*, **5**, 162 (1970).
245. B. A. Trofimov, N. I. Shergina, A. S. Atavin, E. I. Kositsyna, A. V. Gusarov, and G. N. Gavrilova, *Izv. Akad. Nauk S.S.S.R., Ser. Khim.*, 116 (1972).
246. B. A. Trofimov, I. S. Emelyanov, M. E. Yaselman, A. S. Atavin, B. V. Prokopev, A. V. Gusarov, G. N. Vanyukhin, and M. M. Ovchinnikova, *Reakts. Sposobnost Org. Soedin.*, **6**, 934 (1969).
247. O. N. Vylegjanin, V. B. Modonov, and B. A. Trofimov, *Tetrahedron Letters*, 2243 (1972).
248. E. N. Guryanova, I. P. Goldshtein, E. N. Prilezhaeva, and L. V. Tsymbal, *Izv. Akad. Nauk S.S.S.R. Otd. Khim. Nauk*, 810 (1962).
249. E. N. Guryanova, V. I. Snegotskii, V. I. Laba, and E. N. Prilezhaeva, *Izv. Akad. Nauk S.S.S.R., Ser. Khim.*, 1157 (1971).
250. L. M. Kataeva, E. G. Kataev, and T. G. Mannafov, *Zh. Strukt. Khim.*, **10**, 830 (1969).
251. F. D. Verderame and J. G. Miller, *J. Phys. Chem.*, **66**, 2185 (1962).
252. J. R. Rittenhouse, W. Lobunez, D. Swern, and J. G. Miller, *J. Amer. Chem. Soc.*, **80**, 4850 (1958).
253. W. H. T. Davison, *J. Chem. Soc.*, 2456 (1951).
254. D. Belitskus and G. A. Jeffrey, *Acta Cryst.*, **18**, 458 (1965).
255. I. K. Larsen, *Acta Chem. Scand.*, **22**, 843 (1968).
256. M. G. Waite and G. A. Sim, *J. Chem. Soc. (B)*, 752 (1971).
257. B. J. Price and I. O. Sutherland, *Chem. Commun.*, 1070 (1967).
258. S. Göttlicher and P. Ochsenreiter, *Chem. Ber.*, **107**, 398 (1974).
259. S. Winstein and R. E. Wood, *J. Amer. Chem. Soc.*, **62**, 548 (1940).
260. M. Yasumi, *J. Chem. Soc. Japan*, **60**, 1208 (1939).
261. G. F. Longster and E. E. Walker, *Trans. Faraday Soc.*, **49**, 228 (1953).
262. B. Collingwood, H. Lee, and J. K. Wilmshurst, *Aust. J. Chem.*, **19**, 1637 (1966).
263. A. M. Sørensen, *Acta Chem. Scand.*, **25**, 169 (1971).
264. M. Ōki and H. Nakanishi, *Bull. Chem. Soc. Japan*, **44**, 3419 (1971).
265. J. E. Katon and M. D. Cohen, *Can. J. Chem.*, **52**, 1994 (1974).
266. O. Exner and V. Jehlička, unpublished results.
267. L. Radom, W. A. Lathan, W. J. Hehre, and J. A. Pople, *Aust. J. Chem.*, **25**, 1601 (1972).
268. M. Ōki and H. Nakanishi, *Bull. Chem. Soc. Japan*, **45**, 1993 (1972).
269. P. Rosmus, R. Mayer, K. Herzog, and E. Steger, *Tetrahedron Letters*, 4495 (1967).
270. D. Leibfritz and H. Kessler, *Chem. Commun.*, 655 (1970).
271. B. Ekelund, *Acta Acad. Aboensis, Math. Phys.*, **19**, 1 (1954).
272. J. H. Markgraf, G. A. Lee, and J. F. Skinner, *J. Phys. Chem.*, **72**, 2276 (1968).
273. O. Exner and V. Jehlička, *Coll. Czech. Chem. Commun.*, **35**, 1514 (1970).

274. P. A. Hopkins and R. J. W. Le Fèvre, *J. Chem. Soc.* (*B*), 338 (1971).
275. H. J. Vledder, F. C. Mijlhoff, J. C. Leyte, and C. Romers, *J. Mol. Struct.*, 7, 421 (1971).
276. A. J. de Kok and C. Romers, *Rec. Trav. Chim.*, 88, 625 (1969).
277. A. Boogaard, H. J. Geise, and F. C. Mijlhoff, *J. Mol. Struct.*, 13, 53 (1972).
278. W. Lobunez, J. R. Rittenhouse, and J. G. Miller, *J. Amer. Chem. Soc.*, 80, 3505 (1958).
279. P. F. Oesper and C. P Smyth, *J. Amer. Chem. Soc.*, 64, 768 (1942).
280. V. N. Vasileva and E. N. Guryanova, *Zh. Fiz. Khim.*, 33, 1976 (1959).
281. C. G. Le Fèvre and R. J. W. Le Fèvre, *J. Chem. Soc.*, 3373 (1950).
282. O. Exner and B. Plesničar, *J. Org. Chem.*, 39, 2812 (1974).
283. S. Wolfe, *Accounts Chem. Res.*, 5, 102 (1972).
284. W. E. Stewart and T. H. Siddal, *Chem. Rev.*, 70, 517 (1970).
285. J. E. Worsham and M. E. Hobbs, *J. Amer. Chem. Soc.*, 76, 206 (1954).
286. A. Kotera, S. Shibata, and K. Sone, *J. Amer. Chem. Soc.*, 77, 6183 (1955).
287. S. Mizushima, T. Simanouti, S. Nagakura, K. Kuratani, M. Tsuboi, H. Baba, and O. Fujioka, *J. Amer. Chem. Soc.*, 72, 349, (1950).
288. I. Suzuki, M. Tsuboi, T. Shimanouchi, and S. Mizushima, *Spectrochim. Acta*, 16, 471 (1960).
289. H. Lumbroso, *Bull. Soc. Chim. France*, 2132 (1970).
290. H. Lumbroso, C. Pigenet, and P. Reynaud, *C.R. Acad. Sci. Paris, Ser. C.*, 264, 732 (1967).
291. M. Kitano, T. Fukuyama, and K. Kuchitsu, *Bull. Chem. Soc. Japan*, 46, 384 (1973).
292. M. Kitano and K. Kuchitsu, *Bull. Chem. Soc. Japan*, 47, 631 (1974).
293. Y. Koyama, T. Shimanouchi, and Y. Iitaka, *Acta Cryst.*, 27B. 940 (1971).
294. C. J. Brown, *Acta Cryst.*, 21, 442 (1966).
295. T. H. Siddall, W. E. Stewart, and A. L. Marston, *J. Phys. Chem.*, 72, 2135 (1968).
296. J. W. Smith, *J. Chem. Soc.*, 4700 (1961).
297. M. J. Aroney, R. J. W. Le Fèvre, and A. N. Singh, *J. Chem. Soc.* (*B*), 1183 (1966).
298. K. G. Rao and C. N. R. Rao, *J. Mol. Struct.*, 15, 303 (1973).
299. G. N. Ramachandran, A. V. Lakshminarayanan, and A. S. Kolaskar, *Biochim. Biophys. Acta*, 303, 8 (1973).
300. G. N. Ramachandran and A. S. Kolaskar, *Biochim. Biophys. Acta*, 303, 385 (1973).
301. R. L. Jones and R. E. Smith, *J. Mol. Struct.*, 2, 475 (1968).
302. J. Smolíková, A. Vítek, and K. Bláha, *Coll. Czech. Chem. Commun.*, 38, 548 (1973).
303. J. Smolíková, Z. Koblicová, and K. Bláha, *Coll. Czech. Chem. Commun.*, 38, 532 (1973).
304. M. Tichý, K. Dušková, and K. Bláha, *Tetrahedron Letters*, 237 (1974).
305. O. Exner, H. P. Härter, V. Stauss, and O. Schindler, *Chimia* (*Switz.*), 26, 524 (1972).
306. R. Huisgen and H. Walz, *Chem. Ber.*, 89, 2616 (1956).
307. W. Walter and H. Hühnerfuss, *J. Mol. Struct.*, 4, 435 (1969).
308. C. Pigenet and H. Lumbroso, *Bull. Soc. Chim. France*, 3743 (1972).
309. R. C. Neumann and V. Jonas, *J. Org. Chem.*, 39, 929 (1974).

310. B. H. Bracher and R. W. H. Small, *Acta Cryst.*, **26B** 1705 (1970).
311. V. A. Granzhan, S. F. Malone, N. A. Barba, and S. K. Laktionova, *Izv. Akad. Nauk Mold. S.S.R., Ser. Biol. Khim. Nauk*, **2**, 71 (1971).
312. H. P. Härter, M. Neuenschwander, and O. Schindler, *Helv. Chim. Acta*, **54**, 649 (1971).
313. P. Bouchet, J. Elguero, R. Jacquier, and J.-M. Pereillo, *Bull. Soc. Chim. France*, 2264 (1972).
314. H. Lumbroso and J. Barassin, *Bull. Soc. Chim. France*, 3190 (1964).
315. F. Hansen and R. Grønbaek-Hazell, *Acta Chem. Scand.*, **23**, 1359 (1969).
316. D. Gattegno and A. M. Giuliani, *Tetrahedron*, **30**, 701 (1974).
317. R. Huisgen and J. Reinertshofer, *Liebigs Ann. Chem.*, **575**, 197 (1952).
318. E. A. Shott-Lvova and Ya. K. Syrkin, *Dokl. Akad. Nauk S.S.S.R.*, **87**, 639 (1952).
319. W. Walter and K.-P. Ruess, *Liebigs Ann. Chem.*, **746**, 54 (1971).
320. G. Lepore, S. Migdal, D. E. Blagdon, and M. Goodman, *J. Org. Chem.*, **38**, 2590 (1973).
321. C. Béguin and T. Gäumann, *Helv. Chim. Acta*, **41**, 1971 (1958).
322. A. Braibanti, A. Tiripicchio, and M. Tiripicchio-Camellini, *J. Chem. Soc. Perkin II*, 2116 (1972).
323. F. Grambal, J. Mollin, and M. Hejsek, *Monatsh. Chem.*, **101**, 120 (1970).
324. W. Walter, H. Hühnerfuss, A. Neye, and K.-P. Ruess, *Liebigs. Ann. Chem.*, 821 (1973).
325. G. Tóth, *Acta Chim. Acad. Sci. Hungariae*, **64**, 101 (1970).
326. C. M. Lee and W. D. Kumler, *J. Amer. Chem. Soc.*, **84**, 565, 571 (1962).
327. A. Laurent, E. Laurent, and N. Pellissier, *Tetrahedron Letters*, 2955 (1970).
328. E. A. Noe and M. Raban, *J. Amer. Chem. Soc.*, **96**, 1598 (1974).
329. L. H. Piette, J. D. Ray, and R. A. Ogg, *J. Chem. Phys.*, **26**, 1341 (1957).
330. R. J. W. Le Fèvre, R. K. Pierens, D. V. Radford, and K. D. Steel, *Aust. J. Chem.*, **21**, 1965 (1968).
331. V. M. Csizmadia, S. A. Houlden, G. J. Koves, J. M. Boggs, and I. G. Csizmadia, *J. Org. Chem.*, **38**, 2281 (1973).
332. G. J. Karabatsos and R. A. Taller, *J. Amer. Chem. Soc.*, **86**, 4373 (1964).
333. A. E. Lutskii and B. P. Kondratenko, *Zh. Obshch. Khim.*, **29**, 2077 (1959).
334. V. Všctečka and O. Exner, *Coll. Czech. Chem. Commun.*, **39**, 1140 (1974).
335. P. H. Gore, P. A. Hopkins, R. J. W. Le Fèvre, L. Radom, and G. L. D. Ritchie, *J. Chem. Soc. (B)*, 120 (1971).
336. R. J. W. Le Fèvre and A. J. Williams, *J. Chem. Soc.*, 1825 (1960).
337. M. J. Aroney, R. J. W. Le Fèvre, and S.-S. Chang, *J. Chem. Soc.*, 3173 (1960).
338. J. W. Smith, *J. Chem. Soc.*, 81 (1961).
339. H. Kofod, L. E. Sutton, P. E. Verkade, and B. M. Wepster, *Rec. Trav. Chim.*, **78**, 790 (1959).
340. H. Lumbroso and C. G. Andrieu, *Bull. Soc. Chim. France*, 1575 (1973).
341. V. A. Granzhan, *Zhur. Strukt. Khim.*, **9**, 1107 (1968).
342. K. B. Everard and L. E. Sutton. *J. Chem. Soc.*, 2312 (1949); 16 (1951).
343. C. L. Cheng, P. H. Gore, and G. L. D. Ritchie, *Aust. J. Chem.*, **26**, 867 (1973).
344. A. R. Katritzky, H. J. Keogh, S. Ohlenrott, and R. D. Topsom, *J. Amer. Chem. Soc.*, **92**, 6855 (1970).
345. K. Antoš, A. Martvoň, and P. Kristián, *Coll. Czech. Chem. Commun.*, **31**, 3737 (1966).

346. V. G. Vasilev, V. N. Vasileva, Yu. K. Maksyutin, and A. F. Volkov, *Zh. Org. Khim.,* **6**, 1864 (1970).
347. R. A. Y. Jones, A. R. Katritzky, and A. V. Ochkin, *J. Chem. Soc. (B),* 1795 (1971).
348. A. E. Lutskii, V. T. Alekseeva, and B. P. Kondratenko, *Zh. Fiz. Khim.,* **35**, 1706 (1961).
349. E. Bock, R. Wasylishen, B. E. Gaboury, and E. Tomchuk, *Can. J. Chem.,* **51**, 1906 (1973).
350. B. Klabuhn, E. Clausen, and H. Goetz, *Tetrahedron,* **29**, 1153 (1973).
351. V. Baliah and K. Aparajithan, *Tetrahedron,* **19**, 2177 (1963).
352. H. Lumbroso and R. Passerini, *Bull. Soc. Chim. France,* 311 (1957).
353. M. Sanesi and M. Lazzarone, *Ricerca Sci.,* **33**, (II–A), 299 (1963).
354. M. Aroney, R. J. W. Le Fèvre, and J. Saxby, *J. Chem. Soc.,* 1167 (1963).
355. G. C. Pappalardo and S. Pistarà, *Tetrahedron,* **28**, 1611 (1972).
356. G. C. Pappalardo and G. Ronsisvalle, *J. Mol. Struct.,* **16**, 167 (1973).
357. M. Grimaud, M. Loudet, R. Royer, and G. Pfister-Guillouzo, *Bull. Soc. Chim. France,* 1169 (1974).
358. G. Montaudo, P. Finocchiaro, E. Trivellone, F. Bottino, and P. Maravigna, *Tetrahedron,* **27**, 2125 (1971).
359. R. F. Rekker and W. T. Nauta, *Rec. Trav. Chim.,* **80**, 747 (1961).
360. G. Montaudo, P. Finocchiaro, and S. Caccamese, *J. Org. Chem.,* **38**, 170 (1973).

CHAPTER **2**

Liquid crystals with X=Y groups

J. P. Van Meter
Eastman Kodak Company, Rochester, New York, U.S.A.

I. INTRODUCTION

Research in the field of liquid crystals has gone through various periods of activity since the time the liquid crystalline state was identified as such

93

by Reinitzer[1] in 1888. During the last few years, interest in liquid crystals has intensified once again, primarily as a result of the wide variety of possible commercial applications. Such applications include electronic windows which go from clear to frosted with the flip of a switch, numeric displays for wrist watches, digital voltmeters, and computer terminals, point-of-sale advertising panels, flat-screen television, temperature indicators, non-destructive testing, and iridescent inks.

The term 'liquid crystal' has been used for all materials that are capable of forming the liquid crystalline or mesomorphic state, although most of these substances are actually crystalline solids at room temperature. The unusual properties that are normally associated with this intermediate state of matter are evident only when the material is in its mesomorphic state. A large part of the recent synthetic work involving liquid crystals was aimed at the preparation of liquid crystals spanning room temperature (ideally -54 to $71°C$), which is a primary requirement for most practical applications. As the understanding of the physical chemistry of these materials developed, and as the various applications became more sophisticated, liquid crystals were designed and prepared in such a way as to meet these new requirements. These structural changes had to be made in such a way that mesomorphism was still maintained in addition to satisfying the usual room temperature constraint. Hence, the preparation of new liquid crystals, although conceptually simple, has been a challenge for the chemist who must delicately balance a large number of physical properties into one molecule.

In contrast to the approach in the previous volumes of this series, where the primary subject has been the functional group, this chapter, of necessity, will concern itself primarily with the properties of the whole molecule as influenced by the functional group. In the case of liquid crystals the main purpose of the $X=Y$ functional group, the main functional group usually linking aromatic rings, is to help provide the geometric anisotropy which is a necessary condition for liquid crystallinity. A change in the functional group is then reflected in a change in the chemical and physical properties of the molecule as a whole. The purpose of this chapter is to provide a general overview with emphasis on recent work in the field. An extensive discussion of the preparation of $X=Y$ groups is not given, but rather, typical and recent methods that have been applied to liquid crystal preparation. Most of the early work in the field has been adequately covered by Gray[2]. Kast[3] has tabulated all known liquid crystals up to 1959. More recent tables[4] and now books edited by Gray[5] are now available. Other reviews covering all aspects of liquid crystals can be found in References 6–28. The line notation for thermotropic liquid crystalline phase transitions devised by Verbit[29] will be used throughout this chapter. In this notation K, S, N, C, I represent the solid, smectic, nematic, cholesteric, and isotropic phases, respectively. For example, K129N1931 signifies a crystal-to-nematic transition at 129 °C, the compound persists in the nematic phase to 193 °C which is the nematic-to-isotropic transition temperature. Monotropic transitions are recorded in parentheses.

II. CLASSIFICATION AND STRUCTURE OF THE LIQUID CRYSTALLINE STATE

The liquid crystalline state is intermediate between that of the crystalline solid and the liquid phase, the associated order being less than that of the solid state but more than the liquid phase. These materials have some of the optical properties of crystalline solids while physically appearing as turbid fluids with a range of viscosities. Compounds capable of forming the liquid crystalline phase are long, flat, and fairly rigid about the long axis of the molecule. Liquid crystals are subdivided into two general types: those that are formed by the action of heat (thermotropic), and those formed by the action of solvents (lyotropic) on certain compounds. Both of these processes may be considered as a stepwise breakdown of the crystal lattice leading to the liquid crystalline state. Increasing amounts of heat or solvent result in the completely disordered state of an isotropic liquid or a true solution for the thermotropic and lyotropic liquid crystals, respectively. Thermotropic liquid crystals can be classified as either *enantiotropic*, for materials which form the liquid crystalline state both on heating the crystalline solid or cooling the isotropic liquid, or *monotropic*, for those that are formed only upon supercooling the isotropic liquid to a temperature below the melting point of the solid. A monotropic mesophase is therefore metastable with respect to the solid. The ordering in the thermotropic and lyotropic phases can be classified as either smectic, nematic, or cholesteric (chiral nematic) as discussed below. This chapter will deal exclusively with thermotropic systems.

A. Smectic Phase

The most ordered liquid crystalline phase is known as the smectic phase. In Figure 1 is a schematic representation of the main features of this structure which is a parallel arrangement of the long axes of the molecules in layers. At least eight different modifications of the smectic structure have been identified or proposed[30]. The work of Sackman and Demus[31]

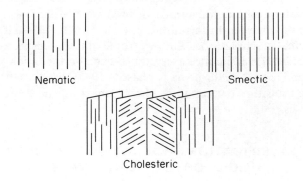

FIGURE 1. Molecular arrangement of liquid crystalline phases.

and of de Vries[32] has revealed some of the details of the molecular arrangement of these phases. The most common smectic phases are designated as
A, B, C. The smectic A modification corresponds to a parallel arrangement
of the molecules perpendicular to the plane of the layers with a random
arrangement of the molecules within the layer. The smectic C is similar to
the A modification, except that the molecules are tilted with respect to the
plane of the layers. The tilt angle may be temperature dependent[33]. The
smectic B structure has long-range order within the layers, and the long
axes of the molecules may be perpendicular or at an angle with respect
to the planes. De Gennes has proposed a notation B_A and B_C for these
variations[34]. For compounds that contain all three of these phases, the
order of occurrence with increasing temperature is B, C, A.

B. Nematic Phase

In the nematic phase the long axes of the molecules are still oriented
in a parallel fashion; however, the centres of gravity of the molecules are
now randomly distributed (Figure 1). In contrast to the smectic phase
there is generally considered to be only one nematic phase, although
de Vries[35], on the basis of an X-ray photographic study of bis-(4'-n-
octyloxybenzal)-2-chloro-1,4-phenylenediamine, has proposed the existence of a cybotactic nematic phase in which there is additional order
within the classical nematic structure. The turbid nematic phase is much
less viscous than the smectic phase.

C. Cholesteric Phase

Although this phase is sometimes classified as a special type of nematic
structure, the author believes that it should be treated as a separate class

because of its unique physical properties. The cholesteric phase (Figure 1) occurs only in optically active substances and corresponds to a twisted nematic layer which results in a helical structure. It consists of a layered structure resembling the smectic phase, but each layer has an order characteristic of a nematic phase. The pitch of the helix is the thickness corresponding to a rotation through 360° and is typically 0·2–20 µm. The characteristic iridescent colour of the cholesteric phase results from the selective reflection of one circularly-polarized component of light. The wavelength of the reflection band is determined by the pitch of the helical structure. Temperature, mechanical disturbances such as pressure or shear, and traces of organic vapour affect the pitch of the helix, which in turn results in various colour changes. The cholesteric phase is found in derivatives of cholesterol, but not in cholesterol itself. Materials that possess the general shape of a typical nematic structure but contain an asymmetric carbon atom are also cholesteric[36]. This general class of liquid crystals is referred to as 'chiral nematic' to distinguish them from those derived from cholesterol for which the adjective 'cholesteric' is used[37]. At one time, the helical structure of the cholesteric phase was attributed to steric interactions in cholesteryl molecules which resulted in a slight displacement of one layer from the next. This is now known to be untrue, as the chiral nematic materials indicate. Gray[38] has provided an example of the limiting case in which an optically active deuterated compound was shown to be cholesteric. The addition of optically active compounds to nematic materials also forms a cholesteric phase[39]. Although thermal transformations between the nematic and cholesteric phase do not occur, cholesteric materials with the opposite screw sense can be mixed in suitable proportions to give a compensated nematic liquid crystal[40]. Electric and magnetic fields can force a cholesteric liquid crystal into the nematic arrangement.

D. Forces and Molecular Order in the Liquid Crystalline State

The forces that are responsible for the long-range order in the liquid crystalline state have been the subject of many recent papers[41-51]. These theories deal with the involvement and non-involvement of dipole–dipole attractions, induced dipole attractions, and repulsion and dispersion interactions in maintaining the order in the liquid crystalline state; they are at present in a state of flux, as the theories are being refined and changed to accommodate new experimental data. Maier and Saupe[41,42] have described the degree of order within the nematic liquid crystal by an orientational order parameter S (equation 1), where θ is the angle

$$S = 1/2\langle 3\cos^2\theta - 1\rangle \tag{1}$$

between the long axis of a molecule and some preferred direction. S can be determined experimentally by several methods[19]. For a perfectly parallel orientation, $S = 1$ and for the isotropic phase, $S = 0$. Typical values for nematic liquids near the isotropic point are between 0·3 and 0·5. The thermal motion of the molecules tends to counteract this parallel orientation, which therefore results in a temperature dependence for the order parameter[52]. To a first approximation, the temperature dependence of the order parameter is reflected in the temperature dependence of most physical properties of the nematic liquid crystal[53].

III. DETECTION AND IDENTIFICATION OF LIQUID CRYSTALS

There are several methods that have been used to detect and identify liquid crystalline phases and their thermal transitions. These range from the simple observations of the melting behaviour in a capillary to the use of differential thermal analysis. In addition, in the course of studying other physical parameters, for example the dielectric anisotropy, phase transitions sometimes are indicated by a sudden change in the value of the measured physical property[54].

A. Visual Observation

The existence of liquid crystalline character in a given substance can usually be detected by observing the melting characteristics of the material. If the substance is an enantiotropic liquid crystal, the material is converted at the melting point into a turbid fluid which persists throughout the mesomorphic range until it becomes clear at the isotropic transition temperature. For materials with both a smectic and a nematic phase, the smectic phase, because of its higher viscosity, adheres to the capillary walls, and at the smectic-to-nematic transition the sudden change in viscosity is readily apparent as the less viscous nematic phase flows to the bottom of the capillary. In many cases, the transition temperatures obtained in this fashion for the smectic-to-nematic transition are very reproducible. For monotropic liquid crystals, the substance will exhibit the usual sharp melting characteristics, becoming a clear isotropic liquid. Upon supercooling to a temperature below the melting point, a turbid fluid is obtained. Some materials supercool more readily than others, and the mesomorphic phase is easily detected while other materials crystallize so easily that the monotropic phase goes undetected. Cholesteric materials are often indicated by the appearance of the characteristic iridescent colours somewhere within its liquid crystalline range. The shortcomings of this method are obvious, since it is normally impossible to detect any of the smectic-to-smectic thermal transitions, and in some cases the solid-to-smectic transition is not clearly defined. It should be noted, however, that for preliminary investigations of a large number of compounds, this method offers much in the way of efficiency.

Another visual or optical method of detecting mesophases is by the microscopic study of their textures[2]. Most recently Sackman and Demus[31] have used this method in connection with their miscibility studies of polymorphic liquid crystals. The preparation and identification of these textures can be difficult, and the optical method is usually used in conjunction with the thermal analytical methods discussed in the next section[55].

B. Thermal Methods of Liquid Crystal Detection

The availability of precision instruments for use in quantitative thermal investigations has provided the researcher with another means for the detection and identification of liquid crystalline phases[56]. Differential thermal analysis or differential scanning calorimetry are the most frequently employed. The thermodynamic order of liquid crystalline systems has been reviewed by Porter[57]. It has been generally assumed that the largest heat of transition was associated with the crystalline solid-to-mesomorphic transition, and that the remaining mesophase-to-mesophase and mesophase-to-isotropic transitions involved much smaller enthalpy changes[22,57]. Recently, several liquid crystalline compounds have been investigated in which the heat of transition for a mesophase-to-mesophase transition is larger than that of the crystal-to-mesophase transition. This fact is no longer surprising in view of the high degree of order associated with some smectic phases. One such compound is 4-butyloxybenzal-4'-ethylaniline[30].

IV. PREPARATION OF LIQUID CRYSTALS

Since a detailed study of the chemistry and formation of the various functional groups has already been given in previous volumes in this series, we shall consider only the chemistry which has been used in the liquid crystal field. We shall restrict our attention to the recent synthetic

work, most of which has been in the area of relating mesomorphic properties with changing molecular structure. Usually the work involves the
preparation of a homologous series of liquid crystals containing a given
functional group, or making a systematic change in the basic molecular
structure, particular attention being paid to the accompanying change in
the physical properties such as transition temperatures, dielectric anisotropy, dipole moments, or heats of transition. As mentioned in the
preceding sections, liquid crystals with low crystal-to-mesomorphic
transition temperatures have been the primary goal of organic chemists
in recent years. Rather than giving an extensive list of the transition
temperatures, we shall give some representative examples in each class of
liquid crystals, which will indicate some of the lowest-melting materials
that have been prepared. Other important physical properties will be cited
where appropriate.

In order to limit the length of this review, we shall not consider liquid
crystals containing heterocyclic rings here. Recent work in this area is
described in References 58–60.

A necessary condition for the formation of the liquid crystalline state
is a rod-shaped molecule with some degree of unsaturation which usually
takes the form of two or more aromatic rings connected by an even-
numbered functional group. This basic structure has various terminal
substituents which range from small polar groups such as halogen atoms,
cyano, or nitro groups to long-chain alkyl or alkoxy groups. The following
generalizations can be made with respect to the terminal substituents of
liquid crystals and their mesomorphic transition temperatures. In general,
predictions as to the crystal-to-mesomorphic transition temperature
cannot be made; however, as the carbon chain is lengthened, the tendency
towards lower-melting materials is increased. Liquid crystals of moderate
chain length are normally purely nematic; increasing the length of the
carbon chain leads to both smectic and nematic mesophases, and very
long chain lengths usually give only smectic phases. As the polarizability
of the terminal substituent is increased, the mesomorphic-to-isotropic
transition temperature is also increased. Branching in the carbon chain
may or may not lead to lower-melting materials; however, it is found that
the nematic-to-isotropic transition temperature is always decreased as a
result of the increase in the width of the molecule, which therefore decreases the lateral attractive forces which are responsible for liquid crystal
formation. Lateral ring substitution, which increases the width of the
molecule, also reduces the mesomorphic-to-isotropic transition temperature. Lateral substitution has been found to lower the crystal-to-mesomorphic transition temperature in some systems. These generalizations

are found to apply to most liquid crystalline systems, regardless of the functional group present in the molecule.

The following sections will give a general description of the preparation and properties of the various liquid crystals organized according to their main functional group. The main functional group is defined as the linkage between the aromatic rings.

A. Liquid Crystals Containing the Imine Group

Liquid crystals containing the imine function are prepared by the condensation of the aromatic aldehyde and an aromatic amine under a variety of conditions, which include heating the reactants with or without solvent or catalyst[61], refluxing in ethanol[62,63], or azeotropically removing the water in refluxing benzene with a sulphonic acid catalyst[64]. In general, the desired liquid crystal is formed directly by this reaction. In some cases, however, a benzylideneaniline intermediate is first prepared which is then converted to the desired liquid crystal by a subsequent reaction. This is illustrated by equation (2) for the preparation of p-alkoxybenzylidine p-aminophenols (1), which are then converted to the ester or carbonate derivative with the appropriate acid chloride or alkyl

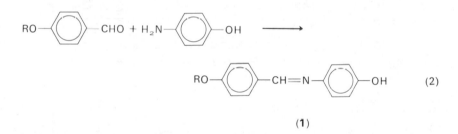

$$\tag{2}$$

(1)

chloroformate in the presence of pyridine or triethylamine[65,66]. Steinsträsser and Pohl[67] have prepared some ester derivatives of 1 by first forming the potassium salt.

For applications requiring liquid crystals with very high resistivity, it may be desirable to avoid the use of the sulphonic acid in the preparation of the liquid crystal, since the final traces of the catalyst may be difficult to remove.

Representative examples of terminally substituted benzylideneanilines are given in Table 1. Most notable is N-(p-methoxybenzylidene)-p-n-butylaniline (MBBA)[68], which is a relatively stable, room-temperature nematic material that has a negative dielectric anisotropy[70].

TABLE 1. Substituted benzylideneanilines

R—⟨benzene⟩—CH=N—⟨benzene⟩—R′

Compound	R	R′	Transition temperature(°C)	Reference
1	CH_3O	C_4H_9	K22N48I	68
2	C_2H_5O	C_4H_9	K37N80I	68
3	C_4H_9O	C_4H_9	K7S46N75I	62
4	$C_8H_{17}O$	C_2H_5	K65S71N75I	61
5	$C_{16}H_{33}O$	C_2H_5	K71S75·5I	61
6	C_3H_7O	C_3H_7O	K107N133I	63
7	C_3H_7O	C_4H_9	K41N59I	63
8	$CH_3OCH_2CH_2O$	C_4H_9	K34·5N56·6I	63
9	$CH_3\!-\!CH\!=\!CH\!-\!CH_2\!-\!O$	C_4H_9	K62·7N94·2I	69
10	$C_4H_9O\!-\!\overset{O}{\overset{\|}{C}}\!-\!O$	$O\!-\!\overset{O}{\overset{\|}{C}}\!-\!OCH_3$	K69N115I	66
11	$CH_3\!-\!\overset{O}{\overset{\|}{C}}\!-\!O$	OC_3H_7	K53N112I	67
12	CN	OC_8H_{17}	K73·2S82·8N107·5I	71

One important correlation of structure versus transition temperature can be made by comparing compounds 6 and 7 in Table 1, where it is found that the replacement of an oxygen atom adjacent to the phenyl ring with a methylene group results in a decrease in all the transition temperatures for the resulting liquid crystal. In this example the crystal-to-nematic transition temperature has been decreased by 66 °C as a result of this structure modification. Thus, terminal alkyl substitution can lead to lower-melting liquid crystals.

In addition to MBBA, another material which has stimulated a considerable amount of theoretical and experimental interest is N-p-cyano-benzylidene-p-n-octyloxyaniline (12 in Table 1). In particular, this compound was thought to have a second order nematic–smectic A phase transition[71]. Recently Torza and Cladis[72] have provided evidence that this transition is weakly first order.

The preparation of laterally substituted benzylideneanilines has been described by van der Veen[73] and others[74,75]. Most of these materials were found to melt at a lower temperature than the corresponding unsubstituted materials; however, the accompanying decrease in the nematic thermal stability was such that most of the materials resulted in very narrow enantiotropic nematic phases or monotropic character (2).

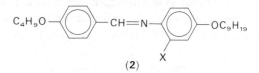

(2)

X – H, K101N113I; X = CH₃, K61N64I

One of the problems associated with the use of benzylideneanilines in practical applications has been the relative instability of the imine linkage. The preparation of some o-hydroxy-substituted liquid crystals, which are stabilized through hydrogen bonding[76], has been described. These materials appear to be chemically more stable than the corresponding unsubstituted materials. In general, o-hydroxy substitution increases the temperature of both the crystal-to-nematic and the nematic-to-isotropic transition, as indicated for the hydroxy analogue of MBBA (3).

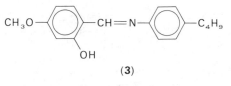

(3)

K45N65I

This class of liquid crystals appears to be an exception to the general rule that increasing the width of the molecule decreases the nematic-to-isotropic transition temperature. However, this should be considered as a special case, since it would be expected that internal hydrogen bonding will increase the overall polarizability of the molecule, which accounts for the increased thermal stability of the mesophase.

A series of bis(4'-n-alkoxybenzal)1,4-phenylenediamines (4) was prepared by Gray and coworkers[77] in 1955 and more recently reinvestigated by Arora and coworkers[78] who have detected a multiplicity of smectic phases in these materials. The heptyloxy derivative is reported to have eight first-order phase transitions. Arora[79] has also prepared a similar series derived from 2-chloro- and 2-methyl-1,4-phenylenediamine which exhibits low transition temperatures as a result of the lateral substituent.

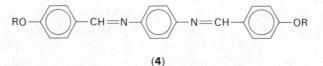

(4)

Terephthal-bis-butylaniline (TBBA) (5) is of interest since it has been shown to have the smectic A, C, B, and the nematic phase[80]. The tilt angle of the smectic C phase is temperature-dependent. This material has been studied by means of electron[81] and nuclear magnetic resonance[82]. Additional metastable phases of TBBA have recently been described[83].

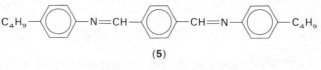

(5)

$$K113S_B144S_C172S_A200N236I$$

The preparation of several low-melting chiral nematic benzylideneanilines has been described by Dolphin and coworkers[84]. The active form of p-ethoxybenzal-p-(β-methylbutyl)aniline (6) is mesomorphic at room temperature. These materials are of interest since they can be combined with the corresponding racemic compounds to give mixtures with varying pitches while all the other thermodynamic properties remain constant.

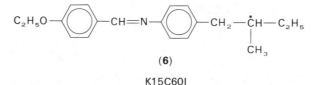

(6)

$$K15C60I$$

The preparation of the first optically active smectic liquid crystal (7) was described by Helfrich and Oh[85].

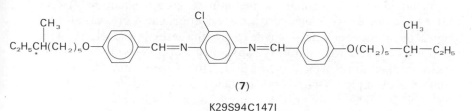

(7)

K29S94C147I

Since the occurrence of single-component, room-temperature liquid crystals is extremely rare, it has been necessary to mix two or more low-melting materials which may result in a room-temperature mesomorphic mixture. In most practical applications, this has been the general procedure for obtaining broad-ranged liquid crystalline materials spanning room temperature. In the case of some Schiff-base mixtures, it has been found that the physical properties varied from preparation to preparation. Sorkin and Denny[86] attributed this to the formation of new compounds by an exchange reaction (equation 3).

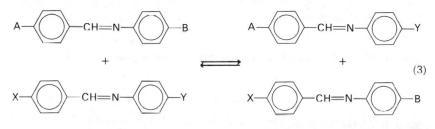

(3)

This reaction had been studied previously by Ingold and Piggott[87] and more recently by Toth and coworkers[88] who postulated an exchange reaction involving the interaction of a neutral and a protonated molecule, rather than a mechanism involving the free aldehyde and amine as suggested by Sorkin and Denny. This reaction leads to an equilibrium that is practically independent of the substituents due to the small differences in the change in free energy of the reactants and products. The reaction is catalysed by protons and other cations capable of quaternizing the azomethine group.

Some of the disadvantages of using liquid crystal materials containing the imine function have been their chemical and electrochemical instability. In this regard, several workers[89-92] have carried out detailed studies using MBBA as a model system.

B. Liquid Crystals Containing the Nitrone Group

Young and coworkers[93] have described the mesomorphic properties of some liquid crystals containing the nitrone group (equation 4). The intermediate hydroxylamine derivative was not isolated but was allowed to react with the aromatic aldehyde to give the desired product. All of the nitrones described were dialkoxy substituted and possessed very narrow mesomorphic ranges which melted above 100 °C with some decomposition, for example, N-(4-n-octyloxyphenyl)-α-(4-methoxyphenyl)nitrone (K112N128I). This study provided enough data to indicate that this class of materials appears to be of little practical value.

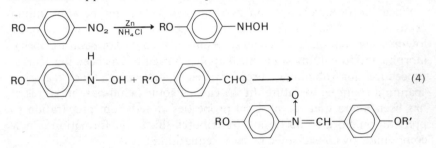

$$\qquad\qquad (4)$$

C. Liquid Crystals Containing the Imidoyl Cyanide Group

Van der Veen and de Jeu[94] have prepared some alkyl- and alkoxy-substituted imidoyl cyanides (equation 5). These materials were yellow

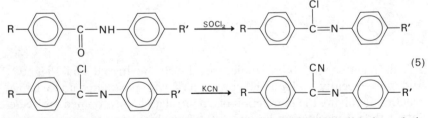

$$\qquad\qquad (5)$$

crystalline solids which formed monotropic mesophases well below their crystal-to-isotropic transition temperatures. The melting points were higher than those of the corresponding liquid crystals without the cyano group. The dielectric anisotropy of these materials was found to be slightly positive (+0·6).

D. Liquid Crystals Containing the Acetylenic Group

Verbit and Tuggy[95] have prepared a series of symmetrical p-alkoxy-phenyl acetylene dicarboxylates (8), which had narrow nematic ranges

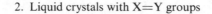
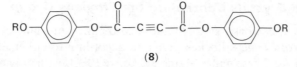

(8)

or were monotropic. A comparison of these compounds with the bis-*p*-alkoxyterephthalates revealed that the replacement of a phenyl ring by an acetylene group resulted in lower crystal-to-mesomorphic transition temperatures.

Some *p,p'*-disubstituted diphenyl acetylenes have been prepared by Malthete and coworkers[96] by the rearrangement of some halogenated diaryl ethanes or ethylenes in the presence of base (equation 6). Most of the

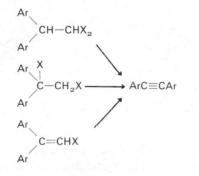

(6)

alkyl-alkoxyphenyl-substituted materials that were prepared melted below 50 °C. A comparison of *p,p'*-diethoxyphenylacetylene (K162I) with *p*-ethoxyphenyl-*p'*-propylphenyl acetylene (K90N98·5I) is another example of the significantly lower transition temperatures accompanying the substitution of a methylene group for an oxygen atom adjacent to the phenyl ring. The dielectric anisotropies of these materials were found to be negative for the dialkoxyphenyl-substituted materials and positive for the alkylphenyl alkoxyphenyl materials.

These workers were also the first to successfully apply the Schröder–Van Laar equation to the calculation of the eutectic composition and transition temperature in binary liquid crystalline systems[97]. Other examples of the use of this equation in mesophase systems can be found in the more recent literature[98,99].

An electrochemical investigation of some of these diphenylacetylenes indicated that they are electrically more stable than MBBA[100].

E. Liquid Crystals Containing the Ethylene Group

Young and coworkers[101] have studied the mesomorphic properties of some dialkoxy-*trans*-stilbenes prepared according to either equation (7) or equation (8). These materials melted above 100 °C with only monotropic mesophases observed. In this connection they devised the so-called

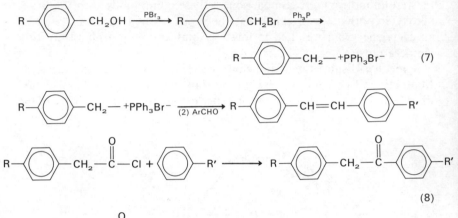

$$(7)$$

$$(8)$$

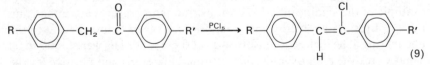

powder method for detecting monotropic mesophases. This consisted in preparing a fine powder of the crystalline material on a cover slide. After melting, the probability of super-cooling one of the small isotropic droplets to its mesomorphic state was thus increased. The same group[102] continued their work in the stilbene class by preparing some α-chloro-*trans*-stilbene derivatives (equation 9), some of which melted below 35 °C.

$$(9)$$

dl-4-(2-Methylhexyl)-4′-methoxy-α-chloro-*trans*-stilbene (K22N35I) falls into the rare class of single-component, room-temperature materials. Unfortunately, these materials were found to be light-sensitive, thus detracting from their usefulness in practical applications[38].

Van der Veen and Hegge[103] have prepared *p*-ethoxy-*p*′-hexyloxy-α-cyano-*trans*-stilbene (K54N80·51) by the condensation of the corresponding benzyl cyanide and benzaldehyde. This material has a highly negative dielectric anisotropy ($\Delta\varepsilon = -5\cdot5$ at 55°C)[104].

F. Liquid Crystals Containing the Azo Group

Steinsträsser and Pohl[105] have prepared some low-melting *p*-alkyl-*p*'-alkoxy and *n-p*-alkyl-*p*'-acyloxyazobenzenes (equation 10). Murase and Watanabe[106] have prepared some *p*-alkyl-*p*'-alkoxy azobenzenes via

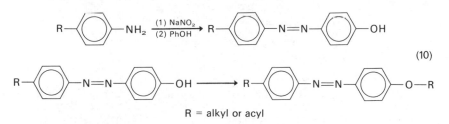

(10)

R = alkyl or acyl

a different route. The reduction of a *p*-alkylnitrobenzene gave the corresponding hydroxylamine in a 10–20% yield. Subsequent oxidation with ferric chloride gave the nitrosobenzene, which was then condensed with a *p*-alkoxyaniline to give the desired liquid crystal (equation 11).

$$ArNO_2 \xrightarrow[Zn]{NH_4Cl} ArNHOH \xrightarrow{FeCl_3} ArN=O$$

(11)

$$ArN=O + ArNH_2 \xrightarrow{HOAc} ArN=NAr$$

Van der Veen and his coworkers[107] have prepared some symmetrical *p,p*'-dialkylazobenzenes by the oxidation of *p*-alkylanilines with MnO$_2$. Most of these dialkyl substituted materials were low-melting, however, they exhibited only monotropic mesophases, with the exception of the diheptyl derivative (K40N47I) and some longer-chained smectic derivatives.

A comparison of the nematic-to-isotropic transition temperature with increasing carbon chain lengths for a series of azobenzenes with dialkoxy, alkoxy alkyl, and dialkyl substituents has been made[108] which illustrates the decrease in the transition temperature as the oxygen content of the molecule decreases. The dielectric anisotropy of the dialkylazobenzenes is positive. As the alkyl groups are replaced with alkoxy groups, $\Delta\varepsilon$ becomes smaller but never reaches a negative value[109].

Table 2 contains some representative examples of substituted azobenzenes: as the crystal-to-mesomorphic transition temperatures are lowered by virtue of alkyl substitution, the resulting mesophase becomes less thermally stable. This behaviour is typical of all liquid crystalline systems, in that as the intermolecular forces responsible for holding the

J. P. Van Meter

crystal lattice together become weaker, so do the forces necessary for maintaining liquid crystallinity.

TABLE 2. Substituted azo and azoxybenzenes

$$R-\bigcirc-X-\bigcirc-R'$$

R	R'	—N=N—	Reference	$\overset{\overset{\text{O}}{\uparrow}}{\text{—N=N—}}$	Reference
OC_4H_9	OC_4H_9	K135I(124N)	3	K107N134I	3
C_5H_{11}	OC_4H_9	K66N86I	105	K40N102I[a]	110
C_5H_{11}	C_5H_{11}	K48I(38N)	107	K22N65I	107

[a] Isomeric mixture.

G. Liquid Crystals Containing the Azoxy Group

Steinsträsser and Pohl[111] have prepared some p-alkyl-p'-alkoxy and p-alkyl-p'-acyloxyazoxybenzenes, which melt below 40 °C, by the oxidation of the corresponding azo compounds with hydrogen peroxide in acetic acid. Owing to the lack of stereospecificity in the oxidation of the unsymmetrical azo compounds, the resulting products were isomeric mixtures. The percent composition of the mixture was determined by n.m.r. spectroscopy. The azoxybenzene with butyl and methoxy substituents can be separated chromatographically[68] or by fractional crystallization[112] to give the individual isomers. The structure of the individual isomers was determined by spectroscopy using shift reagents[112]. As would be expected, the mixtures obtained from the oxidation melt at lower temperatures than the corresponding pure components. The group of van der Veen[107] has prepared some p,p'-dialkylazoxybenzenes by the oxidation of the corresponding alkylaniline with hydrogen peroxide under basic conditions. The dibutyl derivative is a room-temperature nematic material (K14N28I). McCaffrey and Castellano have prepared some p-alkoxy-p'-acyloxy-azoxybenzenes[113] by the performic acid oxidation of the corresponding azo compounds. The transition temperatures of the intermediate azo compounds, along with some thermodynamic properties, of the azoxy compounds are given in this paper. Additional p-alkyl-p'-alkoxyazoxybenzenes have been described by Murase and

Watanabe[106]. The melting points of the azoxy compounds are generally lower and the thermal stability of the mesophase is higher than the azo materials (Table 2).

Van der Veen has prepared some *o*-hydroxyazobenzenes by employing a coupling reaction (equation 12), followed by the alkylation via the potassium salt, to give the desired liquid crystal[114]. One of these materials (**9**) was found to be a room-temperature nematic material (K17N82I).

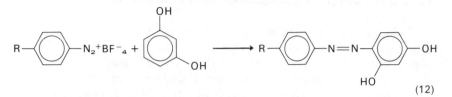

(12)

The alkyl-alkoxy derivatives were oxidized to the corresponding azoxy compounds, and in this case only one isomer was obtained (**10**). The

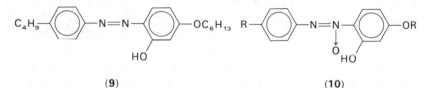

(**9**)   (**10**)

resulting azoxy compounds have higher crystal-to-nematic and nematic-to-isotropic transition temperatures than the corresponding azo compounds.

The measured dielectric anisotropies of the dialkyl azoxybenzenes are positive, the dialkoxy derivatives are negative[109]. De Jeu and collaborators[115] have measured the dielectric anisotropy of two mixtures of the isomeric methoxybutylazoxybenzenes, from which they have determined the value of the individual components. 4′-Butyl-4-methoxyazoxybenzene (**11**) was found to have a positive dielectric anisotropy, but that of the

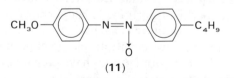

(**11**)

other isomer was negative. The dielectric properties of di-*n*-heptylazoxy-benzene, which has both a nematic and smectic phase, have been investigated[116]. Near the smectic-to-nematic transition, the positive dielectric anisotropy decreases and results in a change of sign as the material passes into the smectic phase.

A calorimetric study of the dialkylazoxybenzenes[117] has revealed a second-order nematic-to-smectic phase transition for the di-*n*-hexyl derivative[118].

H. Liquid Crystals Containing the Ester Group

Although cholesteric esters deviate from the previously defined description of the main functional group, in that the X=Y group links aliphatic moieties, this class of materials has been studied extensively. This work has involved preparation and study of various homologous series containing the basic cholesteryl structure[119] or a modified steroid derivative. Recent work in the area of steroid modification and references to early work on cholesterics in general can be found in a paper by Leder[120]. Cholesteric liquid crystals are relatively unstable to ambient conditions and the lack of long-term stability has caused problems in various practical applications such as temperature sensing. The stability can be improved by the use of u.v. absorbers and polymeric protecting media[121]. Most of the early work on cholesteric liquid crystals has been in the area of detection and determination of the transition temperatures; more recent work has dealt with the determination of the pitch[122] and rotatory sense[123] of the material.

Extensive investigations of liquid crystalline phenyl benzoate derivatives have appeared recently. The yellow colour of the azoxy compounds and the instability associated with Schiff bases have prompted work on liquid crystals containing the ester function. Castellano and coworkers[124] have prepared some *p*-alkoxyphenyl *p'*-alkylcarbonatobenzoates (equation 13). Several of these materials melted below 50 °C. Gray[38] has reinvestigated some of these materials and found different transition temperatures for his materials which he has attributed to impurities in the Castellano preparations. Steinsträsser[125] and Van Meter and Klanderman[126] have prepared a variety of *p,p'*-disubstituted phenyl benzoate derivatives which include some dialkyl-substituted materials that are isotropic liquids at room temperature but have enantiotropic nematic phases below room temperature (2 in Table 3). Both groups have prepared room-temperature nematic mixtures of these colourless esters. It is known that the dielectric anisotropy depends upon the terminal substituents in the phenyl benzoate system[127,128]. Thus, mixtures with either positive or negative dielectric character can be prepared. Table 3 contains a list of some representative phenyl benzoate liquid crystals.

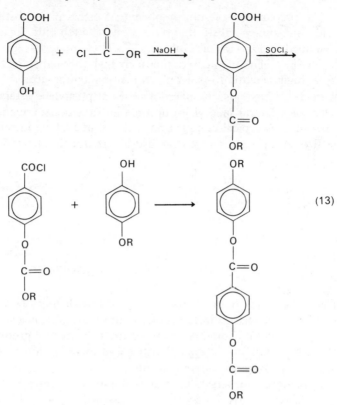

(13)

TABLE 3. Liquid crystalline phenyl benzoates

Compound number	R	R′	Transition temperature(°C)	Reference
1	$C_5H_{11}O-\overset{O}{\overset{\|}{C}}-O$	OC_6H_{13}	K46N74I	103
2	C_7H_{15}	C_4H_9	K9N15I	125
3	C_8H_{17}	$O-\overset{O}{\overset{\|}{C}}-C_5H_{11}$	K45N60I	125
4	OCH_3	OC_4H_9	K78N82I	126
5	OCH_3	C_5H_{11}	K30N44I	126

The preparation of some *p*-phenylene esters of *p*-substituted benzoic acids and some *p*-substituted phenyl esters of terephthalic acid have been described by Dewar and Goldberg[129]. These materials are characterized by having a nematic phase of relatively high thermal stability as indicated by a high nematic-to-isotropic transition temperature. Arora and co-workers[130] have also prepared a series of phenylene esters in which the effect of a lateral methyl group on the transition temperatures was investigated. A typical comparison is found in **12**. The introduction of the methyl group results in a decrease in the transition temperatures of all

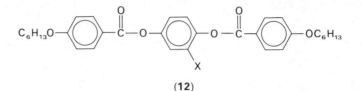

(**12**)

X = H, K124N213I; X = CH$_3$, K88N172I

phases; this has been attributed to the broadening of the molecule, as well as to the change in molecular conformation. It is also assumed that the increased dissymmetry, resulting from the methyl group, leads to less efficient packing in the crystal lattice and therefore lowers the crystal-to-mesomorphic transition temperature.

Haut and coworkers[131] have prepared some unsymmetrical *p*-phenylene di-*p*-alkoxybenzoates according to equation (14). Although the pre-

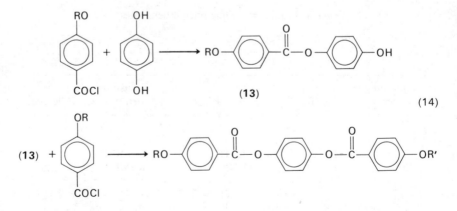

(14)

paration of the *p*-hydroxyphenyl *p*-alkoxybenzoates (**13**) involved the use of a large excess of hydroquinone, some symmetrical ester was usually produced which had to be removed prior to the formation of the un-symmetrical liquid crystalline esters. An alternative method for the preparation of these intermediates comes to mind, which involves the reaction of a *p*-alkoxybenzoyl chloride with *p*-benzyloxyphenol, a com-mercially available compound, to give a protected phenyl benzoate ester. Catalytic hydrogenolysis would give the desired intermediate **(13)** with toluene as the only by-product. The results of this work indicate that increasing the dissymmetry of the molecule by unsymmetrical terminal substituents is not as effective in lowering the crystal-to-mesomorphic transition temperatures as the introduction of a lateral substituent into the aromatic ring.

Several other investigators have studied the dibenzoate esters exten-sively in the hope of obtaining nematic liquid crystalline materials with low crystal-to-nematic transition temperatures. Young and collabora-tors[132] have prepared some ring-methylated phenyl benzoyloxybenzoates (equation 15). The preparation of these materials involved the formation

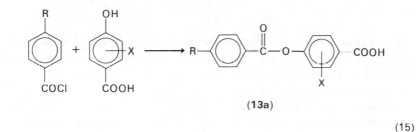

(13a)

(15)

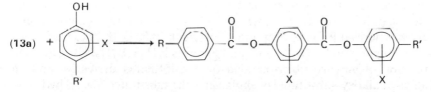

of intermediate 4-(4′-substituted benzoyloxy)benzoic acids **(13a)**. These materials were obtained in 50–80% yields by a procedure that involved a 4-h reflux in aqueous dioxane to hydrolyse the acid anhydride impurity formed in the reaction. The subsequent reaction of these intermediates via the acid chloride gave the desired liquid crystals. The materials prepared consisted of terminal butyl and ethoxy groups with one or more lateral methyl groups *ortho* to the oxygen of the ester linkage. The lowest-melting material in this study had a crystal-to-nematic transition tempera-ture of 91°C (1 in Table 4).

TABLE 4. Substituted phenyl p-benzoyloxybenzoates

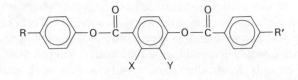

Compound number	R	X	Y	R'	Transition temperature	Reference
1	C_4H_9	H	CH_3	OC_2H_5	K91N179I	132
2	C_4H_9	H	H	C_4H_9	K89N183I	133
3	C_5H_{11}	H	Cl	C_5H_{11}	K67N130I	134
4	C_5H_{11}	Cl	H	C_5H_{11}	K39N122I	134
5	C_8H_{17}	Cl	H	C_7H_{15}	K39N105I	134
6	C_7H_{15}	Cl	H	C_5H_{11}	K33N114I	135
7	CN	Cl	H	C_7H_{15}	K69N160I	136

Steinsträsser[133] has also prepared some substituted phenyl benzoyloxy-benzoate liquid crystals. He prepared a variety of terminally substituted materials without lateral substitution; the lowest-melting material he obtained was the dibutyl derivative (2 in Table 4).

Van Meter and Klanderman[134,137] have also investigated the phenyl p-benzoyloxybenzoate system. It was shown that the crystal-to-meso-morphic transition temperatures of the unsymmetrical materials were significantly lower than those of the corresponding isomeric liquid crystals derived from hydroquinone or terephthalic acid. These materials were prepared in two steps: the first involves a H_2SO_4/H_3BO_3 catalysed reaction of p-hydroxybenzoic acid or a chlorinated derivative and an alkyl- or alkoxy-substituted phenol to give the substituted phenyl hydroxy-benzoate intermediate (equation 16). This combination of catalysts was discovered by Lowrance[138] as a new method for the direct esterification

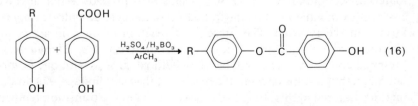

(16)

(14)

of phenols, and was applied with remarkable success to the preparation of these liquid crystal intermediates (14). The reaction was found to be very selective and free of polymer, the reaction occurring exclusively with the hydroxyl group on the phenol. The reaction of 14 and a benzoyl chloride gave the desired liquid crystal. From this study it was concluded that the combination of terminal dialkyl substituents and lateral chloro groups *ortho* to the carbonyl function of the ester provided low-melting materials. Mixtures of these materials have provided room-temperature compositions; for example, a 1:1 by weight mixture of compounds 4 and 5 in Table 4 has a nematic range of K5N111I[139]. Other dialkyl-substituted phenyl benzoyloxybenzoates with lateral chloro substitution were prepared by Van Meter and Seidel[135]. The lowest-melting compound obtained in this work was 6 (Table 4), melting at 33 °C.

A dielectric study[136] has indicated that the inherent dielectric anisotropy (measured at 10 kHz) for the liquid crystals derived from hydroquinone or terephthalic acid is negative, whereas it is positive for the phenyl benzoyloxybenzoate system. As would be expected, structural changes of the basic system can modify or change the sign of the dielectric anisotropy. The dielectric anisotropy of 7 in Table 4 has been found to be +35 at 97 °C[136].

Klanderman and Criswell[140] have prepared some chiral nematic liquid crystalline esters derived from optically active 4-(2-methylbutyl) benzoic acid. Low-melting materials with broad nematic ranges were found for the compounds in the phenyl *p*-benzoyloxybenzoate series. These liquid crystals are expected to have improved stability compared to conventional cholesteric compounds.

I. Liquid Crystals Containing the Biphenyl Nucleus

Before leaving the preparative chemistry of liquid crystals, mention should be made of the recent work of Gray and coworkers[141] who have prepared some low-melting cyanobiphenyls, for example, 4'-*n*-pentyl-4-cyanobiphenyl (K23N35I). In theory, the stability problems associated with the central linkages could be eliminated by removing the linkage as in the biphenyls. However, it has yet to be demonstrated whether or not these materials will be the ultimate in liquid crystals as far as stability in various applications is concerned. There has been one report[142] of erratic changes in the resistivity of these materials. Gray and coworkers[143] have also prepared an optically active biphenyl which is not mesomorphic; however, it does induce cholesteric properties when combined with mixtures of the other liquid crystalline biphenyls.

V. COMMENTS ON MESOPHASE THERMAL STABILITY

Of the many empirical generalizations that have been made in the course of studying liquid crystals, one of the more useful is a comparison of the relative thermal stability of the mesophases: a high thermal stability is reflected in a high mesophase-to-isotropic transition temperature[144]. The *length* of the mesomorphic range is *not* an indication of high thermal stability, as had been reported in one paper[145]. It has recently been pointed out that mesophase-to-isotropic transition temperature is not a reliable guide to the intermolecular forces associated with the liquid crystalline state[146]. However, considerations of the mesophase-to-isotropic transition temperature in relation to structural changes can give some insight into the anticipated thermal properties to be expected from a given liquid crystalline material. One example of this is the oxygen–methylene exchange which has been discussed in the previous sections. Another example is a comparison of the influence that various functional groups have on mesomorphic thermal stability. This comparison has been made by several workers[101,113,126,147], the most comprehensive being that of Rondeau and collaborators[112], who have prepared a series of diaryl derivatives all with methoxy-butyl terminal substituents. The order obtained, which is in agreement with the other workers, is as follows:

$$-O-\overset{\overset{\displaystyle O}{\|}}{C}- \; < \; -C\equiv C- \; < \; -CH=N- \; < \; -N=N- \; < \; CH=\overset{\overset{\displaystyle O}{\uparrow}}{N}- \; < \; -N=\overset{\overset{\displaystyle O}{\uparrow}}{N}- \; <$$

$$-CH=CH-$$

   This order is of interest in that one is able to predict the relative mesophase thermal stability to be expected with respect to a change in the central linkage. For example, the fact that the dialkyl-substituted phenyl benzoates are very low-melting isotropic liquids is understandable as we have incorporated two structural changes, dialkyl substituents and an ester linkage, both of which have weak intermolecular cohesive forces.

   No discussion of liquid crystals would be complete without mention of the well-known even–odd effect in the nematic-to-isotropic transition temperatures and entropies, which is found for homologous series of liquid crystals. As a carbon chain is lengthened, the nematic-to-isotropic transition temperature may either increase or decrease depending on the number of atoms in the terminal substituents. The amplitude of this change decreases with increasing carbon chain lengths[108]. A theoretical discussion of these facts in terms of the ordering of the molecular end chains has been given by Marcelja[148].

VI. APPLICATIONS OF LIQUID CRYSTALS

The driving force behind the intensive interest in liquid crystals is the utilization of their unique physical properties, in particular in the area of electrooptical applications, which include seven-segment display panels, television screens, and point-of-sale advertising panels. Both black-and-white and colour imaging systems are theoretically possible[149]. Some of the advantages of liquid crystal displays are low operating voltages, low power consumption, ability to withstand high levels of ambient light (liquid crystals modulate existing light rather than generate their own light). Some of the problems associated with liquid crystal cells include difficulty in cell fabrication, molecular surface alignment problems, lifetime, response times, and contrast ratio and viewing angle considerations.

Liquid crystals can also be used in temperature sensing, chromatography, as ultrasonic image detectors, and as solvents for spectroscopy.

A. Electrooptical Applications

Electric and magnetic fields are capable of inducing molecular re-orientation in liquid crystals which are reflected in a variety of optical changes. For practical applications, electric fields are generally employed. Most applications involve the use of nematic liquid crystals, although there are some electrooptical effects reported involving smectic phases. Cholesteric liquid crystals can be used in conjunction with nematics to give systems capable of storing information. The liquid crystal cell is essentially a parallel-plate capacitor composed of two conducting electrodes with the liquid crystal as a dielectric, and at least one of the electrodes is transparent. The liquid crystal layer is usually 10–20 μm thick and is separated by suitable spacers. The liquid crystal material is normally purified to a state of high purity and resistivity (10^{10}–10^{11} ohm cm) by means of recrystallization, distillation, chromatography, or zone refining[150], or by the use of electrodialytic membranes[151]. The material can then be doped to change its conductivity[152] or its alignment properties[22] depending on the type of electrooptical effect for which it is intended.

The construction of a liquid crystal cell requires surface preparation for uniform orientation of the liquid crystal. This is required in order to give the maximum transparency in the off or unexcited state as well as for the overall quality of the device. Although many methods of surface preparation have been developed, there appears to be no universal method that applies to all liquid crystalline systems. Usually the optimum surface treatment must be determined for each liquid crystal composition[153]. There are basically two types of surface orientation. One results in an alignment such that the long axes of the molecules are perpendicular to the cell surface (homeotropic), and the other is one in which the long axes of the molecules are parallel to the surface (homogeneous).

A recent review[23] adequately covers some of the early technology, which includes rubbing and chemical cleaning. Newer methods include the absorption of a mono-layer of a quaternary ammonium salt on the cell surface[154,155], vacuum depositing materials such as gold, silicon monoxide, or copper[156], and polymeric films[157]. Kahn[158] has described some silane coupling agents that interact with the surface to produce a permanent bond to the substrate. Silane coupling agents are available that will aid both types of surface alignment.

There are two main mechanisms by which an electric field interacts with the liquid crystal to produce the various electrooptical effects. One is purely dielectric and the other involves a condition-induced electrohydrodynamic interaction. The most common dielectric effects are induced birefringence, twisted nematic effect, the guest–host interaction, and the cholesteric-to-nematic phase change[15,21,159]. Dynamic scattering and the storage effect involve electronic conduction.

The dielectric anisotropy ($\Delta\varepsilon$) is the material parameter that determines the molecular orientation of the liquid crystal with respect to an electric field. It is defined as $\varepsilon_\parallel - \varepsilon_\perp$ where $\varepsilon_\parallel$ and $\varepsilon_\perp$ are the dielectric constants measured parallel and perpendicular to the long axis of the molecule. The dielectric anisotropy may be either positive or negative depending on whether the direction of maximum polarizability is parallel or perpendicular to the long axis of the molecule[160]. This can usually be determined by consideration of the magnitude and direction of the permanent dipole moment. In particular, the sign is always positive when the dipole moment is parallel to the long axis of the molecule[160,161]. In the case of p,p'-di-n-heptyloxyazobenzene, the dielectric anisotropy is positive even though the molecular dipole moment is perpendicular to the long molecular axis. For this compound, the direction of maximum polarizability is parallel to the long axis of the molecule, thus accounting for the positive anisotropy[160]. The dielectric anisotropy is dependent on temperature and frequency[161].

Once the dielectric anisotropy of a given compound is known, the molecular reorientations under the influence of an electric field can be understood easily and the mechanisms of the various electrooptical effects can be qualitatively visualized. For a material with a positive dielectric anisotropy and an initial homogeneous alignment, application of an electric field perpendicular to the electrode will result in a re-orientation such that the direction of maximum polarizability of the molecule will be parallel to the field (Figure 2). A similar rearrangement is possible with negative materials and an initial homeotropic alignment. These effects are known as Freedericksz transitions[38,159].

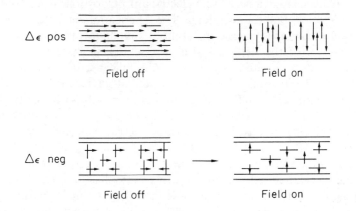

$\Delta\epsilon$ pos

Field off   Field on

$\Delta\epsilon$ neg

Field off   Field on

FIGURE 2. Electric field orientations.

Two representative electrooptical effects that require materials of different dielectric anisotropy and operate by different mechanisms will be given as illustrative examples. Other electrooptical effects along with their associated optics have been adequately discussed elsewhere[15,21]. One of the most widely investigated effects is the twisted nematic field effect[162], which requires a material with a positive dielectric anisotropy and a high resistivity. The basic cell configuration consists of a homo-geneously aligned liquid crystal cell that has one cell wall rotated 90° from the other with respect to the alignment direction. In a transmissive device between cross-polarizers the cell is transparent. The application of an electric field perpendicular to the cell surface forces the molecules away from the cell surfaces to assume a homeotropic alignment. The transmitted light is now extinguished. Removal of the electric field allows the molecules to relax back to their initial state, and the cell becomes transparent again. This effect is purely dielectric in nature.

Dynamic scattering[163] is another widely studied electrooptical effect. This effect requires a material with a negative dielectric anisotropy and a certain amount of conductivity. For a cell with an initial homogeneous alignment, the application of an electric field will cause the molecules to change from this dielectrically stable orientation as a result of an electro-hydrodynamic interaction. This conduction-induced turbulence causes birefringent regions which strongly scatter light. The liquid crystal cell goes from clear to frosted upon application of the electric field.

An application involving the use of cholesteric materials is known as the storage effect[164]. The composition used for this effect usually contains a nematic material with a negative dielectric anisotropy to which about 10% of a cholesteric material is added. The application of an electric field induces electrohydrodynamic instabilities similar to dynamic scattering. After removal of the field, the resulting scattering state remains. This scattering texture is caused by the random helical structures induced by the cholesteric material. Upon application of an electric field of high frequency, a dielectric effect takes over and unwinds the helix to the initial, clear, non-scattering state. This same storage effect can also be obtained by the addition of optically active materials to nematic materials[165].

Primarily as a result of the inherently higher viscosity of the smectic phase, electrooptical effects utilizing this mesophase are rare. An interesting use of a smectic liquid crystal described by Kahn[166] involves rapid cooling from the isotropic phase of a smectic material with a positive dielectric anisotropy. The rapid cooling freezes in the disorder characteristic of the isotropic state and the resulting texture is therefore capable of light scattering. Local erasure can be accomplished by heating to the nematic or isotropic state, followed by slow cooling to the non-scattering homeotropic texture. A faster response is obtained by the simultaneous application of an electric field, which dielectrically assists in forming the homeotropic texture. The latter field-induced reorientation results in high-speed erase capabilities.

In order to interface with today's microsecond technology, the response time of the liquid crystal device is of critical importance. Some recent observations will be mentioned. Response times can be improved or changed by several different methods. The doping of materials used in dynamic scattering improves the rise time[152]. The initial alignment of dynamic scattering cells affects the response times[167]. Homeotropically aligned cells are slower than homogeneously aligned cells, apparently because for homeotropic alignment the molecules must first realign in keeping with the negative dielectric anisotropy. The turn-off times of some devices have been observed to be dependent upon the type of treatment given the cell surface prior to cell construction[142].

The turn-on time of the device can be increased by increasing the applied field. On the other hand, the decay time is dependent on the elastic constants and viscosity of the material. There are some ester-containing liquid crystals available[139,168] that show a dielectric loss in $\varepsilon_\parallel$ at frequencies of several kHz. For materials with a positive dielectric anisotropy, this results in a change of the sign of the dielectric anisotropy with a change in frequency. Use of this fact can result in a decrease in the decay time of the liquid crystal device[139,169,170]. This simply involves a change in frequency, such that the dielectric anisotropy changes sign. The return to the initially aligned state is therefore dielectrically assisted by the high-frequency electric field, which decreases the decay time. This effect should find wide use in the practical application of liquid crystal devices.

B. Spectroscopic Studies with Liquid Crystals

The use of liquid crystals in u.v., i.r., e.s.r., and n.m.r. spectroscopy has become widespread in recent years[18,171,172]. N.m.r. and e.s.r. studies have been the most prevalent. Information concerning the nature of the liquid crystalline phase itself can be obtained[81,82], or the liquid crystalline phase can be used as a solvent which can provide information such as conformational preferences and bond lengths for the dissolved solute[173].

C. Liquid Crystals as Reaction Solvents

The possibility of altering the course of chemical reactions by conducting them in liquid crystalline solvents has been considered for some time. It has been demonstrated recently by Dewar and Nahlovsky[174] that for the Claisen rearrangement of cinnamyl phenyl ether in a nematic solvent, no change of the reaction products or rate occurred, even under the influence of a magnetic field which would increase the orientation of the nematic phase[178].

D. Temperature and Solvent Sensing

Depending on the pitch of the helix, cholesteric liquid crystals may selectively reflect light in the visible region of the spectrum. The pitch is temperature dependent and therefore, may be used to monitor temperature changes via colour changes. This technique has been applied to industrial uses, such as detecting short circuits on circuit boards and in the biomedical area for the detection of skin tumours. Low levels of solvent vapour, such as chloroform or benzene, can also change the pitch, which provides a means of solvent detection[175].

E. Liquid Crystals in Chromatographic Separations

The use of liquid crystals as stationary phases in chromatographic separation of structural isomers has been studied extensively[176,177]. The selectivity of liquid crystalline phases is a consequence of the molecular order. The observed chromatographic selectivity is a result of several factors including molecular shape, polarity, polarizability, and flexibility of the solute molecule. Work in this area is still in the early stages. The optimum conditions must be determined for each type of separation to be studied.

VII. CONCLUSION

Interest in liquid crystals has grown considerably in the last several years. At present there are approximately 5000 references to liquid crystals and their uses. The author has attempted to cover most of the areas of interest, citing the most recent references. The interested reader can refer to the references included in these papers to locate most of the prior work in any given area.

Many promises and predictions of a bright future for liquid crystals have been made during the last 10 years. The application of liquid crystals has been difficult to implement. There are probably two reasons for this. The first is due to the very reason they are useful in the first place; that is, the liquid crystalline state is very sensitive to all types of stimuli, and therefore, it is sometimes difficult to isolate the desired effect from extraneous interactions that may interfere with the desired application. The second involves the concept of technological coupling, which results from the fact that most of the interesting liquid crystal effects, especially in the electrooptical field, are relatively new and to a certain extent undeveloped; and the necessary coupling to more sophisticated technology is, as expected, difficult. The future in liquid crystals will show continued effort to bring liquid crystal devices to the market place and an increase in our fundamental understanding of the structure and forces involved in the liquid crystalline state.

VIII. REFERENCES

1. F. Reinitzer, *Monatsh. Chem.*, **9**, 421 (1888).
2. G. W. Gray, *Molecular Structure and the Properties of Liquid Crystals*, Academic Press Inc., London, 1962.

3. W. Kast, in *Landolt-Börnstein Zahlenwerte und Funktionen aus Physik, Chemie, Astronomie, Geophysik, und Technik*, 6th Ed., Springer, Berlin, 1960, Vol. II, Part 2a, p. 266.
4. D. Demus, H. Demus, and H. Zaschke, *Flüssige Kristalle in Tabellen*, VEB Deutscher Verlag für Grundstoffindustrie, Leipzig, 1974.
5. G. W. Gray, *Liquid Crystals and Plastic Crystals*, Vol. I and II, Wiley-Interscience, New York, 1974.
6. G. H. Brown and W. G. Shaw, *Chem. Rev.*, **57**, 1049 (1957).
7. I. G. Chistyakov, *Soviet Phys. Uspekhi*, **9**, 551 (1967).
8. G. Elliott, *Chem. Brit.*, **9**, 213 (1973).
9. R. Steinsträsser, *Chem. Ztg*, **95**, 661 (1971).
10. H. Liebig and K. Wagner, *Chem. Ztg*, **95**, 733 (1971).
11. D. B. DuPre, E. T. Samulski, and A. V. Tobolsky in *Polymer Science and Materials* (Ed. A. V. Tobolsky and H. F. Mark), Wiley-Interscience, 1971, Chapter 7.
12. H. Baessler, *Festkörperprobleme XI*, 99 (1971).
13. J. G. Grabmaier and H. H. Krueger, *Ver. Deut. Ing.* (VDI)Z., **115**, 629 (1973).
14. L. T. Creagh, *Proc. IEEE*, **61**, 814 (1973).
15. L. A. Goodman, *J. Vac. Sci. Technol.*, **10**, 804 (1973).
16. J. A. Castellano, *RCA Rev.*, **33**, 296 (1972).
17. J. L. Fergason, T. R. Taylor, and T. B. Harsch, *Electro-Technol.* (New York), **85** (1), 41 (1971).
18. R. Steinsträsser and L. Pohl, *Angew. Chem. Int. Ed.*, **12**, 617 (1973).
19. A. Saupe, *Angew. Chem. Int. Ed.*, **7**, 97 (1968).
20. G. H. Brown, J. W. Doane, and V. D. Neff, *Crit. Rev. Solid State Sci.*, 1(**3**), 303 (1970).
21. A. Sussman, *IEEE Trans. Parts, Hybrids, Packag.*, **8**, 24 (1972).
22. S. E. B. Petrie, H. K. Bücher, R. T. Klingbiel, and P. I. Rose, *Eastman Org. Chem. Bull.*, **45**, No. 2 (1973).
23. F. J. Kahn, G. W. Taylor, and H. Schonhorn, *Proc. IEEE*, **61**, 823 (1973).
24. L. L. Bonne and J. P. Cummings, *IEEE Trans. Electron Devices*, **ED-20**, 962 (1973).
25. A. R. Kmetz, *IEEE Trans. Electron Devices*, **ED 20**, 954 (1973).
26. G. Durand and J. D. Litster in *Annual Review of Materials Science* (Ed. R. A. Huggins, R. H. Bube, and R. W. Roberts), Annual Reviews Inc., Vol. 3, (1973).
27. A. Saupe in *Annual Review of Physical Chemistry* (Ed. H. Eyring, C. J. Christensen, and H. S. Johnston), Annual Reviews Inc., Vol. 24 (1973).
28. G. H. Brown and J. W. Doane, *Appl. Phys.*, **4**, 1 (1974).
29. L. Verbit, *Mol. Cryst. Liq. Cryst.*, **15**, 89 (1971).
30. A. de Vries and D. L. Fishel, *Mol. Cryst. Liq. Cryst.*, **16**, 311 (1972).
31. H. Sackman and D. Demus, *Mol. Cryst. Liq. Cryst.*, **21**, 239 (1973).
32. A. de Vries, *Mol. Cryst. Liq. Cryst.*, **20**, 119 (1973).
33. T. R. Taylor, S. L. Arora, and J. L. Fergason, *Phys. Rev. Lett.*, **25**, 722 (1970).
34. P. G. de Gennes, *Mol. Cryst. Liq. Cryst.*, **21**, 49 (1973).
35. A. de Vries, *Mol. Cryst. Liq. Cryst.*, **10**, 219 (1970).
36. G. W. Gary, *Mol. Cryst. Liq. Cryst.*, **7**, 127 (1969).
37. D. Dolphin, Z. Muljiani, J. Cheng, and R. B. Meyer, *J. Chem. Phys.*, **58**, 413 (1973).
38. G. W. Gray, *Mol. Cryst. Liq. Cryst.*, **21**, 161 (1973).

39. A. D. Buckingham, G. P. Ceaser, and M. B. Dunn, *Chem. Phys. Lett.*, **3**, 540 (1969).
40. E. Sackman, S. Meiboom, L. C. Snyder, A. E. Meixner, and R. E. Dietz, *J. Amer. Chem. Soc.*, **90**, 3567 (1968).
41. W. Maier and A. Saupe, *Z. Naturforsch.*, **14a**, 882 (1959).
42. W. Maier and A. Saupe, *Z. Naturforsch.*, **15a**, 287 (1960).
43. S. Chandrasekhar and N. V. Madhusudana, *Acta Cryst.*, **A27**, 303 (1971).
44. S. Chandrasekhar and N. V. Madhusudana, *Mol. Cryst. Liq. Cryst.*, **17**, 37 (1972).
45. R. Alben, *Mol. Cryst. Liq. Cryst.*, **13**, 193 (1971).
46. R. Alben, J. R. McColl, and C. S. Shih, *Solid State Commun.*, **11**, 1081 (1972).
47. W. L. McMillan, *Phys. Rev.* **A8**, 1921 (1973).
48. R. J. Meyer and W. L. McMillan, *Phys. Rev.* **A9**, 899 (1974).
49. J. R. McColl and C. S. Shih, *Phys. Rev. Lett.*, **29**, 85 (1972).
50. P. J. Wojkowicz, *RCA Rev.*, **35**, 105 (1974).
51. E. B. Priestly, *RCA Rev.*, **35**, 144 (1974).
52. W. G. F. Ford, *J. Chem. Phys.*, **56**, 6270 (1972).
53. I. Haller, H. A. Huggins, H. R. Lilienthal, and T. R. McGuire, *J. Phys. Chem.*, **77**, 950 (1973).
54. R. T. Klingbiel, D. J. Genova, and H. K. Bücher, *Mol. Cryst. Liq. Cryst.*, **27**, 1 (1974).
55. G. W. Smith and Z. G. Gardlund, *J. Chem. Phys.*, **59**, 3214 (1973).
56. J. F. Johnson and G. W. Miller, *Thermochim. Acta*, **1**, 373 (1970).
57. R. S. Porter, E. M. Barrell II, and J. F. Johnson, *Accounts Chem. Res.*, **2**, 53 (1969).
58. H. Schubert, Wiss. *Z. Univ. Halle*, **19**, 1 (1970).
59. H. Schubert and H. Zaschke, *J. Prakt. Chem.*, **312**, 494 (1970).
60. J. A. Nash and G. W. Gray, *Mol. Cryst. Liq. Cryst.*, **25**, 299 (1974).
61. K. Murase, *Bull. Chem. Soc. Japan.*, **45**, 1772 (1972).
62. J. B. Flannery, Jr., and W. Haas, *J. Phys. Chem.*, **74**, 3611 (1970).
63. H. J. Dietrich and E. L. Steiger, *Mol. Cryst. Liq. Cryst.*, **16**, 263 (1972).
64. J. A. Castellano, J. E. Goldmacher, L. A. Barton, and J. S. Kane, *J. Org. Chem.*, **33**, 3501 (1968).
65. M. J. Rafuse and R. A. Soref, *Mol. Cryst. Liq. Cryst.*, **18**, 95 (1972).
66. T. R. Criswell, B. H. Klanderman, and D. C. Batesky, *Mol. Cryst. Liq. Cryst.*, **22**, 211 (1973).
67. R. Steinsträsser and L. Pohl, *Z. Naturforsch.*, **26b**, 87 (1971).
68. H. Kelker, B. Scheurle, R. Hatz, and W. Bartsch, *Angew. Chem. Int. Ed.*, **9**, 962 (1970).
69. E. L. Steiger and H. J. Dietrich, *Mol. Cryst. Liq. Cryst.*, **16**, 279 (1972).
70. F. Rondelez, D. Diguet, and G. Durand, *Mol. Cryst. Liq. Cryst.*, **15**, 183 (1971).
71. W. L. McMillan, *Phys. Rev.*, **A4**, 1419 (1973).
72. S. Torza and P. E. Cladis, *Phys. Rev. Lett.*, **32**, 1406 (1974).
73. J. van der Veen and A. H. Grobben, *Mol. Cryst. Liq. Cryst.*, **15**, 239 (1971).
74. K. Murase, *Chem. Lett.*, 471 (1972).
75. Z. G. Gardlund, R. J. Curtis, and G. W. Smith, *J. Chem. Soc. Chem. Commun.*, 202 (1973).
76. H. Hirata, S. N. Waxman, I. Teucher, and M. M. Labes, *Mol. Cryst. Liq. Cryst.*, **20**, 343 (1973).

77. G. W. Gray, J. B. Hartley, A. Ibbotson, and B. Jones, *J. Chem. Soc.*, 4359 (1955).
78. S. L. Arora, T. R. Taylor, J. L. Fergason, and A. Saupe, *J. Amer. Chem. Soc.*, **91**, 3671 (1969).
79. S. L. Arora, J. L. Fergason, and A. Saupe, *Mol. Cryst. Liq. Cryst.*, **10**, 243 (1970).
80. T. R. Taylor, S. L. Arora, and J. L. Fergason, *Phys. Rev. Lett.*, **25**, 722 (1970).
81. G. R. Luckhurst and F. Sundholm, *Mol. Phys.*, **21**, 349 (1971).
82. Z. Luz and S. Meiboom, *J. Chem. Phys.*, **59**, 275 (1973).
83. J. Doucet, A. M. Levelut, and M. Lambert, *Phys. Rev. Lett.*, **32**, 301 (1974).
84. D. Dolphin, Z. Muljiani, J. Cheng, and R. B. Meyer, *J. Chem. Phys.*, **58**, 413 (1973).
85. W. Helfrich and C. S. Oh, *Mol. Cryst. Liq. Cryst.*, **14**, 289 (1970).
86. H. Sorkin and A. Denny, *RCA Rev.*, **34**, 308 (1973).
87. C. K. Ingold and H. A. Piggott, *J. Chem. Soc.*, **121**, 2793 (1922).
88. G. Toth, I. Pinter, and A. Messmer, *Tetrahedron Lett.*, 735 (1974).
89. A. Lomax, R. Hirasawa, and A. J. Bard, *J. Electrochem. Soc: Electrochem. Sci. Technol.*, **119**, 1679 (1972).
90. A Denat, B. Gosse, and J. P. Gosse, *Chem. Phys. Lett.*, **18**, 235 (1973).
91. A. Denat, B. Gosse, and J. P. Gosse, *J. Chim. Phys. Physiochim. Biol.*, **70**, 319 (1973).
92. A. Denat, B. Gosse, and J. P. Gosse, *J. Chim. Phys. Physiochim. Biol.*, **70**, 327 (1973).
93. W. R. Young, I. Haller, and A. Aviram, *Mol. Cryst. Liq. Cryst.*, **13**, 357 (1971).
94. J. van der Veen and W. H. de Jeu, *Mol. Cryst. Liq. Cryst.*, **27**, 251 (1974).
95. L. Verbit and R. L. Tuggy, *Mol. Cryst. Liq. Cryst.*, **17**, 49 (1972).
96. J. Malthete, M. Leclercq, M. Dvolaitzky, J. Gabard, J. Billard, V. Pontikis, and J. Jacques, *Mol. Cryst. Liq. Cryst.*, **23**, 233 (1973).
97. J. Malthete, M. Leclercq, J. Gabard, J. Billard, and J. Jacques, *C. R. Acad. Sci. Paris Ser. C.*, **273**, 265 (1971).
98. E. C.-H. Hsu and J. F. Johnson, *Mol. Cryst. Liq. Cryst.*, **20**, 177 (1973).
99. D. Demus, Ch. Fietkau, R. Schubert, and H. Kehlen, *Mol. Cryst. Liq. Cryst.*, **25**, 215 (1974).
100. A. Denat and B. Gosse, *Chem. Phys. Lett.*, **22**, 91 (1973).
101. W. R. Young, I. Haller and A. Aviram, *Mol. Cryst. Liq. Cryst.*, **15**, 311 (1972).
102. W. R. Young, A. Aviram, and R. J. Cox, *J. Amer. Chem. Soc.*, **94**, 3976 (1972).
103. J. van der Veen and Th. C. J. M. Hegge, to be published.
104. W. H. de Jeu and J. van der Veen, *Phys. Lett.*, **44A**, 277 (1973).
105. R. Steinsträsser and L. Pohl, *Z. Naturforsch.*, **26b**, 577 (1971).
106. K. Murase and H. Watanabe, *Bull. Chem. Soc. Japan*, **46**, 3142 (1973).
107. J. van der Veen, W. H. de Jeu, A. H. Grobben, and J. Boven, *Mol. Cryst. Liq. Cryst.*, **17**, 291 (1972).
108. W. H. de Jeu, J. van der Veen, and W. J. A. Goosens, *Solid State Commun.*, **12**, 405 (1973).
109. W. H. de Jeu and Th. W. Lathouwers, *Z. Naturforsch*, **29a**, 905 (1974).
110. British Patent Specification, 1285388.
111. R. Steinsträsser and L. Pohl, *Tetrahedron Lett.*, **22**, 1921 (1971).
112. R. E. Rondeau, M. A. Berwick, R. N. Steppel, and M. P. Servé, *J. Amer. Chem. Soc.*, **94**, 1096 (1972).

113. M. T. McCaffrey and J. A. Castellano, *Mol. Cryst. Liq. Cryst.,* **18**, 209 (1972).
114. J. van der Veen and Th. C. J. M. Hegge, *Angew. Chem. Int. Ed.,* **13**, 344 (1974).
115. W. H. de Jeu and Th. W. Lathouwers, *Chem. Phys. Lett.,* **28**, 239 (1974).
116. W. H. de Jeu, Th. W. Lathouwers, and P. Bordewijk, *Phys. Rev. Lett.,* **32**, 40 (1974).
117. J. van der Veen, W. H. de Jeu, M. W. M. Wanninkhof, and C. A. M. Tienhoven, *J. Phys. Chem.,* **77**, 2153 (1973).
118. W. H. de Jeu, *Solid State Commun.,* **13**, 1521 (1973).
119. G. J. Davis, R. S. Porter, and E. M. Barrell, II, *Mol. Cryst. Liq. Cryst.,* **10**, 1 (1970).
120. L. B. Leder, *J. Phys. Chem.,* **58**, 1118 (1973).
121. L. C. Scala and G. D. Dixon, *Mol. Cryst. Liq. Cryst.,* **10**, 411 (1970).
122. L. B. Leder, *J. Phys. Chem.,* **54**, 4671 (1971).
123. F. D. Saeva, *Mol. Cryst. Liq. Cryst.,* **18**, 375 (1972).
124. J. A. Castellano, M. T. McCaffrey, and J. E. Goldmacher, *Mol. Cryst. Liq. Cryst.,* **12**, 345 (1971).
125. R. Steinsträsser, *Z. Naturforsch.,* **27b**, 774 (1972).
126. J. P. Van Meter and B. H. Klanderman, *Mol. Cryst. Liq. Cryst.,* **22**, 271 (1973).
127. W. H. de Jeu and Th. W. Lathouwers, *Mol. Cryst. Liq. Cryst.,* **26**, 225 (1974).
128. R. T. Klingbiel, D. J. Genova, T. R. Criswell, and J. P. Van Meter, *J. Amer. Chem. Soc.,* **96**, 7651 (1974).
129. M. J. S. Dewar and R. S. Goldberg, *J. Org. Chem.,* **35**, 2711 (1970).
130. S. L. Arora, J. L. Fergason, and T. R. Taylor, *J. Org. Chem.,* **35**, 4055 (1970).
131. S. A. Haut, D. C. Schroeder, and J. P. Schroeder, *J. Org. Chem.,* **37**, 1425 (1972).
132. W. R. Young, I. Haller, and D. C. Green, *J. Org. Chem.,* **37**, 3707 (1972).
133. R. Steinsträsser, *Angew. Chem. Int. Ed.,* **11**, 634 (1972).
134. J. P. Van Meter and B. H. Klanderman, *Mol. Cryst. Liq. Cryst.,* **22**, 285 (1973).
135. J. P. Van Meter and A. K. Seidel, *5th International Liquid Crystal Conference,* Stockholm, Sweden, 1974.
136. J. P. Van Meter, R. T. Klingbiel, and D. J. Genova, *Solid State Commun.,* **16**, 315 (1975).
137. J. P. Van Meter and B. H. Klanderman, *J. Amer. Chem. Soc.,* **95**, 626 (1973).
138. W. W. Lowrance, Jr., *Tetrahedron Lett.,* 3453 (1971).
139. H. K. Bücher, R. T. Klingbiel, and J. P. Van Meter, *Appl. Phys. Lett.,* **25**, 186 (1974).
140. B. H. Klanderman and T. R. Criswell, *J. Amer. Chem. Soc.,* **97**, 1585 (1975).
141. G. W. Gray, K. J. Harrison, and J. A. Nash, *Electron. Lett.,* **9**, 130 (1973).
142. A. Ashford, J. Constant, J. Kirton, and E. P. Raynes, *Electron. Lett.,* **9**, 118 (1973).
143. G. W. Gray, K. J. Harrison, and J. A. Nash, *Electron. Lett.,* **9**, 616 (1973).
144. G. W. Gray, *Mol. Cryst.,* **1**, 333 (1966).
145. R. F. Tarvin and F. W. Neetzow, *J. Chem. Soc. Chem. Commun.,* 396 (1973).
146. M. J. S. Dewar, A. Griffin, and R. M. Riddle in *Liquid Crystals and Ordered Fluids*, Vol. 2 (Ed. J. F. Johnson and R. S. Porter), Plenum Press, New York, 1974, p. 733ff.
147. L. E. Knaak, H. M. Rosenberg, and M. P. Servé, *Mol. Cryst. Liq. Cryst.,* **17**, 171 (1972).
148. S. Marcelja, *J. Chem. Phys.* **60**, 3599 (1974).
149. T. J. Scheffer, *J. Appl. Phys.,* **44**, 4799 (1973).

150. A. W. Neumann and L. J. Klementowski, *J. Thermal Anal.*, **6**, 67 (1974).
151. J.-C. Lacroix and R. Tobazeon, *Appl. Phys. Lett.*, **20**, 251 (1972).
152. L. T. Creagh and A. R. Kmetz, *J. Electron. Mat.*, **1**, 350 (1972).
153. I. Haller, *Appl. Phys. Lett.*, **24**, 349 (1974).
154. J. E. Proust, L. Ter-Minassian-Saraga, and E. Guyon, *Solid State Commun.*, **11**, 1227 (1972).
155. I. Haller, *J. Chem. Phys.*, **57**, 1400 (1972).
156. J. L. Janning, *Appl. Phys. Lett.*, **21**, 173 (1972).
157. J. C. Dubois, M. Gazard, and A. Zann, *Appl. Phys. Lett.*, **24**, 297 (1974).
158. F. J. Kahn, *Appl. Phys. Lett.*, **22**, 386 (1973).
159. W. Helfrich, *Mol. Cryst. Liq. Cryst.*, **21**, 187 (1973).
160. H. Gruler and G. Meier, *Mol. Cryst. Liq. Cryst.*, **12**, 289 (1971).
161. M. Schadt, *J. Chem. Phys.*, **56**, 1494 (1972).
162. M. Schadt and W. Helfrich, *Appl. Phys. Lett.*, **18**, 127 (1971).
163. G. H. Heilmeier, L. A. Zanoni, and L. A. Barton, *Appl. Phys. Lett.*, **13**, 46 (1968).
164. G. H. Heilmeir and J. E. Goldmacher, *Appl. Phys. Lett.*, **13**, 132 (1968).
165. H. Haas, J. Adams, and G. Dir, *Chem. Phys. Lett.*, **14**, 95 (1972).
166. F. J. Kahn, *Appl. Phys. Lett.*, **22**, 111 (1973).
167. C. H. Gooch and H. A. Tarry, *J. Phys. D*, **5**, L25 (1972).
168. W. H. de Jeu, C. J. Gerritsma, P. van Zanten, and W. J. A. Goossens, *Phys. Lett.*, **39A**, 355 (1972).
169. G. Baur, A. Stieb, and G. Meier in *Liquid Crystals and Ordered Fluids*, Vol. 2 (Ed. J. F. Johnson and R. S. Porter), Plenum Press, New York, 1974, p. 645ff.
170. T. S. Chang and E. E. Loebner, *Appl. Phys. Lett.*, **25**, 1, (1974).
171. A. Saupe, *Mol. Cryst. Liq. Cryst.*, **16**, 87 (1972).
172. S. Meiboom and L. C. Snyder, *Accounts Chem. Res.*, **4**, 81 (1971).
173. A. d'Annibale, L. Lunazzi, G. Fronza, R. Mondelli, and S. Bradamante, *J. Chem. Soc. Perkin II*, 1908 (1973).
174. M. J. S. Dewar and B. D. Nahlovsky, *J. Amer. Chem. Soc.*, **96**, 460 (1974).
175. W. H. Toliver, Sr., J. L. Fergason, E. Sharpless, and P. E. Hoffman, *Aerospace Med.*, **41**, 18 (1970).
176. E. Grushka and J. F. Solsky, *Anal. Chem.*, **45**, 1836 (1973).
177. L. E. Cook and R. C. Spangelo, *Anal. Chem.*, **46**, 122 (1974).
178. In a recent study F. D. Saeva, P. E. Sharpe, and G. R. Olin, *J. Amer. Chem. Soc.*, **97**, 204 (1975), it was found that γ-methylallyl-*p*-tolyl ether undergoes a Claisen rearrangement in a cholesteric liquid crystalline solvent to produce a chiral product.

CHAPTER 3

Thermochemistry of X=Y groups

Robert Shaw

Physical Sciences Division, Stanford Research Institute,
Menlo Park, California 94025, U.S.A.

> The more extensive a man's knowledge of what has been done, the greater
> will be his power of knowing what to do.   Disraeli

I. INTRODUCTION

This chapter is concerned with the recent advances that have occurred
in the thermochemistry of X=Y groups. By 'recent' is meant since the
publication around 1970 of the two excellent monographs[1,2], Stuhl,
Westrum, and Sinke's, *The Chemical Thermodynamics of Organic Com-*
pounds, and Cox and Pilcher's, *Thermochemistry of Organic and Organo-*
metallic Compounds, and of the *Chemical Reviews* paper[3] on the estimation of chemical thermodynamic properties of organic compounds by
group additivity. The thermochemistry of C=O compounds has recently
been reviewed by Benson and Eigenmann[4]. The thermochemistry of
C=N and N=N compounds have been treated in chapters in this
series[5,6]. A check of one of the primary sources of thermochemical
information, IUPAC's annual *Bulletin of Thermochemistry and Thermo-*
dynamics shows that only a few papers have been published in the field
of C=C compounds.

The main thrust of this review will therefore be on two important unpublished papers[7,8] on *Electrostatics and the Chemical Bond*, on a collection of the latest group values for estimating thermochemical properties, and on some recent experimental work. Only ground state molecules will be considered.

Following earlier reviews in this series, the main thermochemical quantity of interest is the heat of formation for the ideal gas state at 298 K (25 °C). IUPAC[9] have recommended that the symbol for this quantity be denoted $\Delta_f H^\theta_m$ (chemical formula, g, 298·15 K). All the values discussed here will be for the ideal gas state at 298·15 K, so the symbol for heat of formation will be abbreviated to $\Delta_f H^\theta_m$ (chemical formula). IUPAC have also recommended that the units should be joules/mole, abbreviated J/mol. In keeping with these recommendations, all heats in this review will be in units of kJ/mol, with the previously accepted unit of kcal/mol in parentheses. The conversion factor is cal = 4·18 J.

II. ELECTROSTATICS AND THERMOCHEMISTRY

What appears to be the most significant development of the decade in understanding the thermochemistry of organic molecules has just been made by Benson and Luria[7,8] and by Palm[10], working independently. Benson and Luria have proposed simple electrostatic models that account for the heats of formation of all *n*-alkanes, for branched alkanes up to C_7, and for the unsaturated hydrocarbons, alkenes, alkynes, and aromatic compounds. For the present purposes we will be concerned only with the alkanes and the alkenes.

A. Electrostatic Model for Alkanes

In the electrostatic model[7], the C—H bond is polarized so that the hydrogen atom carries a positive charge, $+y$, and the carbon atom has a neutralizing negative charge, $-y$. In this model methane has four hydrogen atoms, each with charge $+y$, and the central carbon atom therefore carries a neutralizing charge, $-4y$. The alkanes, methane, ethane, propane, 2-methylpropane, and 2,2-dimethylpropane are shown in Figures 1 to 5.

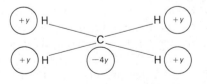

FIGURE 1. The electrostatic model for methane.

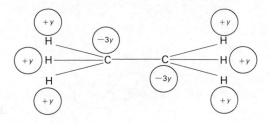

FIGURE 2. The electrostatic model for ethane.

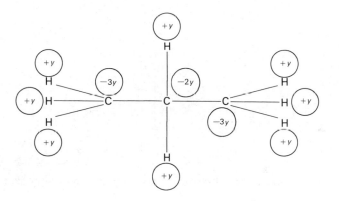

FIGURE 3. The electrostatic model for propane.

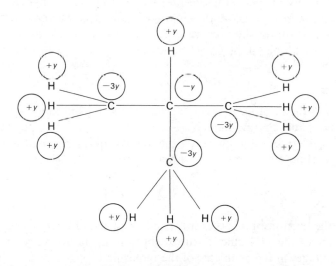

FIGURE 4. The electrostatic model for 2-methylpropane.

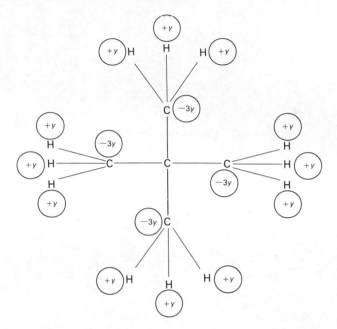

FIGURE 5. The electrostatic model for 2,2-dimethylpropane.

In 2,2-dimethylpropane, the central carbon atom has no hydrogen atoms bonded to it, and therefore the central carbon atom has no charge.

The electrostatic static energy, E_{el}, for any molecule arises from the interactions along all the C—H bonds in the molecule. In general,

$$E_{el} = \sum_{i<j} q_i q_j / r_{ij} \tag{1}$$

where r_{ij} is the distance between charges q_i and q_j. Since all the hydrogens have the same charge, namely $+y$, and all the carbon atoms have a balancing charge, $-ny$, where n is the number of hydrogen atoms bonded to that carbon atom, equation (1) reduces to,

$$E_{el} = y^2 \left[\sum_{i<j} n_i n_j / r_{ij} \right] \tag{2}$$

where the term in brackets is independent of y and depends only on the geometry of the molecule. It turns out that E_{el} is not very sensitive to small changes in the geometry of the molecule.

The value for y was found by attributing the heat of isomerization of alkanes to differences in electrostatic energies. The result was $|y| = 2 \cdot 78 \pm 0 \cdot 5 \times 10^{-11}$ esu $= 5 \cdot 81 \pm 1 \times 10^{-2}$ electronic charge.

Benson and Luria[7] then showed that the heats of formation of the alkanes could be represented by the simple formula:

$$\Delta_f H_m^\theta(C_n H_{2n+2}, \text{g, 298 K}) = -2(n+1) - 0 \cdot 5 + E_{el}(C_n H_{2n+2}) \qquad (3)$$

Heats of formation of alkanes calculated by Benson and Luria, using the electrostatic method from equation (3), are compared with observed values in Table 1.

TABLE 1. Heats of formation of alkanes $[\Delta_f H_m^\theta(\text{g, 298 K})]$ calculated by the electrostatic method compared with observed values[a]

Compound	Observed		Calculated		Δ(obs–calc)	
	kJ/mol	kcal/mol	kJ/mol	kcal/mol	kJ/mol	kcal/mol
CH_4	$- 74 \cdot 8$	$(-17 \cdot 9)$	$- 71 \cdot 9$	$(-17 \cdot 2)$	$-2 \cdot 9$	$(-0 \cdot 7)$
C_2H_6	$- 84 \cdot 4$	$(-20 \cdot 2)$	$- 84 \cdot 4$	$(-20 \cdot 2)$	$0 \cdot 0$	$(0 \cdot 0)$
C_3H_8	$-103 \cdot 7$	$(-24 \cdot 8)$	$-104 \cdot 5$	$(-25 \cdot 0)$	$0 \cdot 8$	$(0 \cdot 2)$
$n\text{-}C_4H_{10}$	$-125 \cdot 8$	$(-30 \cdot 1)$	$-125 \cdot 4$	$(-30 \cdot 0)$	$-0 \cdot 4$	$(0 \cdot 1)$
$i\text{-}C_4H_{10}$	$-134 \cdot 2$	$(-32 \cdot 1)$	$-132 \cdot 9$	$(-31 \cdot 8)$	$-1 \cdot 3$	$(-0 \cdot 3)$
$n\text{-}C_5H_{12}$	$-146 \cdot 7$	$(-35 \cdot 1)$	$-146 \cdot 3$	$(-35 \cdot 0)$	$-0 \cdot 4$	$(-0 \cdot 1)$
$i\text{-}C_5H_{12}$	$-154 \cdot 2$	$(-36 \cdot 9)$	$-153 \cdot 4$	$(-36 \cdot 7)$	$-0 \cdot 8$	$(-0 \cdot 2)$
$neo\text{-}C_5H_{12}$	$-168 \cdot 4$	$(-40 \cdot 3)$	$-169 \cdot 3$	$(-40 \cdot 5)$	$0 \cdot 8$	$(0 \cdot 2)$
$n\text{-}C_6H_{14}$	$-167 \cdot 2$	$(-40 \cdot 0)$	$-166 \cdot 8$	$(-39 \cdot 9)$	$-0 \cdot 4$	$(-0 \cdot 1)$
$n\text{-}C_7H_{16}$	$-187 \cdot 7$	$(-44 \cdot 9)$	$-187 \cdot 8$	$(-44 \cdot 9)$	$0 \cdot 0$	$(0 \cdot 0)$

[a] Units are kJ/mol; in parentheses, kcal/mol. The equation for calculating heats of formation of alkanes is $\Delta_f H_m^\theta(C_n H_{2n+2}, \text{g, 298 K}) = -2(n+1) - 0 \cdot 5 + E_{el}(C_n H_{2n+2})$ from Reference 7.

B. Electrostatic Model for Alkenes

The electrostatic model for the alkenes is slightly more complex because there are two types of carbon atoms; namely, the sp^3 carbon atoms that we met in the alkanes, and the sp^2 carbon atoms. For brevity, an sp^3 carbon atom is denoted by C and an sp^2 carbon atom is denoted by C_d. The nucleus of a C_d atom is less shielded by the compressed pi bond electrons than the nucleus of a C atom which is shielded by the two sigma bond electrons. Thus one would expect that the hydrogen atom in a C_d—H bond will donate more of its electron cloud to the C_d atom than the hydrogen atom in a C—H bond will donate to the C atom. The result is that the C_d—H bond will be more polarized than the C—H bond. The charges on the H and C_d in the C_d—H bond are denoted $+y'$ and $-y'$, respectively, and it follows from the foregoing that $|y'| > |y|$.

The electrostatic model of ethene is in Figure 6.

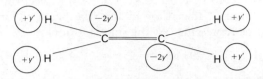

FIGURE 6. The electrostatic model for ethene.

In propene, the differences in the electronegativity of the two carbon atoms results in an additional formal charge of δ' across the $C—C_d$ bond, with $+\delta'$ being on the C atom and $-\delta'$ being on the C_d atom.

The electrostatic model for propene is shown in Figure 7 and that for 2-methylpropene in Figure 8.

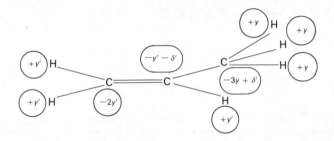

FIGURE 7. The electrostatic model for propene.

Benson and Luria derived the formulas for the electrostatic energies of the alkenes from the known geometries; for example, for ethene:

$$E_{el}/(esu^2/\text{Å}) = -5.67y'^2 \qquad (4)$$

For propene:

$$E_{el}/(esu^2/\text{Å}) = -6.57y^2 - 3.82y'^2 + 0.88yy' - 0.67\delta'^2$$
$$+ 3.35y\delta' - 0.76y'\delta' \qquad (5)$$

and so on for the other alkenes.

The value of the parameter y is the same as that determined from the alkanes; namely, $y = 2.78 \times 10^{11}$ esu. The isomers, 2-methylpropene and $trans$-but-2-ene, have the same number of types of bonds. Benson and Luria[7] therefore assigned their difference in heat of formation to their

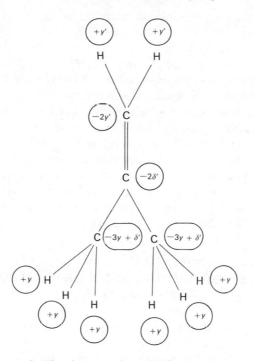

FIGURE 8. The electrostatic model for 2-methylpropene.

difference in electrostatic energies. They were then able to assign values of y' and δ', specifically:

$$|y'| = 3 \cdot 2 \times 10^{11}\ \text{esu} \quad \text{and} \quad |\delta'| = 1 \cdot 2 \times 10^{11}\ \text{esu.}$$

Note that $|y'| > |y|$, which is consistent with the earlier discussion of the model. Substituting the values for y, y', and δ' into the equations for E_{el}, such as (4) and (5), the values of E_{el} for some representative alkenes were found (see Table 2).

The electrostatic model by itself is not enough to account for the heats of formation of an alkene. The heat of formation may therefore be separated into electrostatic and non-electrostatic energies:

$$\Delta_{\text{f}}H_{\text{m}}^{\theta}(\text{alkene, g, 298 K}) = \sum \Delta_{\text{f}}H_{\text{m}}^{\theta}(\text{bond contributions}) + E_{\text{el}} \quad (6)$$

where $\sum \Delta_{\text{f}}H_{\text{m}}^{\theta}$ (bond contribution) is the sum of the contributions of bonds, such as $(C_{\text{d}}{-}H)$, $(C_{\text{d}}{-}C)$, and $(C_{\text{d}}{-}C_{\text{d}})$ to the non-electrostatic energy. By considering the heats of formation of ethene, propene, and allene, together with previously derived values for (C—C) and (C—H), Benson and Luria derived the following values for the bond contributions: $\Delta_f H_m^\theta(C_d—H) = -15\cdot1$ kJ/mol $(-3\cdot6$ kcal/mol), $\Delta_f H_m^\theta(C_d—C) =$ $145\cdot0$ kJ/mol $(34\cdot7$ kcal/mol)   and $\Delta_f H_m^\theta(C_d—C_d) = -23\cdot0$ kJ/mol $(-5\cdot5$ kcal/mol). Substituting in equation (6), Benson and Luria calculated the heats of formation shown for the alkenes in Table 2.

Clearly, this new method of estimating heats of formation is more realistic, chemically speaking, than the more straightforward additivity schemes that have previously been developed. As such, it is a big advance in our understanding of the factors that are important in determining the heat of formation of organic molecules. However, from a practical point of view of estimating heats of formation of not-very-polar hydrocarbons, such as the alkenes, it remains to be seen whether this new development will replace conventional group additivity methods[3].

III. LATEST GROUP VALUES FOR COMPOUNDS CONTAINING X=Y

Estimation of heats of formation of organic compounds by group additivity has been extensively discussed in previous reviews in this series and elsewhere[3,5,6]. Suffice to say that group additivity turns out to be an excellent compromise between simplicity and accuracy, and that it is rapidly becoming accepted as the prime method of estimation. As mentioned in the Introduction, separate papers have recently been published giving new values for groups required for compounds containing C=C, C=O, C=N, and N=N. It seems appropriate to gather these group values in one place for easier access (see Table 3).

The following notation is used: C_d is a vinyl or benzyl carbon atom; C_t is a carbon atom in a triple bond; C_B is a carbon atom in a benzene ring; CO is treated as a single functional entity, as though it were an atom; N_I is an imide nitrogen atom, as in C=N; and N_A is an azo nitrogen atom, as in N=N.

IV. SOME RECENT EXPERIMENTAL WORK

The choice of material for this section is very subjective. I am indebted to many authors who have communicated their results in advance of publication. Heats of formation of stable molecules are from Cox and

TABLE 2. Electrostatic energy contributions (E_{el}) and comparison of heats of formation of alkenes [$\Delta_f H_m^\theta$(alkene, g, 298 K)] calculated by the electrostatic method compared with observed values[a]

Compound	E_{el}		Observed		Calculated		Δ(obs−calc)	
	kJ/mol	kcal/mol	kJ/mol	kcal/mol	kJ/mol	kcal/mol	kJ/mol	kcal/mol
Ethene	−34.9	(−8.36)	51.8	(12.4)	49.7	(11.9)	2.1	(0.5)
Allene	−38.3	(−9.16)	190.6	(45.6)	191.4	(45.8)	− 0.8	(−0.2)
Propene	−46.0	(−11.01)	20.5	(4.9)	19.6	(4.7)	0.8	(0.2)
But-1-ene	−63.0	(−15.07)	− 0.8	(−0.2)	− 3.8	(−0.9)	2.9	(0.7)
2-Methylpropene	−63.6	(−15.21)	−18.0	(−4.3)	−17.1	(−4.1)	− 0.8	(−0.2)
cis-But-2-ene	−58.6	(−14.01)	− 7.9	(−1.9)	−11.3	(−2.7)	3.3	(0.8)
					[− 7.1]	[(−1.7)	− 0.8	(−0.2)][b]
trans-But-2-ene	−58.4	(−13.98)	−12.5	(−3.0)	−11.3	(−2.7)	− 1.2	(−0.3)
2-Methyl-but-2-ene	−75.2	(−17.98)	−42.2	(−10.1)	−49.7	(−11.9)	7.5	(1.8)
					[−44.3]	[(−10.6)	2.1	(0.5)][b]
2,3-Dimethyl-but-2-ene	−91.8	(−21.95)	−68.6	(−16.4)	−82.4	(−19.7)	13.8	(3.3)
					[−71.5]	[(−17.1)	2.9	(0.7)][b]

[a] Units are kJ/mol; in parentheses, kcal/mol. The equation for calculating heats of formation is $\Delta_f H_m^\theta$ (alkene, g, 298 K) = $\sum \Delta_f H_m^\theta$ (bond contribution) + E_{el} (from Reference 8).

[b] These calculated values have an additional 4.2 kJ/mol (1.0 kcal/mol) per 1–4 non-bonded as H ⋯ H repulsion and 1.2 kJ/mol (0.3 kcal/mol) per 1–5 non-bonded H ⋯ H repulsion.

TABLE 3. Revised or recently obtained values of $\Delta_f H_m^\theta$(group, g, 298 K) for groups containing C_d, CO, N_A, or N_I[a]

Group	$\Delta_f H_m^\theta$(g, 298 K) kJ/mol	kcal/mol	Ref.	Group	$\Delta_f H_m^\theta$(g, 298 K) kJ/mol	kcal/mol	Ref.
C—(Cl)(H)$_2$(N$_A$)	−25.1	(−6.0)	5	C$_d$—(Cl)$_2$(N$_I$)	43.0	(10.3)	6
C—(Cl)(CO)(H)$_2$	−21.7	(−5.2)	4	C$_d$—(C$_B$)(Br)(N$_I$)	29.3	(7.0)	6
C—(Cl)(CO)$_2$(H)	−22.6	(−5.4)	4	C$_d$—(C$_B$)(Cl)(N$_I$)	−39.3	(−9.4)	6
C—(Cl)$_2$(CO)(H)	−7.1	(−1.7)	4	C$_d$—(C$_B$)(F)(N$_I$)	−184.3	(−44.1)	6
C—(Cl)$_2$(H)(N$_A$)	−14.2	(−3.4)	5	C$_d$—(C$_B$)(I)(N$_I$)	133.8	(32.0)	6
C—(Cl)$_3$(CO)	5.8	(1.4)	4	C$_d$—(C$_B$)(N)(N$_I$)	−5.8	(−1.4)	6
C—(Cl)$_3$(N$_A$)	−12.5	(−3.0)	5	C$_d$—(C$_B$)(N$_I$)(O)	−13.0	(−3.1)	6
C—(C$_B$)(CO)(H)$_2$	−22.6	(−5.4)	4	C$_d$—(C$_d$)(O)	37.0	(8.9)	4
C—(C$_d$)(CO)(H)$_2$	−15.9	(−3.8)	4	C$_d$—(CO)(H)	20.9	(5.0)	4
C—(C$_d$)(H)$_2$(O)	−27.2	(−6.5)	4	C$_d$—(CO)(O)	48.5	(11.6)	4
C—(C$_I$)(CO)(H)$_2$	−22.6	(−5.4)	4	C$_d$—(H)(Br)(N$_I$)	−12.5	(−3.0)	6
C—(CO)(H)$_3$	−42.2	(−10.1)	4	C$_d$—(H)(Cl)(N$_I$)	−27.2	(−6.5)	6
C—(CO)$_2$(H)$_2$	−31.8	(−7.6)	4	C$_d$—(H)(F)(N$_I$)	−226.1	(−54.1)	6
C—(H)$_3$(N$_A$)	−42.2	(−10.1)	5	C$_d$—(H)(I)(N$_I$)	92.0	(22.0)	6
C—(H)$_3$(N$_I$)	−42.2	(−10.1)	6	C$_d$—(H)(N)(N$_I$)	−47.6	(−11.4)	6
C$_B$—(CO)	15.5	(3.7)	4	C$_d$—(H)(N$_I$)(O)	−54.6	(−13.1)	6
C$_d$—(Cl)(CO)	31.4	(7.5)	4	C$_d$—(H)(O)	36.0	(8.6)	4
C$_d$—(Cl)(H)(N$_I$)	36.0	(8.6)	6	C$_d$—(H)$_2$(N$_I$)	26.3	(6.3)	6
C$_d$—(Cl)(Br)(N$_I$)	−24.7	(−5.9)	6	CO—(C)(C$_B$)	−129.2	(−30.9)	4
C$_d$—(Cl)(Cl)(N$_I$)	−39.3	(−9.4)	6	CO—(C)(CO)	−122.1	(−29.2)	4
C$_d$—(Cl)(F)(N$_I$)	−238.3	(−57.0)	6	CO—(C)(H)	−121.6	(−29.1)	4
C$_d$—(Cl)(I)(N$_I$)	79.8	(19.1)	6	CO—(C)(O)	−146.7	(−35.1)	4

Group	kJ/mol	(kcal/mol)	n	Group	kJ/mol	(kcal/mol)	n
C_d—(C)(N)(N_I)	−59.8	(−14.3)	6	CO—(C)_2	−131.3	(−31.4)	4
C_d—(C)(N_I)(O)	−66.9	(−16.0)	6	CO—(C_B)(CO)	−112.0	(−26.8)	4
C_d—(C)(O)	43.0	(10.3)	4	CO—(C_B)(H)	−121.6	(−29.1)	4
CO—(C_B)(O)	−153.0	(−36.6)	4	N—(C)(C_d)(H)	64.4	(15.4)	6
CO—(C_B)_2	−107.8	(−25.8)	4	N—(C)(C_d)(N)	124.6	(29.8)	6
CO—(C_d)(H)	−121.6	(−29.1)	4	N—(C)(H)(N_I)	87.4	(20.9)	6
CO—(C_d)(O)	−133.8	(−32.0)	4	N—(C)_2(C_d)	102.0	(24.4)	6
CO—(C_I)(H)	−121.6	(−29.1)	4	N—(C)_2(N_I)	122.1	(29.2)	6
CO—(CO)(H)	−105.8	(−25.3)	4	N—(C_d)(H)(N)	89.9	(21.5)	6
CO—(CO)(O)	−122.5	(−29.3)	4	N—(C_d)(H)_2	20.1	(4.8)	6
CO—(H)(O)	−134.2	(−32.1)	4	N—(H)_2(N_I)	47.6	(11.4)	6
CO—(H)_2	−108.7	(−26.0)	4	N_I—(C_d)(H)	50.2	(12.0)	6
CO—(O)_2	−125.0	(−29.9)	4	N_I—(C_d)(N)	104.1	(24.9)	6
CO—(CH)_3 ring strain	94.5	(22.6)	4	N_A—(C)	112.9	(27.0)	5
CO—(CH_2)_4 ring strain	21.7	(5.2)	4	O—(C)(C_d)	−127.5	(−30.5)	4
CO—(CH_2)_5 ring strain	9.2	(2.2)	4	O—(C)(CO)	−180.2	(−43.1)	4
				O—(C_B)(CO)	−153.4	(−36.7)	4
				O—(C_d)(CO)	−188.9	(−45.2)	4
				O—(C_d)(H)	−158.4	(−37.9)	4
				O—(C_d)_2	−138.0	(−33.0)	4
				O—(CO)(H)	−242.9	(−58.1)	4
				O—(CO)_2	−194.4	(−46.5)	4

^{a} The order of listing is C, C_B, C_d, C_I, CO, H, Br, Cl, F, I, N, N_A, N_I, O, and ring strain. The units are kJ/mol; in parentheses, kcal/mol.

Pilcher[2] and bond strength and heats of formation of radicals are from Kerr, Parsonage, and Trotman-Dickenson[11], unless otherwise noted.

A. Pi-bond Strengths

The pi-bond strength is defined as the heat of reaction (7)

$$X{=}Y \longrightarrow \ ^{\cdot}X{-}Y^{\cdot} \tag{7}$$

The heat of reaction (7) may be regarded as the sum of the heats of reactions (8) and (9):

$$W^{\cdot} + X{=}Y \longrightarrow WX{-}Y^{\cdot} \tag{8}$$

$$WX{-}Y^{\cdot} \longrightarrow W^{\cdot} + \ ^{\cdot}X{-}Y^{\cdot} \tag{9}$$

The heat of reaction (9) may be taken to be the same as the heat of reaction (10):

$$WX{-}YZ \longrightarrow W^{\cdot} + \ ^{\cdot}X{-}YZ \tag{10}$$

That is, pi-bond strength $(X{=}Y)$

$$= \Delta H^{\theta}[\text{reaction (7)}]$$

$$= \Delta H^{\theta}[\text{reaction (8)}] + \Delta H^{\theta}[\text{reaction (9)}]$$

$$= \Delta H^{\theta}[\text{reaction (8)}] + \Delta H^{\theta}[\text{reaction (10)}]$$

Now, $\quad \Delta H^{\theta}[\text{reaction (8)}] = \Delta_f H^{\theta}(WX{-}Y^{\cdot}) - \Delta_f H^{\theta}(W^{\cdot}) - \Delta_f H^{\theta}(X{=}Y)$

$$\Delta H^{\theta}[\text{reaction(10)}] = \Delta_f H^{\theta}(W^{\cdot}) + \Delta_f H^{\theta}(^{\cdot}X{-}YZ) - \Delta_f H^{\theta}(WX{-}YZ)$$

$\therefore$ pi-bond strength $(X{=}Y) = \Delta_f H^{\theta}(XY{-}Y^{\cdot}) + \Delta_f H^{\theta}(^{\cdot}X{-}YZ)$
$$- \Delta_f H^{\theta}(X{=}Y) - \Delta_f H^{\theta}(WX{-}YX)$$

For example, consider the pi-bond strength in ethylene. In this case, X and Y are both CH_2. Suppose W and Z are both H. Then, $WX{-}Y$ and $^{\cdot}X{-}YZ$ are the same radical, namely the ethyl radical; hence,

pi-bond strength $(CH_2{=}CH_2)$

$$= 2\Delta_f H^{\theta}(C_2H_5^{\cdot}) - \Delta_f H^{\theta}(C_2H_4) - \Delta_f H^{\theta}(C_2H_6)$$

$$= 216 \cdot 5(51 \cdot 8) - 51 \cdot 8(-12 \cdot 4) + 84 \cdot 4(20 \cdot 2) \ \text{kJ/mol (kcal/mol)}$$

$$= 249 \cdot 1 \ \text{kJ/mol} \ (59 \cdot 6 \ \text{kcal/mol})$$

To take a more general example, consider the pi-bond strength in formaldehyde, $H_2C{=}O$. X is CH_2, Y is O, and let W and Z be H.

pi-bond strength ($H_2C=O$)

$$= \Delta_f H^\theta(CH_3O^\bullet) + \Delta_f H^\theta(^\bullet CH_2OH) - \Delta_f H^\theta(CH_2O) - \Delta_f H^\theta(CH_3OH)$$

$$= 14 \cdot 6(3 \cdot 5) - 21 \cdot 7(-5 \cdot 2) + 108 \cdot 7(26 \cdot 0) + 201 \cdot 1(48 \cdot 1) \text{ kJ/mol (kcal/mol)}$$

$$= 302 \cdot 7 \text{ kJ/mol} (72 \cdot 4 \text{ kcal/mol})$$

An alternative way of calculating the pi-bond strength comes about because the heat of reaction (10) is $D(W-XYZ)$, which is the $W-X$ bond dissociation energy in WXYZ, and the heat of reaction (8) is $-D(W-XY^\bullet)$, which is minus the $W-X$ bond dissociation energy in $WXY^\bullet$. If values of $D(W-XY^\bullet)$ are not tabulated directly, they can be readily calculated from $D(W-XY^\bullet) = \Delta_f H^\theta(W^\bullet) + \Delta_f H^\theta(XY) - \Delta_f H^\theta(WXY^\bullet)$.

Thus, in general:

pi-bond strength (WX=YZ)

$$= D(W-XYZ) - D(W-XY^\bullet)$$
$$= D(W-XYZ) - \Delta_f H^\theta(W^\bullet) - \Delta_f H^\theta(XY) + \Delta_f H^\theta(WXY^\bullet)$$

Taking ethene as an example, X and Y are both CH_2, and let W and Z be H,

pi-bond strength ($CH_2=CH_2$)

$$= D(H-CH_2CH_3) - \Delta_f H^\theta(H^\bullet) - \Delta_f H^\theta(CH_2CH_2) + \Delta_f H^\theta(CH_3CH_2^\bullet)$$

$$= 409 \cdot 6(98) - 217 \cdot 8(-52 \cdot 1) - 51 \cdot 8(-12 \cdot 4)$$

$$+ 108 \cdot 3(25 \cdot 9) \text{ kJ/mol (kcal/mol)}$$

$$= 248 \cdot 3 \text{ kJ/mol} (59 \cdot 4 \text{ kcal/mol})$$

For the pi-bond strength in formaldehyde we can use either $D(H-CH_2OH)$ or $D(H-OCH_3)$. Taking $D(H-CH_2OH)$ first, we have W and Z are H, X is CH_2, and Y is O,

pi-bond strength ($CH_2=O$)

$$= D(H-CH_2OH) - \Delta_f H^\theta(H^\bullet) - \Delta_f H^\theta(CH_2O) + \Delta_f H^\theta(CH_3O^\bullet)$$

$$= 397 \cdot 1(95) - 217 \cdot 8(-52 \cdot 1) + 108 \cdot 7(26 \cdot 0)$$

$$+ 14 \cdot 6(3 \cdot 5) \text{ kJ/mol (kcal/mol)}$$

$$= 302 \cdot 6 \text{ kJ/mol} (72 \cdot 4 \text{ kcal/mol})$$

On the other hand, using $D(H-OCH_3)$ W and Z are both H, X is O, and Y is CH_2, pi-bond strength (CH$_2$=O)

$$= D(H-OCH_3) - \Delta_f H^\theta(H^\cdot) - \Delta_f H^\theta(CH_2O) + \Delta_f H^\theta(HOCH_2^\cdot)$$

$$= 433 \cdot 5(103 \cdot 7) - 217 \cdot 8(-52 \cdot 1) + 108 \cdot 7(26 \cdot 0)$$

$$-21 \cdot 7(-5 \cdot 2) \text{ kJ/mol (kcal/mol)}$$

$$= 302 \cdot 6 \text{ kJ/mol } (72 \cdot 4 \text{ kcal/mol})$$

Some representative pi-bond energies are listed in Table 4, including some unpublished results by Rodgers on the fluorinated ethenes. The factors that determine pi-bond strengths are not generally well understood. Solly, Golden, and Benson[12] have attributed the lack of stabilization energy in the acetonyl radical to the strong pi-bond strength of the carbonyl group, but (see section on stabilization energy of acetonyl radical) this is still something of an open question.

B. Stabilization Energy of the Acetonyl Radical

The extent of delocalization of the unpaired electron in radicals containing double bonds has been the subject of a great deal of controversy over the years. The delocalization in the allyl radical is measured by comparing reactions (11) and (12):

$$CH_3CH_2CH_3 \longrightarrow CH_3CH_2CH_2 + H - 410 \text{ kJ/mol (98 kcal/mol)} \quad (11)$$

$$CH_2CHCH_3 \longrightarrow CH_2CHCH_2 + H - 372 \text{ kJ/mol (89 kcal/mol)} \quad (12)$$

The difference in C—H bond strengths of 38 kJ/mol (9 kcal/mol) is the stabilization energy or resonance energy of the allyl radical. For comparison, the stabilization energy of the methylallyl radical is 63 kJ/mol (15 kcal/mol).

After extensive work, the values given above for the allyl and methylallyl radicals are now generally agreed. However, the value for the stabilization energy of the acetonyl radical is open to question. The stabilization energy is measured by comparing the C—H bond strength in acetone with that in propane:

$$CH_3CH_2CH_3 \longrightarrow CH_3CH_2CH_2 + H - 410 \text{ kJ/mol (98 kcal/mol)} \quad (11)$$

$$CH_3COCH_3 \longrightarrow CH_3COCH_2 + H - X \text{ kJ/mol } (X' \text{ kcal/mol}) \quad (13)$$

The difference $410 - X$ kJ/mol ($98 - X'$ kcal/mol) is then the stabilization energy of the acetonyl radical. In the first place there is the question of whether or not reaction (11) is a good yardstick with which to compare reaction (13). Arguments presented by Solly, Golden, and Benson (SGB)[12] suggest that the comparison is valid only when there is no stabilization

TABLE 4. Calculation of pi-bond strengths for some representative X=Y compounds [The units are all kJ/mol (kcal/mol)]

X=Y	WXYZ	D(W—XYZ)	$\Delta_f H^\theta$(W')	$\Delta_f H^\theta$(XY)	$\Delta_f H^\theta$(WXY')	Pi-bond strength
$CH_2=CH_2$	CH_3CH_3	404·6 (98)	217·8 (52·1)	31·8 (12·4)	108·3 (25·9)	248·3 (59·4)
$CH_3CH=CH_2$	$CH_3CH_2CH_3$	397·1 (95)	217·8 (52·1)	20·5 (4·9)	94·5 (22·6)	253·3 (60·6)
$CH_3CH=CHCH_3$	$CH_3(CH_2)_2CH_3$	397·1 (95)	217·8 (52·1)	−12·5 (−3·0)	54·3 (13·0)	246·2 (58·9)
$CH_2=CF_2$						271·7 (65)[a]
$CF_2=CF_2$						219·4 (52·5)[a]
$CH_2=O$	CH_3OH	397·1 (95)	217·8 (52·1)	−108·7 (−26·0)	14·6 (3·5)	302·6 (72·4)
$CH_3CH=O$	CH_3CH_2OH	388·7 (93)	217·8 (52·1)	−166 (−39·7)	−18·4 (−4·4)	318·5 (76·2)
$(CH_3)_2C=O$	$(CH_3)_2CHOH$	380·4 (91)	217·8 (52·1)	−217 (−51·9)	−92·4 (−22·1)	287·2 (68·7)
$H_2C=NH$	CH_3NH_2	405·5 (97)[b]	217·8 (52·1)	66·9 (16)[c]	189·8 (45·4)	310·6 (74·3)
$NH=NH$	NH_2NH_2	347 ± 20 (83 ± 5)	217·8 (52·1)	209 (50)[d]	297 ± 20 (71 ± 5)	217 ± 40 (51·9 ± 10)
$NH=O$	H_2NOH	397 ± 20 (95 ± 5)	217·8 (52·1)	99·5 (23·8)	121 ± 20 (29 ± 5)	201 ± 40 (48·1 ± 10)

[a] Unpublished results by A. S. Rodgers, 1974.
[b] Estimated by D. M. Golden from K. A. W. Parry and P. J. Robinson, *Int. J. Chem. Kinetics*, **5**, 27 (1973).
[c] Assuming[6] the heat of hydrogenation is 89.9 kJ/mol (21·5 kcal/mol).
[d] Reference 6.

of the acetonyl radical, i.e. if $X = 410$ kJ/mol (98 kcal/mol). Grzechowiak, Kerr, and Trotman-Dickenson (GKTD)[13] proposed that $X = 385$ kJ/mol (92.1 kcal/mol), giving a stabilization energy of 25 kJ/mol (6 kcal/mol). The value obtained by GKTD was not based on experimental results but on an empirical, Polanyi relationship that had some inconsistencies[12]. The conclusion by SGB was that the experimental value is more firmly based.

Now the plot thickens. Tsang[14] has studied the pyrolysis reactions (14) and (15) by his widely-acclaimed, competitive, shock-tube technique:

$$CH_3COCH(CH_3)CH_2CH_3 \longrightarrow$$

$$CH_3CO + CH(CH_3)CH_2CH_3 - 318 \text{ kJ/mol (76 kcal/mol)} \qquad (14)$$

$$\longrightarrow CH_3COCH(CH_3) + CH_2CH_3 - 297 \text{ kJ/mol}$$

$$(71 \text{ kcal/mol}) \qquad (15)$$

These experiments suggest that the methylacetonyl radical has a stabilization energy of 21 kJ/mol (5 kcal/mol) or more. Tsang plans experiments on the pyrolysis of methyl n-alkyl ketones that should provide direct values for the stabilization energy of the acetonyl radical.

C. Olefin Isomerization

Recent papers on the isomerization of butenes[15–17] have given heats and entropies of isomerization over a temperature range of 250 to 900 K. The main conclusion[17] from these studies is that the entropy of but-1-ene at 298.16 K is 311.4 J/(mol K) (74.5 cal/(mol K)) compared with the API value of 305.3 J/(mol K) (73.0 cal/(mol K)). In addition, the API data for cis- and trans-but-2-ene are all right at 298 K, but are slightly off at temperatures up to 900 K.

V. ACKNOWLEDGEMENTS

It is a pleasure to acknowledge the helpful discussions with Sidney W. Benson, David M. Golden, Menachem Luria, J. Alistair Kerr, Alan S. Rodgers, and Wing Tsang. These authors kindly gave me permission to quote their unpublished results. I am indebted to Mrs. Elaine Adkins for typing the manuscript.

VI. REFERENCES

1. D. R. Stull, E. F. Westrum, and G. C. Sinke, *The Chemical Thermodynamics of Organic Compounds,* Wiley, New York, 1969.
2. J. D. Cox and G. Pilcher, *Thermochemistry of Organic and Organometallic Compounds,* Academic Press, New York, 1970.
3. S. W. Benson, F. R. Cruickshank, D. M. Golden, G. R. Haugen, H. E. O'Neal, A. S. Rodgers, R. Shaw, and R. Walsh, *Chem. Rev.,* **69**, 279 (1969).
4. H. K. Eigenmann, D. M. Golden, and S. W. Benson, *J. Phys. Chem.,* **77**, 1687 (1973).
5. R. Shaw, in *Chemistry of Hydrazo, Azo, and Azoxy Groups* (Ed. S. Patai), Wiley, New York, 1974.
6. R. Shaw, in *Chemistry of Amidines and Imidates* (Ed. S. Patai), Wiley, New York, 1975.
7. S. W. Benson and M. Luria, *J. Amer. Chem. Soc.,* **97**, 704 (1975).
8. S. W. Benson and M. Luria, *J. Amer. Chem. Soc.,* **97**, 3337 (1975).
9. M. L. McGlashan, *Ann. Rev. Phys. Chem.,* **24**, 51 (1973).
10. V. Palm, *Reakts. Sposobnost Org. Soedin.,* **10**, 413 (1973).
11. J. A. Kerr, M. J. Parsonage, and A. F. Trotman-Dickenson, in *Handbook of Chemistry and Physics,* 55th Edition 1974–1975, Chemical Rubber Company, 1974, p. F 204.
12. R. K. Solly, D. M. Golden, and S. W. Benson, *Int. J. Chem. Kinetics,* **2**, 11 (1970).
13. J. Grzechowiak, J. A. Kerr, and A. F. Trotman-Dickenson, *J. Chem. Soc.,* 5080 (1965).
14. W. Tsang, personal communication, 1974.
15. J. Happel, M. A. Hnatow, and R. Mezaki, *J. Chem. Eng. Data,* **16**, 206 (1971).
16. H. Akimoto, J. L. Sprung, and J. N. Pitts, *J. Amer. Chem. Soc.,* **94**, 4850 (1972).
17. E. F. Meyer and D. G. Stroz, *J. Amer. Chem. Soc.,* **94**, 6344 (1972).

CHAPTER **4**

Mechanisms of elimination and addition reactions involving the X=Y group

ANTHONY F. COCKERILL and
ROGER G. HARRISON

Lilly Research Centre Limited, Erl Wood Manor, Windlesham, Surrey, U.K.

I. SCOPE

There have been many reports concerning mechanisms of elimination and addition reactions involving X=Y groups since some of these aspects were discussed in previous chapters in this series of books on the *Chemistry of Functional Groups*. Space prevents a comprehensive coverage, and, consequently, we therefore confine our discussion to the salient changes in our understanding of the reaction mechanisms which have occurred during the last few years. Our subject matter can be conveniently categorized into four sections:

(i) Formation of X=Y in elimination reactions of X—Y: Typical examples, which include alkene, carbonyl, thiocarbonyl and imine forming eliminations, constitute the major portion of this chapter.

(ii) Formation of X=Y in addition reactions of X≡Y: Nucleophilic and electrophilic additions to alkynes and nitriles fall into this classification. We omit electrophilic additions to alkynes and our discussion in this section as a whole is very brief.

(iii) Addition reactions of X=Y to give X—Y: Our discussion concerning alkenes is limited to reactions involving nucleophilic 1,2-additions. Electrophilic additions, 1,3-dipolar additions and carbene insertions and additions are discussed in other chapters. Heterolytic additions to carbonyls and imines comprise the major part of this section.

(iv) Elimination reactions of X=Y to give X≡Y: Alkyne and nitrile forming eliminations typify this section.

The examples mentioned briefly in these sections are of paramount importance in synthetic organic chemistry. Hence, although the major theme of this chapter concerns reaction mechanisms, we comment where relevant, on the relative synthetic merits of certain reactions. The reactivity of the group X=Y is mainly determined by the nature of the constituent atoms X and Y, and to a lesser extent its molecular environment. Even so, very similar mechanistic descriptions may apply to widely differing reactions. For example, in both alkene or carbonyl-forming elimination reactions, the problem may narrow to attempting to distinguish between concerted bimolecular elimination or stepwise carbanion elimination. Hence, we could approach our discussion in two ways; (i) either by categorizing under reaction mechanisms, or (ii) by the functional group. We prefer the latter approach, as this is consistent with presentations in earlier volumes, and in general, the chemist, who is interested in synthesis, is more concerned with the particular reactions of a functional group rather than in a comparison of reaction mechanisms of differing groups. We do however, attempt where possible to draw analogies and emphasize differences in the behaviour of different functional groups under comparable reaction conditions.

Our sections on the chemistry of each functional group begin with a brief summary of the material discussed in the original volumes. This is followed by a description of the major reviews, which have appeared in recent years, then the more specific details of mechanism.

II. ALKENE-FORMING ELIMINATIONS

A. General Comments

The Chemistry of Alkenes contains two chapters concerning the mechanisms of β-eliminations giving alkenes. Maccoll's discussion[1] describes the unimolecular gas phase pyrolyses of alkyl halides and esters. Under homogeneous conditions, these eliminations occur with *syn*-stereospecificity via a cyclic transition state in which varying degrees of carbonium-ion character develop at the carbon bearing the more electronegative leaving group. Our mechanistic concept of these reactions has not changed significantly during the last decade, although a wider range of experimental techniques are now available and more attractive thermal eliminations for synthetic purposes have been developed. However, dramatic changes have occurred in our understanding of mechanisms of solvolytic and base catalysed eliminations since this subject was reviewed by Saunders in 1963[2].

Saunders' chapter concerns mainly the mechanisms of β-elimination of alkyl halides, esters and 'onium salts in protic solvents containing their lyate ions. Under these conditions the E2 and to a lesser extent, the E1 mechanism are the dominant processes. Examples of the E1cB mechanism were rare and founded on uncertain evidence, and the α'–β elimination was only observed under exceptional conditions. Saunders outlined the concept of the variable transition state for concerted E2 reactions, in which bond breaking and making processes need not be completely synchronous[2,3]. This concept was used to explain changes in kinetic parameters, orientation, and stereochemical features of olefin-forming eliminations. In general, most E2 reactions were considered as occurring with *anti*-stereospecificity, *syn*-stereospecificity being encountered only in activated systems in which the acidity of the β-hydrogen was enhanced by electron-withdrawing groups, or when *anti*-elimination was precluded. In dubious cases, *syn*-elimination was attributed to the E1cB mechanism. Both steric and electronic factors were proposed to explain orientational behaviour, different groups of workers providing evidence to support their own preference. Another unsettled problem concerned the mechanism of elimination of alkyl halides and esters with weak bases such as thiolate and halide ions in aprotic solvents. These reagents were regarded as nucleophiles towards carbon rather than hydrogen, and to explain their 'unexpected' effectiveness at causing elimination reactions, a merged mechanism for E2 and S_N2 reactions was proposed by some authors. Others preferred explanations in terms of polarizability factors and the variable transition state theory.

During the last decade the considerable improvement in analytical equipment (chromatographic, spectrometric, stop-flow, etc.) has enabled much more definitive kinetic analysis. In addition a much wider range of reaction conditions have been used to effect elimination reactions. As a result, numerous examples of concerted *syn*-elimination have been uncovered, and in some cases *syn*-elimination has been shown to occur in preference to *anti*-elimination. A considerable number of variants of the E1cB mechanism have been demonstrated, although nearly all of them are of little consequence in the synthesis of alkenes. The concept of a common transition state for E2 and S_N2 reactions has been disproved, but has been replaced by a modified transition state for bimolecular elimination termed the E2C mechanism. The E2C mechanism implies an important role for nucleophile–C_α interaction in the transition state, and is invoked for the reactions of alkyl halides and esters with nucleophiles in aprotic media. Both steric and electronic factors are now accepted as influential under certain conditions in controlling the stereochemistry and orientation of elimination. The role of ion pairs in elimination reactions involving either cationic or anionic intermediates has been realized. Considerably more evidence has been provided to support the concept of the variable transition state for bimolecular E2 reactions. However, some controversy exists concerning reactions which exhibit transition states with considerable ionic character. Some authors regard these as variants of the E1 and E1cB mechanisms, in which stepwise cleavage of bonds occurs, with the initial cleavage being rate determining and essentially non-reversible. We shall discuss all of these aspects in the subsequent sections.

In some of the more recent volumes in this series, the reactions of carbonium ions, generated in reactions other than solvolyses of alkyl halides and esters have been discussed. Alkene yields are generally very low in deaminations[4,5] and we therefore exclude them from our discussion. On the other hand, we devote a small section to the discussion of dehydration reactions[6], which are more relevant synthetically.

The mechanisms of a wide variety of β-eliminations giving alkenes have been discussed in detail in two monographs[7,8] and one review[9]. Specific reviews on the following aspects have also appeared in the last 10 years; the variable transition state for concerted elimination[10], the stereochemistry of base-catalysed β-eliminations[11], the carbanion mechanism in olefin formation[12,13,14], mechanisms of pyrolytic eliminations[15,16], dehydration reactions using solid catalysts[17] and mechanisms of dehalogenations[18].

B. Terminology for Solvolytic and Base Induced β-Eliminations

Most heterolytic β-eliminations in solution involve hydrogen as the less electronegative eliminating fragment. X is usually a halide, an ester or an 'onium salt in synthetic applications, although a wider range of leaving groups have been studied kinetically. Most of the eliminations are described by one of three possibilities, the E1, E2 and E1cB mechanisms. This is basically the terminology devised over 40 years ago by Hughes and Ingold[19]. It has withstood the test of time, although as we shall see in subsequent sections, a multitude of minor modifications are now used.

The difference between the three mechanisms rests on the relative timing of the breaking of the C_β—H and C_α—H bonds. The unimolecular elimination, E1, requires only that the substrate is dissolved in a solvent which is capable of assisting ionization of X with its bonding electrons, prior to the cleavage of the C_β—H bond (equation 1). Depending on the relative magnitudes of $k_{-1}(X^-)$ and k_2, either C_α—X or C_β—H cleavage may be rate determining.

$$H-\overset{\diagdown}{\underset{\diagup}{C}}_\beta-\overset{\diagup}{\underset{\diagdown}{C}}_\alpha-X \;\underset{k_{-1}}{\overset{k_1}{\rightleftharpoons}}\; H-\overset{\diagdown}{\underset{\diagup}{C}}-\overset{\diagup}{\underset{\diagdown}{C}}+ \; + \; \bar{X} \;\overset{k_2}{\longrightarrow}\; H^+ \; + \; \overset{\diagdown}{\underset{\diagup}{C}}=\overset{\diagup}{\underset{\diagdown}{C}} \; + \; \bar{X} \qquad (1)$$

In both cases, the rate is first order in the substrate, but under pre-equilibrium conditions $(k_{-1}(X^-) \gg k_2)$ an inverse dependence on X^- is expected if the concentration of the intermediate carbonium ion is low (i.e. allowing application of the steady state hypothesis). An additional complication concerns the mode of removal of the β-hydrogen. If assistance is provided by the leaving group X, then overall *syn*-elimination is anticipated, whereas if the solvent acts as the base, *anti*-elimination is more probable unless the stereochemical arrangement of the leaving group in the substrate is not lost by rapid rotation about the C_α—C_β bond in the carbonium ion[20]. This intermediate carbonium ion can clearly capture a solvent molecule to give an S_N1 product, rather than lose a β-hydrogen, and the ease of this process makes E1 reactions undesirable as a good synthetic method for alkenes.

Basic reagents are most often used to induce β-eliminations giving alkenes. The E2 mechanism (equation 2) describes the situation in which both the C_β—H and C_α—X bonds cleave simultaneously via a single transition state. Although this reaction is essentially concerted, the bond making and breaking processes need not occur in a synchronous fashion. Transition states with carbanion character at C_β or carbonium ion character at C_α, and varying degrees of double-bond character can be depicted. This is essentially the concept of the variable transition state for concerted bimolecular eliminations, and we shall discuss more specific terminology to cater for numerous variants subsequently (see Section II.D.1).

$$B + H-\overset{\diagdown}{\underset{\diagup}{C}}-\overset{\diagup}{\underset{\diagdown}{C}}-X \;\overset{slow}{\longrightarrow}\; [\overset{\delta+}{B}---H---\overset{\diagdown}{\underset{\diagup}{C}}\cdots\overset{\diagup}{\underset{\diagdown}{C}}\overset{\delta-}{---X}] \;\longrightarrow\; \overset{+}{B}H \; + \; \overset{\diagdown}{\underset{\diagup}{C}}=\overset{\diagup}{\underset{\diagdown}{C}} \; + \; \bar{X} \qquad (2)$$

The stereoselectivity of E2 reactions can be strictly controlled as the stereochemical identity of the leaving groups with respect to each other is maintained in the transition state. For this reason E2 reactions are the methods of choice in alkene synthesis. For eliminations from rigid small alicyclic or bicyclic compounds in which the dihedral angle between the C_α—X and C_β—H bonds is clearly defined in the substrate, the following terminology for the stereochemistry of elimination is used in this text: (dihedral angle, 0–30°) *syn*-periplanar elimination; (60 ± 30°) *syn*-clinal; (120 ± 30°) *anti*-clinal; (150–180°) *anti*-periplanar. In acyclic molecules, free rotation allows all orientations of the eliminating fragments to be attained in the transition state. Diastereoisomers can be used to delineate the stereochemistry of elimination, but since it is impossible to distinguish between the periplanar and clinal orientations, the general terms *syn* and *anti* are used to describe the stereochemistry. In line with terminology adopted in all previous reviews, we shall use the historic terms Hofmann and Saytzev products to describe the less and more substituted alkene when elimination may involve more than one β-hydrogen atom (equation 3).

$$CH_3CH_2CHXCH_3 \longrightarrow CH_3CH=CHCH_3 + CH_3CH_2CH=CH_2 \qquad (3)$$

Saytzev product   Hofmann product

The E1cB mechanism describes the situation in which the $C_\beta—H$ bond cleaves prior to the $C_\alpha—X$ bond (equation 4). Depending on the relative magnitudes of $k_{-1}(\overset{+}{B}H)$ and k_2, either leaving group may be cleaved in the slow step.

$$H—\overset{|}{C}—\overset{|}{C}—X + B \underset{k_{-1}}{\overset{k_1}{\rightleftharpoons}} \overset{+}{B}H + \overset{|}{\bar{C}}—\overset{|}{C}—X \overset{k_2}{\longrightarrow} \overset{+}{B}H + \overset{|}{C}=\overset{|}{C} + \bar{X} \quad (4)$$

In most cases, both the E1cB and E2 mechanisms should exhibit first-order kinetics in both the substrate and the base. In fact, one of the most challenging problems facing the kineticist has been to develop clear-cut methods for distinguishing the E2 mechanism from all of the variants of the E1cB mechanism. We discuss the success achieved during the last 10 years in the next section.

For leaving groups which contain α'-hydrogens (e.g., $X = \overset{+}{N}Me_3$, $\overset{+}{S}Me_2$), the ylid mechanism (also termed $\alpha'-\beta$ elimination) has to be considered (equation 5)[21]. Analysis of the deuterium content of the eliminated X allows this mechanism to be distinguished from the E2 and E1cB routes. The ylid mechanism is *syn*-stereospecific and consequently should exhibit features in common with pyrolytic eliminations (e.g., the Cope elimination of amine-oxides, Section II.H.2).

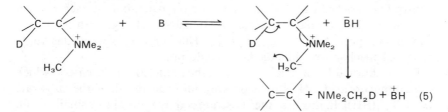

C. E1cB Mechanisms

1. Mathematical possibilities and kinetic differences from the E2 mechanism

There are a number of mathematical solutions to the general equation for the E1cB mechanism (equations 6 and 7)[14,22-24].

$$H-\overset{|}{\underset{|}{C}}-\overset{|}{\underset{|}{C}}-X + B^- \underset{k_{-1}}{\overset{k_1}{\rightleftharpoons}} \overset{|}{\underset{|}{\bar{C}}}-\overset{|}{\underset{|}{C}}-X + BH \qquad (6)$$

$$\overset{|}{\underset{|}{\bar{C}}}-\overset{|}{\underset{|}{C}}-X \xrightarrow{k_2} \overset{|}{\underset{|}{C}}=\overset{|}{\underset{|}{C}} + X^- \qquad (7)$$

If the substrate and conjugate acid, BH, are of similar acidity, and the leaving group X is not very labile ($k_{-1}(BH) > k_2$), then in presence of excess base, the substrate will be converted extensively into its conjugate base, which then undergoes a rate-determining unimolecular decomposition to the alkene product. Under these conditions, a further increase in the base concentration can cause little change in the carbanion concentration and the rate of reaction should exhibit first-order kinetics in the substrate but zero-order in the base (equation 8). This mechanism is termed the $(E1)_{anion}$ mechanism[14,22], although other authors have described it as 'E1cB of the second type'[23,24].

$$\text{Rate}((E1)_{anion}) = k_2(RX) \qquad \text{if } k_1 \gg k_{-1}(BH) > k_2 \qquad (8)$$

Kinetically the $(E1)_{anion}$ mechanism may exhibit many features in common with the E1 mechanism. The rate is determined by the cleavage of the $C_\alpha-X$ bond. Thus, as long as the substrate remains extensively converted into its conjugate base with changes in the substrate structure, the following kinetic predictions can be made: (i) a significant leaving group isotope effect or element effect (e.g., k_{Br}/k_{Cl}) is expected; (ii) electron-releasing substituents at C_α should enhance the rate of elimination; (iii) electron-withdrawing substituents at C_β should retard the elimination, by causing further delocalization of the electron pair at C_β, thereby reducing its effectiveness in aiding cleavage of the $C_\alpha-X$ bond; (iv) if the substrate is labelled with β-deuterium. and the reaction is quenched when little alkene has formed, complete or extensive exchange with the protic solvent is expected, i.e. $k_{exchange} \gg k_{elim}$. The first two expectations apply to the E1 mechanism while the last two do not.

Two additional E1cB mechanisms can be depicted for the case in which the conjugate acid BH is considerably more acidic than the substrate ($k_1 \ll k_{-1}(BH)$). In this case the substrate is only ionized to a small extent, enabling application of the steady state treatment to the concentration of the carbanion. Equation 9 shows that the rate law is determined by the relative magnitudes of $k_{-1}(BH)$ and k_2. If $k_{-1}(BH) \gg k_2$, the elimination involves a rapid pre-equilibrium followed by a slow unimolecular elimination from the carbanion and the rate is described by equation (10). This situation is called the reversible anion mechanism, $(E1cB)_R$[14], although the description 'pre-equilibrium E1cB' has often been used [12,24]. The rate in this case should exhibit first-order kinetics in both the substrate and the base if BH is the solvent. However if BH is not the solvent (e.g., $\bar{B} = NR_3$, $BH = \overset{+}{N}HR_3$, solvent $= EtOH$), then the reaction should reflect specific base catalysis as the rate is dependent on the buffer ratio, $\bar{B}/BH$, and not simply on $\bar{B}$. As befits a rapid pre-equilibrium, the rate of β-deuterium exchange in the substrate with a protic solvent should occur much more rapidly than the rate of elimination. However,

$$\text{Rate} = k_2(\text{carbanion}) = \frac{k_1 k_2 (RX)(B^-)}{k_{-1}(BH) + k_2} \qquad (9)$$

$$\text{Rate } ((E1cB)_R) = \frac{k_1 k_2 (RX)(B^-)}{k_{-1}(BH)} \qquad (10)$$

as Breslow has pointed out[25], although this is a necessary requisite of an $(E1cB)_R$ mechanism, it cannot alone be used as evidence for this mechanism rather than an E2 mechanism, in which exchange cannot occur, since the exchange could be an irrelevant side reaction.

Substituent effects on rates of $(E1cB)_R$ reactions are more difficult to predict than for the $(E1)_{anion}$ mechanism, as they may exert their influence on either the position of the pre-equilibrium or the subsequent cleavage. Electron-withdrawing substituents at C_β should enhance the rate of elimination by increasing the steady-state concentration of the carbanion, even though they may retard k_2. Similarly, electron-releasing substituents at C_α may effect k_2 more than k_1/k_{-1} and hence slightly promote the rate. Rate-determining C_α—X cleavage should reflect a significant isotope effect or element effect.

Another solution of equation 9 results if $k_2 \gg k_{-1}(BH)$ (equation 11). In this case, proton removal is rate limiting and the resulting carbanion decomposes unimolecularly to alkene more rapidly than it undergoes deprotonation by the solvent. This situation is called the irreversible carbanion mechanism, $(E1cB)_{irr}$, and is the most difficult to distinguish from the E2 reaction in which carbanion character is extensive. Both reactions follow second-order kinetics, involve rate-determining proton removal and should exhibit negligible or small α-leaving group effects. However, in one case proton removal is coupled with other bond changes, albeit small ones, whereas in the $(E1cB)_{irr}$, the isotope effect for proton removal should simulate simple carbon acid ionizations. This difference appears to be significant in the eliminations of 1,1-diaryl-2,2,2-trichloro-ethanes (DDT) with various bases as we see subsequently (Section II.C.4.a)[26]. Another ploy concerns modifications in the concentration of BH in order to increase $k_{-1}(BH)$ relative to k_2, and cause a change over to an $(E1cB)_R$ mechanism. This should not be possible with an E2 reaction. Both the $(E1cB)_{irr}$ and E2 routes are subject to general base rather than specific base catalysis as is anticipated for the $(E1cB)_R$ mechanism.

$$\text{Rate } ((E1cB)_{irr}) = k_1(RX)(B^-) \qquad \text{if } k_2 \gg k_{-1}(BH) \qquad (11)$$

Another variant involves ion pairs, and was originally invoked for the elimination of *cis*-dibromoethylene with amines in dimethyl formamide[27]. Neither protium exchange with labelled solvent, nor a hydrogen isotope effect is observed. However, the element effect k_{Br}/k_{Cl} is large. These facts accord with Scheme 1, which depicts the $(E1cB)_{ip}$ mechanism. The intimate ion pair collapses to reactant or product without equilibration with the solvent, thereby avoiding β-protium exchange. $k_2 \ll k_{-1}$, restricting the influence of β-protium substitution to a secondary isotope effect on k_2 and an equilibrium isotope effect on k_1/k_{-1}, both of which approximate to unity[28]. Ion pair mechanisms are most likely in solvents of low ionizing and solvating power, especially of the aprotic type in which anion solvation is poor[29]. Table 1 compares the predicted kinetic effects for the ElcB mechanisms with those for an E2 reaction with a transition state with considerable carbanion character.

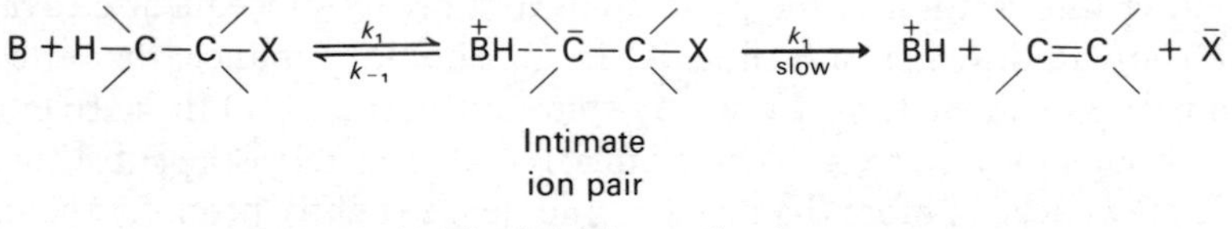

$$\text{B} + \text{H} - \overset{|}{\underset{|}{\text{C}}} - \overset{|}{\underset{|}{\text{C}}} - \text{X} \; \underset{k_{-1}}{\overset{k_1}{\rightleftharpoons}} \; \overset{+}{\text{B}}\text{H} - - - \bar{\text{C}} - \overset{|}{\underset{|}{\text{C}}} - \text{X} \; \underset{\text{slow}}{\overset{k_1}{\longrightarrow}} \; \overset{+}{\text{B}}\text{H} + \; \text{C} = \text{C} + \bar{\text{X}}$$

Intimate
ion pair

SCHEME 1. The $(E1cB)_{ip}$ mechanism

Hine[30] predicted that the following requirements to encourage E1cB rather than an E2 reaction: (i) a leaving group X which is not easily displaced with its bonding electrons; (ii) an intermediate carbanion which is stable relative to the reactants and products under the reaction conditions, and (iii) an unsaturated product which is comparatively unstable relative to the reactants. All of these factors are apparent in the examples of the E1cB mechanism which have been uncovered in recent years. The alkene is seldom isolated as it undergoes rapid nucleophilic addition in the presence

TABLE 1. Kinetic predictions for base-induced β-eliminations

$$\bar{B} + (D)H{-}\overset{\displaystyle /}{\underset{\displaystyle \backslash}{C}}_\beta{-}\overset{\displaystyle /}{\underset{\displaystyle \backslash}{C}}_\alpha{-}X \longrightarrow BH + \overset{\displaystyle \backslash}{\underset{\displaystyle /}{C}}{=}\overset{\displaystyle /}{\underset{\displaystyle \backslash}{C}} + \bar{X}$$

Mechanism	Kinetic[a] order	β-Protium exchange faster than elimination	General or specific base catalysis	k_H/k_D	Electron withdrawal at C_β[d]	Electron release at C_α[d]	Leaving group isotope effect or element effect
$(E1)_{anion}$	1	Yes	General[c]	1·0	Rate decrease	Rate increase	Substantial
$(E1cB)_R$	2	Yes	Specific	1·0	Small rate increase	Small rate increase	Substantial
$(E1cB)_{ip}$	2	No	General[e]	1·0 → 1·2	Small rate increase	Small rate increase	Substantial
$(E1cB)_{irr}$	2	No	General	2→8	Rate increase	Little effect	Small to negligible
$E2$[b]	2	No	General	2→8	Rate increase	Small rate increase	Small

[a] All mechanisms exhibit first-order kinetics in substrate.
[b] Only transition states with considerable carbanion character considered in this table.
[c] Specific base catalysis predicted if extent of substrate ionization reduced from almost complete.
[d] Effect on rate assuming no change in mechanism is caused; steric factors upon substitution at C_α and C_β have not been considered. The rate predictions are geared to substituent effects such as these giving rise to Hammett reaction constants on β- and α-aryl substitution.
[e] Depends on whether ion pair assists in removal of leaving group.

of the basic reagents to give a product of apparent substitution. This problem limits the synthetic potential and raises the need to prove kinetically the absence of a direct substitution reaction on the substrate.

2. (E1)$_{anion}$ mechanisms

This mechanism is more prevalent in carbonyl and imine forming eliminations in which the β-hydrogen if attached to the heteroatom is acidic. An early example concerns the synthesis of nitromethane (equation 12)[31]. As long as the substrate is at least 2 pK units more acidic than the conjugate acid of the base, the reaction should follow first-order kinetics[14]. As carbon is less electronegative than the heteroatoms, it is necessary for the β-hydrogen to be activated by a strongly electron-withdrawing group, to enable observation of (E1)$_{anion}$ mechanisms giving alkenes.

$$\bar{O}H + OH-\overset{O}{\overset{\|}{C}}-CH_2NO_2 \longrightarrow H_2O + \bar{O}-\overset{O}{\overset{\|}{C}}-CH_2NO_2 \longrightarrow CO_2 + \bar{C}H_2NO_2 \quad (12)$$

$$(E_{co}1)_{anion} \quad \text{mechanism}$$

Observation of the carbanion intermediate provides good evidence for the (E1)$_{anion}$ mechanism. Berndt[32] in fact isolated the carbanion of 3-t-butyl-4,4-dimethyl-2,3-dinitropentane as its ammonium salt. This salt decomposed slowly in the solid state, but in methanol above 0 °C underwent rapid elimination of nitrite to give an unstable alkene (Scheme 2).

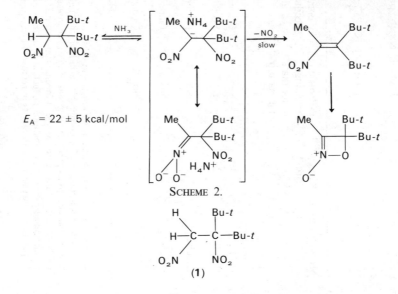

SCHEME 2.

The remarkable stability of this carbanion probably derives from steric inhibition to internal displacement of nitrite by the electron pair, because the slightly less encumbered structure **1** cannot be isolated as its ammonium salt since it undergoes rapid elimination.

The nitro group has also been used as the activating substituent in the $(E1)_{anion}$ eliminations of *cis*- and *trans*-**2** with methoxide ion (Scheme 3)[22]. The reaction was monitored by u.v. spectroscopy and spectra consistent with **2**, **3** and **5** were obtained during the course of the reaction. The rates of ionization of the two isomers (measured by stop-flow techniques) differ slightly because of differences in steric inhibition to proton abstraction, and the large primary isotope effect accords with a symmetrical transition state for proton removal[33]. Anion formation is essentially complete shortly after the reactants are mixed. Olefin **4**, formed in the slow step, is unstable and is converted rapidly into the more conjugated alkene anion **5**, which is isolated on neutralization of the reaction. The rate of formation of **5** is independent of the base concentration and the stereochemistry of the reactant **2**. The composite isotope effect for the conversion of 1,3,3-trideutero-**2** into **5** of 1·7 reflects mainly a secondary isotope effect.

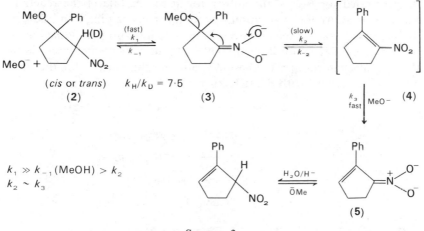

SCHEME 3.

The six-membered analogue of **2** is about 2 pK units less acidic and the rate of its elimination with methoxide ion increases with base concentration, consistent with an $(E1cB)_R$ pathway[34]. A change to the more basic medium of *t*-BuOK/*t*-BuOH causes increased substrate ionization and incursion of the $(E1)_{anion}$ mechanism. The induction period of about 0·8 to 1 sec in the rate of appearance of phenoxide from 2-phenoxynitroethane on treatment with ethoxide or hydroxide ion, and the sensitivity of elimination rate to solvent changes are consistent with the operation of the $(E1)_{anion}$ mechanism[35]. Bordwell has stated that in protic solvents, the only single electron withdrawing group which is capable of sustaining the $(E1)_{anion}$ mechanism is the nitro group. For the less-electron-withdrawing C=O and C≡N and MeSO$_2$ groups, a stronger base is required[14].

3. (E1cB)$_R$ mechanisms

a. *Eliminations of 2-substituted aryloxyethanes.* Crosby and Stirling[36] monitored the rates of elimination of a series of β-activated phenoxyethanes by following the appearance of phenoxide ion spectrophotometrically. The elimination rates vary by a factor of 10^{11} with changes in the activating group in the two basic systems used (see Table 2). The quantitative yields of phenoxide ion and 2-substituted ethoxyethane (for the ethoxide induced reactions) exclude eliminations involving the activating substituent (e.g., Hofmann elimination etc.). To confirm the observed reactions were eliminations, it was necessary to show that the ethers did not arise via a direct substitution. To this end, the intermediacy of the alkene was demonstrated for two of the substrates in the middle of the reactivity range (X = SOAr, CN) by trapping experiments using piperidine (Scheme 4). Needless to say, in the absence of the lyate ions, no piperidine adduct was obtained. Except for 2-phenoxynitroethane, which eliminates via the $(E1)_{anion}$ mechanism in the ethanolic solution, the kinetics for the other eliminations accord with the (E1cB)$_R$ route.

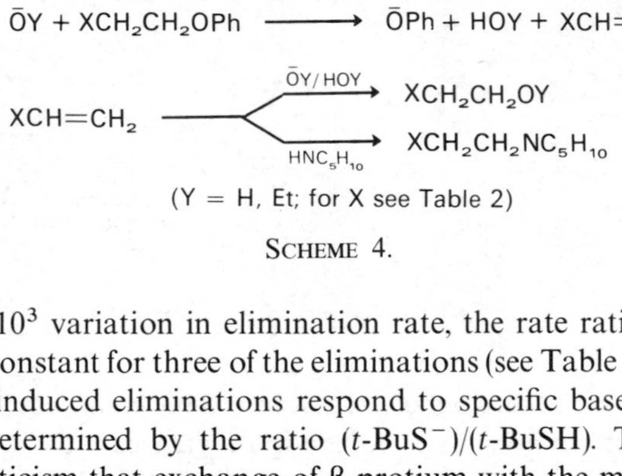

$$\bar{O}Y + XCH_2CH_2OPh \longrightarrow \bar{O}Ph + HOY + XCH=CH_2$$

$$XCH=CH_2 \underset{HNC_5H_{10}}{\overset{\bar{O}Y/HOY}{\diagdown}} \begin{array}{l} XCH_2CH_2OY \\ XCH_2CH_2NC_5H_{10} \end{array}$$

(Y = H, Et; for X see Table 2)

SCHEME 4.

Despite a 10^3 variation in elimination rate, the rate ratio $k_{t\text{-}Bu\bar{S}}/k_{\bar{O}Et}$ is remarkably constant for three of the eliminations (see Table 2). In addition, the thiolate induced eliminations respond to specific base catalysis, the rate being determined by the ratio $(t\text{-}BuS^-)/(t\text{-}BuSH)$. To circumvent Breslow's criticism that exchange of β-protium with the medium may be an irrelevant side reaction to the β-elimination, Crosby and Stirling employed a more involved method to demonstrate a rapid pre-equilibrium involving proton removal. They compared the rates of elimination of

TABLE 2. Rate coefficients and products in eliminations of
β-substituted phenoxyethanes

$$\text{XCH}_2\text{CH}_2\text{OPh} + \bar{\text{O}}\text{Y} \xrightarrow{\text{slow}} \text{XCH}=\text{CH}_2 + \bar{\text{O}}\text{Ph} \xrightarrow[\text{fast}]{\bar{\text{O}}\text{Y}} \text{XCH}_2\text{CH}_2\text{OY}$$

	I		II		III

	$(\bar{\text{O}}\text{Y} = \bar{\text{O}}\text{Et}/\text{EtOH})$				$(\bar{\text{O}}\text{Y} = \bar{\text{O}}\text{H}/\text{H}_2\text{O}^e)$
X	$k_{\text{elim}}{}^a$	%III	%II	$k_{t-\text{Bu}\bar{\text{S}}}/k_{\bar{\text{O}}\text{Et}}$	$k_{\text{elim}}{}^a$
NO_2	b	99	91		3.2×10^2
$\overset{+}{\text{PPh}}_3$	$6.0 \times 10^3{}^c$	84	99		2.0
$\overset{+}{\text{SMe}}_2$	1.9×10^2	85	97		$4.8 \times 10^{-2}{}^f$
COMe	3.2×10	95	92	0.23	3.1
CHO	2.6×10	d	93		
SO_2OEt	7.3×10^{-1}	97	96		4.5×10^{-1}
SO_2Ph	3.5×10^{-1}	99	100	0.26	
SO_2Me	1.5×10^{-1}	92	94		
CN	9.4×10^{-2}	87	95		
CO_2Et	6.1×10^{-2}	104	100	0.24	1.5×10^{-2}
$\text{SO}_2\text{N}(\text{CH}_2\text{Ph})_2$	1.6×10^{-2}	99	100		
SOPh	1.2×10^{-3}	99	85		
CONH_2	1.1×10^{-3}	95	98		1.4×10^{-3}
SOMe	4.3×10^{-4}	91	105		$9.2 \times 10^{-5}{}^f$
CONEt_2	2.6×10^{-4}	98	97		
$\overset{+}{\text{NMe}}_3$	3.1×10^{-7}	d	88		
COO^-	7.0×10^{-8}	91	94		

^{a} In litre mol^{-1} sec^{-1} at 25 °C;
^{b} El$_{\text{anion}}$ mechanism operates;
^{c} Estimated from data in $\bar{\text{O}}\text{H}/\text{H}_2\text{O}$;
^{d} Not isolated;
^{e} 2% EtOH except results f.

protic substrates in aqueous sodium hydroxide with those for the deuterated substrates in $\text{NaOD}/\text{D}_2\text{O}$. These observed isotope effects, $k_{\text{H}}/k_{\text{D}}$ (0.66 for $\text{Me}_2\overset{+}{\text{S}}\text{CH}_2(\text{D}_2)\text{CH}_2\text{OPh}$ and 0.78 for $\text{MeSOCH}_2(\text{D}_2)\text{CH}_2\text{OPh}$ at 25°C) contain contributions of primary, secondary and solvent origin. After allowances for the larger protolysis constant of H_2O than D_2O^{37}, and the solvent and secondary isotope on both the pre-equilibrium and carbanion decomposition, a primary isotope effect of unity is predicted for the β-proton abstraction. This is as expected for the $(\text{E1cB})_\text{R}$ mechanism. However, considering the assumptions necessary to factorize the various isotope effects, it is doubtful if this approach is any more certain than the normal exchange criterion.

The comprehensive list of substituents merits an analysis as to the origin of the rate variations. These changes may reflect the influence of X (Table 2) on either the pre-equilibrium or the carbanion decomposition (see equations 6, 7, and 10, Section II.C.1). For the ethoxide induced eliminations, the nitro derivative is excluded as its reacts via the $(E1)_{anion}$ route. Rappoport[38] has argued that reprotonation of the remaining carbanions will approach diffusion control, thus reducing substituent effects to changes on k_1 and k_2. As k_1 is enhanced by electron withdrawal and k_2 by electron release, some self cancellation of the gross substituent effect is possible. Stirling and Crosby[36] have demonstrated linear correlations between $\log k_{elim}$ (rate change 10^{11}) and pK_a(CH_3X, variation 15·5 pK units), $\log k_{ion}$ (CH_3X exchange in DO^-/D_2O, rate change 10^9) and $\log k_{add}$ (nucleophilic addition to $RCH=CHX$). The similar magnitude of substituent effects for the first three terms suggests changes in X affect mainly values of k_1 for the ethoxide induced eliminations. The importance of anion delocalization in promoting the elimination rate is demonstrated by the much greater reactivity of the substrates in which $X = \overset{+}{S}Me_2$ and $\overset{+}{P}Ph_3$ than when $X = \overset{+}{N}Me_3$. Although the latter substituent has the greatest electron-withdrawing polar effect, the second row elements of the other 'onium salts possess low lying d-orbitals which can partake effectively in $p\pi$–$d\pi$ overlap, thereby affording greater delocalization of the carbanion electron pair. It is well known that protons exchange more rapidly when situated alpha to a sulphonium than an ammonium group[39,40].

Nuclear substitution in the activating group ($X = ArSO_2$ and $ArSO$) gives Hammett reaction constants of 2·1 and 1·7, respectively for the ethoxide induced eliminations of 2-X-1-phenoxyethanes[36]. The ratio of these reaction constants is similar to the ratios for the same substituents for ionization of β-X-acetic, propionic and trans-acrylic acids. In addition, the reaction constant of 2·1 is very similar to that of 2·32 observed for the γ-E1cB reaction of aryl 3-chlorosulphones with t-BuOK, a reaction in which a carbanion mechanism is indicated[41]. Nuclear substitution in the phenoxy leaving group gives Hammett reaction constants of 1·5 ($X = ArSO_2$) and 1·2 ($X = ArSO$). The reaction constant for the ionization of phenols is 2·23[42], so these values suggest the C_α—O bond is about half broken in the unimolecular decomposition of the carbanion.

b. *Fluorenyl substrates.* In methanol containing methoxide ion, 9-tritio-9-trifluoromethyl-fluorene (**6**) undergoes a faster exchange of tritium than loss of fluoride ion[43]. The dibenzofulvene is unstable under the reaction conditions and undergoes rapid nucleophilic addition (Scheme 5). In view of the poor leaving-group ability of fluoride and the aromatic nature of the fluorenyl carbanion (**7**), it seems highly probable that the elimination follows the (E1cB)$_R$ mechanism, However, although the rapid tritium exchange accords with the formation of a carbanion, its intermediacy in the elimination reaction is not proven.

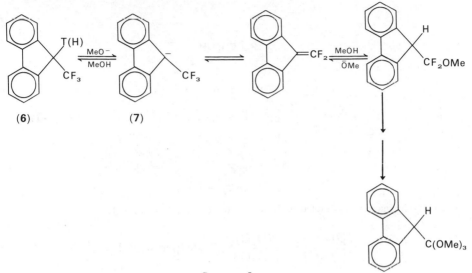

SCHEME 5.

More convincing evidence for the (E1cB)$_R$ mechanism for the dehydration of 9-fluorenylmethanol has been provided by complex isotope studies. More O'Ferrall and Slae[44] determined the rates of elimination of the 9-hydrogen and 9-deuterium compounds spectrophotometrically in water and deuterium oxide, and the rates of exchange of 9-tritium in both solvents containing their lyate ions. The exchange was faster than elimination to give dibenzofulvene, and induction periods were observed in both the elimination of protic substrate in D$_2$O and 9-deutero substrate in H$_2$O. The results again indicate the rapid reversible formation of a carbanion, but do not implicate its intermediacy in the elimination. This latter point was confirmed by a comparison of the experimental isotope effects with the theoretical predictions for the (E1cB)$_R$ and E2 mechanisms (see Scheme 6, equations 13–16). If the reaction is entirely E2, then $k_b \gg \alpha k_1$ and equations (17) and (18) apply. Equation (17) incorporates only a primary isotope effect, whereas equation (18) also incorporates solvent effects arising from the difference between hydroxide and deuteroxide as base and leaving group. These latter effects would have to be unreasonably large for the experimental ratio of 0·92 to incorporate a primary isotope effect of 7·2.

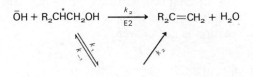

*labelled hydrogen atom

SCHEME 6.

$$k_E^H = k_b^H + \alpha_H k_1^H \tag{13}$$

$$k_E^D = k_b^D + \alpha_D k_1^D \tag{14}$$

$$k_X^D = k_b^D + k_1^D \tag{15}$$

$$\alpha = k_b/(k_2 + k_{-1}) \tag{16}$$

(E = elimination; X = exchange; H,D refer to the labelled hydrogen, $\overset{*}{H}$)

$$\frac{k_E^H(D_2O)}{k_E^D(D_2O)} = \frac{k_b^H(D_2O)}{k_b^D(D_2O)} = 7\cdot2 \text{ (experimental)} \tag{17}$$

$$\frac{k_E^H(H_2O)}{k_E^D(D_2O)} = \frac{k_b^H(H_2O)}{k_b^D(D_2O)} = 0\cdot92 \text{ (experimental)} \tag{18}$$

However, if the $(E1cB)_R$ mechanism operates, $\alpha k_1 \gg k_b$ and equations (19) and (20) apply. Equation (19) constitutes simply a primary isotope effect, consistent with the experimental value of 7·2 for carbanion formation. Equation (20) can be converted into equation (21) as exchange is faster than elimination $(k_{-1} \gg k_2)$. The isotope effect on k_2 should be small as it is of secondary origin. Primary isotope effects on k_1 and k_{-1} should be similar and hence cancel each other, such that the experimental value of 0·92 is consistent with an $(E1cB)_R$ mechanism.

$$\frac{k_E^H(D_2O)}{k_E^D(D_2O)} = \frac{k_1^H(D_2O)}{k_1^D(D_2O)} = 7\cdot2 \text{ (experimental)} \tag{19}$$

$$\frac{k_E^H(H_2O)}{k_E^D(D_2O)} = \frac{\alpha_H k_1^H(D_2O)}{\alpha_D k_1^D(D_2O)} = 0\cdot92 \text{ (experimental)} \tag{20}$$

$$\frac{k_E^H(H_2O)}{k_E^D(D_2O)} = \frac{k_1^H k_{-1}^D k_2(H_2O)}{k_1^D k_{-1}^H k_2(D_2O)} \tag{21}$$

The $(\text{E1cB})_R$ mechanism also operates in methanol, but in t-butyl alcohol, the exchange reaction is suppressed and the isotope effects are consistent with either an E2 or $(\text{E1cB})_{irr}$ elimination, the latter being preferred[45].

c. *Eliminations from β-keto derivatives.* The elimination of methanol from 4-methoxybutan-2-one (equation 22, X = OMe) with a variety of bases in aqueous isopropyl alcohol is specific base catalysed[46,47]. Exchange of β-protium in D_2O is 226-times faster than elimination, and the rate of exchange responds to general base catalysis as expected for a rate-determining proton transfer. The Brönsted coefficient, $β$, for elimination with amine bases is about unity. These facts accord with the operation of the $(\text{E1cB})_R$ mechanism. When the leaving group is the more labile acetoxy, β-protium exchange is no longer faster than elimination, which now follows general base catalysis, consistent with either an E2 or $(\text{E1cB})_{irr}$ mechanism. In an attempt to resolve this point, Fedor and his coworkers have measured the eliminations of a series of 4-substituted benzoyloxy[48] and phenoxy[49] butan-2-ones with a variety of amine bases in aqueous solutions. These substrates possess leaving groups whose conjugate acids have pK_a values spanning the range between methanol ($\simeq 16$) and acetic acid ($\simeq 4$).

$$XCH_2CH_2COCH_3 + B \underset{k_{-1}}{\overset{k_1}{\rightleftharpoons}} XCH_2\overset{-}{C}HCOCH_3 + \overset{+}{B}H \overset{k_2}{\longrightarrow} \overset{-}{X} + H_2C{=}CHCOCH_3$$

$$(22)$$

The β-eliminations of the p-substituted benzoates from 4-(p-substituted benzoyloxy)-2-butanones are general base catalysed and virtually insensitive to the substituent in the leaving group. This insensitivity is unusual for E2 reactions (e.g., arene sulphonates[50,51]) and accords more favourably with an $(\text{E1cB})_{irr}$ reaction in which $k_{-1}(\overset{+}{B}H) > k_2$, and thence k_1 is rate determining (cf. equation 22). The eliminations of the substituted phenoxides support this view. The rates of formation of the but-3-en-2-ones are described by equation 23, with tertiary amines (B) and hydroxide ion as the bases. This equation for the $(\text{E1cB})_R$ mechanism predicts that plots of k_{obs} against (B) should exhibit curvature at high buffer concentration $(k_{-1}(\overset{+}{B}H) \gg k_2)$, if a substantial fraction of the substrate is converted into its enolate anion. In fact, saturation kinetics of this kind are observed for four of the five phenoxy derivatives, only the p-CN derivative, which

$$k_{obs} = \frac{k_1(B)}{(k_{-1}/k_2)(\overset{+}{B}H) + 1} + \frac{k_{\overline{OH}}K_w}{a_{\overset{+}{H}}}$$

$$(23)$$

is by far the best leaving group, failing to show this behaviour. Rearranging equation 23, and using a double reciprocal plot of $1/(k_{obs} - k_{O\bar{H}}K_w/a_H^+)$ against $1/B$ enables determination of the values of k_1, k_{-1} and k_2[49]. Rate coefficient k_2 is markedly affected by the aryl substituent ($\rho = 0.67 \pm 0.08$, relative to $\rho' = 1$ for the ionization of the corresponding phenols), suggesting about 60–70% of the charge resides on the leaving group in the transition state for C_α—O cleavage. This is very similar to the charge distribution in the eliminations of phenoxide from 2-aryloxyethyl sulphones, sulphoxides and sulphonium salts discussed in Section C.3.a. Both k_1 and k_{-1} values change little with the p-substituent, the reaction constants being similar to that observed for the ionization of phenoxyacetic acids.

The hydroxide induced eliminations of the substituted phenoxy and benzoyloxy derivatives and the exchange reaction of 4-methoxybutan-2-one are related by equation (24). The generality of this expression accords

$$\log k_{\bar{O}H} = -1.16 \pm 0.07 \, pK_a + 6.25 \qquad (24)$$

(11 points, pK_a values for methoxyacetic acid, phenoxyacetic acids and benzoyloxyacetic acids)

with proton removal for the whole series being a separate process from C_α—O bond cleavage. The authors concluded that the E1cB mechanism was generally applicable to the eliminations of β-oxyketones possessing leaving groups, whose conjugate acids have pK_a values in the region 4–16. For good leaving groups, k_1 is rate determining, k_2 is rate determining for the poor leaving groups, and for intermediate cases, partitioning of the enolate anion is significant.

An interesting observation arising from the work of Fedor and Glave concerns the sensitivity of k_2 to the conjugate acid of the amine base, suggesting an ion pair association of $\overset{+}{B}H$ and the enolate anion[49]. Ion pairs undergoing continual internal return have been invoked to explain the apparent lack of β-deuterium exchange and interconversion of *threo-* and *erythro-*4,4'-dichlorochalcone dichlorides (8) during elimination induced in ethanol containing various buffers[52]. However, this explanation is not convincing, as the ethoxonium enolate ion pair would not be expected to have an appreciable lifetime in 'acidic' ethanol, and the observed rate depression in acidic solution should only arise if protonation is by an external acid[53]. Other than the exchange requirement, the kinetics of elimination of the chalcone dichlorides are consistent with an $(E1cB)_R$ mechanism[52].

$$p\text{-ClC}_6\text{H}_4\text{COCHClCHClC}_6\text{H}_4\text{Cl-}p$$

(8)

d. *Eliminations from 2-phenylsulphonyl-1-X-ethanes.* The results in Table 3 indicate the delicate balance between rate-determining ionization and elimination with change of α-leaving group[54]. Proton removal is rate-determining and no buffer saturation rate plot is observed for the acetate (either E2 or $(\mathrm{E1cB})_{\mathrm{irr}}$). However, the ionization rate is greater with the more powerful electron withdrawing $\overset{+}{\mathrm{N}}\mathrm{Me}_3$ group, and the 'poorer' departing ability of this group changes k_2/k_{-1} such that at high buffer concentration, the $(\mathrm{E1cB})_{\mathrm{R}}$ mechanism operates. The changes in k_1 accord with steric hindrance to proton abstraction. In contrast the authors point out that the ratio of k_2/k_{-1} is little affected by α- or β-phenyl substitution in the substrate, suggestive of carbanion intermediates with little double-bond character. However, the steric encumbrance in the carbanions would be expected to greatly restrict the orientation of the aryl group and thereby minimize the normal resonance delocalizing property of the aryl ring.

4. $(\mathrm{E1cB})_{\mathrm{irr}}$ mechanisms

In cases for which $k_2 \sim k_{-1}(\overset{+}{\mathrm{B}}\mathrm{H})$, an irreversible carbanion mechanism can be inferred at lower buffer concentration, if buffer saturation (and hence a change to the $(\mathrm{E1cB})_{\mathrm{R}}$ mechanism) is observed in more concentrated solution. However, if $k_2 \gg k_{-1}(\overset{+}{\mathrm{B}}\mathrm{H})$, or if the conjugate acid of the

TABLE 3. Structural effects on base-induced eliminations on the $(\mathrm{E1cB})_{\mathrm{irr}}$ borderline[54]

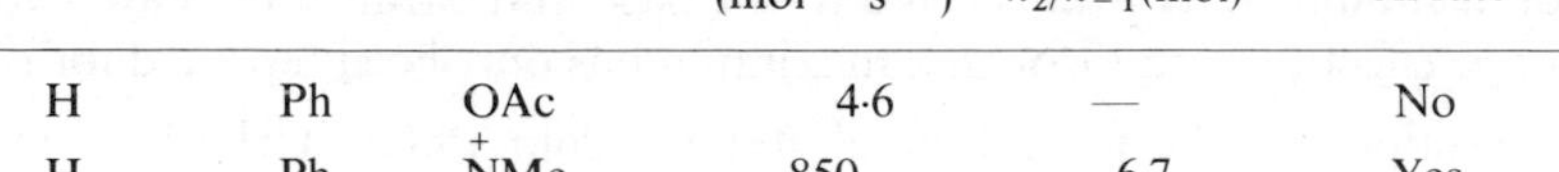

($\mathrm{B} = \mathrm{Et_3N}$, $\overset{+}{\mathrm{B}}\mathrm{H} = \mathrm{Et_3\overset{+}{N}H}$ buffers in EtOH at 25 °C)

R^1	R^2	X	$10^2 k_1$ ($\mathrm{mol^{-1}\,s^{-1}}$)	10^4 k_2/k_{-1}(mol)	Buffer saturation on rate
H	Ph	OAc	4·6	—	No
H	Ph	$\overset{+}{\mathrm{N}}\mathrm{Me}_3$	850	6·7	Yes
H	H	$\overset{+}{\mathrm{N}}\mathrm{Me}_3$	86	—	No
Ph	H	$\overset{+}{\mathrm{N}}\mathrm{Me}_3$	24	10	Yes
Ph	H	$\overset{+}{\mathrm{N}}\mathrm{Me_2Et}$	6·8	19	Yes
H	H	$\overset{+}{\mathrm{N}}\mathrm{Et_2Me}$	53	—	No
H	H	$\overset{+}{\mathrm{N}}\mathrm{Me_2Ph}$	160	18	Yes

base is also the solvent, this method is inoperable. Alternative procedures include comparison of kinetic effects for eliminations with those for proton abstractions (e.g., cf. equation 24, Section II.C.3.c), and estimation of proton abstraction rates for elimination reactions from data for exchange reactions on model compounds. Both of these approaches allow only a preference for, rather than certain proof of, an $(E1cB)_{irr}$ instead of an E2 mechanism.

a. *The elimination of* 1,1-*di*(p-*chlorophenyl*)-2,2,2-*trichloroethane* (*DDT*) *in protic solvents.* The dehydrohalogenation of DDT with various bases in protic solvents (equation 25) was for over two decades regarded as a E2 elimination involving a transition state with considerable carbanion character. The substantial kinetic data available included: (i) second order kinetics[55-59]; (ii) a large positive reaction constant[55] ($\rho = 2.4$, 30 °C, $\bar{O}H/92\%$ EtOH); (iii) low k_H/k_D (e.g., 3.8, 25 °C, $\bar{O}Et$; 3.1, 45 °C, SPh(EtOH))[56]; (iv) very large $k_{\bar{O}Et}/k_{\bar{S}Ph}$ elimination rate ratio (14,000)[56,59]; (v) Brönsted components, $\beta = 0.88$ ($\bar{S}Ar$, EtOH), 0.77 ($\bar{O}Ar$, EtOH)[56,59]; (vi) general rather than specific base catalysis[59] ($\bar{S}Ph/HSPh/EtOH$); and (vii) no isotopic exchange of β-hydrogen[56]. However, although these results clearly exclude the $(E1cB)_R$ mechanism, they accord equally well with an $(E1cB)_{irr}$ mechanism.

$$(p\text{-}ClC_6H_4)_2CHCCl_3 \xrightarrow[ROH]{B} (p\text{-}ClC_6H_4)_2C{=}CCl_2 + \overset{+}{B}H + \bar{C}l \qquad (25)$$

McLennan and Wong have recently measured the deuterium isotope effects for the DDT elimination with several bases in ethanol (Table 4)[26,60,61]. As the base strength is increased, the isotope effect (k_H/k_D) passes through a maximum, the position of which corresponds to the calculated pK_a value of DDT. This is the expected result for an 'uncoupled' proton removal (($E1cB)_{irr}$ mechanism). When the pK_a of the base and the substrate are equal, the maximum isotope effect is predicted because the proton should be half-transferred to the base in the transition state[33]. An isotope effect passing through a maximum has only been reported for one E2 reaction, i.e. the formation of styrene from $PhCD_2CH_2\overset{+}{S}Me_2$ with hydroxide ion in DMSO–water mixtures[62,63]. As the aprotic composition of the solvent is increased, the isotope effect attains a maximum. However, the medium basicity (H_-) at this point is 10–15 pK units away from the predicted pK_a value of the substrate[61]. Thus, it appears that the DDT reaction behaves more like a carbon–acid ionization than an elimination reaction. It is noteworthy that for both of the above reactions, the isotope maxima are clearly defined in contrast to some theoretical predictions[64] and experimental findings for carbon acid ionizations[65], which indicate much broader maxima.

The maximum isotope effect for the elimination of DDT with phenoxide ion suggests a transition state in which the β-hydrogen is about half transferred to the base. This is in conflict with the Brönsted value of 0·88 for the DDT eliminations with substituted phenoxides, which indicates a more product-like transition state. The authors felt that the Brönsted component was the less certain criterion of mechanism, although Bordwell and Boyle have argued that both kinetic parameters are insensitive measures of transition state structure[65].

TABLE 4. Deuterium isotope effects for the elimination of DDT in ethanol with various bases at 45 °C[61]

$$(p\text{-}ClC_6H_4)_2CDCCl_3 \xrightarrow{\text{B}} \overset{+}{B}D + (p\text{-}ClC_6H_4)_2C{=}CCl_2 + Cl^-$$

Base	PhS^-	$p\text{-}NO_2C_6H_4O^-$	PhO^-	MeO^- [a]	EtO^-	$t\text{-}BuO^-$ [b]
k_H/k_D	3·13	4·83	6·21	4·75	3·40	3·38
pK_a BH in EtOH	9·3	13·3	15·8	18·3[a]	20·3	—

[a] in methanol; $k_H/k_D = 5\cdot16$ at 30 °C.
[b] in t-butyl alcohol at 30 °C.

For the DDT eliminations, the isotope effect decreases when the medium is changed from MeOH/NaOMe to the more basic t-BuOK/t-BuOH[61]. This result is in striking contrast to the increase in isotope effect accompanying the same solvent change for a whole series of concerted eliminations (reactions showing significant isotope effects or elements involving both eliminating groups) giving alkenes, alkynes and carbonyl products[61,66].

McLennan and his coworkers prefer the E2 mechanism for the chloride ion induced eliminations of DDT in aprotic solvents as the reaction constants are considerably smaller than those observed for the alkoxide induced eliminations ($\rho = 1\cdot31$ in acetone; 0·99 in DMF)[67]. This result is surprising as aprotic solvents possess poor anion solvating properties for small ions, and relative to protic media, C_β—H bond cleavage should be more facile than to C_α—Cl cleavage. Presumably charge is more effectively dispersed in an E2 transition state than in a delocalized carbanion.

b. *Leaving group effects on elimination rates.* For an (E1cB)$_{irr}$ mechanism the role of the leaving group should be confined to its polar effect on the rate of abstraction of the β-hydrogen. However, in an E2 mechanism, the weakening of the bond to the leaving group decreases the energy of the transition state below that of a stepwise process. Thus, in this latter case, elimination should occur more rapidly than the predicted proton removal.

The problem, therefore, narrows to measuring the proton exchange in a relevant model compound, and making the appropriate allowance for the polar effect of the leaving group. The uncertainties in the method are illustrated by the stereoconvergent eliminations of the isomeric 2-p-tolylsulphonylcyclohexyl tosylates (equation 26). With hydroxide ion as

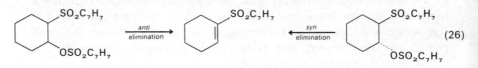

$$\text{(26)}$$

the base, *anti* elimination occurs about 400-times more rapidly than *syn* elimination[68]. In presence of *t*-amine bases, the eliminations exhibit general base catalysis, thus excluding the $(E1cB)_R$ mechanism. As with many stereoconvergent eliminations, which were reported two decades ago[69], Weinstock, Pearson and Bordwell preferred an E2 mechanism for the *anti* elimination, but presumed that the *syn* elimination was $(E1cB)_{irr}$[68], as it was not then realized that E2 reactions may occur with *syn*-stereospecificity (see Section II.E). When the rate of exchange of **9** was found to be 10^{-3}- to 10^{-5}-times slower than the eliminations, their view was modified[70]. They considered the polar effect of the tosylate was too small to account for the differential, and therefore proposed that both eliminations were E2 processes.

(9)   (10)

Hine and Ramsay found that the rate of exchange of **10** was 500-times greater than **9**, and using these two data points to calculate $\rho*$ for the series, they predicted that the exchange rate was slower than the *anti* elimination by a factor of about 100, but slightly faster than the *syn* elimination[71]. Thus, they resurrected the initial interpretation, despite the considerable uncertainty in a two-point extrapolation. Steric inhibition of β-proton abstraction is likely to be greater with the more bulky tosylate than the smaller methoxyl and hydrogen substituents. Thus, simple use of $\sigma*$ alone may over estimate the exchange rate.

Because of the uncertainty in calculation, the predicted rate should be at least two orders of magnitude less than the elimination rate for an E2 mechanism to be assumed. Conversely, proximity of exchange and elimination rates allows an E1cB mechanism to be inferred, but does not constitute proof. In a more recent publication, Bordwell, Weinstock, and Sullivan[72] have stated that an E1cB mechanism is still possible even if the predicted exchange rate is considerably smaller, as exchange may involve considerable internal return, especially with tertiary amine bases. This potential loophole greatly nullifies the use of the rate-of-exchange predictions, as the result can clearly be interpreted as desired.

Bordwell, Weinstock, and Sullivan's recent paper contains more definite proof that both eliminations involve the same mechanism[72]. Both reactions give very similar reaction constants for substituents in the leaving group ($\rho = 0.56$ (ŌH at 0 °C), 0.33 (NMe$_3$ at 25 °C) for *syn* elimination, and $\rho = 0.42$ (ŌH at 0 °C) 0.50 (NMe$_3$ at 25 °C) for *anti* elimination). These reaction constants are considerably smaller than those observed for documented E2 reactions of other arene sulphonates (e.g., 2-pentylarenesulphonates, $\rho = 1.35^{50}$; 2,2-diphenylarenesulphonates, $\rho = 1.1^{73}$; 2-substituted arylethylarenesulphonates, $\rho = 0.94$ to 1.24^{51}). However, whether these values are small enough for an (E1cB)$_{irr}$ rather than an E2 mechanism is debatable. It would be worthwhile investigating the C$_\alpha$–oxygen isotope effect, and the β-deuterium isotope effect with various bases on the line outlined for the DDT elimination.

Kaplan and Burlinson have used an α-fluorine effect to infer an (E1cB)$_{irr}$ mechanism for the elimination of nitrous acid (equation 27)[74].

$$ZC(NO_2)_2CH_2CH_2CN \xrightarrow{\text{OH}} [ZC(NO_2)=CHCH_2CN] \qquad (27)$$

$$Z = NO_2, F, \text{ and } Cl.$$

5. (E1cB)$_{ip}$ mechanisms

The kinetics of elimination of **11** (Scheme 7) accord with the (E1cB)$_R$ mechanism except that the observed rate is not dependent on (B̄H)$^{-1}$. However, in view of the low dielectric constant of the solvent and its poor anion solvating properties, it seems likely that C̄N expulsion from the carbanion is electrophilically assisted by (B̄H), i.e. the (E1cB)$_{R\text{-}ip}$ mechanism[24]. Electrophilic assistance to leaving group cleavage is well-known in nucleophilic aromatic[75] and vinylic[76] substitution reactions. With more basic amines (Et$_3$N and (n-Bu)$_3$N), compound **11** is extensively converted into its conjugate base and the (E1)$_{anion}$ mechanism, involving an ion-paired intermediate, operates. The elimination of **11** is truly 'base catalysed' rather than 'base induced', since only a trace of amine is required

176   Anthony F. Cockerill and Roger G. Harrison

to affect the self-decomposition. The liberated HCN is a weaker acid than
11 (pK_a HCN $\simeq$ 9), and the eliminated $\bar{\text{C}}$N therefore abstracts a proton
from **11**.

$$\text{ArC(CN)}_2\text{CH}^*\text{(CN)}_2 + \text{B} \underset{k_{-1}}{\overset{k_1}{\rightleftharpoons}} \text{ArC(CN)}_2\bar{\text{C}}\text{(CN)}_2 \ \overset{+}{\text{B}}\text{H} \xrightarrow{k_2} \text{Ar(CN)C}{=}\text{C(CN)}_2$$

(11) $+ \bar{\text{C}}\text{N} + \overset{+}{\text{B}}\text{H}$

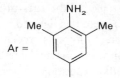

solvent: $CHCl_3$; B = morpholine, substituted anilines or pyridines.
$^*k_H/k_D = 1.00 \pm 0.06$ at 30 °C; $k_{exchange} > k_{elimination}$
$k_{obs} = V_{obs}/(\mathbf{11})(\text{B}) = k_1 k_2/k_{-1}$.

SCHEME 7.

The (E1cB)$_{R\text{-}ip}$ mechanism has also been invoked to explain the second-
order kinetics, primary deuterium isotope effect of unity, and the insen-
sitivity of rate to added $\text{Et}_3\overset{+}{\text{N}}\text{H}\bar{\text{Cl}}$ (only a two-fold variation) in the
elimination of **12** with Et_3N in aprotic solvents[77] (Scheme 8).

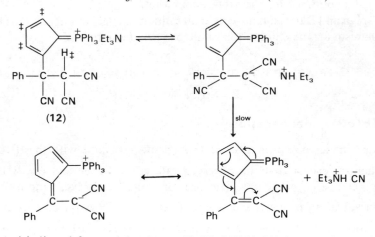

‡ $k_H/k_D = 1.0 \longrightarrow 1.1$, solvents; MeCN; 20% MeCN–benzene; benzene.

SCHEME 8.

The (E1cB)$_{ip}$ mechanism is anticipated to show a preference for *syn*
elimination[78], as is found for the dehydrofluorinations of the diastereoisomers **13** and **14** (equation 28)[79]. The geminal isotope effect, $k_{15}/k_{14} = 1.1$

is of secondary origin, arising from a hybridization change on passing into an extensively delocalized carbanion ion pair, whereas the inverse isotope effect, $k_{15}/k_{13} = 0.8$ reflects predominantly an equilibrium isotope effect.

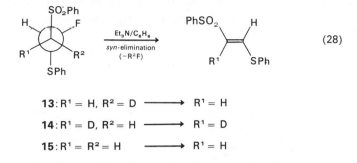

(28)

13: $R^1 = H$, $R^2 = D$ ⟶ $R^1 = H$

14: $R^1 = D$, $R^2 = H$ ⟶ $R^1 = D$

15: $R^1 = R^2 = H$ ⟶ $R^1 = H$

Element effects and primary deuterium isotope effects for the eliminations of 2-phenylsulphonylethyl halides with a variety of bases suggest that the bromide and chloride follow the E2 mechanism, whereas the fluoride in aprotic media, reacts via an $(E1cB)_{ip}$ mechanism[79b].

6. Miscellaneous E1cB reactions

E1cB mechanisms have been proposed (i) for a number of reverse Michael additions[80,81], (ii) for hydrolysis of esters possessing strongly electron-withdrawing aryl groups[82], (iii) to explain isocyanate intermediates in the hydrolysis of carbamates[83], and (iv) for the reaction of 3'-O-(methylsulphonyl)thymidine-5-carboxylate with bases in DMF[84].

7. Concerted versus stepwise eliminations giving alkenes

Bordwell[13] has questioned the existence of concerted reactions involving the formation and cleavage of as many as four bonds. Considering the recent proliferation of examples of the E1cB mechanism and the well-known two step eliminations for the majority of E_{CO}, E_{CS}, E_{CN}, $E_{C=N}$ and $E_{C=C}$ eliminations, he concludes that concerted bimolecular eliminations are the exception rather than the rule. In concerted reactions, the energy released in bond formation is regarded as being used simultaneously to aid bond cleavage. Arguing that elimination with concomitant aromatization should provide a strong driving force for a concerted E2 reaction, Bordwell, Happer and Cooper[85] compared the aromatizing elimination (equation 29) with the non-aromatizing elimination (equation 30). For **16b**, $k^*_{exchange} \gg k_{elim}$ and k_{16a} is only twice as large as k_{16b} exchange. As the elimination of **16b** is clearly $(E1cB)_R$, the authors invoked the $(E1cB)_{irr}$ mechanism for **16a** in view of the similarity in elimination (**16a**) and exchange rate (**16b**). Thus, they conclude that even the potential driving force of aromatization is insufficient to promote a concerted elimination and the reaction follows the principle of least motion. In general they feel that for eliminations involving activated β-hydrogens, concerted eliminations will be rare[14,85].

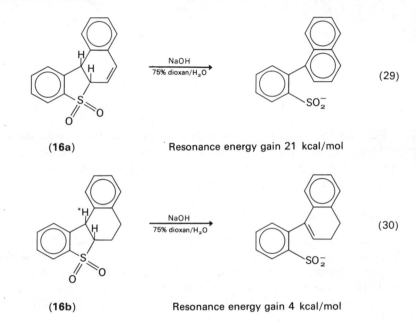

(16a)   Resonance energy gain 21 kcal/mol   (29)

(16b)   Resonance energy gain 4 kcal/mol   (30)

Generalizations based on isolated examples are unsound. Hückel calculations imply the resonance energy gain in passing from **16a** into its conjugate base is similar to the gain on 'aromatization'[86]. In addition, the geometrical alignment of the eliminating fragments is far from coplanar[86,87], and sulphone is a reluctant leaving group[86]. Thus, it is not surprising that an $(E1cB)_{irr}$ mechanism apparently prevails in this case.

D. The Variable Transition State for Concerted Eliminations in Protic Media

1. Terminology

Most recent reviews give detailed coverage of this important aspect[2,7–10,88], so our discussion here is brief. Bunnett's[3] original concept of transition states ranging from 'E1cB-like' to 'central E2' to 'E1-like' has been consolidated and extended to include terms such as 'reactant-like' and 'product-like'. Figure 1 depicts the interrelationship of these variants. This formulation is generally used to predict directional changes in the transition state with changes in the reactant structures, rather than to 'pin-point' the exact position. To this end, the descriptive terms 'more carbanion-like', 'increased carbonium ion character', etc., are used. Ion pair entities are not included in our concept of E2 reactions as these consti-tute variants of the E1cB and E1 mechanism. We consider the role of base–C_α interaction under eliminations in aprotic media (see Section II.G). There have been several theoretical attempts to predict the effect of structural changes in the reactants on the transition state[89,90]. In general these approaches predict that the transition state will move closer in structure to the stabilizing influence. In this context, carbanion stabilizing substituents at C_β should promote an increase in carbanion character, whereas conjugative substituents at C_α and C_β should encourage a transi-tion state with extensive double-bond character, thus falling into the central-E2 or product-like region.

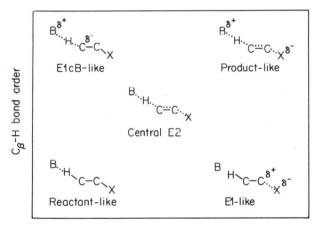

FIGURE 1. E2 transition states.

2. The effect of the leaving group

A change to a more reluctant leaving group should stabilize reactant and carbanion-like transition states, but destabilize product-like and carbonium-ion like structures. Thus an increase in C_β—H cleavage

Anthony F. Cockerill and Roger G. Harrison

relative to C_α—X cleavage is anticipated. Table 5 lists some data on 2-arylethyl derivatives consistent with this prediction.

All of the reactions are clearly of the E2 category. They exhibit significant deuterium isotope effects and either a substantial element effect or leaving-group isotope effect[63,98–100]. The kinetics are clearly second order in all cases, and *anti* elimination is assumed, although it has only been proved in a few cases using substrates specifically labelled with deuterium[101].

For the uncharged leaving groups, the ρ values increase with decreasing reactivity. The Hammett reaction constants reflect mainly charge accumulation at the benzylic carbon, and the increase in ρ is thus interpreted as indicative of increased carbanion character. The 'onium groups do not fit the reactivity sequence, but are generally assumed to be among the poorer leaving groups on account of the large ρ values. Some degree of caution is necessary in assuming the ρ values as listed reflect a continual increase in carbanion character, since without a change in transition state structure, a change of leaving group should modify the reaction constant. The poorer leaving groups are generally more electron withdrawing and hence the need for stabilization of the carbanion character at C_β by the aryl substituent is reduced. The increased electron withdrawing effect of X down the series may mask, in part, a substantial change in transition state, as ρ barely doubles in passing from the least to most carbanion-like structure. In any case, except for the sulphoxide, the ρ values are much

TABLE 5. E2 eliminations of 2-arylethyl derivatives at 30°C

$$\text{Ar—CH}_2\text{CH}_2\text{X} + \bar{\text{B}} \longrightarrow \text{BH} + \text{Ar—CH=CH}_2 + \bar{\text{X}}$$

	BH	EtOH		i-PrOH		t-BuOH	
X	k_{rel}	ρ	k_H/k_D [c]	ρ	k_H/k_D [k]	ρ	k_H/k_D [k]
I	26,600	2·07[e]	—			1·88[f]	—
Br	4100	2·14[e]	7·1[c]			2·08[f]	7·9[c]
OTs	392	2·27[f]	5·7[c]			2·49[i]	8·0[c]
Cl	68	2·61[g]	—			—	—
$\overset{+}{\text{S}}\text{Me}_2$	37,900	2·75[e]	5·1[c]	3·5[h]	6·7[h]	—	—
F	1	3·12[g]	—	—	—	—	—
$\overset{+}{\text{N}}\text{Me}_3$	760	3·77[d]	3·5[a,c]	3·9[h]	3·8[h]	3·04[d]	6·9[d]
SOMe	—	—	—			4·4[j]	2·7[b]

[a] Extrapolated from data at higher temperatures.
[b] Solvent contained 2·18 M-DMSO at 52 °C– References as follows [c] 91; [d] 92; [e] 93; [f] 94; [g] 95; [h] 96; [i] 97; [j] 51 (at 40 °C).
[k] Ratio of 2-arylethyl-X/2,2-dideutero-2-arylethyl-X.

smaller than those observed in the formation of carbanions from carbon acids ($\rho = 4$ to 7^{102}).

For all three solvent systems, the deuterium isotope effects decrease with increasing ρ value. The value of 7–8 for the bromide is about the maximum predicted on theoretical grounds[33] for half transference of the proton in the transition state, and the other decreasing values are considered to reflect gradually greater extents of proton transfer to the base. This interpretation clearly accords with the increasing carbanion character predicted from the ρ values and is supported additionally by solvent isotope studies[103], which indicate proton transfer is greater for the $\overset{+}{N}Me_3$ than $\overset{+}{S}Me_2$ case. These observations, coupled with the greater extent of $C_\alpha-S^+$ than $C_\alpha-N^+$ cleavage, as indicated by the leaving group isotope effects[63,98–100] all accord with the hypothesis that the carbanion character of the transition state should be increased by a change to a poorer leaving group.

Elimination reactions of substituted arenesulphonates (Table 6) enable separation of steric and electronic factors. The increase in ρ_y with decreasing electron withdrawing power of substituent Z implies greater carbanion character in the transition state for the poorer leaving group. As the acidity of the β-hydrogen is enhanced by greater electron-withdrawing substituents Y, the reaction constant ρ_z decreases, consistent again with an increase in carbanion character[51].

3. The influence of stereochemistry

More positive reaction constants, smaller primary deuterium isotope effects and leaving group isotope effects are generally found for *syn* than for

TABLE 6. Hammett reaction constants for elimination of substituted 2-arylethyl arene sulphonates with *t*-BuOK/*t*-BuOH at 40 °C[51]

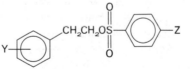

Y	*p*-OMe	*p*-Me	H	*m*-OMe	*p*-Cl	*m*-Cl
ρ_z	1·24	1·24	1·08	1·06	1·01	0·94
Z	*p*-Me	H	*p*-Br	*p*-NO₂		
ρ_y	2·49	2·50	2·36	2·03		

the corresponding *anti* eliminations. These results indicate that *syn* eliminations involve transition states which are more carbanion-like, whereas double-bond character is greater for *anti* eliminations (Table 7). There are several reasons which may explain these facts. In *syn* elimination, base-leaving group steric interaction is greater, and this may weaken the $B-H_\beta$ bond, necessitating a greater extent of cleavage. Alternatively, this greater proximity encourages a cyclic ion-pair mechanism with electrophilic assistance by the base cation aiding $C_\alpha-X$ cleavage, thereby diminishing the demand for assistance from the developing $C_\beta-H$ electron pair. In *syn* elimination, the electron pair developing at C_β is incorrectly situated to exert a rearside displacement in encouraging $C_\alpha-X$ cleavage and consequently this may require a greater concentration of charge at C_β to provide a similar displacing momentum to the *anti* elimination. Another explanation based on quantum mechanical calculations argues that overlap of developing orbitals is considerably less favourable for a *syn* orientation[106].

4. The influence of base and solvent

It is difficult to separate the influence of base and solvent as both are usually changed simultaneously in elimination reactions. Even if the base is maintained nominally constant, changes in the solvent can influence the association and solvation behaviour of the base, and thus change its intrinsic activity. Most often alkoxide–alcohol media are employed and we shall mainly confine our discussion to these.

Along the series $\bar{O}Et \rightarrow i\text{-}PrO^- \rightarrow t\text{-}BuO^-$, the base strength increases[107] whilst the solvent ionizing power decreases[108]. Hence, for eliminations involving uncharged groups, the solvent change should retard $C_\alpha-X$ cleavage, and promote greater charge dispersal in the transition state, and possibly more double-bond character. However, a stronger base should encourage a more reactant-like transition state. The combination of these factors is likely to produce a decrease in $C_\beta-H$ bond breaking, a slightly greater decrease in $C_\alpha-X$ bond cleavage and a marginal increase in double-bond character. Thus, k_H/k_D should increase and ρ should change little. These trends are consistent with the eliminations of 2-arylethyl bromides and tosylates (Table 5). It should be remembered however, that ρ usually varies inversely with the solvent dielectric constant[109], so a slight increase, without change of carbanion character at C_β, in passing from ethanol to *t*-butyl alcohol is expected. An indication that double-bond character is slightly greater in the less polar solvent is provided by the eliminations of 1,2-diphenyl-1-propyl halides[110].

TABLE 7 Some kinetic parameters for *anti* and *syn* eliminations

Substrate	Stereochemistry	Dihedral angle	*syn/anti*	$\rho(50\,^\circ\text{C})^a$
	exo–syn-periplanar	0°		3·12
			30·7	
	exo–anti-clinal	120°		2·56
	endo–syn-periplanar	0°		—
			0.58	
	endo–anti-clinal	120°		—
	syn	0–30°		2·76
			0·10	
	anti	150°		1.48

TABLE 7 (continued)

Substrate	Stereochemistry	Dihedral angle	syn/anti	$\rho(50°C)^a$
	syn	0–25°		2·90
			0·39	
	anti			2·18
			$k_{14}/k_{15}{}^c$	$k_H/k_D{}^b$
	anti-periplanar	180°	1·0108	5·40
	syn	0–60°	1·0064	2·60

[a] Base: t-BuOK/t-BuOH, at 50 °C, reference 104.
[b] Base: NaOEt/EtOH at 60 °C, reference 105.
[c] Nitrogen isotope effect.

Following *anti* stereospecificity, the *erythro* isomer eliminates to the less stable *cis* stilbene, whereas the *threo* isomer gives α-methyl-*trans*-stilbene. As the reactants have similar stability, the rate ratio reflects the extent of double-bond character, a value of 50 being predictable under thermodynamic conditions. Under kinetic control, $k_{threo}/k_{erythro}$ is almost unity with ethoxide in ethanol, but attains 5·4 for the bromide and 15 for the chloride in t-butyl alcohol containing t-butoxide. The rate ratio, k_{OTs}/k_{Br} is less in $\bar{O}Et/EtOH$ than in t-BuO^-/t-BuOH (cf. 0·10 vs 0·22) for the 2-arylethyl eliminations. This result was assumed to indicate less C—X cleavage in the more ionizing solvent[111]. However, although large values of k_{OTs}/k_{Br} certainly accord with extensive C—X cleavage, interpretations of small ratios are uncertain as k_{OTs}/k_{Br} is most probably not a linear function of C—X cleavage for small extents of bond rupture[51,112]. Clearly, there is a need for a more definitive measure of C—X cleavage for uncharged leaving groups.

For charged leaving groups the base effect is predicted as above, but a less polar solvent should encourage C_α—X^+ cleavage in the transition state. Table 5 shows that proton transfer does decrease with a change to a stronger base as k_H/k_D increases. However, whereas ρ changes little for the $\overset{+}{N}Me_3$ elimination, (EtOH → i-PrOH), and clearly decreases in the most basic medium (t-BuOH), an abrupt rise in ρ is noted for the $\overset{+}{S}Me_2$ elimination (EtOH → i-PrOH). These results imply an increase in carbanion character in one case and a decrease in the other as the base strength is increased and the solvent polarity is reduced. There are obvious limitations in assuming all substrates with charged leaving groups will respond similarly to changes in reaction conditions.

Mixed protic–aprotic solvents have gained favour as media for base induced reactions. Addition of DMSO to t-BuOK/t-BuOH causes no change in the ρ or k_H/k_D values for the elimination of 2-arylethyl bromides[113]. It should be noted that this is one of the few reactions in which the rate rises faster than the medium basicity as measured by the ionization of nitroanilines. Conversely, the ρ values rise sharply for a similar solvent change in the eliminations of the substituted 2-arylethyl arenesulphonates[114]. This result again illustrates the uncertainty in general predictions. Presumably hydrogen bonding interactions are more important for the sulphonate leaving group, and consequently as the aprotic content of the solvent is increased, C_α—O cleavage is retarded more than C—Br cleavage.

The rate of elimination of 2-arylethyldimethylsulphonium ion with hydroxide ion increases by 10^6 as the solvent is changed from water to 85 mol-% dimethyl sulphoxide[62]. This dramatic rate increase reflects an increase in medium basicity and a decrease in water activity. The activation energy decreases as might be expected for eliminations involving gradually less solvated and more reactive base. This change obviously contains a significant ground state contribution. At low DMSO concentrations (0–20%), the sulphur isotope effect decreases markedly[62,63]. The ρ value increases during this region and thereafter remains essentially unchanged. The k_H/k_D value changes little initially, then rises, passes through a maximum, and then declines with increasing DMSO concentration[62]. The ylid mechanism was not excluded and the results were interpreted as an increase in carbanion character and a gradual shift to a more reactant-like transition state with increasing aprotic content of the medium. This reaction in fact was the first example of a maximum primary isotope effect to be observed experimentally, in agreement with the theoretical interpretation, by just changing the solvent. Subsequently, similar behaviour has been reported for the elimination of the corresponding ammonium salt[115] and a number of simple proton transfers of carbon acids[116].

5. The effect of substituents at C_α and C_β

It would seem reasonable to expect electron-withdrawing substituents at C_β to facilitate C_β—H cleavage and retard C_α—X cleavage, thereby encouraging an increase in carbanion character of the transition state[3]. However, interpreting this expectation in terms of extent of bond cleavage is an uncertain occupation. In the 2-arylethyltrimethylammonium ion eliminations, electron-withdrawing substituents accelerate the rate ($\rho = 3.77$, Table 5) and cause an increase in k_H/k_D but a decrease in the nitrogen isotope effect (Table 8). Hence, in line with the Hammond postulate[118], a more reactant-like transition state is observed for the more reactive substrate. Whether this represents an increase in carbanion character is debatable, as it is difficult to judge which leaving-group cleavage is diminished more. Theoretical calculations indicate the C—X isotope effect should increase almost linearly with the extent of cleavage[119]. However, various shaped profiles, exhibiting both well-defined and very flat maxima have been obtained from theoretical considerations for k_H/k_D[64,120]. Experimentally, similarly contrasting profiles have also been reported[62,65,116,121,122].

In the 1-arylethyltrimethylammonium ion, the influence of the aryl substituent appears to be mainly concerned with its effect on C_β—H cleavage as electron-withdrawing substituents promote the rate ($\rho = 0.95$) and reduce slightly the extent of C_β—H cleavage, if it is assumed that the transition states have extensive carbanion character, with the proton always more than half transferred. The isotope effect as measured of course contains a secondary contribution and this, rather than the primary effect, could be responsible for the small change observed. The secondary α-D isotope effects do not show a regular variation with the electronic nature of the aryl substituent. Secondary isotope effects may be of steric[123], inductive[124], hyperconjugative[125] or hybridization change origin[126]. Usually in eliminations, the latter explanation is preferred for α-D effects, although for small extents of C—X cleavage, and hence little hybridization change at C_α, the other factors may be more important. The authors[117] explained their results in terms of a shift to a more reactant-like transition state with greater carbanion character as the electron-withdrawing power of the aryl substituent is increased.

A decrease in the α-deuterium isotope effect with increased electron withdrawal is observed in the eliminations of 2-arylethyl tosylates

TABLE 8. The influence of aryl substituents on E2 reactions of 1- and 2-arylethyl derivatives

Substrate	$ArCD_2CH_2NMe_3^+$[99] EtOH 40°C		$ArCH(CD_3)NMe_3^+$[117] EtOH 70°C	$ArCD(Me)NMe_3^+$[117] EtOH 70°C	$ArCH_2CD_2OTs$[112] t-BuOH 30°C
Substituent in Ar	k_H/k_D	k_{14}/k_{15}[a]	k_H/k_D	k_H/k_D	k_H/k_D
p-OMe	2.64	1.0137	4.35	1.04	1.047
H	3.23	1.0133	4.45	1.02	1.043
p-Cl	3.48	1.0114			1.017
p-CF$_3$	4.16	1.0088	4.86	1.07	

[a] Leaving group isotope effect.

(Table 8)[112]. This result suggests successively less rehybridization at C_α in the transition state and is in agreement with the predictions based on the Hammett ρ_z values of Table 6. In line with the ammonium ion eliminations a reduction in the extent of C_β—H cleavage with increased electron withdrawal is predicted. This prediction, however, is at variance with the Brönsted components, β, for the elimination of 2-arylethyl bromides with substituted phenoxides. Values of $0.67(p\text{-}NO_2C_6H_4CH_2CH_2Br)$ and $0.54(C_6H_5CH_2CH_2Br)$ for β suggest that proton transfer is more extensive with the more strongly electron withdrawing nitro group[127]. Recent results of Bordwell's suggest that β is a rather unreliable measure of transition state structure for slow proton transfers[65].

Despite the combination of more than one kinetic method, the results in Table 8 do not allow a definitive conclusion as to the variation in carbanion character at C_β with the electronic effect of α- and β-aryl substituents. However, it can be concluded that the transition state becomes more reactant-like with increased electron withdrawal.

6. Concluding remarks

To restrict the size of our discussion in this section, we have limited most of our examples to the 2-arylethyl series. This series is in fact the most well documented, but it may not be a truly representative model for eliminations of simple aliphatic molecules because all of the transition states lie in the carbanion region. Despite the large volume of results, many gaps are apparent, and often only tentative proposals are possible. The following conclusions can be drawn, but these should be used only as guidelines:

(i) A change to a poorer leaving group increases the carbanion character of the transition state.

(ii) Weaker bases, in more ionizing solvents promote a decrease in carbanion character.

(iii) *Anti* eliminations are more synchronous than *syn* eliminations, which exhibit greater carbanion character.

(iv) Electron-withdrawing groups, particularly at C_β, and to a lesser extent at C_α, promote a more reactant-like transition state, which may possess a greater degree of carbanion character.

At the present time, more experimental results are required, but more importantly, a deeper understanding of the theoretical variations of these kinetic effects as measures of transition state structure is needed. Some of these kinetic parameters are complex for simple single-bond breaking processes. There is no guarantee that kinetic parameters should behave in a similar manner in more complex reactions, such as eliminations in which several bond changes occur simultaneously. Allowance may have to be made for the effect of one bond change on the kinetic parameters being assessed for an adjacent bond. Calculations of this kind have only reached the initial stages of development[119], and are to be encouraged in the future.

E. The Stereochemistry of Concerted Eliminations in Protic Media

This aspect has been reviewed in detail[8,9,11]. *Anti* elimination is most common, but numerous examples of *syn* elimination, in some cases as the preferred mode, have been uncovered. The terminology is described earlier in this chapter (Section II.B).

1. Theoretical explanations and hypotheses

No single explanation accommodates all of the facts, and it seems certain that several features control the stereochemistry. For effective overlap the orbitals developing at C_α and C_β should be aligned in parallel. For this reason periplanar eliminations should be preferred to clinal arrangements. This reasoning of course does not explain the preference of *anti* over *syn* elimination.

An electrostatic factor was originally proposed by Hückel[128] to account for the preference for *anti* elimination in dehydrohalogenation with alkoxide ions. He regarded base-leaving group repulsion as unfavourable in *syn* elimination (**17**). However, model calculations predicted this factor was too small to account for *anti/syn* rate ratios of 7000–24,000 for the benzene hexachlorides[129]. In addition, *anti* stereospecificity is observed when X is an 'onium salt and if both B and X are uncharged. Clearly, in such cases electrostatic attraction should promote *syn* elimination.

(**17**)

Syn elimination requires eclipsing of groups at C_α and C_β. If the minimum eclipsing effect is modelled on the 3 kcal mol^{-1} barrier to rotation in ethane, a rate differential of 160 for $k_{anti}/_{syn}$ is predicted[7,12]. Of course during an elimination reaction, bonds other than those being made and broken undergo rehybridization. Using ethyl chloride as a model, the principle of least motion has been used to show that the energy for rehybridization of non-eliminating bonds increases along the series; *anti*-periplanar < *syn*-periplanar < *anti*-clinal < *syn*-clinal[130,131].

Both quantum mechanical calculations[106] and frontier orbital theory[132] indicate a preference for *anti* over *syn* elimination. A correlation between spin–spin coupling in n.m.r. spectra and the relative rates and stereochemistries of eliminations has been demonstrated[133]. This arises because the ease of delocalization of spin density and of charge run parallel. CNDO/2 calculations were used to extend this work[134]. They showed for model systems that the site having strong positive spin coupling to a given proton is the site to which negative charge is preferentially transferred. Hence, except when B and X are both uncharged, charge transfer is more effective to the *anti* than to the *syn* position. However, *syn* transition states are expected to possess extensive carbanion character.

In one of the most favoured arguments, bimolecular elimination is compared with bimolecular nucleophilic substitution. The electron pair from the C_β—H bond is considered as equivalent to the nucleophile and *anti* orientation is envisaged as correctly aligned for a displacement with inversion at C_α (18). However, as originally highlighted by Ingold[135], if the transition state possesses extensive carbanion character, the p-orbital at C_β may form a well-developed lobe of electron density on the side away from the C_β—H bond. In this case, displacement with inversion at C_α now requires an overall *syn* elimination (19)[136]. In both cases, displacement with inversion minimizes electron repulsion effects.

(18)   (19)

From these various theories and hypotheses a clear preference for *anti* elimination emerges. However, *syn* elimination is expected to become more competitive for transition states with extensive carbanion character. Additionally, steric or conformational restrictions which preclude periplanar orientations may markedly alter the *syn/anti* rate ratio. As we shall see, these predictions are borne out experimentally.

2. Experimental methodology

All the stereochemical investigations concern comparative analyses of the products or rates of elimination of diastereoisomeric substrates. In many cases, the relative positions of X and the β-hydrogen are fixed by the presence of a β-alkyl substituent. Thus, if elimination occurs from a chair conformation, the cyclohexyl series can be used to compare *anti*-periplanar and *syn*-clinal stereochemistry. *Anti* elimination from **20** can give both **22** and **23**, whereas *syn* elimination yields the less substituted alkene **23**. On the other hand, *anti* elimination from diastereoisomer **21** can give only **23**. Thus, if it is shown that **20** and **21** do not epimerize prior to elimination (e.g., excluding an E1cB mechanism) and that the alkene products do not interconvert under the reaction conditions, product analysis enables deduction of the stereochemistry of elimination. The limitation in this type of study is that the 2-R substituent may influence elimination into the substituted branch. Hence, the result reflects a combination of stereochemical and orientational factors.

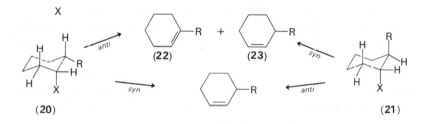

The above problem is minimized if the R substituent is deuterium. As shown for the acyclic series, *anti* elimination from diastereoisomer **24** gives the *trans*-alkene **26** containing one atom equivalent of deuterium, whereas the *syn* elimination gives the unlabelled alkene. For the isomeric **25**, the stereochemical and labelled assignments are reversed. Both substrates may also yield the *cis*-ene **27**, the proportion of which may be influenced by the kinetic isotope effect. Thus, the stereochemistry of elimination can be deduced from a knowledge of the proportions of the geometric alkenes and their specific deuterium content. This kind of analysis is a considerable improvement in the technique for determining stereochemistry of elimination. It has only been made possible in recent years by developments in gas chromatography, which enables facile separation of the alkenes, and mass spectroscopy, which is used to determine the deuterium content precisely.

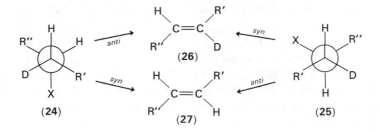

3. Experimental results

a. *Alicyclic rings.* A major advance in our understanding of the stereo-chemistry of elimination reactions stemmed from Sicher and Závada's work on medium sized ring systems[137,138]. From their data in Table 9, virtually no isotope effect is observed for the formation of *cis*- and *trans*-**29** in runs 2 and 5, but substantial isotope effects are apparent for runs 3 and 6. These results demonstrate that the same hydrogen is eliminated to give both alkenes from a given substrate, i.e. *anti* elimination → *cis*-ene, *syn* elimination → *trans*-ene. The *syn* elimination of the 'onium salt does not follow the ylid mechanism as unlabelled trimethylamine was isolated[139]. This divergent stereochemical behaviour is now termed the *syn–anti* dichotomy[11].

In the original studies, the Czech workers showed the dichotomy was general for the elimination reactions of all medium sized ring trimethyl-ammonium salts under a variety of basic conditions[139,141]. They developed an alternative kinetic criterion dependent on rate profiles. Similar profiles (rate versus ring size) are predicted for different reactions which possess similar salient steric features. The Cope elimination was adopted as the reference for the *syn* process (Figure 2A) and the S_N2 reaction of cycloalkyl bromides, which involves displacement with inversion at C_α (Figure 2B) was adopted as the model for *anti* elimination. For the trimethyl-ammonium ions, the rates of formation of the *trans*-enes clearly parallel the behaviour of the Cope elimination (cf. Figure 2A, C) whereas the rate changes for *cis*-ene formation are very similar to the S_N2 variations (cf. Figure 2A, B).

The *syn–anti* dichotomy is not so firmly followed in the tosylate elimina-tions[140]. In the cyclododecyl case, isotopic labelling indicates the *cis*-ene is always formed almost exclusively by *anti* elimination, but both *syn* and *anti* elimination can give rise to the *trans*-ene. In fact, with potassium *t*-butoxide as the base, the *syn* contribution decreases with increasing dielectric constant of the solvent; viz: 95%, benzene; 85%, *t*-butyl

TABLE 9. Olefins formed in the bimolecular eliminations of some 1,1,4,4-tetramethyl-7-X-cyclodecanes[138–140]

No.[a]	Substrate	(28)				(29)	
		% trans	% cis	% trans	k_H/k_D[b]	% cis	k_H/k_D[b]
1.	NMe₃	35·9	9·20	49·3	—	5·60	—
2.	NMe₃ (D)	38·4	9·70	45·8	1·1	6·10	1·0
3.	NMe₃ (D)	55·9	13·2	28·7	2·7	2·20	3·6
4	OTs	10·8	7·8	74·7	—	6·7	—
5.	OTs (D)	10·3	8·4	75·4	0·95	5·9	1·1
6.	OTs (D)	17·0	14·0	63·0	1·9	6·0	1·9

[a] Base MeOK–MeOK at 40 °C (runs 1–3); t-BuOK–t-BuOH at 100 °C (runs 4–6).
[b] Calculated from alkene composition as follows:

$$(k_H/k_D)\text{-}trans \text{ (run 2)} = \frac{\{\% \, trans\text{-}29 \text{ (run 1)}\}\{\% \, trans\text{-}28 \text{ (run 2)}\}}{\{\% \, trans\text{-}28 \text{ (run 1)}\}\{\% \, trans\text{-}29 \text{ (run 2)}\}}$$

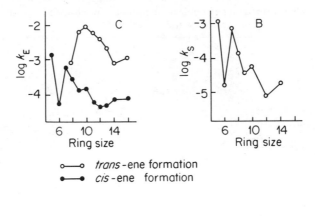

o——o *trans*-ene formation
●——● *cis*-ene formation

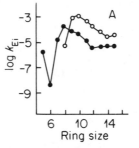

FIGURE 2. Rate profiles for elimination and substitution reactions of cycloalkyl derivatives. A: Ei rates of cycloalkyldimethylamine oxides in *t*-BuOH at 70·6 °C[137]; B: S_N2 rates of cycloalkyl bromides with KI in acetone at 60 °C[142]; C: E2 rates of cycloalkyltrimethylammonium ions with *t*-BuOH at 55 °C[141].

alcohol; 30%, DMF. This observation suggests a role for ion association in *syn* elimination.

Using the rate profile approach, the bimolecular eliminations of cycloalkyl bromides were shown to follow the *syn–anti* dichotomy with *t*-BuOK/*t*-BuOH[143]. However, with the more polar but less basic medium of KOEt–EtOH, the rate profile changes imply that both alkenes are formed by mainly *anti* elimination. With *t*-BuOLi in DMF, *anti* elimination is again dominant. Finally for cyclodecyl bromide[143], elimination with *t*-BuOK/*t*-BuOH gives a predominance of the *trans*-ene (5:1), whereas the *cis*-ene is the dominant product (4:1) when the more basic and less associated base, tetramethylammonium *t*-butoxide is used[144].

Thus, the following generalizations can be made concerning the stereo-chemistry of bimolecular elimination of medium sized rings. (i) The *syn–anti* dichotomy is always followed in the trimethylammonium ion eliminations, but the *syn → trans* route becomes less competitive with better leaving groups. (ii) For the better leaving groups, the *syn → trans* route is promoted by associated bases in weakly ionizing solvents. Thus, it is possible to select conditions to optimize the yield of either geometric cycloalkene[11,145]. (iii) The *syn → trans* route is most dominant for the $C_{10} \to C_{12}$ rings. The factors favouring *syn* elimination are such that even for the $C_8 \to C_{10}$ series, in which the *trans*-ene is the thermodynamically less stable alkene[146], it is clearly the dominant product in the Hofmann eliminations. We shall consider a theoretical explanation of these facts in a subsequent section (Section II.E.4).

We have not so far commented on the *syn → cis*-ene route. In general, for bimolecular eliminations of medium sized rings, this pathway is much less important than the corresponding *anti*-elimination. For example, in the Hofmann elimination, *cis*-cyclo-octene arises 85% by *anti* and only 15% by *syn* elimination, whereas the *trans*-ene is exclusively a *syn* elimination product[147]. However, the unimolecular solvolyses of cyclodecyl tosylates, in a variety of solvents, give both alkenes predominantly by a *syn* elimination, in which the departing tosylate probably aids removal of the β-hydrogen (cf. **30**)[148].

(**30**)

The initial assumption that E2 reactions should exhibit *anti* stereo-specificity arose in part from the accumulation of results from the cyclo-hexyl series in which this statement is generally correct. Deviations are usually limited to cases in which the β-hydrogen is activated by a strongly electron-withdrawing group. *Anti* elimination of course requires that the leaving groups occupy axial conformations, and for bulky groups such as trimethylammonium ion, steric factors may be prohibitive in some cases. Thus, when bulky or ion associated bases are employed, *syn* elimination and the ylid mechanism may become competitive with the *anti* pathway[149]. A recent example concerns the elimination of menthyltrimethylammonium ion which gives 25% of the more stable alkene **33** on treatment with butoxide ion. Presumably axial proton abstraction from the more stable

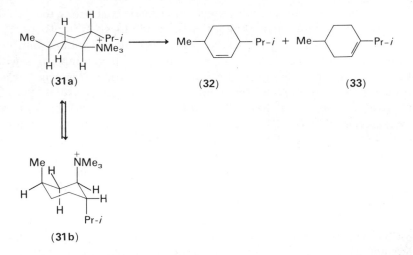

conformer **31a** is easier than from **31b** in which all the bulky groups are axially orientated.

   b. *Acyclic systems.* The discovery of the *syn–anti* dichotomy for elimination reactions of medium sized alicyclic rings, prompted investigations into a wider range of acyclic reactant structures and reaction conditions. The result has been the finding of several examples of the *syn–anti* dichotomy, and although apparently not as preponderant as in the alicyclic series, even the *syn* → *cis*-ene route is operative in some cases. The stereochemical experiments in all cases have relied on deuterium labelling techniques.

   2-Butyl bromide[150] and tosylate[151] undergo exclusive *anti* elimination in forming 2-butene with a wide variety of base-solvent systems. Similar stereospecificity is found for the 2-butyltrimethylammonium ion eliminations[152,153]. 2-Phenylethyltrimethylammonium ion eliminates with *anti* stereospecificity in protic solvents[101], but the corresponding chloride and tosylate undergo *syn* elimination, via an ion pair mechanism (cf. **34**), on reaction with *t*-BuOK in benzene[154]. The separate roles of free and associated bases in elimination reactions have been realized. It is well known that activated *syn* elimination occurs more readily than unactivated *anti* elimination in cyclopentyl tosylates (equation 31)[155]. However, in the presence of dicyclohexyl-18-crown-6 ether (**35**), the *anti* elimination becomes dominant[156]. Clearly, the activated *syn* elimination involves an associated butoxide, and possibly a cyclic ion pair mechanism (cf. **34**). Similar variations in product proportions can be induced by changing the counterion from an alkali metal to the more sterically

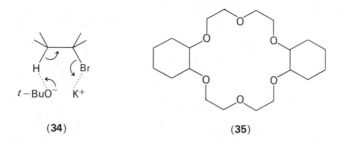

(34)   **(35)**

hindered tetraalkylammonium cation, which shows a greatly reduced tendency to associate with anions in poorly ionizing media[157].

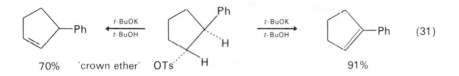

70%   'crown ether'   OTs   91%

Figure 3 depicts the variation in *syn* elimination contribution to give the *trans*-ene. Most of the results concern the $\overset{+}{N}Me_3$ leaving group, and with a given base, the *syn* contribution increases with the molecular size and complexity of the alkyl framework. For all substrates, *syn* elimination is enhanced relative to *anti* elimination by an increase in base strength and decrease in medium polarity. For the 5-decyl series, the *syn* contribution decreases along the order: $\overset{+}{N}Me_3 > I^- > Cl > OTs$. It is interesting that this is the order of decreasing carbanion character as indicated by the Hammett reaction constants for the 2-phenylethyl eliminations (see Table 5, II.D.2). The relationship, however, may be purely fortuitous.

In the majority of the cases, the corresponding *cis*-enes are formed predominantly, and occasionally exclusively, by *anti* elimination.

4. The origin of the *syn–anti* dichotomy

Saunders and his coworkers have attributed the *syn–anti* dichotomy for Hofmann eliminations to steric inhibition of proton abstraction being greater for the *anti* → *trans* than *anti* → *cis* route[8,152,160]. The bulky trimethylammonium ion forces the alkyl substituents into the conformations shown (**36–38**) with the result that the β-hydrogen becomes

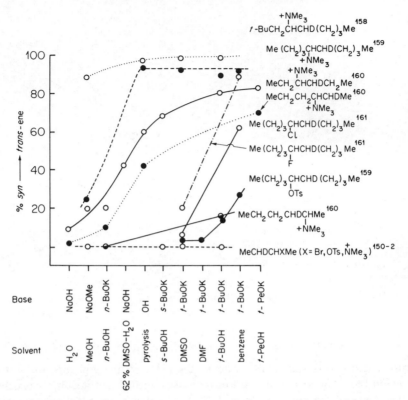

FIGURE 3. Variation in the *syn → trans*-ene elimination pathway with structural features of the substrate and base in acyclic systems. (The % refer to the fraction of *trans*-ene formed by *syn* elimination in the unlabelled substrate. The D label is merely to indicate the direction of elimination in each case. The lines joining the points for a given substrate are not intended to portray specific slope changes. Reference numbers are situated by each structure.)

'enveloped' particularly by substituents on the β^1 carbon and to a lesser extent by those on the γ-carbon (38). *Syn* elimination is promoted by

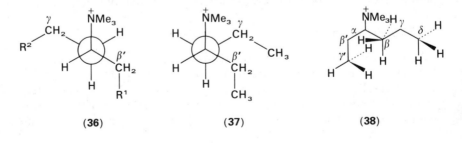

(36)   (37)   (38)

stronger bases in poorer ionizing solvents, conditions which should encourage a more reactant-like transition state[92]. Thus the argument is
based on transition states with conformations similar to the ground state.
In the alkoxide series, stronger bases also tend to be larger, and steric
inhibition to proton abstraction for the *anti* → *trans* pathway becomes
greater. On the other hand, approach to the *syn* β-hydrogen is less hindered,
and conformation **37** shows that for the *anti* → *cis* path, one side of the
molecule is unimpeded to base approach. Finally, the low preference for
the *syn* → *cis* route is attributed to unfavourable ordinary eclipsing effects
(cf. **39**), which develop as the small amount of double-bond character
constrains the transition state towards a periplanar conformation. For
medium sized ring systems, the carbon framework of the ring provides a
more formidable steric barrier, **40**, and the propensity of the *syn–anti*
dichotomy is greater, being extended even to smaller leaving groups such
as tosylate. Steric effects alone cannot account for the general phenomenon
of the *syn–anti* dichotomy. Small leaving groups such as fluoride and
chloride in the 5-decyl series exhibit substantial *syn* → *trans*-ene pathway

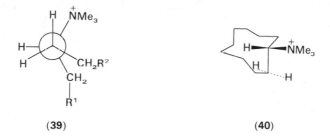

(**39**)   (**40**)

in solvents encouraging ion association[161]. Arguments in terms of transition states with extensive carbanion character seem unlikely, as carbanion
character is expected to be considerably less in the alkyl and cycloalkyl
series than in the 2-arylethyl series, and *anti* elimination predominates in
the latter[101]. Závada and his coworkers have proposed a critical role for
attractive interactions between substrate and the base to account for the
dichotomy[162]. As the *syn* elimination contribution is enhanced by increased base concentration (e.g., *t*-BuOK/*t*-BuOH)[163], and by a reduction
in medium polarity (*t*-BuOK in *t*-BuOH → benzene)[164], but decreased by
the addition of crown ethers and more polar solvents (*t*-BuOK in
t-BuOH → DMF)[164], it appears that associated bases (e.g., *t*-BuOK)
favour *syn* elimination, but free ions (*t*-BuO$^-$) promote *anti* elimination
preferentially for uncharged leaving groups. For $\overset{+}{N}Me_3$ eliminations free
ions appear to be responsible for the dichotomy[161], as association between

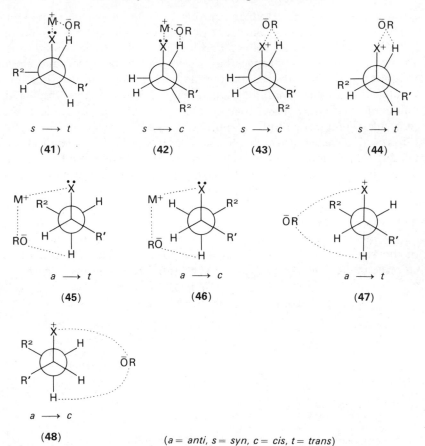

$(a = anti, s = syn, c = cis, t = trans)$

the 'onium ion and alkoxide produces a more reactive form of the dissociated base[144]. Structures **41** → **48** depict the eight possible routes available for eliminations of uncharged and charged leaving groups[162]. For the *syn* series, eclipsing interactions are least in **41** and **44**, conformations leading to *trans*-ene. For the *anti* series, the sideways approach is cumbersome for the *trans*-ene production, but less impeded for *cis*-ene formation, although the dual interaction for the 'onium salt (**48**) seems 'geometrically stretched' to the present authors. Ion pairing alone cannot account for the *syn–anti* dichotomy, as small substrates such as 2-butyl exhibit *anti*-stereospecificity. Both the ion association and steric explanation have certain attractive features. It is probable that both operate in all systems to some extent, molecular complexity and reaction media causing changes in the relative importance of the two hypotheses.

F. Orientation of Elimination in Protic Media

There have been numerous reports of experimental factors promoting changes in both positional and geometric orientation during the last decade. Most of the details are summarized in reviews[3,7,8-10]. In the search for the causative factors, investigators have generally measured product ratios for a series of reaction conditions gradually increasing in molecular complexity. Such changes in both substrate and base are the very factors which are now known to encourage the *syn–anti* dichotomy. Unfortunately, few combined stereochemical and orientational studies have been reported and arguments have generally been built upon *anti*-stereospecificity. As a result much of the data will remain of limited theoretical significance until the stereospecificity is known. We shall, therefore, restrict our coverage to simple substrates, in which *anti* stereospecificity is very probable, or reactions in which the stereochemistry was simultaneously reported.

1. The effect of the leaving group

The synthetic chemist is generally confronted with the preparation of a specific structure. Thus, in optimizing the yield of a specific alkene, the problem narrows to variations in the leaving group and reaction conditions. The magnitude of their effect however, is primarily controlled by the substrate framework.

Brown and Wheeler[165] considered that the increased steric size of the leaving group was responsible for increased Hofmann orientation in the elimination of some 2-pentyl derivatives (Table 10). However, a similar trend is predictable from the ease of heterolysis of the C_α—X bond. The choice of leaving groups is somewhat unfortunate in that only some are top-symmetric, and conformational preferences for the others add a complicating factor. Using a more precise gas chromatographic procedure, Saunders and coworkers measured the products of elimination of 2-pentyl halides[166] and Bunnett and Bartsch those for 2-hexyl halides[167]. Their results show that for halides, the propensity of Hofmann orientation does not accord with the steric size of the leaving group. The percentage Hofmann product increases and the *trans/cis*-2-ene ratio decreases with increasing difficulty of cleavage of the C_α—X bond. This trend accords with the variable transition state theory, poorer leaving groups encouraging greater carbanion and less double-bond character. In this context, it is interesting that the rates of elimination and Saytzev/Hofmann ratios correlate with the Hammett reaction constants for the 2-arylethyl eliminations (Figure 4). However, the rates and product ratios do not

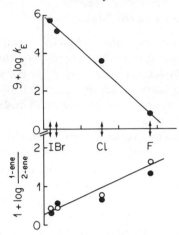

(2-arylethyl-X eliminations)

FIGURE 4. Correlation of elimination rates of 2-pentyl and 2-hexyl halides with Hammett reaction constants for corresponding eliminations of 2-arylethyl derivatives in ethanol–ethoxide. [For ρ values see Table 5; for product ratios and eliminations rates, see Table 10; solvent: NaOH/MeOH for 2-hexyl eliminations (●); NaOEt/EtOH for the 2-pentyl eliminations (○)].

correlate with the σ^* values of the halogens[167], suggesting the inherent acidity of the β-hydrogen is not a critical factor determining the orientation of elimination, as has previously been proposed by Ingold and Hughes for Hofmann eliminations[168]. Although fluoride is a good choice for Hofmann product formation, the substrate reactivity is considerably depressed and alkene isomerization to the Saytzev product may become competitive with elimination. The generality of these results may be limited to simple alkyl halides as the *trans/cis*-ene ratios show an irregular trend in the eliminations of 1-phenyl-2-X-propanes (Table 10A)[169]. In this case, the authors suggested an increase in carbanion character is also accompanied by increased double-bond character, encouraged by the conjugative phenyl. Hence, eclipsing interactions become most significant for the fluoride which gives the highest *trans/cis* ratio. A large *trans/cis* ratio could arise with the incursion of a *syn → trans* elimination for the poorest leaving group (i.e., ion pair elimination) and it would be worth checking this possibility.

2. The effect of base and solvent

As solvent and base are generally changed simultaneously, it is seldom possible to separate the contribution of each. For the 2-substituted alkanes, with uncharged leaving groups, the combination of increased

TABLE 10. Eliminations of 2-pentyl and 2-hexyl derivatives

	I		II		III
	Saytzev Products				Hofmann Product

(a) $\overset{+}{M}\overset{-}{OR}$ = KOEt[165]

X	Br	I	OTs	$\overset{+}{S}Me_2$	SO_2Me	$\overset{+}{N}Me_3$
%III	31	30	47	87	89	98

(b) $\overset{+}{M}\overset{-}{OR}$ = NaOEt[166]

X	F	Cl	Br	I
%III	82	35	25	20
I/II	2·6	3·5	3·8	4·1

(c) $\overset{+}{M}\overset{-}{OR}$ = NaOMe[167] (2-hexyl series)

X	F	Cl	Br	I
%III	70	33	28	19
I/II	2·3	2·9	3·0	3·6
9 + log k	0·74	3·56	5·16	5·75

(d) t-BuOK (2-hexyl series)[170]

X	F	Cl	Br	I
%III	97	88	80	69
I/II	1·2	1·1	1·4	1·8

TABLE 10A. trans/cis-β-Methyl styrene ratios in elimination of 1-phenyl-2-X-propanes with ethoxide in ethanol at 60°C[169]

X	F	Cl	Br	I
trans/cis	112·4 ± 45	25·0 ± 0·1	24·7 ± 1·3	28·3 ± 1·7

base size, less dissociating solvent and ion-paired rather than dissociated base, promotes a shift towards Hofmann behaviour (Figure 5). In fact, even 2-butyl iodide only just follows the Saytzev rule on treatment with basic triethylcarbinol. In most cases, the decrease in Saytzev/Hofmann ratio is parallelled by a decrease in trans/cis-Saytzev-ene[175]. For eliminations in polar media in the previous section, we interpreted these variations in terms of a decrease in double-bond character and an increase in carbanion character of the transition state. These factors are clearly unsatisfactory in accounting for elimination behaviour in less dissociating media, in which trans/cis ratios may be less than unity or exceed the thermodynamic predictions[176,177]. In some cases, a change of stereochemistry is responsible. However, this explanation does not hold for 2-butyl tosylate eliminations (base: % 1-ene; trans/cis-2-ene: (i) t-BuOK–t-BuOH, 64, 0·58; (ii) t-BuOK–DMSO; 61, 2·53)[177]. Brown and

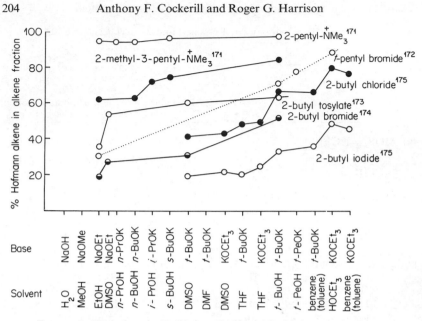

FIGURE 5. Variations in Hofmann alkene yield with base and solvent.

Klimisch[176] proposed steric interactions with bulky bases cause conformation **49** to predominate over **50**. Arguing in terms of more reactant-like transition states in the strongly basic butoxide media, Froemsdorf and coworkers also assumed a preference for the same conformation **49**[177]. However, a change of solvent to DMSO hardly affects the positional orientation, but markedly alters the geometric orientation. Subsequently, it has been shown that addition of 'crown ether' to *t*-butoxide–*t*-butyl alcohol solutions gives a similar *trans/cis*-ene ratio to the DMSO result[157]. Thus, it seems *cis*-ene formation is favoured by ion-paired base, whereas *trans*-ene formation is promoted more effectively by dissociated ions. Possibly the answer still involves conformation **49** as ion-paired base can

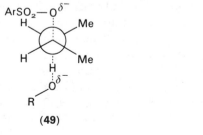

(49)

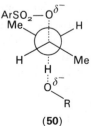

(50)

approach unimpeded, side on, giving *cis*-ene, whereas an alkyl group–base interaction is necessary for conformation **37** which gives the *trans*-ene (cf. **46**).

For the 'onium salts, Hofmann orientation is often so extreme in polar solvents, that little change in orientation is noted in more basic media (Figure 5). However, for both the 'onium salts in Figure 5, the increase in Hofmann orientation is accompanied by an increase (abnormal compared with alkyl halides) in the *trans/cis*-ene ratios[171]. In the 2-pentyl case, the Saytzev-ene *trans/cis* ratios are continually below unity. However, the Hofmann-ene *trans/cis* ratios vary from 5 to 11 in the elimination of 2-methyl-3-pentyltrimethylammonium ion. Saunders has used this abnormal behaviour as an indication that the *syn* → *trans*-ene process is becoming competitive in more basic media[152,171]. The explanation is based on steric inhibition of proton abstraction and is discussed in detail in Section II.E.4. Substitution at the β'-carbon has a greater impact than at the γ-carbon as shown by the more rapid decline in *cis*-ene yield with increasing base strength in the eliminations of 5-nonyl → 4-ene, 4-heptyl → 3-ene and 3-hexyl → 3-ene than the eliminations of 3-pentyl, 2-hexyl and 2-pentyl all to give 2-enes (Figure 6). However, as the *cis*-ene yield declines with increased base strength, base–alkyl interactions in conformations leading to *trans*-ene cannot be as severe as in those giving *cis*-enes. Clearly, the variation cannot alone be explained by a reduction in the rate of the *anti* → *trans*-ene. The slight increase in *cis*-2-butene yield with medium basicity was assumed to imply continual operation of *anti* elimination for all the basic systems.

Eliminations of 2- and 3-pentyldimethylsulphonium ions with *t*-butoxide ion give *trans/cis* ene ratios of 3·5 → 5·0[178]. Initially assumed to be indicative of a *syn*-E2 mechanism, subsequent labelling studies have confirmed that the yield mechanism accounts for most of the reaction[179].

Nearly 20 years ago, Brown showed that the yield of Hofmann product in the eliminations of alkyl bromides increased with the steric size of a number of pyridine bases[180]. A steric interpretation was also assumed to explain the increase in Hofmann orientation along the alkoxide series; ethoxide → *t*-butoxide → *t*-pentoxide → *t*-heptoxide (see Figure 5)[172]. However, in both series the base and solvent were simultaneously changed and the stereochemistry of elimination was not examined. More recently, Froemsdorf has shown that with substituted phenoxide bases in dimethyl sulphoxide, the yield of 1-butene from 2-butyl tosylate increases with the basicity of the medium[173]. Admittedly, he used a considerably less encumbered substrate than Brown, but even so it would seem reasonable to attribute the previous results to both steric and electronic effects and not

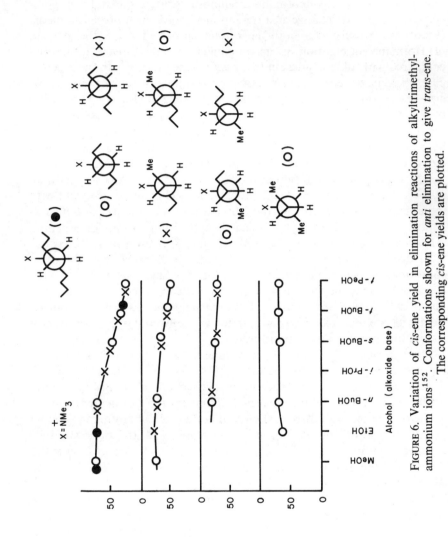

FIGURE 6. Variation of cis-ene yield in elimination reactions of alkyltrimethyl-
ammonium ions[152]. Conformations shown for anti elimination to give trans-ene.
The corresponding cis-ene yields are plotted.

just the former alone. Very recently, it has been shown that the sensitivity of orientation of elimination of 2-butyl iodide in DMSO to changes in the base strength decreases markedly as the base is changed from oxy anion to aniline anion to carbanion[180a]. In conclusion, careful choice of the base can now be used to achieve as large a variation in the Hofmann–Saytzev or *cis/trans* olefin ratios for a given molecular structure as can be obtained by varying the leaving group. The use of a halide leaving group and *t*-alkoxide base can often offer a more convenient synthesis of the less substituted alkene than the more classical Hofmann eliminations. A recent synthetic example concerns the synthesis of β-cedrene[181] (equation 32).

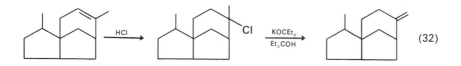

$$(32)$$

3. The effect of alkyl structure

The bias towards Hofmann orientation as the alkyl substitution at C_α and C_β was increased in the elimination of alkyl halides and 'onium salts was interpreted in terms of a balance of hyperconjugative and inductive effects by Hughes and Ingold[168] and hyperconjugative and steric effect by Brown[182]. However, the complex alkyl structures employed in the investigations are now known to exhibit the *syn–anti* dichotomy and a re-evaluation of the data with a knowledge of the contribution of the various stereochemical requirements is needed. A recent evaluation of the eliminations of some acyclic ammonium salts illustrates the complexity of the problem (see Figure 7)[183].

The various stereochemical orientations leading to alkenes A and B exhibit quite different rate variations with changes in the alkyl substituent R. For elimination away from R to give the A-alkenes, the authors[183] argued that the influence of R should be confined to a steric role, R being too distant significantly to alter the inherent acidity of the β-hydrogen. For the *syn* → *trans* A, the rate increases continually with the bulk of R, consistent with relief of steric interaction between R and the $\overset{+}{N}Me_3$ group (or the *n*-butyl fragment) in the ground state being relieved in the transition state (51, 52). The two *anti* pathways to give *cis*-A (53) and *trans*-A (54)

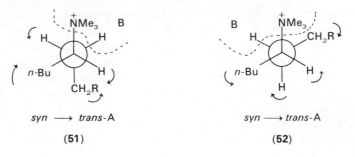

syn ⟶ trans-A   syn ⟶ trans-A

(51)   **(52)**

show little rate variation with R. Presumably in these cases, the reduction in $R \leftrightarrow \overset{+}{N}Me_3$ steric interaction is counterbalanced by an increase in

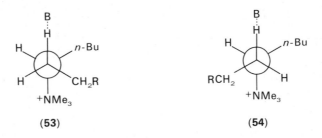

(53)   **(54)**

$B \leftrightarrow n\text{-Bu}$ interaction in passing into the transition state. It would be interesting to see if the same rate trend is found for a smaller base. Clearly, in the formation of alkene A, the *syn* → *trans* route becomes more competitive with increasing bulk of R, not because of a rate reduction in the *anti* → *trans* route, but because of rate acceleration in the *syn* pathway (cf. II.E.4).

For formation of alkenes B, the R substituent can exert both steric and polar effects. Ingold[168] had previously argued that steric effects should build up very rapidly with increasing material bulk, whereas the polar factor should exhibit a gentler rise. The significance of this interpretation is however in doubt, as steric effects can manifest themselves in the form of the polar variation. The gentle rise in rate with increasing R for the *syn* → *trans*-A series is an example. The *anti* → *cis*-B may reflect a significant steric interaction, as the rate declines markedly from R = *n*-Pr to *t*-Bu. The more gentle decreases in rate for the other pathways could originate from either steric or polar effects. All three reaction profiles for B-alkene formation have been extrapolated through the composite rate consistent with relief of steric interaction between R and the $\overset{+}{N}Me_3$ group (or the *n*-butyl fragment) in the ground state being relieved in the transition the recent development of tritium-n.m.r. should alleviate this problem[184], as the necessary chirality can be introduced into the substrate using all three isotopes of hydrogen and the products identified by proton and tritium-n.m.r. (equation 33). Presumably tritium will be eliminated considerably less easily than hydrogen and to a lesser extent, deuterium.

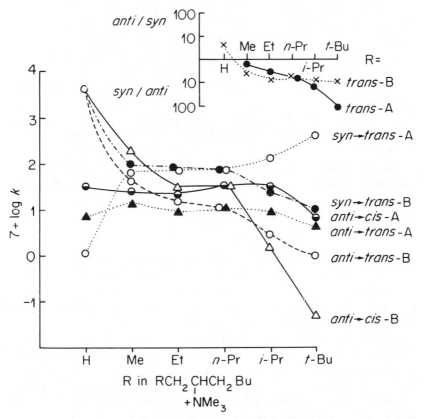

FIGURE 7. Approximate rate constants for the elimination pathways in the E2 reactions of quaternary ammonium salts with *t*-BuOK in *t*-BuOH at 35 °C.

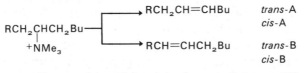

For experimental determinations, see Reference 183.

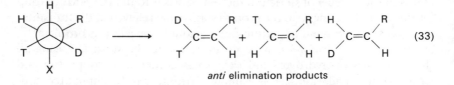

$$anti \text{ elimination products} \tag{33}$$

Final comment is warranted on the *syn/anti* ratios to give *trans* A and B (see Figure 7). Only when R = H, is the *anti* → *trans*-A the dominant mode. *Syn* elimination becomes considerably more dominant in alkene-A formation than alkene-B formation with increasing bulk of R.

In conclusion, for the eliminations of simple alkyl halides in polar media, the Hofmann–Saytzev and *cis–trans* olefin ratios can be explained using the variable transition state theory. However, for more complex acyclic compounds possessing a wider range of leaving groups, in a wider range of basic media, the stereochemistry of elimination should be carefully evaluated before explanations of varying orientation are contemplated. From a synthetic view, considerable progress has been made, but more investigations are required before a general theory to account for orientation under a wider range of reaction conditions is clearly established.

G. Eliminations in Aprotic Media

1. Synthetic results

In 1956, Winstein, Darwish and Holness reported the facile elimination of *cis-* and *trans*-4-*t*-butylcyclohexyl tosylate with halide ions in acetone[185]. Since that time, more polar aprotic solvents such as dimethyl sulphoxide (DMSO), dimethyl formamide (DMF), methyl cyanide and hexamethylphosphoramide (HMPA) have gained favour as media for reactions involving nucleophilic anions[186]. The dipole in all of these solvents is exposed at the negative end, but sterically shielded at the positive end. Thus, aprotic solvents effectively solvate cations but leave anions rather 'bare'. As a result, small anions, such as halide, possess enhanced nucleophilicity towards carbon and hydrogen compared to the situation in protic media. Mixtures of protic and aprotic media can often be used to attain the required medium basicity to affect a reaction, and some examples have been illustrated in the previous sections. However, in this section we shall confine our discussions mainly to aprotic media. The absence of a hydroxyl group precludes the formation of solvolytic products. In some cases it is possible to cause elimination with base and leaving group the same, and thereby minimize the problem of competing substitution and elimination. Most eliminations in aprotic media concern reactions of tertiary, secondary and to a lesser extent primary halides and arene sulphonates with various halide and other inorganic anions. The eliminations are characterized by a greater adherence to the Saytzev rule and *anti*-stereospecificity than corresponding reactions with alkoxide bases (e.g., equations 34–36)[187,188]. Generally the ratio of geometric alkenes approaches the thermodynamic ratio[187]. This behaviour is complementary to eliminations with *t*-alkoxides which give preferentially the Hofmann alkene and the *cis*-olefin. Cation–anion association is minimal with the sterically shielded tetra-*n*-butyl ammonium ion, and eliminations are consequently much faster using this than the corresponding alkali metal salts[189].

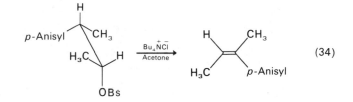

(34)

$$CH_3CH_2\underset{\underset{OTs}{|}}{C}HCH(CH_3)_2 \longrightarrow CH_3CH_2CH{=}C(CH_3)_2 + CH_3CH{=}CHCH(CH_3)_2$$

t-BuOK–*t*-BuOH	51·1%	48·9%	(35)
Bu₄ṄC̄l–acetone	92·5%	7·5%	(36)

2. Mechanistic proposals

At the time of writing, there is no satisfactory explanation of all the mechanistic details of eliminations in aprotic media[190]. Originally Winstein[185] suggested a common intermediate (**55**) was involved in both substitution and elimination reactions. This idea of a merged mechanism was furthered by Eliel and coworkers[191] and Csapilla[192] reviewed the whole of elimination reactions from this standpoint. Subsequent work by Parker, Winstein and coworkers[187] rules against a common intermediate or transition state. The *erythro* and *threo* isomers of 3-*p*-anisyl-2-butyl

(**55**)

chloride on treatment with chloride ion give vastly different elimination products. Microscopic reversibility requires the same transition state for conversion of *threo* → *erythro* and *vice versa*. Since the isomers lead to the same substitution transition state, the different elimination products exclude a merged intermediate. However, this does not exclude a common reaction profile for elimination and substitution prior to formation of the transition state[9,190].

Some aspects of the merged mechanism have been retained in Parker's formulation, termed the E2C mechanism, which constitutes an extension to the variable transition state theory (Scheme 9)[193]. Strong bases are

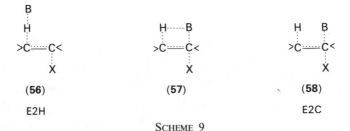

(56)   (57)   (58)

E2H   E2C

SCHEME 9

expected to confine their interaction with the substrate to the β-hydrogen (E2H, **56**), whereas weak bases (e.g., halide ion in aprotic media) with greater carbon nucleophilicity attack mainly at the α-carbon (E2C, **58**). In most eliminations in aprotic media, intermediate transition state **57** is proposed. The written formulation of the E2C mechanism clearly resembles an S_N2 process, but it must be remembered that this is regarded as a separate feature.

The main support for $B—C_\alpha$ interaction is provided by linear correlations between S_N2 and E2 rates (e.g., for cyclohexyl tosylates) with a variety of weak bases, but lack of correlation with strong alkoxide bases[193,194]. However, when it is recollected that rates of elimination of alicyclic ring ammonium salts giving *cis*-enes (typical E2H reactions, cf. II.E.3.a) correlate linearly with S_N2 reactions of the corresponding bromides, some doubt is cast on the validity of these correlations as evidence for $B—C_\alpha$ interaction. In addition, S_N2 and E2 rates do not always run parallel for all substrates under apparently E2C conditions[190,194]

Element effects (k_{Br}/k_{Cl}) are large in halide-ion promoted eliminations in aprotic media, and often exceed 1000 (e.g., **59**, **60**) for tertiary substrates, but are considerably smaller for secondary substrates (e.g., **61**). The large values imply considerable $C_\alpha—X$ cleavage whereas the results for the secondary series are similar to those of about 60 observed for ethoxide-induced eliminations in the 2-arylethyl series. Parker suggests that the

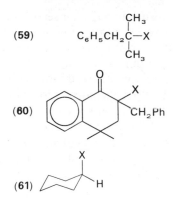

(59) $C_6H_5CH_2\overset{\overset{\displaystyle CH_3}{|}}{\underset{\underset{\displaystyle CH_3}{|}}{C}}-X$ $k_{Br}/k_{Cl} = 3400$ ($\bar{Cl}$ in acetone)[195]

(60) CH_2Ph $k_{Br}/k_{Cl} = 1160$ ($\bar{Br}$ in MeCN)[196]

                              1500 ($\bar{Cl}$ in MeCN)[196]

(61)   H $k_{Br}/k_{Cl} = 23$ ($\bar{Cl}$ in DMF).

tertiary substrates and cyclohexyl tosylate react via E2C-like transition states (termed loose substrates) whereas cyclohexyl bromide and chloride are more like E2-H substrates[198]. As yet, no leaving group isotope effects have been measured for these aprotic eliminations and these may provide a better measure of C_α—X cleavage than the element effects, the magnitude of which may be 'telescoped' in aprotic relative to protic media[186].

For eliminations of simple acyclic halides and tosylates in aprotic media, the deuterium isotope effects are generally small, ranging from about 2 to 4[196,199]. Values of this size are consistent with little or extensive C_β—H cleavage if the base approaches the proton to give a linear C_β—H—B bond in the transition state. In terms of the E2C concept, side approach of the base would influence the bending rather than stretching modes and therefore a maximal isotope effect of about 2·5—3 is expected even for half transference in the transition state[200].

Recently Ford[190] has suggested that a preformed tight ion pair may be attacked by the base in aprotic eliminations in the rate-determining step. This mechanism is equivalent to the $(E2)_{ip}$ notation of Bordwell[14] and requires carbonium ion character to develop at C_α and consequently should be favoured by electron-releasing substituents at C_α. In fact, the elimination of 1-aryl-1-bromopropanes with $n\text{-}Bu_4\overset{+}{N}\bar{Br}$ in DMF proceeds faster with increased electron donation from the aryl substituent[201]. The rate variation is much smaller than is observed for the corresponding solvolysis in 90% acetone–water ($\rho = -4.7$ at 75 °C and a regular Hammett plot with either σ or σ^+ is not observed. This result appears more consistent with variable E2C transition states rather than an $(E2)_{ip}$ mechanism. However, it is possible that the poorly solvated chloride ion in the aprotic media provides most of the stabilization of the incipient double bond in a tight ion pair, whereas in the solvolysis reaction, the halide is effectively solvated by the protic solvent. This results in a greater separation of anion from the cation and both sterically and electronically may allow greater stabilization by the aryl substituents. Substitution of a second halogen at C_α generally shifts various kinetic parameters for the entire substrate to values expected for transition states for E2 reactions with extensive carbanion character (cf. Ar_2CHCCl_3[67], 1,1-dichlorocyclohexanes[202]).

Substituent effects at C_β suggest little carbanion character develops at this site, unlike the situation for most E2 eliminations in protic media. For instance, β-methyl substitution causes a slight rate enhancement, β-phenyl has a similar effect on rate, and a small reaction constant (0·42) is found for elimination of p-substituted 2-benzyl-2-bromo-1-indanones (**62**)[203].

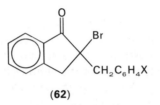

(**62**)

The picture of the transition state emerging from the eliminations of substrates with good leaving groups on treatment with weak bases in aprotic media is thus one in which C_α—X cleavage appears to be the lead process, followed by significant development of double-bond character (*anti* stereospecificity and Saytzev orientation) with the least change occurring in the C_β—H bond. These eliminations may be representative of the long-sought-after examples of E2 reactions exhibiting carbonium ion character. Certainly, there are no clear-cut examples of this type in eliminations induced in protic media with alkoxide bases[8]. Whether they are best described by the present E2C concept or an ion pair is debatable. The nature of the interaction, if any, between the base and C_α is not clearly understood. It could be partial bonding or a through-space interaction. Hopefully future studies will clarify the situation.

3. Eliminations with sulphur nucleophiles

There has been a tendency to develop a common transition state approach to explain all the eliminations induced by weak bases (strong carbon nucleophiles). In addition to inorganic ions in aprotic media, this category has often contained sulphur nucleophiles, such as ethanethiolate and thiophenoxide ion, generally in alcoholic solvents. These latter basic media involve a polarizable nucleophile in a protic solvent and are clearly somewhat different from the combination of hard nucleophile and aprotic solvent. We, therefore, prefer to consider these eliminations in a separate category.

1956 was also the year in which it was originally discovered that for elimination of t-butyl chloride, thiophenoxide was a more effective reagent than the stronger ethoxide[204]. This somewhat surprising result was subsequently attributed to a possible merged mechanism and is clearly a candidate reaction for the E2C category. For a number of additional cyclohexyl derivatives, McLennan[202] demonstrated an inverse relation between the Brönsted component, β(substituted thiophenoxides) and the relative rates of elimination, k_{SPh}^-/k_{OEt}^-[9]. Thus, for eliminations in which the β-proton is extensively transferred to the base, the stronger ethoxide is more effective. However, if proton transfer is considerably less, and counterbalanced by increased C_α—X cleavage, then thiophenoxide is more efficient. This could be because of its greater polarizability and carbon nucleophilicity. The rate variations clearly can be interpreted in terms of variable transition states with different carbanion and carbonium ion character or in terms of E2H and E2C concepts. Bunnett invoked the decrease in k_{SEt}^-/k_{OMe}^- ratios with a change to a poorer leaving group in terms of the variable transition state theory and a reduction in the effectiveness of the polarizability of the thiolate as the substrate leaving group became less labile ($C_6H_5CH_2CMe_2X$; k_{SEt}^-/k_{OMe}^-; X: 6·5, Cl; 0·8, $\overset{+}{S}Me_2$; 0·05, SO_2Me)[205]. He regarded base-C_α interaction as minimal as rates of elimination were insensitive to steric effects at C_α. For example, isopropyl bromide eliminates only about twice as fast as 2-bromo-3,3-dimethylbutane upon treatment with ethane thiolate, after suitable statistical correction[206]. Parker[207] replied by stating that an E2C transition state is looser than an S_N2 transition state and this greater separation between C_α and the base lessens the expected rate-retarding effect of bulky substituents based on our knowledge of bimolecular displacements. Clearly, as for the eliminations in aprotic media, a more definite measure of C_α-base interaction is required than the uncertain rate comparisons which have been used so far.

H. Mechanisms with Carbonium Ion Character

Carbonium ion intermediates or transition states with considerable carbonium ion character are involved in a number of alkene-forming eliminations. Typical of the latter category are the homogeneous unimolecular pyrolytic eliminations. More discrete carbonium ions are involved in solvolytic reactions of alkyl halides and esters and in dehydration reactions, for all of which substitution is the major competing pathway.

However, carbonium ions formed in deaminations, or deoxidations of alcohols with dihalogeno carbenes, are less selective and in addition undergo unimolecular rearrangements[8]. Space limits us to only a cursory discussion in this text. The chemistry of carbonium ions is described in great detail in a series of recent monographs[207a].

1. Ion-pair mechanisms as alternatives to concerted elimination

Sneen and Robbins[208] suggested that the decrease in the second-order rate coefficient for the elimination of 1-phenylethyl bromide with increasing base concentration of sodium ethoxide in ethanol was consistent with an ion pair mechanism. However, McLennan[209] has challenged this interpretation, and shown that the decrease in rate coefficient can be attributed to a variation in the concentration of ion-paired and dissociated ethoxide ion. The role of ion pairs in both base and substrate species clearly merits more detailed consideration in future studies.

2. Mechanisms of pyrolytic eliminations

Maccoll[210,211] and others[212] have reviewed mechanisms of pyrolytic eliminations since his original article in an earlier volume in this series[1]. For gas phase pyrolysis, single-pulse shock tube and chemical activation methods have been developed to supplement the earlier flow and static procedures[211]. The shock tube approach enables a wide temperature study, the heating is homogeneous due to very short reaction times, which in turn minimize heterogeneous reactions. The chemical activation method has been used particularly in the study of alkyl fluoride pyrolysis. The activated fluorides, which can be generated by a variety of procedures, undergo deactivation by either molecular collision or unimolecular elimination. The activation parameters for the latter can be calculated using the Rice–Ramsperger–Kassel–Marcus theory of unimolecular reactions[213]. In general, excellent agreement is obtained between the various experimental procedures for a given elimination[211].

Detailed kinetic analyses of the eliminations of aliphatic halides, acetates, xanthates, isothiocyanates and thiocyanates have been reported. The kinetics of all these eliminations accord with the unimolecular Ei-type transition state with carbonium ion character developing at C_α. As indicated by the decreasing effect of additional α-methyl substitution on rate, carbonium ion character decreases as C_α—X changes as follows; C—Cl, C—OAc, C—N=C=S, to C—S—C≡N[212]. As shown earlier for the halide eliminations, the activation energies for the pyrolysis of various alkyl isothiocyanates are linearly related to the heterolytic and not homolytic bond dissociation energies of the C_α—X bond[214]. Not surprisingly the rates of elimination for a series of alkyl chlorides are linearly related in a logarithmic manner to the solvolysis rate coefficients for the corresponding halides in acetonitrile[215]. All of these results confirm the close analogy between pyrolytic and solvolytic eliminations.

For pyrolysis in solution, the Cope elimination of amine oxides remains the most *syn* stereospecific. The pyrolysis of simple aliphatic sulphoxides is also highly *syn* stereospecific, but for more complex sulphoxides, a duality of mechanism is apparent. At low temperatures (70–80 °C) the diastereoisomeric 1,2-diphenyl-1-propyl phenylsulphoxide eliminations are *syn* stereoselective. However, at higher temperatures, the stereoselectivity decreases, and curved Arrhenius plots are obtained, consistent with the incursion of a radical pair mechanism in competition with the normal Ei process[216]. Sulphoxide pyrolyses should give rise to sulphenic acids which have generally avoided detection due to inherent instability. However, direct spectroscopic evidence[217], trapping procedures[218], and in one case an actual isolation via crystallization during the reaction, have provided firm evidence to support their intermediacy (equation 37)[219]. Selenoxide eliminations are very similar to sulphoxide pyrolyses[220].

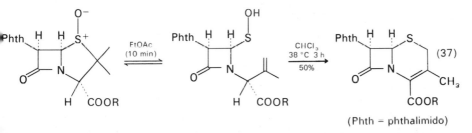

$$(37)$$

(Phth = phthalimido)

Radical intermediates have been invoked to account for the array of products in the Chugaev reaction of primary xanthates[221], and examples of both acetate[222] and xanthate pyrolyses[223] exhibiting *anti*-stereospecificity have been reported. This behaviour, however, is atypical of most ester pyrolyses[212]. The alkene yield in Chugaev reactions is improved if potassium xanthates rather than sodium derivatives are used[224] and in nematic liquid-crystalline solvents[225]. The kinetics of pyrolysis of the following derivatives also accord with the *syn*-Ei transition state: dialkyl carbonates[226], *N*-sulphonyl sulphilimines[227], and amine–imides[228]. Both ionic and concerted eliminations have been suggested for the pyrolysis of alkyldiphenylphosphinates[229]. Other interesting reactions include the pyrolysis of tosyl carbonates[230] (equation 38) and cleavage of *vic*-diols as their carbonate tosylhydrazone salts (Scheme 10)[231].

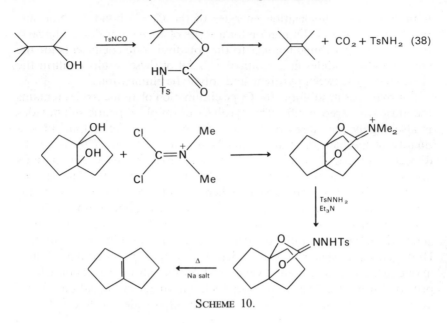

SCHEME 10.

3. Dehydration reactions

There have been a number of reviews[9,232] since Knozinger's chapter in this series on dehydration reactions[6]. For mineral acid dehydrations and hydration of alkenes, Knozinger preferred the π-complex hypothesis developed initially by Taft in which a rapid and reversibly formed association complex between the alkene and the hydronium ion follows or precedes, respectively, the carbonium ion. Noyce has shown from a combination of substituent effects and isotope effects, that the π-complex idea is an unnecessary elaboration in the dehydration/hydration of 1,2-diphenylethanol/*cis*- and *trans*-stilbenes[233].

The affinity of aprotic solvents for hydroxyl groups renders them as potentially good dehydrating catalysts. Complexing of the hydroxyl improves its departing ability and encourages carbonium-ion mechanisms in solvents such as DMSO[234] and HMPA[235].

Sulphur tetrafluoride[236] has been used in addition to more common 'halogenating' reagents such as $SOCl_2$ and $POCl_3$ with pyridine bases. For these latter reagents, Kirk and Shaw have shown that orientation of dehydration is very dependent on the choice of base[237]. In the absence of the knowledge of orientation of elimination of both epimers of steroidal alcohols, stereochemical assignments based on alkene proportions are uncertain. The inner salt **63** is a convenient reagent to affect *syn* dehydration

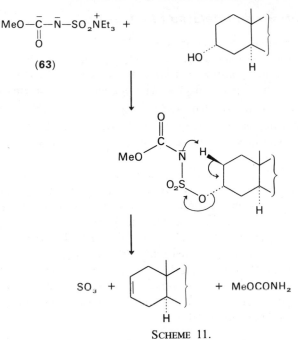

$MeO-\overset{-}{\underset{\underset{O}{\|}}{C}}-\overset{-}{N}-SO_2\overset{+}{N}Et_3$ +

(63)

SO$_3$ +   + MeOCONH$_2$

SCHEME 11.

(Scheme 11)[238], and selective dehydrations have also been accomplished with methyltriphenoxyphosphonium iodide in HMPA[239]. This reagent selectively dehydrates secondary alcohols in presence of tertiary alcohols, and gives predominantly the Saytzev and *trans*-alkene. Another novel reagent is shown in Scheme 12[240]. The sulphurane induces E1-type dehydrations of tertiary alcohols and E2-type dehydrations with secondary alcohols as indicated by the preference for *anti* stereospecificity. On the other hand, primary alcohols form ethers unless the β-hydrogen acidity is enhanced by electron-withdrawing substituents.

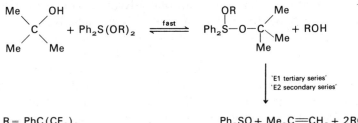

R = PhC(CF$_3$)$_2$ Ph$_2$SO + Me$_2$C=CH$_2$ + 2ROH

SCHEME 12.

I. Dehalogenations and Related Eliminations

Dehalogenations are the most common of alkene-forming eliminations, which do not involve hydrogen as one of the leaving groups. The leaving groups may be chlorine, bromine or iodine, fluorine being inert in most cases. The eliminations can be affected with nucleophilic reagents (e.g., $\bar{\text{I}}$, $\bar{\text{S}}\text{Ph}$) or reducing agents ($\text{Cr}^{\text{II}}\text{en}$, SnCl_2, Na, Zn). An excellent comparison of the efficiency of many reagents in the dehalogenation of *meso*- and *dl*-stilbene dibromides[241] and reviews on the general mechanisms have appeared in recent years[242].

Dehalogenations with nucleophiles such as iodide ion generally occur with *anti*-stereospecificity. Alkene yields are usually greater if aprotic solvents are used, and the eliminations occur faster because of the greater nucleophilicity of the halide[243]. The leaving group mobility declines markedly along the series, I → Br → Cl[244]. Second-order kinetics coupled with the above observations are consistent with the operation of an E2 type mechanism (**63a**)[244], although other types of transition states have also been invoked (**63b–63d**)[245]. In some cases, an initial S_N2 reaction followed by an E2-type elimination may occur, invalidating deductions from the product stereochemistry alone.

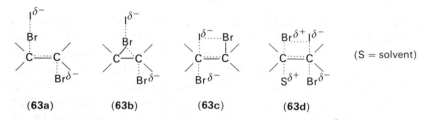

(63a)   (63b)   (63c)   (63d)

Miller has made excellent use of product analysis and the relative rates of dehalogenation of *meso* and *dl*-stilbene dibromides as a guide to reaction mechanism[241–243]. The major elimination components using nucleophilic reagents such as the halide ions in aprotic solvents are the *trans*- and *cis*-stilbene, respectively. However, both substrates give mainly *trans*-stilbene with a variety of metal and metal ion reductants. This suggests the intermediacy of radicals in which free rotation about the C_α—C_β bond allows loss of the original asymmetry and formation of the more stable alkene.

A number of very reactive reductants have been used to affect dehalogenation at low temperatures, enabling the isolation of thermally unstable alkenes. Typical of this category is the facile preparation of cyclobutenes using sodium dihydronaphthylide, sodium biphenylide or disodium phenanthrene in dimethyl ether (cf. equation 39)[246].

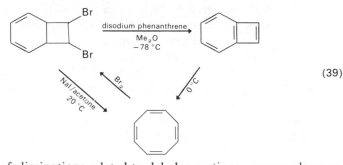

(39)

A number of eliminations related to dehalogenations concern cleavage of a halogen and various other β-related groups such as hydroxyl, tosylate, acetate, amine and alkyl ether. These groups can be effectively cleaved with a variety of metals, the main side reaction being partial reduction (only of the halogen[242]). Chromous ion and its ethylene diamine complex, Cr^{II}en, are also effective reductants, inducing mechanisms involving radical or chromium-organic complexes[247].

III. THE REACTIONS OF NUCLEOPHILES WITH VINYL HALIDES AND ESTERS

We devote only brief comment to nucleophilic addition reactions of alkenes as this subject has been covered in detail previously[248], and has been the subject of recent reviews[249]. Most of the reactions concern addition to alkenes activated by electron-withdrawing groups such as nitro, cyano, acetyl, sulphonyl and aryl groups. The majority of these reactions can be considered as typical Michael additions, the nucleophiles including alkoxides[250], thiolates[251], amines[252], metal alkyls[253] and phosphoranes[254]. A number of the most common mechanisms is shown in Scheme 13. Minor variants not shown concern the stereochemistry and position of apparent substitution (either α or β to the activating group Y). Hydrogen isotope studies, stereochemical analysis and excellent use of the element effects are generally the main kinetic criteria[249,255]. Substitution products may arise directly or more often via either an addition–elimination or elimination–addition sequence. If the latter is involved, then both geometric isomeric substrates give the same product of apparent substitution. The variable transition state theory has been invoked to explain changes in the kinetic parameters in the eliminations of β-halostyrenes with alkoxides[256]. The effects of basicity parallel those observed for E2 reactions, but as expected for a less labile vinylic carbon–halogen bond, the transition states appear to be more carbanionic. Unless ion pairing is prominent, most eliminations exhibit *anti* stereospecificity. *Syn*-elimination is particularly prevalent with metal alkyls in poorly dissociating media[253,257]. These reaction media have also enabled the recognition of the $E_{C=C}2cB$ mechanism[258], in which metal–hydrogen exchange, alpha to the halogen, precedes either bimolecular (thermal process) or second order (base catalysed process, first order in base and in substrate) elimination (Scheme 14). The metallated intermediate was isolated at low temperature and decomposed thermally and in the presence of excess base.

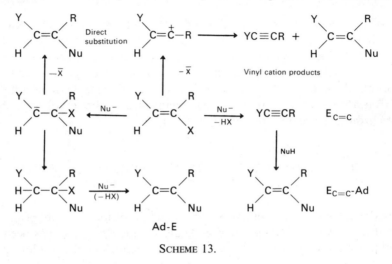

SCHEME 13.

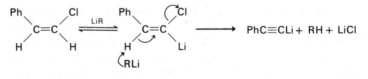

SCHEME 14.

Although cationic intermediates are common in saturated systems, they are seldom encountered in reactions of vinyl halides and esters. Mesomeric interaction with the adjacent π-bond imparts partial double-bond character on the C_α—X bond, thereby opposing its unimolecular fragmentation[259]. However, when X is an exceptionally good leaving group[260] (e.g., OSO_2CF_3) or the vinyl cation is stabilized by appropriately located cyclopropyl[261] or aryl groups[262], the existence of vinyl cation intermediates has been demonstrated[263].

IV. ADDITION AND ELIMINATION REACTIONS INVOLVING THE CARBONYL GROUP

A. General Comments

The most frequently encountered addition reaction of the carbonyl group is a nucleophilic attack at the positive end of the carbonyl dipole. The additions are typically reversible and are often complicated by secondary processes. Such secondary processes may lead to alkenes (e.g., aldol, Wittig reactions). Elimination reactions leading to a carbonyl group are of an oxidative type (e.g., chromic acid oxidation) or the reverse of an addition reaction (e.g., acetal hydrolysis). Compared with addition and elimination reactions involving alkenes, mechanistic studies in this area are less developed. The formation and hydrolysis of acetals has recently been the subject of intense investigation owing to its relevance to biological systems and will constitute a large part of this section, as will the condensation reactions which are of synthetic importance.

B. Condensation Reactions of the Carbonyl Group

This section encompasses the aldol condensation, Knoevenagel, Darzens, Wittig and related reactions. Reviews have appeared on the aldol condensation[264] and the Knoevenagel reaction[265]. The Wittig reaction has been reviewed by Maercker in 1965[266] and by Reucroft and Sammes in 1971[267]. The preparative aspects of these reactions are covered by House's recent book [268]. In most cases, the essential mechanism is well understood and recently investigators have concentrated on stereochemical aspects of the reactions.

1. The aldol condensation

a. *Mechanism.* The aldol condensation will be defined in its strictest sense in this section as the condensation of an α-carbon of an aldehyde or ketone with the carbonyl carbon of another molecule. The reaction is catalysed by either acids or bases. In the latter case (Scheme 15) the rate-limiting step may be proton removal[264,269] or the formation of the new carbon–carbon bond. Where a reactive carbonyl compound is the acceptor, for example an unhindered aldehyde, the proton removal is rate limiting. In this case general base catalysis is significant[270]. The condensation is often followed by an E1cB reaction leading to an α,β-unsaturated carbonyl compound (See Section II.C.3.c). In the acid-catalysed reaction (Scheme 16) the enol form of the ketone reacts in a rate-determining step with the conjugate acid of the carbonyl component. The aldol or ketol formed is dehydrated under the acidic conditions. Two mechanisms have been described in considerable detail for the dehydration step[271,272]. Thus, the essential mechanism of the aldol condensation is well understood. Recent interest has centred on stereochemical analysis for the reactions which lead to diastereoisomeric hydroxy-ketones. This is made difficult because of the occurrence of secondary processes, such as retroaldolization, dehydration and equilibration of the isomeric products via their enolate anions.

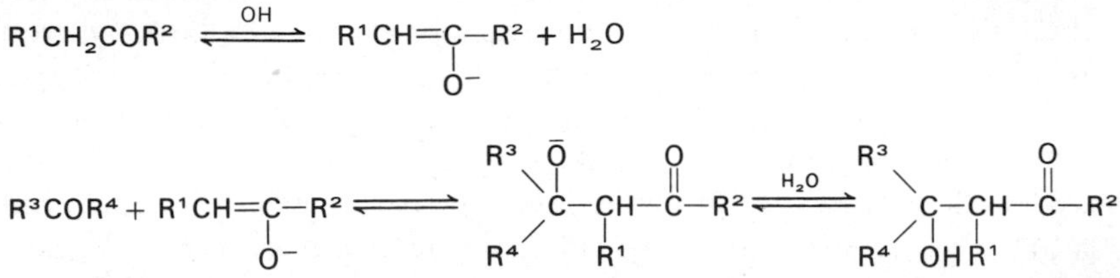

$$R^1CH_2COR^2 \; \underset{}{\overset{OH}{\rightleftharpoons}} \; R^1CH{=}\underset{\underset{O^-}{|}}{C}{-}R^2 + H_2O$$

$$R^3COR^4 + R^1CH{=}\underset{\underset{O^-}{|}}{C}{-}R^2 \; \rightleftharpoons \; \underset{R^4}{\overset{R^3}{\diagdown}}\underset{\underset{R^1}{|}}{\overset{\overset{\bar{O}}{|}}{C}}{-}\underset{}{CH}{-}\underset{}{\overset{O}{\overset{||}{C}}}{-}R^2 \; \underset{}{\overset{H_2O}{\rightleftharpoons}} \; \underset{R^4}{\overset{R^3}{\diagdown}}\underset{\underset{OH\,R^1}{|\ \ |}}{C}{-}CH{-}\overset{\overset{O}{||}}{C}{-}R^2$$

SCHEME 15.

$$R^1CH_2COR^2 \; \underset{}{\overset{\overset{+}{H}}{\rightleftharpoons}} \; R^1CH_2{-}\underset{\underset{OH}{|}}{\overset{\overset{O}{||}}{C}}{-}R^2 \; \underset{}{\overset{-\overset{+}{H}}{\rightleftharpoons}} \; R^1CH{=}\underset{\underset{OH}{|}}{C}{-}R^2$$

$$\left. \begin{array}{l} R^1CH{=}\underset{\underset{\overset{\cdot\cdot}{OH}}{|}}{C}{-}R^2 \\[6pt] R^1CH_2\underset{\underset{\overset{+}{OH}}{\diagup}}{C}{-}R^2 \end{array} \right\} \; \rightleftharpoons \; R^1CH_2{-}\underset{\underset{R^2}{|}}{\overset{\overset{OH}{|}}{C}}{-}\underset{\underset{R^1}{|}}{CH}{-}\overset{\overset{\overset{+}{OH}}{||}}{C}{-}R^2 \; \underset{}{\overset{-\overset{+}{H}}{\rightleftharpoons}} \; R^1CH_2{-}\underset{\underset{R^2}{|}}{\overset{\overset{OH}{|}}{C}}{-}\underset{\underset{R^1}{|}}{CH}{-}\overset{\overset{O}{||}}{C}{-}R^2$$

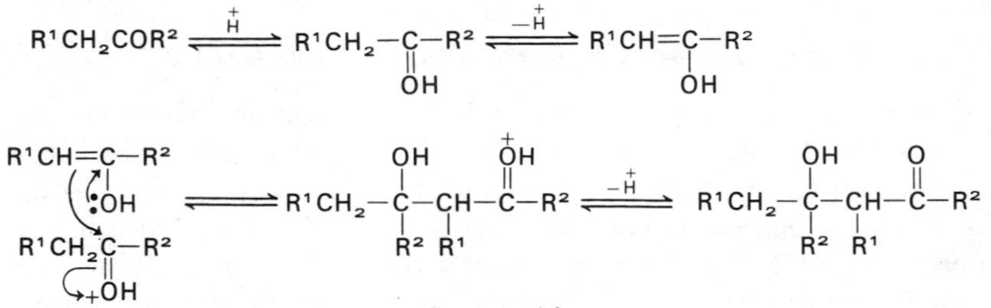

SCHEME 16.

b. *Stereochemical studies.* The stereoselectivity of the aldol condensation is dependent on both the solvent and the cation employed[273,274]. The thermodynamically less stable *threo* isomer can be formed predominantly in the cyclopentanone–acetaldehyde condensation if cyclopentanone is used as solvent and potassium ion as the cation[274]. A similar result is found in the condensation of cyclopentanone and isobutyraldehyde[273]

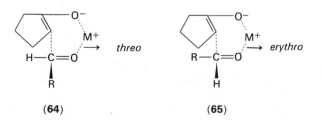

           (64)   (65)

(Table 11). This can be explained on the basis of the transition states **64** and **65**, which lead to the *threo* and *erythro* isomers, respectively. The steric interaction between R and the ring methylenes is less in **64** than in **65**, and hence the *threo* isomer is the kinetically formed product. It would be expected that a solvent of high dielectric constant would decrease the association of the enolate anion with the metal cation. This has been demonstrated by Dubois[273], who showed that in methanol, the reaction of isobutyraldehyde and cyclopentanone led preferentially (68%) to the thermodynamically more stable *erythro* form of 2-(1-hydroxy-2-methyl-propyl)cyclopentanone. In the presence of the cation $\overset{+}{N}Me_4$, the *erythro* form was the major product in both methanol and cyclopentanone, since dissociation of the enolate is expected to be extensive. Miller and his coworkers[275] have examined the effect of cation, concentration and solvent upon the geometry of reaction of enolate anions.

In a generalized situation (Scheme 17), and where the reaction is essentially irreversible ($k_T \gg k_{-T}$ and $k_E \gg k_{-E}$), a measure of the kinetic stereoselectivity, S_{ki}, is given by the *threo*:*erythro* concentration ratio[276,277] ($S_{ki} = [T]/[E]$). Where the reaction is reversible, the thermodynamic and kinetic stereoselectivities, S_{th} and S_{ki}, correspond to the stereochemical composition at equilibrium and at the extrapolated zero reaction time, respectively (equations 40, 41).

$$\text{Ketone + aldehyde} \underset{k_{-T}}{\overset{k_T}{\rightleftharpoons}} \textit{threo}\text{-ketol (T)}$$

$$\overset{k_E}{\searrow}$$

$$\textit{erythro}\text{-ketol (E)}$$

SCHEME 17.

TABLE 11. Percentage of *threo*-isomer from the cyclopentanone–isobutyraldehyde condensation at $-20\,°C$[273]

Solvent	Approximate dielectric constant at $-20\,°C$	Cation	% *threo*
Cyclopentanone	15	Li^+	> 95
		K^+	> 95
		$\overset{+}{N}Me_4$	30 ± 10
Methanol	40	Li^+	30 ± 10
		K^+	30 ± 10
		$\overset{+}{N}Me_4$	30 ± 10

$$S_{ki} = \lim_{t \to 0} \frac{[T]}{[E]} = k_T/k_E \tag{40}$$

$$S_{th} = \lim_{t \to \infty} \frac{[T]}{[E]} = (k_T/k_{-T})/(k_E/k_{-E}) \tag{41}$$

In the reaction of 2,2,5-trimethylcyclopentanone with aliphatic alde-hydes in a weakly polar solvent (THF: MeOH = 9:1), the thermodyna-mically less stable *threo*-ketol is formed predominantly at the beginning of the reaction[276]. A plot of ketol concentration against time (Figure 8) shows the increase and decline in the *threo* form. In a polar solvent (MeOH), the thermodynamically more stable form always predominates.

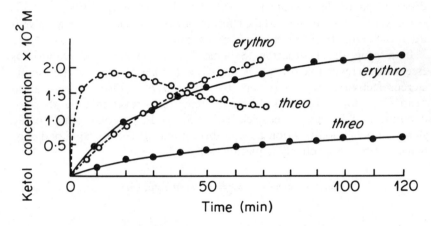

FIGURE 8. The effect of solvent ($\cdots \bigcirc \cdots$, THF/MeOH(90:10);—●—, MeOH) on ketol concentration with reaction time for the condensation of 3-methylbutanal with 2,2,5-trimethylcyclopentanone.

The experimental curves show excellent agreement with the calculated curves from a mathematical model based on equations 42 and 43, where $k_{E'}$ and $k_{T'}$ are complex expressions including base concentration, equilibrium constant and absolute rate constant.

$$\frac{d[T]}{dt} = k_{T'}[\text{ketone}][\text{aldehyde}] - k_{-T'}[T] \tag{42}$$

$$\frac{d[E]}{dt} = k_{E'}[\text{ketone}][\text{aldehyde}] - k_{-E'}[E] \tag{43}$$

A new term, the restoring energy, E_r of the system has been invoked[278] in order to explain the effect whereby equilibrium is attained after an initial increase in the thermodynamically less stable form. The restoring energy is a function of solvent polarity and size of the cationic catalyst, and in fact represents the difference in the activation free energies of the reverse reactions (equation 44).

$$E_r = RT \log (k_{-T}/k_{-E}) = RT \log (S_{ki}/S_{th}) \tag{44}$$

In polar solvents, where the enolate salt is dissociated, the restoring energy is essentially zero. Intermediate effects are examined by varying the percentage of methanol in THF (Figure 9).

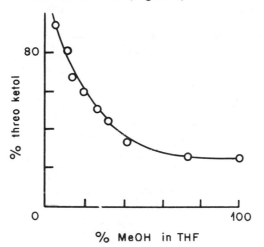

FIGURE 9. Influence of solvent composition on the stereochemical outcome of a condensation between 3,3-dimethylbutanal and 2,2,5-trimethylcyclopentanone at 20 °C with LiOMe as catalyst[278].

On the basis of the above results new nomenclature for the aldol condensation has been proposed[278]; (i) Ad_N2c corresponds to bimolecular nucleophilic addition via a cyclic transition state as found in non-polar solvents, and (ii) Ad_N2o represents the bimolecular addition via an open chain transition state found in polar solvents.

c. *Chelation effects.* House and his coworkers[279] have used metal salts to control the aldol condensation by trapping the intermediate keto alkoxide as a metal chelate (equation 45). In this way an unfavourable equilibrium can be displaced towards products, and side reactions such as polycondensation and dehydration are avoided. Thus when $ZnCl_2$ is added to an ethereal solution of a lithium enolate, subsequent addition of an aldehyde leads to a single aldol product in high yield.

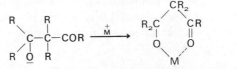

$$\tag{45}$$

A directed aldol condensation has been introduced by Wittig[280] (Scheme 18). The lithio derivative (**66**) of the imine adds to a carbonyl compound with formation of a chelate of the imino alkoxide (**67**). As with the preceding reaction, chelation offers a strong driving force for the process. Hydrolysis then affords the aldol product.

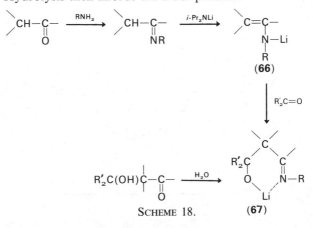

SCHEME 18.   (**67**)

2. Knoevenagel condensation

a. *General comments.* The Knoevenagel condensation[265] is the name given to the reaction between compounds with an acidic methylene group and an aldehyde or ketone. The acidity is conferred on the methylene group by electron-withdrawing groups, for example $CO_2C_2H_5$, NO_2, CN, which stabilize the anionic conjugate base. Weak bases such as amines and their salts are used as catalyst. The presence of a catalytic amount of a carboxylic acid is also important. The reaction is forced to completion by dehydration of the intermediate aldol product and removal of water from the system.

In some cases the amine may act as a nucleophilic catalyst rather than a simple base and an imine or iminium salt is formed as an intermediate[265]. Consequently, primary or secondary amines are often more effective catalysts than tertiary amines. Ionization, addition, or dehydration may be the rate-limiting process.

b. *Solvent effects on anion nucleophilicity.* The equilibrium constants

$$\begin{array}{c} O_2N \\ \diagdown \\ \diagup \quad CH-CH_3 \\ O_2N \end{array}$$

(68)

$$\begin{array}{c} O_2N \\ \diagdown \\ \diagup \quad CH-CH_3 \\ H_3CO_2C \end{array}$$

(69)

and rate constants for nucleophilic addition of the anions from 1,1-dinitroethane (68) and methyl α-nitropropionate (69) to acetaldehyde have been used to characterize the carbon basicities and nucleophilicities of these anions[281–283]. These are found to be highly solvent dependent. For example, in an aprotic medium (DMSO) the more basic anion (α-nitropropionate) was found to react more rapidly than the less basic dinitroethane anion. However, changing to a protic solvent (water) caused a levelling of the basicities and an inversion in nucleophilicity. This result can be attributed to solvation of the anion by hydrogen bonding in protic media[281], and correlates with an earlier observation that in aqueous solvents, the dinitromethane anion abstracts a proton from $H_3\overset{+}{O}$ faster than does the nitromethane anion, although the latter is a 10^6 times stronger base[284].

c. *Kinetic data.* In the addition of nitroalkanes to acetaldehyde (equation 46) the pseudo first-order rate constant, k_{obs}, in a buffer of given composi-

$$\begin{array}{c} O_2N \\ \diagdown \\ \diagup \quad CH-CH_3 + CH_3CHO \rightleftharpoons \\ X \end{array} \quad \begin{array}{c} O_2N \;\; OH \\ | \quad\quad | \\ X-C-CH \\ | \quad\quad | \\ H_3C \;\; CH_3 \end{array} \quad\quad (46)$$

$(X = NO_2, H)$

tion is directly proportional to the acetaldehyde concentration[283]. The second-order rate constant, k''_{obs}, calculated from the slope of a plot of k_{obs} against acetaldehyde concentration, was dependent on the buffer capacity and buffer component ratio. A plot of k''_{obs} against concentration of the buffer component ([HB], [B$^-$]), when their ratio was constant, is linear. The acidic component of the buffer was shown to be kinetically active. The authors claim this represents the first example of general acid catalysis in a carbanion addition to a carbonyl group. A Brönsted coefficient of 0·2 for this catalysis is indicative of a reactant-like transition state.

The importance of base catalysis in the reaction of malononitrile and 3-methylcyclohexanone is shown by the reduction in activation energy in the presence of aniline (11 kcal/mol to 7·6 kcal/mol)[285].

The potassium salt (70) is isolated in high yield when anthraldehyde is condensed with cyanoacetic acid in dimethylformamide (equation 47), and the reaction quenched with potassium hydroxide[286]. Extension of

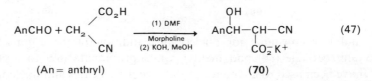

$$
\text{AnCHO} + \text{CH}_2 \begin{array}{c} \diagup \text{CO}_2\text{H} \\ \diagdown \text{CN} \end{array} \xrightarrow[\substack{\text{Morpholine} \\ (2)\ \text{KOH, MeOH}}]{(1)\ \text{DMF}} \text{AnCH}-\underset{\underset{\text{CO}_2^-\text{K}^+}{|}}{\text{CH}}-\text{CN} \qquad (47)
$$

(An = anthryl)   (70)

this reaction to other systems should allow more definitive mechanistic studies of the Knoevenagel condensation without the complication of dehydration.

3. Darzens condensation

The Darzens condensation involves the reaction of an α-halo ester with carbonyl compounds in the presence of a strong base (Scheme 19).

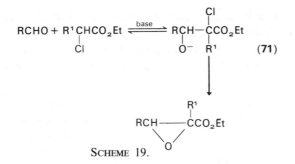

Scheme 19.

Either the aldol condensation or the ring closure may be rate determining[287–289]. Where the ring closure is rate determining, retro-aldolization and hence equilibration of the aldol intermediates (71) may occur. The *erythro* intermediate leads to the *trans* (E)-epoxide and the *threo* intermediate to the *cis*(Z)-epoxide. If the initial aldol condensation is rate determining and the ring closure is rapid relative to the equilibration of the aldol intermediate, the stereochemistry of the glycidate is determined by that of the aldol condensation product[288,289,290]. The structure of the reagents and solvent effects determine which step is rate limiting. For example, in the reaction of ethyl chloroacetate and benzaldehyde using benzene–sodium hydride or ethanol–sodium ethoxide, the reaction is stereoselective, but in the dipolar aprotic solvent, hexamethylphosphoramide (HMPA), the reaction is non-stereoselective (Table 12)[289,291]. HMPA increases the rate of cyclization and the aldol condensation

TABLE 12. Effect of solvent and base on the stereoselectivity of the Darzens condensation.

Solvent	Base	*trans*-Glycidate	*cis*-Glycidate
Benzene	NaH	90	10
Ethanol	NaOEt	90	10
Hexane	NaH	90	10
HMPA	NaH	50	50

becomes rate determining. In HMPA each halohydrin gave the corresponding glycidate, whereas in benzene or ethanol the *erythro*-halohydrin gave the *trans*-glycidate, but the *threo* isomer, in addition to giving the *cis*-glycidate, partly reverted to starting materials. The retro-aldolization was demonstrated by incorporation of C_6D_5CHO into the product when added to the isolated *threo*-halohydrin. This data can be explained in

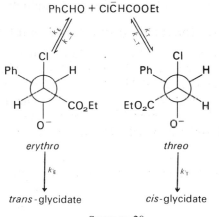

SCHEME 20.

terms of the following rate constants for the reaction (Scheme 20). In HMPA, $k'_E > k'_T \gg k_{-E}$ and k_{-T}; $k_E = k_T$. In ethanol or benzene $k'_E > k_{-E}$ and $k'_E > k_{-T} > k'_T$. Steric or electronic factors in the *erythro* intermediate may govern the more rapid glycidate formation from this isomer. The anion of chloroacetonitrile exhibits no selectivity in competition reactions with various aromatic aldehydes[292]. The slow step is the irreversible formation of the carbanion, followed by a fast aldolization. The transition state is regarded as close to reactants and the reaction as non-stereoselective (Scheme 21). The reaction is irreversible in HMPA,

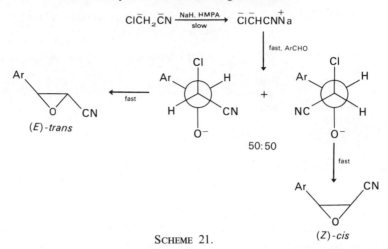

SCHEME 21. (Z)-*cis*

allowing a study of the aldol reaction to be made, since the ratio of the glycidates is representative of that of the aldolization process.

4. The Wittig condensation and related reactions

a. *Introductory comments.* The application of the Wittig reaction in formation of alkenes from carbonyl compounds is established as one of the most important methods available to the synthetic chemist. In recent years it has proved invaluable in several prostaglandin syntheses. The essential features of the reaction[293,294] are summarized in Scheme 22. The quaternary phosphonium salt (72), formed by reaction of a tertiary

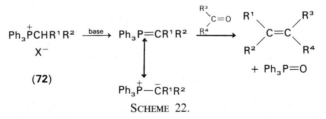

SCHEME 22.

phosphine with an alkyl halide, is treated with base to give a phosphorane which reacts with a carbonyl compound to give an olefin and triphenylphosphine oxide. The mechanism of this step is usually written as involving the formation of a betaine (74), which decomposes via an oxaphosphetane (75) (Scheme 23).

For ylids (73) in which the electron density on the carbanion is effectively delocalized, that is stabilized ylids, the first step is a slow reversible betaine formation followed by a rapid decomposition to products.

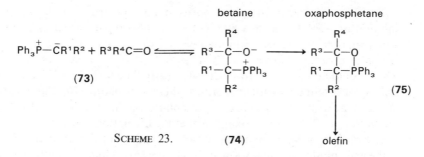

SCHEME 23.   (74)   olefin

With non-stabilized ylids a rapid, reversible first step is followed by a slow decomposition to products, although Schlosser and Christmann[295] have suggested that olefin formation is rapid if the reaction is conducted in a medium in which salts have been rigorously excluded.

b. *Nature of the intermediate.* Several betaines formed from non-stabilized ylids have been isolated[293,296,297], generally as their hydro-bromides or lithium bromide adducts. These methods do not distinguish conclusively between a betaine and an oxaphosphetane intermediate, as the latter could open during the isolation process. Vedejs and Snoble[298] have suggested that the oxaphosphetane may be formed directly from carbonyl compounds and non-stabilized ylids, without the intervention of an ionic betaine intermediate. Evidence for this proposal is based on a study of the reaction of ethylidenetriphenylphosphorane (a non-stabilized ylid) with carbonyl compounds (e.g., cyclohexanone) as illustrated in Scheme 24.

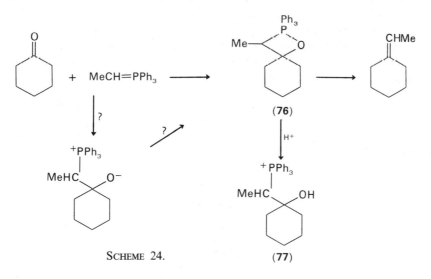

SCHEME 24.   (77)

The Fourier transform phosphorus n.m.r. spectrum of the reaction solution at $-70\,°C$ shows a singlet at δ_{31P} 66·5 p.p.m. which is resolved into a quartet of doublets when subjected to selective decoupling of the aromatic protons. Above $-15\,°C$ a shift to δ-26·1 p.p.m. was observed which could be attributed to $Ph_3P{=}O \cdot LiBr$. The high field ^{31}P chemical shift is consistent only with an oxaphosphetane structure (**76**). The addition of acetic acid at $-70\,°C$ gave a precipitate of the β-hydroxyalkylphosphonium salt (**77**). A direct oxaphosphetane formation can be rationalized in terms of an orthogonal approach of the ylid and the carbonyl π-bonds in the least hindered orientation. A π(2s) + π(2a) cycloaddition then would give the most hindered phosphetane, decomposition of which would lead to a *cis*-olefin (Scheme 25).

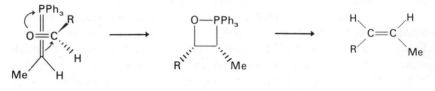

SCHEME 25.

This mechanism accounts for the *cis* stereoselectivity observed in Wittig reactions between hindered aldehydes and non-stabilized ylids (see Section IV.B.4.d). Ramirez and his coworkers[299] have developed an indirect synthesis of an oxaphosphetane (Scheme 26). The 1,3,2-dioxaphospholane (**78**) derived from trimethylphosphine and hexafluoroacetone is converted into the 1,2-oxaphosphetane (**79**) when heated at $80\,°C$ in benzene. Further heating causes decomposition to olefin and a phosphinate ester. In a subsequent paper[300], the structure of the oxaphosphetane (**80**) was confirmed by X-ray analysis.

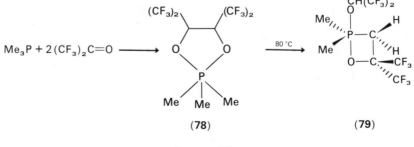

$$Me_3P + 2\,(CF_3)_2C{=}O$$

(78)   (79)

SCHEME 26.

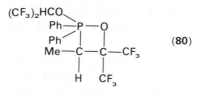

$$(80)$$

In the reaction of stabilized ylids in alcoholic solution, Schweizer and his coworkers[301] have demonstrated that a vinyl phosphonium salt (**81**) rather than an oxaphosphetane is an intermediate (Scheme 27).

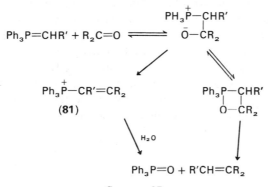

SCHEME 27.

The vinyl phosphonium salt **82** was isolated by acidification of the reaction product from salicylaldehyde and the ylid from allyltriphenylphosphonium bromide in ethanol (equation 48). This reaction would proceed with inversion of configuration at phosphorus.

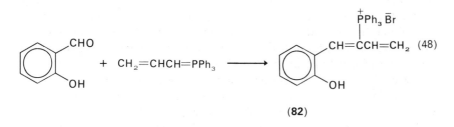

Trippett[302,303] has shown that the above mechanism is not exclusive in the reactions of stabilized ylids in ethanolic solution, since in the reaction of (+)-benzylethylmethylphenylphosphonium iodide and benzaldehyde with ethanolic sodium ethoxide, olefin synthesis occurs with retention of configuration at phosphorus. Trippett[302] suggests that reaction via the oxaphosphetane or vinylphosphonium salt may be competitive.

A CNDO-MO(complete neglect of differential overlap–molecular orbital) investigation of the Wittig reaction and the related Petersen reaction[304] has been made by Carey and his coworkers[305] in an attempt to deduce whether the betaine intermediate decomposes in a concerted manner or in a stepwise process via a four-membered ring intermediate. Plotting energy diagrams based on CNDO-MO calculations suggest that in the Wittig reaction, a rapid, reversible dihydrooxaphosphetane formation is followed by a fragmentation in which cleavage of the P—C bond is advanced in comparison to that of the C—O bond. In the Petersen reaction (equation 49), in which the base catalysed decomposition of a β-hydroxysilane yields an olefin, it is shown from similar calculations that the cleavage of the Si—C bond is more progressed than the C—O bond cleavage. Carey also calculated that in the Wittig reaction, the dihydro-oxaphosphetane is 66–89 kcal/mol more stable than the corresponding betaine, whilst the dihydrooxasiletanide anion is only 18–25 kcal/mol more stable than its betaine and may, therefore, not be a true intermediate[305].

$$\text{OH} \quad \text{SiR}_3 \qquad \longrightarrow \qquad \qquad \qquad (49)$$

c. *Solvent effects.* In a study of solvent effects on the reaction between carboethoxymethylene triphenylphosphorane and benzaldehyde, Rüchardt and coworkers observed a rate decrease in changing the solvent from benzene to the more polar aprotic media (DMF, DMSO)[306]. A similar rate decrease was reported by Frøyen[307] for the reaction of trialkyl-phosphine fluorenylide and triphenylphosphine fluorenylide with *p*-nitrobenzaldehyde. The rate decrease is not compatible with the reversible formation of a betaine in the rate-determining step of the reaction, but can be explained if the transition state is of a less polar character, and may therefore indicate a concerted formation of the P—O bond along with the C—C bond, in agreement with Vedejs's proposal[298].

d. *Stereochemistry and salt effects* Stabilized ylids react with aldehydes to give almost exclusively the *trans*-alkene. This result is explained in terms of preferred formation and decomposition of the less sterically hindered *threo*-betaine (Scheme 28)[293]. With non-stabilized ylids, the thermodynamically less stable *cis*-alkene, formed via the *erythro*-betaine often predominates[295,308]. Changes in solvent or the addition of salts are known to influence the stereochemistry of the Wittig reaction[295,309]. In an investigation of the reaction of a non-stabilized ylid (**83**), and a semi-stabilized ylid (**84**), with aldehydes, Bergelson, Barsukov and Shemyakim[310] found the reaction of **83** in benzene, in the complete absence of salts, yielded exclusively the *cis*-alkenes. Under the same conditions the semi-stabilized ylid (**84**) gave *trans*-alkenes. In the presence of LiI an increased amount of *trans*-alkene was formed from the non-stabilized

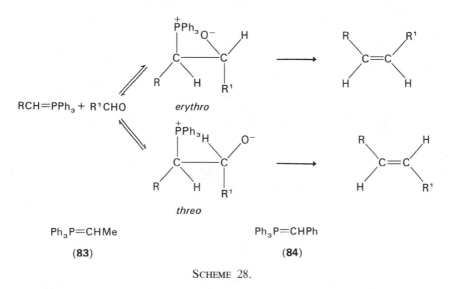

$$RCH{=}PPh_3 + R^1CHO$$

erythro

threo

Ph$_3$P=CHMe   Ph$_3$P=CHPh

(**83**)   (**84**)

Scheme 28.

ylid, and more *cis*-alkene from the stabilized ylid. Thus, in a salt free medium, both reactions are subject to kinetic control and betaine formation is rate determining. In the presence of LiI, stabilization of the *erythro*-betaine derived from the semi-stabilized ylid, promotes formation of the *cis*-alkene. Increased *trans*-alkene from the reaction of the non-stabilized ylid may be explained by coordination of LiI and the betaine, reducing its rate of decomposition to such an extent that this step becomes rate determining and the reaction becomes subject to thermodynamic control. In DMF the reaction is independent of the presence of salts. The ylid **83** then gives predominantly *cis*-alkenes, and the reaction of the semi-stabilized ylid **84** is non-stereoselective. The nature of the added lithium halide (i.e., LiCl, LiBr, LiI) also affects the stereochemical outcome of the reaction, and reflects the coordinating ability of the halide (Table 13).

Schneider[311] has suggested the formation of a trigonal–bipyramidal coordination complex **85** from the ylid and aldehyde to explain the preferential formation of the *erythro*-betaine from non-stabilized phos-phoranes in salt-free media. To form the *erythro*-oxaphosphetane inter-mediate **86**, a small anti-clockwise rotation about the C—O bond is necessary. In this configuration there is steric hindrance between R^2 and a phenyl group, but in the alternative *threo*-oxaphosphetane (**87**) both R^1 and R^2 are close to adjacent phenyl groups. This explanation also accounts for the increased *cis*-stereoselectivity as the size of R^1 increases[312].

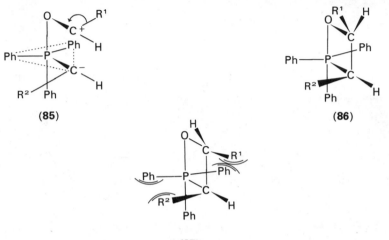

(85)   (86)

(87)

e. *Modified Wittig reactions.* The mechanism and scope of the phos-phonate modification of the Wittig reaction (also called the Horner or Wadsworth–Emmons reaction) has recently been reviewed[313] and will not be discussed further here.

α-Lithio derivatives of phosphonic acid bisamides have been used in olefin synthesis[314–316] (Scheme 29). The intermediate β-hydroxyphos-

TABLE 13. Effect of added salt on the *cis/trans* alkene ratio for the reaction of $Ph_3P{=}CHMe$ with PhCHO in benzene[310]

Added salt	% *cis*-Alkene	% *trans*-Alkene
None	87	13
LiCl	81	19
LiBr	61	39
LiI	58	42

phonamides are crystalline compounds and may be purified prior to thermal decomposition to alkenes.

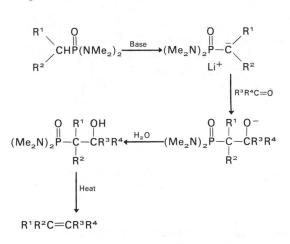

SCHEME 29.

β-Sultines (**88**) have been demonstrated to be intermediates in the sulphur analogue of the Wittig reaction[317] (Scheme 30). The extrusion of sulphur dioxide from the β-sultine is stereospecific. The reaction of

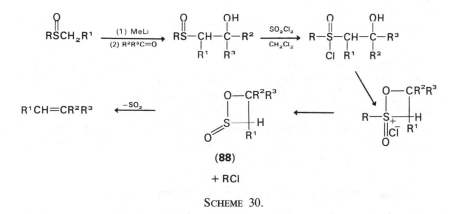

SCHEME 30.

iminophosphoranes with carbonyl compounds has been investigated[318] (Scheme 31). In the reaction of *p*-nitrobenzaldehyde a concave Hammett plot for variation of the *N*-phenyl substituent was obtained. This indicates a change in the rate-determining step with $k_1 \simeq k_2$, the relative magnitude depending on the substituent X. This observation was confirmed with other substituted benzaldehydes. Where X is an electron-donating group,

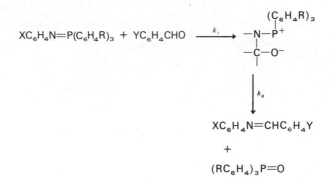

SCHEME 31.

the nucleophilicity of the nitrogen is increased (k_1 increases), but a concurrent increase in the electron density on phosphorus decreases its susceptibility to oxyanion attack (k_2 decreases). The reverse concept applies to the effect of electron-withdrawing groups (i.e., k_1 decreases, k_2 increases). The introduction of a methyl group (X = Me) is sufficient to cause betaine decomposition to become rate determining (k_1 increases) and indicates an efficient transmission of electronic effect through a phenyl ring to nitrogen.

Arsonium ylids have properties intermediate to those of sulphur and phosphorus. The reaction of the stabilized ylid, $Ph_3\overset{+}{As}-\overset{-}{C}HCOPh$ with aldehydes yields the expected olefins[319], but with α,β-unsaturated ketones, a Michael addition leading to cyclopropanes occurs. Other stabilized ylids, such as phenacylid[320], carbomethoxymethylid[320], fluorenylid[321], and cyclopentadienylid[322] afford olefinic products on reaction with carbonyl compounds. Non-stabilized arsonium ylids, such as methylid[323] and ethylid[324] yield mainly epoxides. Semi-stabilized ylids, e.g. benzylid, give mixtures of olefin and epoxide.

C. Organometallic Additions to Carbonyl Compounds

Within the scope of this section are the Grignard reactions, alkyl lithium, alkyl cadmium, and alkyl zinc addition reactions. The former constitute the major part of the section. The stereochemistry of additions to carbonyl groups is discussed in a subsequent section on reduction (see Section IV.D.1.c).

1. The Grignard reaction

Since the chapter by Eicher in this series[325], a comprehensive review of the Grignard reaction has appeared[326]. The addition of Grignard reagents to carbonyl compounds may result in a normal 1,2-addition, a two or one electron reduction, and, in the case of conjugated carbonyl compounds, a 1,4-addition. Enolization of the carbonyl group may also compete which leads to recovered starting material. Reduction may become competitive with sterically hindered ketones. To appreciate the mechanism, an understanding of the structure of the Grignard reagent is necessary.

a. *Structure of the reagent.* For many years, the Schlenk equilibria[327] (Scheme 32) were accepted as a simple explanation of the nature of Grignard reagents in ethereal solvents. Ashby and his coworkers demonstrated

$$2RMgX \rightleftharpoons R_2Mg + MgX_2$$

$$R_2Mg + MgX_2 \rightleftharpoons R_2Mg \cdot MgX_2$$

SCHEME 32.

the existence of the species $RMgX$ in tetrahydrofuran[328] and a concentration effect on structure was also shown[329]. A modified set of equilibria were then suggested[329] (Scheme 33). The ionic nature of the $C-Mg$ bond

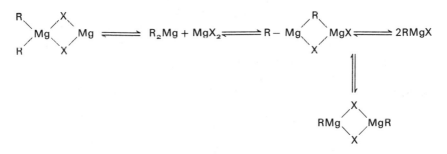

SCHEME 33.

has been established by n.m.r. studies[330]. The degree of association of Grignard reagents in ethereal solvents may vary from unity in dilute solutions to four in more concentrated solution[331,332]. The equilibrium association in tetrahydrofuran may be represented by Scheme 34[333]. The kinetics of the reaction of 4-methylmercaptoacetophenone with cyclopentylmagnesium reagents has been studied by stop flow u.v. and

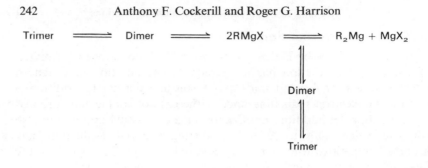

SCHEME 34.

i.r. spectroscopy[334]. The data indicate that the reaction is first order in organomagnesium in dilute solution, but zero order at high concentration. It is suggested that both $RMgX$ and R_2Mg contribute to the reaction, the relative contribution being dependent on the concentration of magnesium bromide. Evidence that the dialkyl magnesium species is more reactive than the Grignard reagent has been presented[335,336].

Methylmagnesium fluoride has recently been prepared[337] and shown to be dimeric in tetrahydrofuran, in contrast to the other methylmagnesium halides.

n-Alkyl- and aryl-magnesium halides have been prepared in hydrocarbon solvents in place of the conventional ethers[338]. The properties of the unsolvated and solvated reagents are generally comparable in the reaction with carbonyl compounds. Aromatic hydrocarbons may also be used as solvents[339]. Hexamethylphosphoramide (HMPA) is believed to enhance the carbanionic character of the Grignard reagent[340].

b. *1,2-Addition*. The 1,2-addition reactions of organometallic compounds with ketones may be discussed in terms of three mechanisms. The first step of all these involves a fast equilibrium formation of a carbonyl–metal complex (equation 50)[341,342]. Subsequent product forma-

$$R_2C{=}O + CH_3M \underset{}{\overset{fast}{\rightleftharpoons}} R_2C{=}O{\cdots}MCH_3 \tag{50}$$

tion may then occur in one of three ways (equations 51, 52, 53). Support

$$R_2C{=}O{\cdots}MCH_3 \longrightarrow R_2\overset{\bullet}{C}{-}OM + \overset{\bullet}{C}H_3 \longrightarrow \underset{\overset{|}{CH_3}}{R_2C{-}OM}$$

(53)

for mechanism (52) has been given by several authors[343,344,345] whilst evidence for mechanism (51) has also been presented[335a,335b,336b,346–348]. Mechanism (53) is not accepted as a major pathway[348], though if Grignard reagents capable of a facile electron transfer are used, for example t-butyl magnesium bromide, this mechanism may contribute significantly[349].

Allyl Grignard reagents exhibit a greater propensity for axial attack on 4-t-butylcyclohexanone than does n-propyl magnesium bromide[350,351]. The allyl reagent yields 52% of the equatorial alcohol, whereas n-propyl magnesium bromide yields 26%. The difference in stereochemical outcome may be attributable to a change from a four-centre transition state for the n-propyl reagent to a six-centre transition state (**89**) for the allyl reagent.

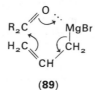

(**89**)

Since this transition state is presumably more flexible and of lower energy than the four-membered case, Ashby[352] has suggested that cyclo-hexanone may exist in the boat form in the transition state. Attack would then take place at the side opposite to the 'flagpole' hydrogen, and on reversion to the chair form the alkyl group would be in the axial position. An alternative S_E2' like mechanism (Scheme 35), involving antarafacial attack by the carbonyl C atom on the allyl reagent is preferred by other authors[353,354]. Mechanisms proceeding via an initial one-electron transfer[349,355,356] leading to a caged radical–radical anion pair, may also be envisaged. The reaction of t-butylallyl magnesium bromide with carbonyl compounds has been used to assess steric hindrance in the neighbourhood of the carbonyl group (Scheme 35, R = t-Bu)[357]. Thus the ratio of (**91**) and (**93**) depends on the steric environment of the carbonyl group; that is, the transition state **90** is more sterically crowded than **92** when R is large.

2-α-Hydroxyalkyl (or aryl) cyclopentanones and cyclohexanones react with Grignard reagents to yield pure diols[358]. cis-2-Acylcyclopentanols and $trans$-2-acylcyclohexanols also undergo stereospecific additions. In

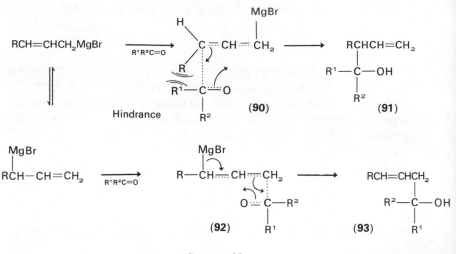

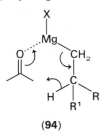

SCHEME 35.

contrast *trans*-2-acylcyclopentanols react in poor yield and low stereo-selectivity. These data can be correlated with the ability of the substrates to form chelates.

c. *Reduction.* The two electron reduction by Grignard reagents in a generalized case is shown in Scheme 36. Reduction may compete with

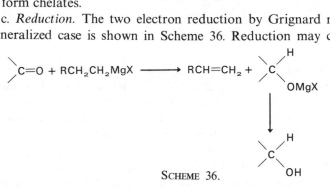

SCHEME 36.

addition when the substrate is sterically hindered. A cyclic mechanism (**94**) involving transfer of the β-hydrogen has been confirmed by deuterium

(**94**)

labelling studies[359]. The reaction is the basis of several asymmetric reductions with Grignard reagents[360]. The preferred transition states for an asymmetric Grignard reduction may be represented by **95**. The larger the difference in steric size between R_L and R_S the greater would be the

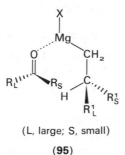

(L, large; S, small)

(95)

expected asymmetric induction. In many cases the consideration of the model transition state (**95**) allows a prediction of the configuration of the product. However, in some reactions electronic factors may be important and the simple approach represented by **95** is no longer valid.

    d. *Enolization.* Under normal conditions, the enolization of the carbonyl group is not a competitive process. However, in hexamethylphosphoramide (HMPA), diisopropylketone is exclusively enolized by *n*-butylmagnesium bromide at the expense of the normal addition reaction (Scheme 37)[361]. Reaction of the enol with an alkyl halide leads to alkylation of the α-carbon atom in high yield.

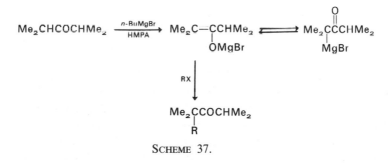

SCHEME 37.

2. Alkyl lithium reactions

    Buhler has shown that the highest yields of addition products of *n*-butyl lithium with carbonyl compounds are obtained by adding a hexane or ether solution of the substrate to the reagent at $-78\,^{\circ}\text{C}$[362]. Reduction or self-condensation of the carbonyl compound is not significant at this temperature. Some enolization occurs leading to recovered starting material after hydrolysis. With t-butyllithium more enolization and lower alcohol yields are found.

In a convenient preparative procedure, the carbonyl compound and alkyl halide are added to lithium pieces in tetrahydrofuran at $<0\,^{\circ}\text{C}$[363].

3. Organoaluminium, -zinc and -cadmium additions

These reagents will be discussed together and their reactions compared with Grignard reagents. Detailed studies on the nature of these reagents have not been made but dimethylzinc and dimethylcadmium are known to be monomeric in ether solution[364], and the stereochemistry of their addition to aldehydes has been discussed[365]. Jones and his colleagues have compared organomagnesium and organocadmium addition reactions[366]. The addition of methylmagnesium compounds (MeMgF, MeMgCl, MeMgBr, and Me_2Mg) and trimethylaluminium to 4-t-butylcyclohexanone in various solvents (hexane, benzene, diethyl ether, tetrahydrofuran, diphenyl ether, and triethylamine) has been investigated (equation 54)[352]. In ether and THF the major product is the axial alcohol

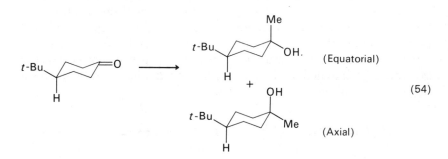

(54)

formed by an equatorial addition, and is independent of the ratio of the reactants. With trimethylaluminium in a hydrocarbon solvent, when reactant to substrate ratio was 1:1, the axial alcohol again predominated. However, when the ratio approaches 2:1, the stereochemistry of the product was reversed and the equatorial alcohol was the major product ($c.\ 90\%$). At a 1:1 ratio, a four-centre transition state was proposed and the reaction was first order in $(CH_3)_3Al$, whereas at a 2:1 ratio a six-centre transition state was preferred, the reaction being second order in $(CH_3)_3Al$[342,352,367]. Steric compression of the 2 and 6 equatorial hydrogens of cyclohexanones, and with the 2-methyl group of 2-methylcyclopentanones may be envisaged to destabilize the six-centre transition state leading to the axial alcohol. No such compression is present with a four-centre transition state[368]. Alternatively conformational changes in the complexed ketone have been invoked to explain the change in stereochemistry. However, a study of the n.m.r. spectrum of the reaction of 4-*t*-butylcyclohexanone[368] offered no evidence for the postulated change from the ketone chair form to a half-chair for ketone–aluminium alkyl complex.

In the reduction of isopropyl phenyl ketone by optically active dialkyl zinc compounds the stereoselectivity is dependent on the distance of the chiral centre from the metal atom[369].

4. The Reformatsky reaction

a. *Introductory remarks.* The Reformatsky reaction has been reviewed in 1972[370] and consequently this section will discuss only the mechanistic consideration about which there is still some disagreement.

A modification of the preparative procedure of the Reformatsky reaction, in which trimethyl borate is used to provide a mildly acidic medium, allows the reaction to proceed at room temperature with high yields[371].

In the Reformatsky reaction an α-halo-ester is reduced by zinc to form a halo-zinc enolate. This reacts with an aldehyde or ketone to give, after hydrolysis, a β-hydroxy ester (Scheme 38).

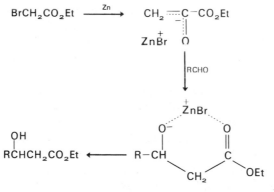

SCHEME 38.

b. *Stereochemical studies.* Recently particular interest has centred on whether the formation of diastereoisomeric β-hydroxy esters is under kinetic or thermodynamic control[370,372–378]. The reaction of various Reformatsky reagents with benzaldehyde in benzene (Scheme 39) has been studied by Canceill and his coworkers[372], who suggested that formation of the intermediates **96** and **97** was practically irreversible.

However, interesting solvent effects are found in the reaction between benzaldehyde and methyl α-bromophenylacetate[375]. In dimethoxyethane,

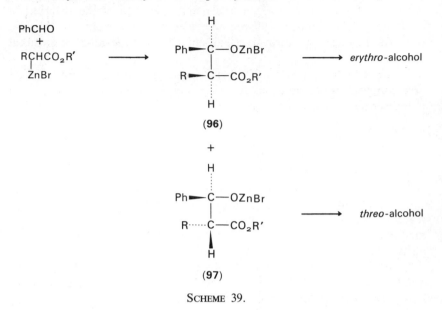

SCHEME 39.

tetrahydrofuran, and dioxan the proportion of the *threo* isomer in the product increased with time. In benzene and dimethoxymethane the reaction was not stereoselective, and in dimethylsulphoxide the *erythro* to *threo* product ratio was independent of time. When the pure *erythro* product was heated at 45 °C for 1 h with $BrZnCH_2CO_2Et$ in benzene, 2·9% of methylphenyl acetate was recovered, whilst in dimethoxyethane 14% of the starting material was recovered. This indicates some reversibility in the reaction although in benzene it is relatively small. Also whilst discussing the question of reversibility of the Reformatsky reaction it is of interest to consider the addition of the reagents derived from methyl α-bromoisovalerate and methyl *t*-butylacetate to benzalaniline at −18 °C which gives the *erythro* amino esters (**98**) (Scheme 40)[376]. At reflux in ether/benzene, the *cis*- and *trans*-β-lactams (**99** and **100**) are formed. Three explanations for this observation are possible: (i) epimerization of **99**; (ii) that the reaction is irreversible and of different stereoselectivity to the low temperature reaction; and (iii) the reaction is reversible but the position of equilibrium is highly temperature dependent.

No epimerization of the isolated β-lactams was observed and when optically active methyl α-bromopropionate reacted with benzalaniline in

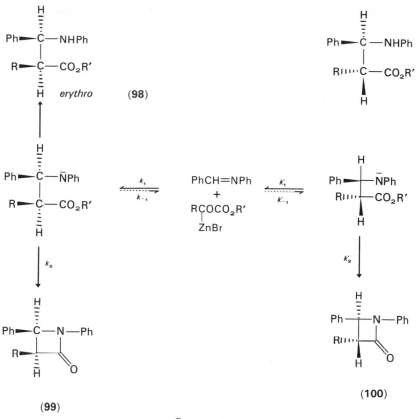

SCHEME 40.

boiling benzene, a *cis* optically active lactam and a racemic *trans* lactam were obtained. If (ii) were true, both lactams would be expected to be optically active, hence a reversible reaction is indicated. This reaction has been extended to other imines and different Reformatsky reagents[379]. A prolonged reaction time at low temperature favours formation of the amino ester (Table 14).

The proportion of the *erythro* hydroxyester formed in the reaction of ethyl α-bromopropionate and acetophenone in benzene increases with time until a constant *erythro* to *threo* product is found[378]. With p-methoxy- and p-chloroacetophenone analogous behaviour is observed. The rate of reaction increases with electron-releasing power of the substituent on the aromatic ring. After a reaction time of only 10 min, complete conversion into the *erythro*- and *threo* β-hydroxyesters had occurred in all the cases investigated, whereas equilibrium was attained only after times greater than 1 h.

TABLE 14. Temperature effect on product ratio in the reaction of benzalaniline and ethyl bromoacetate.

Temperature (°C)	% Ph-azetidinone	% Ph—CH—CH$_2$CO$_2$Et, Ph—NH
44	85	0
20	18	37
0	8	75
−10	0	85

D. Reduction of C=O

Wheeler's earlier chapter in this series[380] concerns catalytic reduction, reduction by dissolving metals, electrochemical reduction, reductive amination, reduction by organic reagents, and metal hydride reductions. In this section we limit our discussion to reduction by organic reagents and metal hydrides. Reductive amination involves hydrogenation of an imine intermediate, the mechanism of formation of which is discussed elsewhere (see Section V.A). Catalytic reduction (i.e., hydrogenation), reduction by dissolving metals, and electrochemical reductions are beyond the scope of this chapter.

1. Reduction by metal hydrides

A discussion of the scope and mechanism of hydride reductions is included in House's book[381]. In its simplest terms it may be regarded as nucleophilic attack of hydride ion at the carbonyl carbon leading to a metal alkoxide. The carbinol is then liberated by addition of acid, base or water.

a. *New reagents.* In the past decade considerable effort has been devoted to the development of new metal hydrides which will allow the stereoselective reduction of the carbonyl group in cyclic ketones. Several such reagents are now available and, although the mechanistic considerations are similar in all cases, some of these are mentioned briefly here. Lithium tri-s-butylborohydride[382] reduces ketones with high stereoselectivity. Hindered ketones such as 2-methylcyclohexanone, 3,3,5-trimethylcyclohexanone, and camphor give over 99·5% of the less stable

TABLE 15. Stereoselectivity of ketone reduction by
lithium-tri-s-butylborohydride at 0 °C[382]

Ketone	Major isomer (%)
2-Methylcyclopentanone	*cis* (98)
2-Methylcyclohexanone	*cis* (99·3)
3-Methylcyclohexanone	*trans* (85)
4-Methylcyclohexanone	*cis* (80·5)
4-*t*-Butylcyclohexanone	*cis* (93)
3,3,5-Trimethylcyclohexanone	*trans* (99·8)
Norcamphor	*endo* (99·6)
Camphor	*exo* (99·6)

epimer (Table 15). Potassium triisopropoxyborohydride[383], and lithium trimesitylborohydride bis(dimethoxyethane) also reduce cyclic ketones stereoselectively[384]. Tetrabutylammonium cyanoborohydride has been used to reduce selectively aldehydes to alcohols in the presence of ketones[385]. Another reagent, lithium perhydro-9b-boraphenalylhydride offers high stereoselectivity in reduction[386] and has recent application in prostaglandin synthesis[387,388]. In contrast to lithium trialkoxyalumino-hydrides[389] an increase in the size of alkyl substituents in boron reagents enhances the stereoselectivity of addition to the carbonyl group.

b. *Asymmetric reductions.* Červinka[390] first demonstrated asymmetric reduction of ketones using complexes derived from lithium aluminium hydride (LAH) and quinine, quinidine, cinchonine and ephedrine.

Further papers in which use of these reagents is described have appeared[391–393]. The reagent derived from LAH and dihydroxy-monosaccharides (e.g., 3-O-benzyl-1,2-cyclohexylidene-α-D-glucofuranose) also effect asymmetric reduction of ketones with high optical yields[394,395]. Similar results are obtained using LAH complexes with monohydroxy sugar derivatives[396].

A new class of chiral reducing agents (**101, 102**) have been reported by Giongo and his coworkers[397]. A high dependence of optical yield on

(**101**)   (**102**)

solvent, temperature and substrate was found. An aromatic group in the substrate leads to an increased optical yield as has previously been observed with chiral lithium aluminium alkoxides[391].

c. *Stereochemistry of reduction.* Most investigations of the stereochemistry of carbonyl reduction with metal hydrides have concerned the reaction of cycloalkanones and this will be the main topic of this section. However, a correlation linking the reduction of cyclic ketones and acyclic ketones has been proposed and is discussed. Several theories have been proposed to explain the stereochemistry of reduction and whilst some of these have since been disproved, there is as yet no general agreement.

(i) Cycloalkanones. In 1956 Dauben and his coworkers[398] suggested that the steric course of reduction of unhindered ketones with metal hydride reducing agents was governed by 'product development control'; that is, product-like transition states leading to the predominant formation of the thermodynamically more stable alcohol. A classic example is the reduction of 4-*t*-butylcyclohexanone which gives *trans*-4-*t*-butylcyclohexanol as the major product. However, with hindered ketones reduction was suggested to be governed by 'steric approach control', with a reactant-like transition state, where the reducing agent approaches from the less hindered plane of the carbonyl group. Modified terminology was proposed by Brown and Deck[389] who suggested that 'steric strain control' and 'product stability control' were more appropriate. An argument against product development control lies in the observation that in the reaction of 4-*t*-butylcyclohexanone and LAH, the product ratio (*c.* 10:1) differs from the equilibrium ratio of the epimeric alcohols[399] and also from the position of equilibrium of alkoxide derivatives of 4-*t*-butylcyclohexanols[400,401] (*c.* 4:1 in THF). This indicates that the equatorial alcohol is a kinetically formed product. An alternative explanation for the reduction of unhindered cyclic ketones places importance on steric hindrance to an equatorial approach caused by the axial hydrogens in the 2 and 6 positions[402,403]. However, replacement of the axial hydrogen at $C_{(2)}$ by a methyl group does not lead to the expected marked reduction in the rate of attack leading to the axial alcohol with LAH[404,405].

An attractive theory suggested by Felkin and Chérest[351] proposes that torsional strain between partial bonds is an important factor in nucleophilic addition to cyclic ketones. Hence, considering the transition states **103** and **104** the steric outcome of reaction will depend on the relative magnitude of strain in **103** and **104**. When Nu⁻ (nucleophile) and R are small, steric strain in **104** is minimal, whereas torsional strain in **103** is still significant. Reaction therefore proceeds via **104** leading to the equatorial alcohol. As the bulk of the substituent R increases steric strain

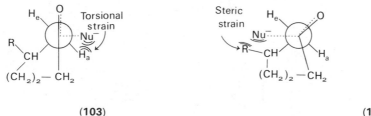

(103)   (104)

Equatorial attack ⟶ Axial   Axial attack ⟶ Equatorial
                    alcohol   alcohol

in **104** will become more severe and the proportion of axial alcohol in the product, obtained via the transition state **103**, will increase. Similarly, an increase in the effective size of the nucleophile will lead to an increase in the production of axial alcohol. In the reaction of Grignard reagents and unhindered ketones, the axial alcohol is the predominant product. Thus this theory dispenses with the concept of 'product development control' and considers a 'reactant-like' transition state in the reduction of both hindered and unhindered cyclic ketones.

Support for this theory has been presented by Chauvière and co-workers[406] and in the kinetic results of Eliel's group[404,405]. Evidence against this approach has been found in a study of the variation in stereo-chemistry of the reduction of 2 isopropylcyclohexanone with LAH as a function of temperature[407]. From the ratio of *cis* to *trans* products the enthalpy difference for the transition states **105** and **106** was calculated. The difference obtained was too large to be reconciled with a reactant-like

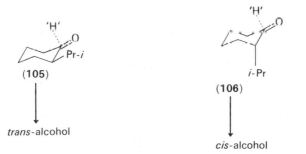

(105)   (106)

trans-alcohol   cis-alcohol

transition state for axial attack and, therefore, a transition state with product-like character was invoked. A similar conclusion has been reached in a study of equatorial attack on 3-ketosteroids by lithium-tri-*t*-butoxyaluminium hydride while varying the substituents at $C_{(5)}$ and $C_{(10)}$ (Scheme 41)[408]. The rate of the reaction was determined by optical rotatory dispersion (ORD). Both the rate constant for axial attack, k_a,

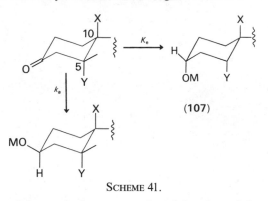

SCHEME 41.

and for equatorial attack, k_e, were reduced by the axial substituents at $C_{(5)}$, the latter markedly, indicating interactions between OM and Y in the transition state leading to **107**. This may be interpreted as indicating product-like character in the transition state for equatorial attack in this case.

Eclipsing of the C—OM bond, where M is the metal, with the equatorial C—H bonds in the 2 and 6 positions of cyclohexanones may also be important[409]. This eclipsing would occur only in the transition state leading to the axial alcohol and would explain the less ready formation of axial alcohols in the case of unhindered ketones.

Recently Klein[410] has proposed an electronic interpretation for the steric course of reductions of cyclohexanones, in which the nucleophile, e.g. hydride ion, attacks by interaction with the LVMO (lowest vacant molecular orbital) of the ketone, which will be easier from the axial direction, unless steric effects are predominant.

On the basis of kinetic data it has been proposed that an important factor in determining the position of the transition state along the reaction coordinate is the nature of the nucleophile[411]. The reaction constant ρ, for a series of nucleophiles is always > 0, indicating an increase in electron density on the central carbon atom as the transition state is approached. The greater the ρ value, the greater the charge variation, and the further the transition state is along the reaction coordinate. Small values of ρ, as found for $PhMgBr$, CH_3MgI, NH_2OH, and SO_3^{2-}, indicate a reactant-like transition state, whilst values for borohydride reduction and cyano-hydrin formation indicate a more product-like transition state (Table 16). Further, in a plot of the free energy of activation, ΔG^*, against the free energy of reaction, ΔG^0, a linear correlation often results. The slope of the line, α, $(\delta\Delta G^* = \alpha\delta\Delta G^0)$, indicates the position of the transition state along the reaction coordinate. Values of α shown in Table 17 again indicate a product-like transition state for cyanohydrin formation. Where the transition state is 'reactant-like', stereoelectronic factors in the substrate may be important in determining reactivity. Conversely when the transition state is 'product-like' steric effects will be important.

TABLE 16. Variation in the reaction constant, ρ, for the reaction of different nucleophiles with carbonyl compounds[411]

Reaction	ρ
$ArCOCH_3 + NaBH_4$	+3·06
$ArCOAr' + LiBH_4$	+2·81
$ArCHO + HCN$	+2·33
$ArCHO + SO_3^{2-}$	+1·27
$ArCOCH_3 + PhMgBr$	+0.41
$ArCOAr' + CH_3MgI$	+0·36
$ArCOCH_3 + NH_2OH$	+0.32

TABLE 17. α-values for the reaction of nucleophiles with carbonyl compounds

Nucleophile	α	Number of ketones
SO_3^{2-}	0·49	12
NH_2OH	0·56	7
$\bar{C}N$	0·74	17

Adamantanone (**108**) has been suggested[412] as a suitable model to deduce the position of the transition state in nucleophilic additions to carbonyl

axial side

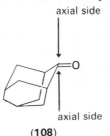

axial side

(**108**)

groups. Both faces of the carbonyl group are identical and are held rigidly (and are equivalent to the axial sides of cyclohexanone). The rate of nucleophilic addition will approximate to $2k_a$, where k_a is the rate constant for axial attack on cyclohexanone, for a 'reactant-like' transition state. For a 'product-like' transition state, $k \simeq 2k_e$. Good correlations for the additions of $\bar{B}H_4$ and $SO_3{}^{2-}$ have been obtained by this simple approach.

(ii) Acyclic ketones. In the consideration of the metal hydride reduction of carbonyl compounds with a chiral centre in the α-position, a prediction as to which diastereoisomer will predominate is given by 'Cram's rule'[413]. In the reduction shown in equation (55), the diastereoisomer **109** is the

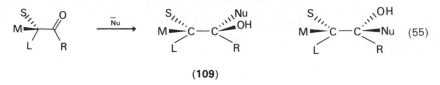

(55)

(109)

(S, M, L ≡ small, medium and large groups)

major product. This is explained by Cram on the basis of a preferred 'reactant-like' transition state **110**, where the incoming nucleophile attacks remote from L and M, and the steric strain between L and R ($\phi \sim 0°$) is less than between L and the metal complexed carbonyl group in the alternative transition state (**111**).

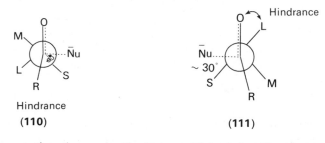

(110)   **(111)**

This interpretation has recently been criticized by Karabatsos[414], who suggests **110** is preferred because when $\phi \sim 30°$, it offers the smallest carbonyl-eclipsed group interactions. However, if either of these proposals were correct, an increase in the size of R, the group attached to the carbonyl carbon, would be expected to destabilize the transition state **110** in favour of **111**, and the reaction would be less stereoselective. Experimentally it has been demonstrated for two series of ketones (L = cyclohexyl or phenyl, M = methyl, S = H; R = Me, Et, i-Pr or t-Bu) that the reaction becomes more stereoselective as the size of R increases[415]. On the basis of these results it has been suggested[415] that the transition state is always reactant-like, that torsional strain involving partial bonds is important, and interactions between $\bar{N}u$ and R are significant steric considerations. In the absence of polar groups the preferred conformation of the transition state is represented by **112**.

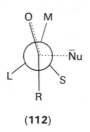

(112)

The stereoselectivity of the reduction of acyclic ketones should increase as L, R or $\overline{\text{Nu}}$ increase in bulk as this further destabilizes the alternative transition states in favour of 112.

An increase in length of the alkyl chain in n-alkyl methyl ketones causes a decrease in rate on changing from ethyl to propyl but further changes have little effect. The interaction of C_γ with oxygen is suggested to reduce the reactivity of the carbonyl group[416].

d. *Metal cation–carbonyl group association.* Although it has long been presumed that the reduction of a carbonyl compound involves association of the carbonyl oxygen atom with the metal cation prior to or concurrent with hydride transfer, no direct evidence for this was available. Ashby and his coworkers[417] have recently presented stereochemical results which indicate that complexation does occur in the reduction of ketones by complex metal hydrides. In the reduction of 2-methylcyclohexanone with LAH an increase in equatorial attack compared with 4-t-butylcyclo-hexanone is observed, which can be explained on the basis of theories

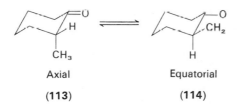

Axial	Equatorial
(113)	(114)

discussed previously. However, in going from LAH to $ClMgAlH_4$ and $Mg(AlH_4)_2$, a substantial increase in *cis*-alcohol is found in the reduction of 2-methylcyclohexanone, whereas no significant variation is observed in the stereochemical outcome of the reduction of 4-t-butylcyclohexanone. The decrease in axial attack on 114 cannot be explained by an increase in steric hindrance towards axial attack as $ClMgAlH_4$ and $Mg(AlH_4)_2$ both give more *exo* attack on camphor and more axial attack on 3,3,5-trimethylcyclohexanone than does LAH. The apparent increase in the

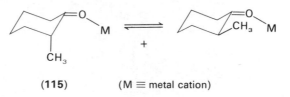

(115)   (M ≡ metal cation)

involvement of conformer **113** in the reduction may be explained if complexation of the carbonyl group does occur during reduction; that is, more reaction occurs via the transition state **115** with the bulky complexing agents $\overset{+}{M}$gCl and $Mg\overset{+}{A}lH_4$ than the smaller $\overset{+}{L}i$. When the conformation of the 2-alkyl substituent is fixed, e.g. 4-t-butyl-2-methylcyclohexanone, the stereochemical outcome is almost the same for all three reducing agents.

e. *Kinetic studies.* Rickborn and Wuesthoff[418] have reviewed the methods available for studying the kinetics of borohydride reduction of ketones. Discrepancies are found between results obtained from iodate titration and those from potentiometric and chromatographic methods. Spectrophotometric results, obtained by following the disappearance of the $n \rightarrow \pi^*$ ketone band at 285 nm, agree with those found by potentiometric and chromatographic methods[419].

Secondary kinetic deuterium isotope effects, k_H/k_D, for nucleophilic addition to ketones are dependent on both the structure of the ketone and the nucleophile[420,421,422].

2. Meerwein–Ponndorf–Verley (MPV) reduction

The overall reaction is shown in equation (56); isopropyl alcohol is often the solvent with aluminium isopropoxide as catalyst. The reaction

$$R_2C{=}O + MeCHOHMe \xrightarrow{\text{Al(OCHMe}_2)_3} R_2CHOH + MeCOMe \qquad (56)$$

is generally believed to involve hydride transfer in the rate-determining step[433,424]. However, in the reduction of α-phenylacetophenone to α-phenylethanol, the ketone disappears more rapidly than the alcohol is formed, indicating that the alcohol exchange step may be rate limiting[425].

Aluminium isopropoxide exists mainly as a tetramer[426], but the reactive species is believed to be a trimer[425,427]. Aluminium t-butoxide exists as a dimer[426]. Both cyclic (Scheme 42, **116**) and non-cyclic transition states have been proposed[428] for the reaction. Contrary to early beliefs, the MPV reduction is very rapid for unhindered ketones, and for cyclohexanone and methylcyclohexanone is essentially instantaneous[429].

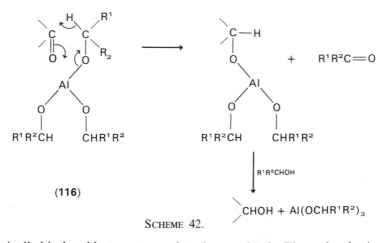

(116)

SCHEME 42.

Sterically hindered ketones are reduced more slowly. The ratio of epimeric alcohols from the reduction of cyclic ketones depends on the substrates and reagent concentration, which may be rationalized in terms of a concentration dependent association of i-PrOH with $(i$-PrO$)_3$Al. Evidence for an association is provided by the increase in rate found upon introduction of an inert solvent (benzene) to the system.

Further examples of asymmetric reduction using chiral alkoxides have been reported[430], but the optical yields are not high. The preferred transition state from steric considerations **117** enables a prediction of the

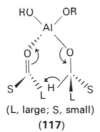

(L, large; S, small)
(**117**)

configuration of the products to be made.

E. Miscellaneous Carbonyl Group Additions

1. Hydration of the carbonyl function

This reaction represents the simplest type of reversible addition to the carbonyl group and may be simply written as equation (57). The reaction

$$R_2C=O + H_2O \rightleftharpoons R_2C(OH)_2 \qquad (57)$$

has been reviewed in 1966[431] and has been discussed in a previous volume of this series[432].

The half-life of the hydration process is short and a study of the mechanism requires techniques developed for studying fast reactions. Ultraviolet, [17]O-n.m.r., p.m.r., polarography, ^{18}O-exchange, chemical scavenging and chemical methods have all been employed. Equilibrium constants calculated by different methods are not always in agreement. The methods used have been reviewed[433].

a. *Mechanism.* The hydration of carbonyl compounds is general acid and general base catalysed and its mechanism is similar to that of the reversible acetal formation. A spontaneous hydration which is, by definition, independent of acid or base also occurs. Eigen[434] has suggested this latter reaction takes place via a cyclic hydrogen-bonded transition state in which two or more water molecules are involved. Values obtained for the entropy of activation for spontaneous hydration of acetaldehyde supports this interpretation[435]. Corroborative evidence is also derived from the kinetic order with respect to water dissolved in a non-aqueous solvent[436–438], and from a study of the isotope effects[438]. In a catalysed reaction one or more of the water molecules is replaced by the catalyst. A recent study[439] of the entropy of activation, over a range of temperatures, for the uncatalysed hydration of 1,3-dichloroacetone in aqueous dioxan suggests that a transition state **118** involving two water molecules in

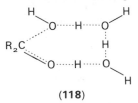

(118)

addition to the reactant molecule fits the data better than one involving one or three additional water molecules. The corresponding transition state **119** is proposed for the same reaction catalysed by benzoic acid.

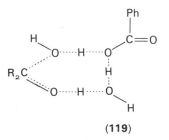

(119)

Stewart and Van Dyke[440] have examined the hydration of ring substituted trifluoroacetophenones in aqueous polar aprotic solvents (dimethyl sulphoxide and sulpholane) by u.v. and n.m.r. spectroscopy. Aqueous sulpholane has a dehydrating effect compared with water, whilst DMSO–water combinations as low as 15 mol-% water are more hydrating than pure water. In the uncatalysed reaction the rate of hydration in water–sulpholane falls as the water percentage decreases. The rate of the acid catalysed reaction is not significantly affected. A hydration factor, W_o, (equation 58) where K_d is the equilibrium constant for the reaction

$$W_o = pK_d + \log[Z]/[Z.H_2O], \qquad (58)$$

(equation 59), may allow the hydration equilibrium constants to be

$$Z + H_2O \rightleftharpoons Z.H_2O \qquad (Z = \text{substrate}) \qquad (59)$$

determined in solutions other than water[440,441]. Thus positive values of W_o indicate a solution less hydrating than water. For negative values of W_o the opposite applies. A plot of $\log[Z]/[Z.H_2O]$ against W_o is linear with unit slope.

An analogy with the acidity function, H_o, may be drawn. W_o, however, may also be a function of substrate and requires further investigation.

b. *Coordination and association effects on hydration.* The ruthenium(III) complex of 4-formylpyridine is greater than 90% hydrated whereas the corresponding ruthenium(II) complex is less than 10% hydrated[442]. This may be attributed to back-bonding which is greater for Ru^{II} than Ru^{III}, and greater when the ligand is in the carbonyl form. Reduction of the hydrated Ru^{III} complex by vanadium(II) gave the Ru^{II} compound in the hydrated form. The rate of dehydration is then reduced by a factor of 70.

In eight- and nine-membered cyclic ketones, the carbonyl group is markedly shielded by transannular interaction with a C_2H_5N group or an oxygen atom. A sulphur atom has no such effect. In consequence the rates of hydration are markedly diminished compared with simple ketones[443]. The reaction was followed by ^{17}O-n.m.r. using ^{17}O-ketones and non-enriched water. The exchange reaction was completely inhibited in N-ethylazocyclodecan-6-one.

2. Addition of alcohols to C=O compounds

In contrast to the reverse reaction, the hydrolysis of acetals (see Section IV.G), few mechanistic studies on the formation of acetals have been made since the article by Ogata and Kawasaki[444] in this series. The mutarotation of glucose and related processes have been reviewed by Capon[445]. The addition of alcohols to C=O compounds is similar mechanistically to the hydration process (Section IV.E.1). Since acetal hydrolysis is to be described in some detail no further discussion of acetal formation is necessary here.

3. The addition of thiols and hydrogen sulphide to C=O compounds

The addition of thiols and hydrogen sulphide to carbonyl compounds[446] has been discussed earlier in this series. Since few mechanistic studies have appeared recently, they are discussed only briefly here.

Addition of a thiol to a carbonyl compound may occur by a direct nucleophilic attack of the thiol anion (equation 61) or by a general acid

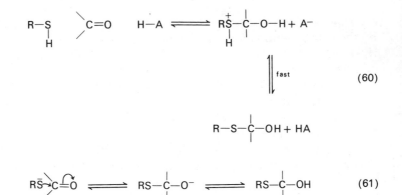

$$(60)$$

$$(61)$$

catalysed reaction (equation 60)[447]. In the latter case a concerted mechanism is possible. A general acid catalysed thiol anion addition ($\alpha = 0.2$) (equation 62) has been proposed for the addition of acidic thiols, such

$$RS\overset{\nearrow}{}C\overset{}{=}\overset{\curvearrowright}{O} \quad H\overset{\curvearrowleft}{-}A \;\rightleftharpoons\; RS-\overset{|}{\underset{|}{C}}-OH + A^- \qquad (62)$$

as thiophenol and thioacetic acid[448a].

Thiazolidine formation from L-cysteine and formaldehyde (Scheme 43) occurs by a rate-determining dehydration of the carbinolamine intermediate in alkaline solution, whilst in acid solution the attack of cysteine on formaldehyde is rate determining. The dehydration in alkaline solution is subject to general acid catalysis ($\alpha = 0.66$)[448b].

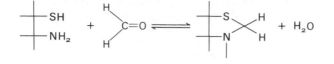

SCHEME 43.

4. The Baeyer–Villiger reaction

The reaction of peracids with ketones offers a preparative method for conversion of cyclic ketones into lactones and acyclic ketones to esters[449]. The relative ease of migration of various groups (Scheme 44) depends partly on the ability of that group to stabilize a partial positive charge in the

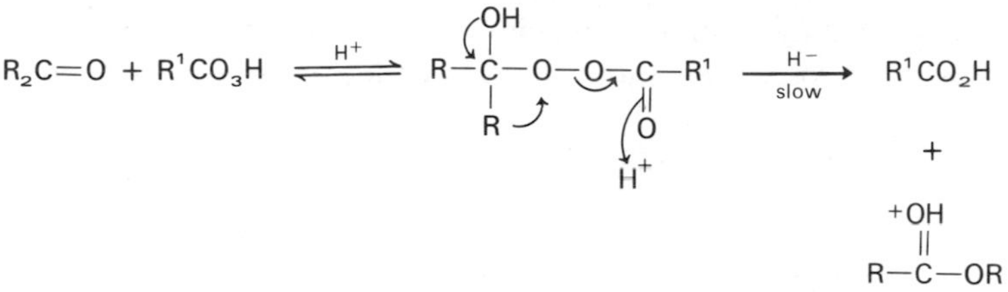

$$R_2C{=}O + R'CO_3H \xrightleftharpoons[\quad]{H^+} R{-}\overset{\overset{\displaystyle OH}{|}}{\underset{\underset{\displaystyle R}{|}}{C}}{-}O{-}O{-}C{-}R' \xrightarrow[\text{slow}]{H^-} R'CO_2H$$

SCHEME 44.

transition state. Thus a tertiary alkyl group migrates readily, and electron-withdrawing substituents in a phenyl group retard its migration. The migrating ability may also be affected by a change in the reaction conditions. Migration is believed to be synchronous with the departure of the leaving group[450]. The rate-limiting process is generally the migration, except for aryl ketones and aldehydes substituted with electron-donating substituents when the addition step often becomes rate determining[450,451].

Recently isotope effects for the Baeyer–Villiger reaction of $PhCD_2COCH_3$, $PhCH_2COCD_3$ and $PhCD_2COCD_3$ have been reported (Table 18)[452]. From these values and the equilibrium isotope effects (K_H/K_D) for hemi-ketal formation, the α- and β-isotope effects for the rate-determining migration step were calculated (equations 63, 64, 65).

$$(k_H/k_D)_{obs}(\text{for } D_2) = (K_H/K_D)_{D_2} \times (k_\alpha)^2 \tag{63}$$

$$(k_H/k_D)_{obs}(\text{for } D_3) = (K_H/K_D)_{D_3} \times (k_\beta)^3 \tag{64}$$

$$(k_H/k_D)_{obs}(\text{for } D_5) = (K_H/K_D)_{D_5} \times (k_\alpha)^2 \times (k_\beta)^3 \tag{65}$$

Thus, $(k_\alpha)^2 = 1\cdot071$ ($1\cdot035$ per deuterium) and $(k_\beta)^3 = 1\cdot166$ ($1\cdot052$ per deuterium). The β-isotope effect for the migration indicates a large degree of carbon–oxygen bond formation in the transition state.

TABLE 18. Observed isotope effects for
the Baeyer–Villiger oxidation of
ketones[452]

Ketone	$(k_H/k_D)_{obs}$
$PhCD_2COCH_3$ (D_2)	0.932 ± 0.004
$PhCH_2COCD_3$ (D_3)	0.921 ± 0.006
$PhCD_2COD_3$ (D_5)	0.858 ± 0.006

F. Elimination Mechanisms for Alcohol Oxidations

1. General considerations

Carbonyl-forming eliminations may be considered as reactions involving either proton (equation 66) or hydride transfer (equation 67).

$$\ddot{B} \quad H-C-O-X \longrightarrow \overset{+}{B}H + \ \ C{=}O + X^- \qquad (66)$$

$$H-C-O-X \longrightarrow H^- + \ \ C{=}O + X^+ \qquad (67)$$

Both mechanisms may also proceed via a cyclic transition state. In this section the reactions will be classified and discussed according to their mechanism. Radical oxidation processes are also briefly considered.

A previous volume in this series[453] dealt in detail with oxidation mechanisms and in 1973 a book was devoted to oxidation[454].

2. Cyclic proton transfer processes

Typical of this reaction type are the thermal eliminations of nitronate esters and oxidation involving sulphonium salt intermediates. The latter are of great synthetic utility and are discussed in detail.

Corey's group[455a] showed that alcohols are oxidized by the addition product from N-chlorosuccinimide and dimethyl sulphide (120, Scheme 45). The decomposition of the sulphonium salt (121) may occur by an $E_{CO}2$ mechanism (equation 68) or a ylid mechanism (equation 69). Support for the latter mechanism is given by the formation of dimethyl sulphide-d_5 (DMS-d_5) from DMS-d_6 in the reaction[455b]. A similar oxidation occurs using $(CH_3)_2\overset{+}{S}Cl{\cdot}\overset{-}{C}l$ prepared from chlorine and dimethyl-sulphide. Yields are high and the reactions are rapid.

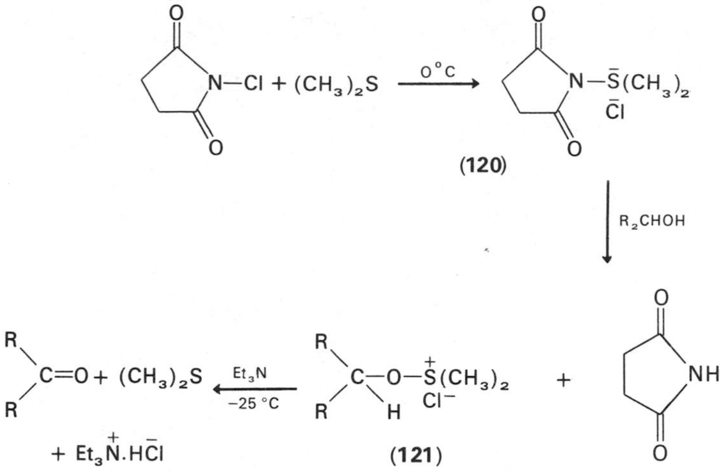

(120)

R₂CHOH

(121)

SCHEME 45.

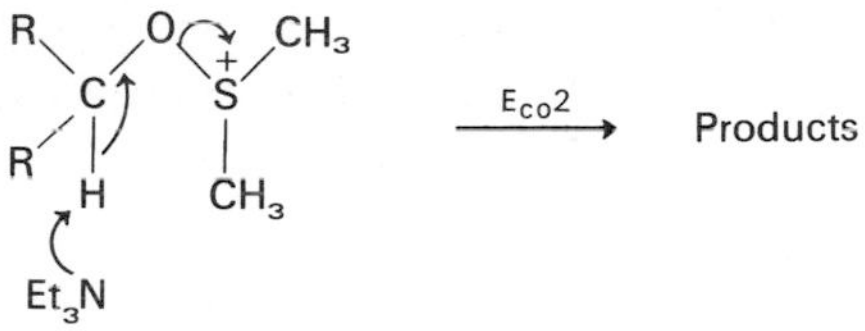

$$\xrightarrow{E_{co2}} \quad \text{Products}$$

(68)

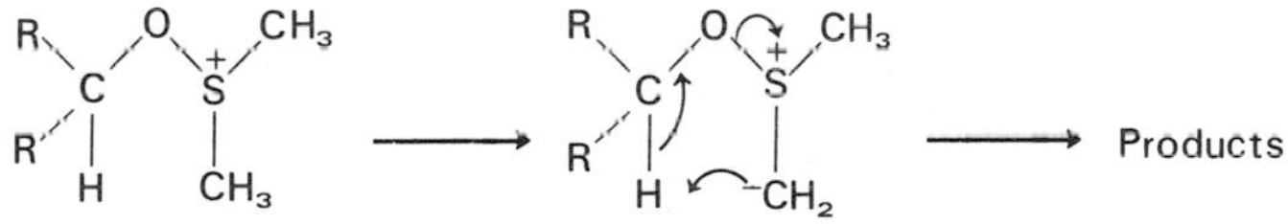

$$\longrightarrow \quad \text{Products} \quad (69)$$

The scope of the Pfitzner–Moffatt[456] oxidation has been reviewed[457]. In the reaction an alcohol is oxidized by a combination of dicyclohexyl-carbodiimide (DCC) and dimethylsulphoxide. Trifluoroacetic acid or phosphoric acid are used as catalysts. Evidence for the proposed mechanism[458] (Scheme 46) is provided by isolation of dicyclohexylurea containing one atom of deuterium from an oxidation using DMSO-d_6, confirming that abstraction of an SCH_3 proton occurs via an intra-molecular mechanism[459]. Nuclear magnetic resonance studies in the absence of an alcohol indicate that the DCC–DMSO adduct (**122**) is formed in low equilibrium concentration[459]. The reaction has recently

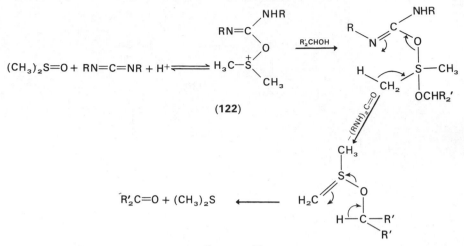

(122)

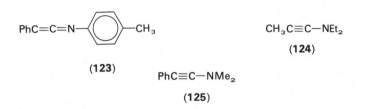

(123)

SCHEME 46.

been conducted using a carbodiimide linked to a cross-linked polystyrene matrix which simplifies product isolation[460]. Diphenylketene-*p*-tolylimine (**123**) and the alkynylamines **124** and **125** have been used in place of the diimide[461].

$$PhC{=}C{=}N{-}\!\!\bigcirc\!\!{-}CH_3$$

(123)

$$CH_3C{\equiv}C{-}NEt_2$$

(124)

$$PhC{\equiv}C{-}NMe_2$$

(125)

The mechanism, illustrated for *N,N*-diethylaminoprop-1-yne (**124**, Scheme 47), involves addition of dimethylsulphoxide to the alkynylamine followed by nucleophilic addition of the alcohol to the resulting sulphoxonium ion. The thermal decomposition of nitronate esters (equation 70) is another example of a cyclic proton transfer mechanism, but has not been investigated in detail.

Corey and Fleet[462] have suggested a proton transfer mechanism for the oxidation of alcohols by a chromium trioxide–3,5-dimethylpyrazole complex (**126**, equation 71). However, no corroborative evidence is available and a hydride transfer process cannot be discounted. The mechanism of the oxidation of alcohols by a similar reagent, chromium

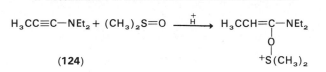

(124)

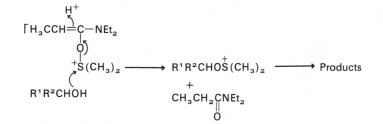

R^1R^2CHOH

SCHEME 47.

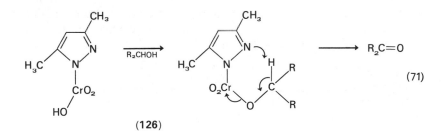

$$\underset{/}{\overset{\backslash}{C}}=\overset{+}{N}\underset{/}{\overset{\backslash}{O}}\overset{O}{\underset{O}{\underset{\backslash}{\overset{H-C}{\overset{\backslash}{/}}}}} \longrightarrow \underset{/}{\overset{\backslash}{C}}=N-OH + \underset{/}{\overset{\backslash}{C}}=O \qquad (70)$$

trioxide–pyridine[463], has not been investigated. The reagent is a powerful synthetic tool in the oxidation of primary alcohols to aldehydes and secondary alcohols to ketones. Isolation of the unstable complex is unnecessary[464] and oxidation is normally complete within 15 min in methylene chloride.

$$ \qquad \xrightarrow{R_2CHOH} \qquad \longrightarrow R_2C=O \qquad (71)$$

(126)

3. Non-cyclic proton transfer mechanisms

This reaction category is exemplified by the $E_{CO}2$ elimination of nitrite esters[465] (equation 72). Hydrolysis to the alcohol and elimination to an alkene may compete, and Baker and his coworkers[466] have investigated

$$R-\underset{\underset{R}{|}}{\overset{\overset{H}{|}}{C}}-O-NO_2 \xrightarrow{\ \bar{O}R\ } \underset{R}{\overset{R}{>}}C=O + NO_2^- + ROH \qquad (72)$$

the effect of substituents on the competing reactions. The observation of significant nitrogen and hydrogen kinetic isotope effects for elimination from benzyl and 9-fluorenyl nitrates are consistent with the $E_{CO}2$ mechanisms[467].

The oxidation of secondary alcohols to ketones has also been accomplished by iodobenzene dichloride in pyridine[468]. A suggested mechanism (equation 73) invokes proton transfer with liberation of iodobenzene.

$$\begin{array}{c} R_2CHOH \\ + \\ PhICl_2 \end{array} \xrightarrow[5-20^\circ C]{} R_2C \overset{H \quad Ph}{\underset{O \quad \quad Cl}{\diagdown \diagup}} \longrightarrow R_2C=O + PhI \qquad (73)$$

Peroxides, hypochlorites, and sulphonate esters also eliminate by proton transfer mechanisms but detailed studies are not available.

4. Non-cyclic hydride transfer processes

Few examples of this category are known. The oxidation of alcohols by triarylcarbonium ion (equation 74) occurs by a hydride transfer mechanism. This reaction also forms the basis of a convenient procedure

$$Ph_3C^+ \overset{\frown}{H}-\overset{|}{\underset{|}{C}}-\overset{\frown}{O}-H \longrightarrow Ph_3CH + \overset{\diagdown}{\underset{\diagup}{}}C=O \qquad (74)$$

for removal of an acetal protecting group (Scheme 48)[469]. Evidence for a mechanism involving intermolecular hydride transfer to the triphenyl-carbonium ion is based on the isolation of deuterated triphenylmethane when a suitably deuterated substrate was used. Isotope effects are indicative of a rate-determining hydride transfer.

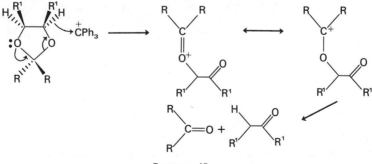

SCHEME 48.

5. Cyclic hydride transfer processes

Cyclic hydride transfer mechanisms are exemplified by chromic acid oxidations and the Oppenhauer oxidation. The latter is the reverse of the Meerwein–Ponndorf–Verley reduction (see Section IV.D.2), and therefore a discussion of the chromic acid oxidation will form the major part of this section.

a. *Chromic acid oxidations.*

(i) General comments. In most of the reactions discussed above only limited mechanistic investigations have been made. In contrast to this, the oxidation of alcohols by chromic acid has been a subject of intense investigation, and the essential features of the mechanism are now well understood[470–472]. With unhindered alcohols initial formation of a chromate ester is followed by a rate-limiting cleavage of the α-carbon–hydrogen bond

$$\begin{array}{c} R \\ \diagdown \\ \diagup \\ R \end{array} CHOH + H_2CrO_4 + H \rightleftharpoons (R_2CHOCrO_3H_2)^+ \qquad (75)$$
$$+ H_2O$$

$$\underset{\underset{R'}{R}}{\overset{\overset{H_2O}{\diagup}}{\diagdown}} \underset{O}{\overset{}{C}} \underset{O}{\overset{OH}{\diagdown}} Cr \overset{}{\underset{O}{\diagdown}} \longrightarrow \underset{R'}{\overset{R}{\diagdown}} C{=}O + H_2CrO_3 \qquad (76)$$

(equations 75, 76) leading to the carbonyl compound. Proton abstraction from the chromate ester could also occur by an intermolecular process. The intermediacy of chromate esters in the oxidation of simple alcohols has been confirmed[472]. Where the chromate ester is sterically crowded the decomposition to products is accelerated and ester formation may become rate limiting[470].

(ii) Fate of the Cr^{IV} species. The chromic acid oxidation of alcohols is a two-electron oxidation in which Cr^{IV} is formed. The fate of the Cr^{IV} species in the reaction has recently been the source of considerable interest[473–477]. The Cr^{IV} species may either reduce Cr^{VI} to Cr^{V} (equation 77) or alternatively, it can oxidize the alcohol in a one- or two-electron process (equations 78, 79). Subsequently, the Cr^{II} species generated in equation (79) could then react with Cr^{VI} as shown in equation (80) and the radical species generated in equation (78) would react further as shown by equation (81). The Cr^{V} moiety formed in all these reactions would function as a two-electron oxidant in converting the organic substrate into ketone[472]. The route in which Cr^{IV} was oxidized to Cr^{V} (equation 77) was generally accepted until Roček and Radkowsky[476,478,479] demonstrated that Cr^{IV} may act as an active oxidant. In this study Cr^{IV} was

$$Cr^{IV} + Cr^{VI} \longrightarrow 2Cr^{V} \tag{77}$$

$$Cr^{IV} + R_2CHOH \longrightarrow Cr^{III} + R_2\overset{\bullet}{C}OH \tag{78}$$

$$Cr^{IV} + R_2CHOH \longrightarrow Cr^{II} + R_2C{=}O \tag{79}$$

$$Cr^{II} + Cr^{VI} \longrightarrow Cr^{III} + Cr^{V} \tag{80}$$

$$Cr^{VI} + R_2COH \longrightarrow R_2C{=}O + Cr^{V} \tag{81}$$

generated from the reaction of chromium(VI) and vanadium(IV) (Scheme 49). Cyclobutanol, a strained small-ring alcohol, is known to be oxidized

$$Cr^{VI} + V^{IV} \underset{k_{-1}}{\overset{k_1}{\rightleftharpoons}} Cr^{V} + V^{V}$$

$$Cr^{V} + V^{IV} \xrightarrow[\text{Rate limiting}]{k_2} Cr^{IV} + V^{V}$$

$$Cr^{IV} + V^{IV} \xrightarrow{k_3} Cr^{III} + V^{V}$$

SCHEME 49.

to a mixture of cyclobutanone and γ-hydroxy butyraldehyde. Roček demonstrated that in the presence of vanadium(IV), γ-hydroxybutyralde-hyde, the result of carbon–carbon bond cleavage, was the sole isolated product. In the oxidation of 1-deuteriocyclobutanol, the γ-hydroxy-butyraldehyde formed retained the deuterium label which was located at the aldehydic carbon. 1-Methylcyclobutanol is oxidized by chromic acid but no simple oxidation products have been isolated[480]. However, in the presence of a secondary alcohol, that is under conditions in which Cr^{IV} is formed, 5-hydroxy-2-pentanone is formed from 1-methylcyclobutanol in good yield[476,479], offering further evidence for the importance of chromium(IV) in chromic acid oxidations. A mechanism consistent with these results is shown in Scheme 50. Roček and Rahman[481] have extended

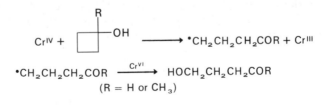

SCHEME 50.

the study of chromium(IV) oxidation to other primary and secondary alcohols in order to ascertain whether cleavage reactions are limited to strained alcohols. Using chromium(IV), produced from chromium(VI) and vanadium(IV), isopropyl alcohol was oxidized to acetone under conditions in which oxidation by vanadium(V) was slow. That the reaction involved radical intermediates (equations 82, 83) was indicated by polymer formation when the reaction was carried out in the presence of acryloni-trile or acrylamide. The possibility of bimolecular radical reactions cannot

$$(CH_3)_2CHOH + Cr^{IV} \longrightarrow (CH_3)_2\dot{C}OH + Cr^{VI} \qquad (82)$$

$$(CH_3)_2\dot{C}OH + V^V \longrightarrow (CH_3)_2C{=}O + V^{IV} \qquad (83)$$

be discounted. A plot of the logarithm of relative rates of chromium(IV) oxidation of a series of primary alcohols against the Taft substituent constant, σ^*, gives a straight line (Figure 10).

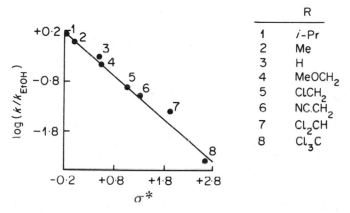

	R
1	i-Pr
2	Me
3	H
4	$MeOCH_2$
5	$ClCH_2$
6	$NC.CH_2$
7	Cl_2CH
8	Cl_3C

FIGURE 10. The dependence of relative rate of oxidation of primary alcohols, RCH_2OH, by chromium(IV) on the σ^* value of the group R[481].

The Taft ρ^* value of -0.85 for primary alcohols suggests that the electronic requirements of the Cr^{IV} oxidation are similar to those of chromium(VI) oxidations. t-Butyl alcohol is unreactive toward chromium(IV).

Phenyl-t-butyl carbinol (127) is known to undergo oxidative cleavage to benzaldehyde[482,483] (equation 84). In the chromic acid oxidation

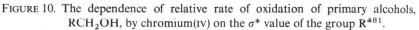

$$C_6H_5CHOHC(CH_3)_3 \longrightarrow C_6H_5COC(CH_3)_3 + C_6H_5CHO + (CH_3)_3COH \quad (84)$$

(127)

of 2-aryl-1-phenylethanols, in aqueous acetic acid containing sodium acetate, oxidative cleavage as well as ketone formation is found[475]. The intermediacy of the benzyl radical is shown by trapping with oxygen. Plots of logarithms of reaction rates against σ for the substituent in the aromatic ring show good Hammett relationships with $\rho = -0.10 \pm 0.02$ for ketone formation and -1.06 ± 0.04 for cleavage. Good correlations with σ^+ were also found ($\rho^+ = -0.09$ for ketone formation and -0.97 for cleavage). The ρ value of -0.1 for ketone formation indicates, as expected, that the remote substituents on the 2-aryl ring have little effect

$$Cr^{VI} + RCHOHR^1 \longrightarrow Cr^{IV} + ketone$$

$$Cr^{IV} + RCHOHR^1 \longrightarrow Cr^{III} + RCHO + \overset{\cdot}{R}{}^1$$

$$Cr^{VI} + R^1 \longrightarrow Cr^{V} + R^1OH$$

$$2Cr^{V} \longrightarrow Cr^{VI} + Cr^{IV}$$

SCHEME 51.

on the rate of ketone formation. A ρ value of -1.06 is not unreasonable for a reaction involving benzyl radicals. A consistent mechanism is shown in Scheme 51.

Cerium(III) or cerium(IV) may be used to remove chromium(IV) from the reaction mixture[483,484,485], and thus inhibit radical processes.

Spectrophotometric observation of chromium(v) in the chromic acid oxidation of isopropanol[472,474] in 97% acetic acid also confirms the role of Cr^{IV} in oxidation. If Cr^{IV} were oxidized by chromium(VI), i.e. $Cr^{IV} + Cr^{VI} \rightarrow 2Cr^{V}$, 2 equivalents of chromium(v) would be accompanied by formation of one equivalent of acetone, whilst if Cr^{IV} functioned as an active oxidant, 2 equivalents of acetone to 1 equivalent of Cr^{V} would be formed. Rate constants for the disappearance of Cr^{VI} and Cr^{V}, determined spectrophotometrically, allow a calculation of the concentration of acetone as a function of time for the two possible mechanisms. The experimental acetone time curve corresponds closely to that calculated for the mechanism involving oxidation of the substrate by chromium(IV).

Thus evidence for an important contribution of Cr^{IV} in oxidation reactions is conclusive, and has led to a significant increase in our understanding of this reaction.

(iii) Chromic acid oxidation of cyclopropanols. A carbonium ion intermediate has been postulated in the oxidation of cyclopropanols by chromic acid[486] (Scheme 52). Oxidation is approximately 2000-times faster than with a 'normal' secondary alcohol and gives β-hydroxypropionaldehyde. Substituents which stabilize the carbonium ion, e.g. 1-alkyl, 2,3-dialkyl, enhance the rate of reaction.

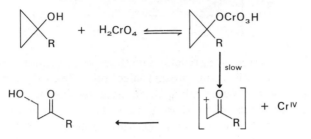

SCHEME 52.

(iv) Chromic acid in a two component substrate system. In an oxidation of a mixture of propan-2-ol and oxalic acid in aqueous perchloric
acid (Scheme 53) by chromium(VI) the rate was shown to exceed that for
either of the components[487]. Cleavage of the C_α—H occurring in the rate-
determining step is indicated by a k_H/k_D effect of 5·8 using 2-deuterio-2-
propanol. Thus the oxidation involves a direct conversion of Cr^{VI} into

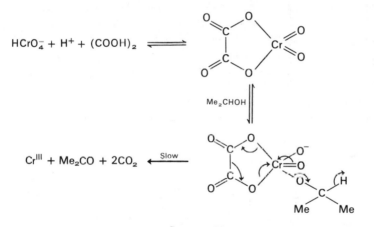

SCHEME 53.

Cr^{III} without the intervention of Cr^{IV} and the possibility of side reactions.
Hydroxy acids and keto acids may be used in place of oxalic acid[488].
The reaction is of potential synthetic utility.

6. Radical mechanisms

a. *Oxidation by silver carbonate.* Kakis[489] has determined the kinetic
isotope effect for the oxidation of several alcohols by silver carbonate on
celite[490] (Table 19). Owing to difficulty in reproducibility, because of the

$$R^{14}CHR^1 + RCDR^1 \xrightarrow[\substack{Celite, \\ benzene \\ or\ heptane}]{Ag_2CO_3} R^{14}CR^1 + R{-}CR^1 \qquad (85)$$
$$\underset{OH}{|}\qquad\underset{OH}{|}\qquad\qquad\underset{O}{\|}\qquad\underset{O}{\|}$$

heterogeneous nature of the reaction, mixtures of non-radioactive deuterated alcohols and radioactive protium alcohols were oxidized in the same
vessel (equation 85). Values of k_H/k_D obtained by this method were of the
same order as those obtained by Eckert and his coworkers[491] for the
oxidation of norbornanols. In a further study[492] the average intermolecular
deuterium isotope effect for primary alcohols was 1·9 and for secondary
alcohols was 3·2. However, in these initial studies the effect of adsorption
on celite was not determined. Recently Kakis and his coworkers[492] have
determined intermolecular and intramolecular isotope effects in the silver
carbonate oxidation in the absence of celite. Intermolecular kinetic isotope
effects were then >6 and provide a more valid criterion of the nature of
the transition state as adsorption effects are minimized. Intramolecular
isotope effects for oxidation of alcohols of the type RCHDOH were an
average of 4·9 and rose to 6·5 when celite was omitted. Normally intermolecular values are greater than intramolecular effects, because of a
contribution of the secondary isotope effect to the former. The difference
here suggests a primary effect due to H versus D adsorption, occurring
prior to elimination. Steric effects support this hypothesis, since in cases
where steric and conformational factors precluded the preferred orientation of the α-hydrogen toward the absorbent surface, the reaction was
slow. The mechanism proposed for the oxidation (Scheme 54) involves a
reversible adsorption of the alcohol, followed by an irreversible, homolytic
cleavage.

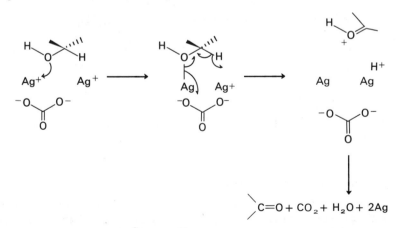

Scheme 54.

TABLE 19. Isotope effects k_H/k_D for
oxidation of alcohols by silver
carbonate/celite[489]

Alcohol	k_H/k_D
1,2,2-Triphenylethanol	3.0
1,2,2-Tri-*p*-anisylethanol	3.1
1,2,2-Tri-*p*-tolylethanol	3.0
1,2-Diphenylethanol	2.9
1,3-Diphenyl-2-propanol	3.1
Benzhydrol	2.9

b. *Oxidation by* 1-*chlorobenzotriazole*. A convenient oxidant for unhindered alcohols is 1-chlorobenzotriazole **128**[493]. An induction period

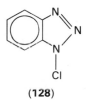

(128)

followed by a rapid reaction is taken as indicative of a radical mechanism[493]
(Scheme 55).

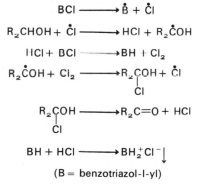

SCHEME 55.

7. Miscellaneous oxidations

a. *Oxidation by* N-*halosuccinimides*. Some confusion exists as to whether
oxidations by *N*-halosuccinimides occur by a radical or ionic mecha-

nism[494]. The fact that the reactions proceed readily in the dark offers support for an ionic process.

In the oxidation of alcohols by N-bromosuccinimide (NBS) in aqueous systems a slow reaction between NBS and the alcohol is followed by a rapid reaction between the alcohol and liberated bromine[495]. A first-order dependence on both NBS and the alcohol was also found. Mercuric salts inhibit the reaction presumably by complexation with Br^-. Added bromide ion or mineral acids accelerate the reaction.

In a study of the oxidation of alcohols by N-chlorosuccinimide[496], while a first-order dependence on NCS was found, a zero-order dependence on the alcohol was determined. The lack of a primary deuterium isotope effect suggests the C—H bond is not cleaved in the rate-limiting step. The difference in reactivity of NBS and NCS can be rationalized in terms of a more facile N—Br bond cleavage due to the difference in electronegativity between the halogens.

b. *Cupric ion oxidations.* The rate law for the oxidation of α-hydroxy-acetophenone by cupric acetate in buffered aqueous pyridine was determined by Wiberg and Nigh (equation 86)[497].

$$\frac{-d[Cu^{II}]}{dt} = k_1[\text{ketol}][B] + k_2[\text{ketol}][B][Cu^{II}] \tag{86}$$

At low concentrations of Cu^{II}, $k_1 \gg k_2$ and the first term in equation 86

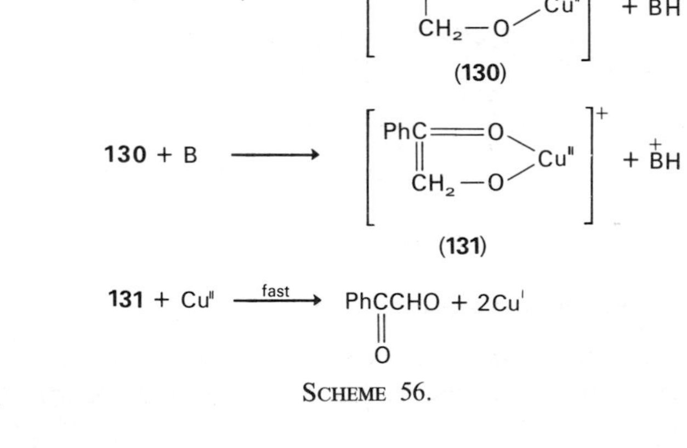

SCHEME 56.

predominates. Where $k_2(Cu^{II}) \gg k_1$, the second term is predominant. A kinetic isotope effect of 7·4 and a Hammett reaction constant, ρ, of $+1·24$ were found, indicating a pathway in which a rate-determining proton removal from the α-methylene of the Cu^{II}-ketol complex occurs (Scheme 56). The reaction exhibits marked steric effects[498], 4-hydroxy-2,2,5,5,-tetramethyl-3-hexanone being unaffected by the reagent. In the transition state **132** large groups at R and R^1 prevent formation of a planar confor-

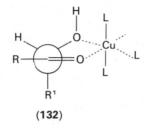

(**132**)

mation required for maximum stability of the chelate. 2-Hydroxycyclo-pentanone is oxidized only slowly as the ring prevents eclipsing of C=O and OH in the transition state. In 2-hydroxycyclohexanone the oxidation is faster.

c. *Oxidation by ceric ions*. Oxidations with ceric ions have recently been reviewed[499]. Ceric ammonium nitrate oxidizes primary alcohols to aldehydes, but fragmentation may occur where an incipient free radical would be stabilized by substituents.

G. Acetal and Related Hydrolyses

Since the section by Salomaa in Volume I of this series[500], reviews by Cordes[501] and a monograph by Fife[502] (1972) have been published. The book by Jencks on *Catalysis in Chemistry and Enzymology*[503] also discusses this topic in some detail. Until recently no example of general acid catalysis had been reported and acetal hydrolysis was believed to proceed by a specific acid catalysed Al mechanism. Since the Al mechanism has been discussed fully[500,501,503] it will be mentioned only briefly here and the majority of this section will be devoted to discussion of the general acid catalysed processes, involving proton transfer in the rate-determining step.

1. Mechanism of acetal hydrolysis

In a generalized equation for acetal hydrolysis (Scheme 57) if $k_2 \ll k_{-1}$, then an Al mechanism operates. If $k_2 > k_{-1}$, then the rate-limiting process is governed by k_1 and the reaction is designated S_E2. This process

Anthony F. Cockerill and Roger G. Harrison

$$X + H_3O^+ \underset{k_{-1}}{\overset{k_1}{\rightleftharpoons}} \overset{+}{X}H + H_2O$$

$$\overset{+}{X}H \xrightarrow{k_2} \text{Products}$$

Scheme 57.

is unlikely as if k_2 were appreciably greater than k_1, then the rate of cleavage of a covalent bond would be greater than that of a diffusion-controlled proton loss. In the more feasible situation where $k_2 \rightleftharpoons k_{-1}$ general acid catalysis is also significant and the mechanism is designated A-S_E2. A general acid catalysed A2 process[504] in which water attacks the protonated acetal is also discussed.

In a typical energy profile (Figure 11) for hydronium-ion catalysed hydrolysis of an acetal (Scheme 58)[502], it can be seen that proton transfer would make a significant contribution to the rate if the peak height of that

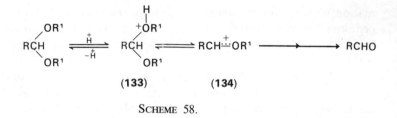

(133)   (134)

Scheme 58.

step is increased, or alternatively if the subsequent bond breaking energy requirement were lowered. That is, in order to observe general acid

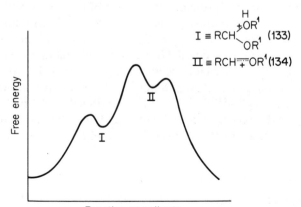

Free energy

Reaction coordinate

Figure 11. Hypothetical energy profile for the A–1 hydrolysis of a simple acetal.

catalysis, there should be a good leaving group or a stable carbonium ion intermediate if the leaving group is poor, and a weakly basic acetal[505].

a. *The A1 mechanism.* The A1 mechanism of acetal hydrolysis (Scheme 59) has been reviewed[501]. The rate-limiting step is regarded as the cleavage of

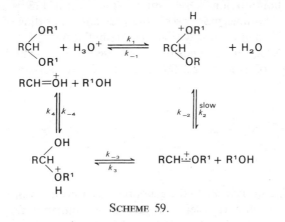

SCHEME 59.

the C—O bond with formation of a stabilized oxocarbonium ion and an alcohol (k_2). The reaction is specific acid catalysed with the Brönsted coefficient, α, equal to unity. It has been suggested that k_4 may be rate-determining in certain cases[505].

b. *The A2 mechanism.* The first report of buffer catalysis in the hydrolysis of an acetal appeared in 1967[504]. The observed buffer catalysis was not consistent with an A1 mechanism in the hydrolysis of 2-(p-methoxyphenyl)-4,4,5,5-tetramethyl-1,3-dioxolane(**135**). An A2 mechanism in which water attacks the protonated acetal (Scheme 60) in the rate-determining step was

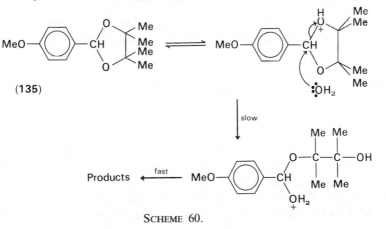

SCHEME 60.

proposed. Supporting this proposal, substitution of a methyl group at the reaction centre decreases the rate of the reaction more than 500-fold[506]. Water participation may be necessary in this reaction as the tetramethyl substituents would be expected to favour reclosure of the ring if the intermediate carbonium ion were formed. The acetal **135** is nevertheless hydrolysed 1000-times more slowly than the corresponding ethylene ketal and 40,000-times more slowly than the diethyl acetal[504].

A similar A2 mechanism has been proposed by Capon and Page[507] for the hydrolysis of isomeric 2,3-*O,O*-benzylidene-norbornane-exo-2-exo-3-diols(**136,137**).

(**136**)   (**137**)

c. *The A-S_E2 mechanism*. In the general acid catalysed A-S_E2 mechanism (Scheme 61), proton transfer is regarded as contributing to the rate-determining step. Values of the Brönsted coefficient, α, are then less than unity.

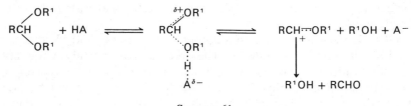

SCHEME 61.

In 2-(*p*-nitrophenoxy)-tetrahydropyran (**138**), the electron-withdrawing nitro group lowers the basicity of the acetal and facilitates cleavage of the

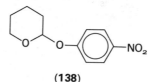

(**138**)

C—O bond, thus fulfilling the conditions necessary for observation of general acid catalysis. General acid catalysis is observed in both water and water–dioxan systems[508,509]. Values of the Brönsted coefficient α are 0·5 for catalysis by chloroacetic, formic and acetic acids in water and 0·65 for dichloroacetic, chloroacetic, formic and acetic acids in 50% water–dioxan.

Second-order rate constants were found to be less in D_2O than water, again consistent with a rate-determining proton transfer.

Since the first example of an A-S_E2 mechanism of acetal hydrolysis, several other cases have been reported[510-518]. Fife and Anderson[518] have attempted to demonstrate general acid catalysis in a series of acetals in which the leaving group was poor but the intermediate carbonium ion was stable. In the hydrolysis reactions of benzophenone diethyl acetal, 2,2-(p-methoxyphenyl)-1,3-dioxolane, 2,3-diphenylcyclopropenone diethyl acetal, ferrocene carboxaldehyde dimethyl acetal and tropone ethylene ketal, general acid catalysis by weak buffer acids (Tris (H)$^+$ and $H_2PO_4^-$) could be detected only with the latter. General acid catalysis in the hydrolysis of tropone diethyl acetal had previously been reported[514]. Thus it appears that when the leaving group is poor, the carbonium ion must be well stabilized before general acid catalysis may be observed.

Benzaldehyde methyl phenyl acetals are subject to general acid catalysed hydrolysis[512,513], but in the corresponding benzaldehyde-methyl S-phenyl thioacetals (**139**), general acid catalysis was not detected[519]. In this sulphur analogue, the C—S bond is cleaved initially but, although the basicity is greatly reduced, the difficulty in bond breaking precludes observation of general acid catalysis.

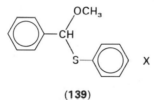

(**139**)

In the hydrolysis of p-substituted benzaldehyde di-t-butyl acetal (**140**) bond breaking is facilitated by relief of steric strain in the ground state, and general acid catalysis has been detected[516].

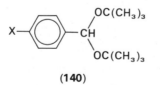

(**140**)

An A-S_E2 mechanism is also proposed for the hydrolysis of O,O'-benzylidenecatechol (**141**)[510]. A Brönsted α value of 0·47 for catalysis by carboxylic acids in water is found.

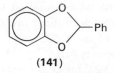

(141)

Kankaanperä and Merilahti[520] have compared solvent effects in A1 and A-S_E2 reactions. The rates of hydrolysis of 2-methyl-1,3-dioxolane (A1 mechanism) and 2-methyl-4-methylene-1,3-dioxolane (A-S_E2 mechanism) in water–dioxan and water–dimethyl sulphoxide were determined. A plot of the logarithm of the relative rate coefficients against the mole fraction of dimethyl sulphoxide (Figure 12) shows that the rates for both the A1 and A-S_E2 reaction pass through minimum where the mole fraction of DMSO is about 0·6. A similar plot is obtained when dioxan is used as solvent.

The same authors[521] have emphasized the need to consider the influence of specific salt effects in experiments to detect general acid catalysis. DMSO–water, rather than the commonly used dioxan–water, is the preferred system for preparing buffer solutions in which specific salt effects are minimized.

d. *pH-independent hydrolysis.* pH-independent hydrolysis of *p*-nitro-phenoxytetrahydropyran[508,509], benzaldehyde *S*-(2,4-dinitrophenyl)

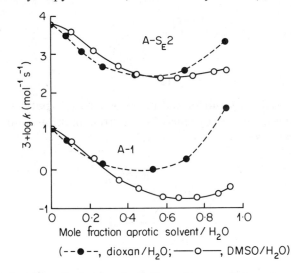

FIGURE 12. The effect of solvent on logarithm of rate coefficients for A-1 hydrolysis of 2-methyl-1,3-dioxolane and A-S_E2 hydrolysis of 2-methyl-4-methylene-1,3-dioxolane at 25 °C[520].

methyl thioacetal[519], and tropone diethyl acetal[514] have been reported. With p-nitrophenoxy tetrahydropyran the pH independent reaction is observed in both 50% aqueous dioxan and water from pH 4 to alkaline pH, and for tropone diethyl acetal at pH values greater than 10. In all these reactions isotope effects which are approximately 1 are found, indicating that water is not acting as a general acid. An uncatalysed ionization is therefore proposed and is reasonable since there is a good leaving group and a relatively stable carbonium ion intermediate.

A different mechanism is proposed for the pH independent hydrolysis of O,O'-benzylidene-catechol (Scheme 62)[510] where $k[H_2O]/k[D_2O] = 1·61$ and $\Delta S^{\neq} = -21·1$ cal deg^{-1} mol^{-1} at 75 °C. A mechanism consistent with these results involves a general acid catalysed hydrolysis as shown in Scheme 62.

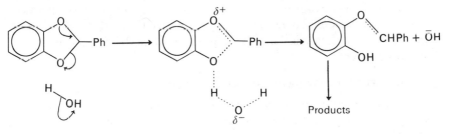

SCHEME 62.

2. Acetal hydrolysis in biological systems

Much of the interest in acetal hydrolysis began in 1965 when the structure of lysozyme, the enzyme which catalyses the hydrolysis of the glycosidic linkages of certain polysaccharides, was elucidated[522]. The catalytic groups are carboxyls of aspartic acid-52 (asp-52) and glutamic acid-35 (glu-35). Glu-35 acts as a general acid catalyst whilst asp-52 stabilizes the positively charged intermediate (equation 87).

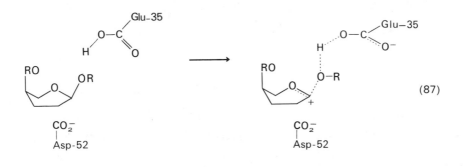

(87)

In this mechanism the sugar ring is held in a strained half-chair conformation and relief of strain in the transition state is suggested to facilitate
bond cleavage. Recently Atkinson and Bruice[523] have shown that the
hydrolysis of 2-methoxy-3,3-dimethyloxetane **142** involves general acid

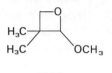

(**142**)

catalysis, so confirming that the proposed ring strain in the lysozyme
catalysed hydrolysis of the glycosidic bond could promote a general acid
catalysed bond cleavage. Catalysis by the carboxyl groups in the hydrolysis
of simple glycosides and acetals has been reviewed recently[524]. Studies of
intramolecular catalysis have been used as a model for the enzyme-
substrate system.

3. Intramolecular general acid catalysis

Intramolecular catalysis by carboxyl groups has been investigated with
salicylic acid derivatives. Rate enhancements of the order of 10^5–10^9 are
observed in comparison with corresponding compounds in which
carboxyl participation is not possible. For compounds in which intermolecular general acid catalysis is observed in the absence of a carboxyl
substituent, intramolecular general acid catalysis has been postulated when
the acidic substituent is present. For example the hydrolysis of 2-(o-
carboxyphenoxy) tetrahydropyran (**143**)[525], benzaldehyde methyl-o-
carboxyphenylacetal (**144**)[525], and the benzaldehyde disalicylacetal

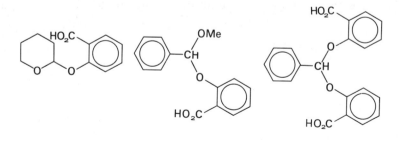

(**143**)   (**144**)   (**145**)

(145)[526] have been studied. In the latter case the pH–rate profile is bell-shaped and a maximum rate enhancement of k_{obs}, in comparison with the dimethyl ester, of 2.7×10^9 is found. Capon[527] has proposed a general acid catalysed mechanism (147) for the hydrolysis of 2-methoxymethoxy benzoic acid (146), whereas Dunn and Bruice[524,528] favour specific acid

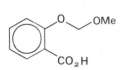 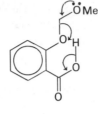 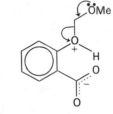

| (146) | (147) T.S. for general acid catalysis | (148) T.S. for specific acid catalysis |

catalysis of the hydrolysis of the anion 148. These mechanisms are kinetically indistinguishable. Recently Kraze and Kirby[529] have determined a Brönsted α value of zero for carboxy-group catalysis ($\rho_{carboxy} = 0.02 \pm 0.08$) in this reaction and a solvent deuterium isotope effect of 1.61 at 39 °C. This data indicates that proton transfer has not occurred to a significant degree in the transition state. A mechanism is proposed in which the solvated carboxyl group is brought into plane as the negative charge develops on the leaving group oxygen. The system is then stabilized by delocalization of the partial negative charge on the *ortho* carboxyl group, and the hydrogen bond of the salicylate anion will form. The suitability of salicylic acid derivates as models for enzyme-substrate reactions is consequently questioned.

4. Hydrolysis of ortho esters

It is worthwhile considering the hydrolysis of ortho esters[501] in context with that of the structurally related acetals. As with the acetals, recent emphasis has been placed on investigation of general acid catalysed hydrolyses[502]. Some of the earlier observations of general acid catalysis[530,531] have been disputed[521,532,533] as the influence of specific salt effects were not considered. Bunton and Wolfe[534] suggested a relatively basic ortho ester was necessary for general acid catalysis to be detected. Brönsted α-coefficients for both the hydrolysis of ethyl orthocarbonate[535] and methyl orthobenzoate[536] are approximately 0.7. No general acid catalysis could be detected in the hydrolysis of triphenyl orthoformate[537], where the stabilization of the intermediate oxocarbonium ion is not great.

In the hydrolysis of diethyl phenyl orthoformate (149), diphenyl ethyl orthoformate (150), and diphenyl ethyl orthoacetate (151) a large general acid catalysis is observed[538]. Brönsted α values for 149, 150 and 151 were 0·47, 0·68 and 0·49, respectively. The value of 0·47 is lower than any previously recorded. In diethyl phenyl orthoformate (149) the leaving group is good and the intermediate carbonium ion is well stabilized. With 150 the leaving group is the same but the intermediate carbonium ion is less stable and thus proton transfer is occurring to a greater extent in the transition state.

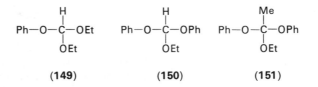

$$\begin{array}{ccc} & \text{H} & \text{H} & \text{Me} \\ & | & | & | \\ \text{Ph}-\text{O}-\text{C}-\text{OEt} & \text{Ph}-\text{O}-\text{C}-\text{OPh} & \text{Ph}-\text{O}-\text{C}-\text{OPh} \\ & | & | & | \\ & \text{OEt} & \text{OEt} & \text{OEt} \end{array}$$

(149)   (150)   (151)

5. Vinyl ether hydrolysis

a. *Mechanism.* Vinyl ether hydrolysis has been reported recently by several groups[539–543]. As represented by a simplified scheme (Scheme 63),

$$RCH{=}CH{-}OR^1 \xrightarrow[\text{slow}]{H^+} RCH_2CH{=}\overset{+}{O}R^1$$

$$\downarrow \text{fast} \quad H_2O$$

$$RCHO + R^1OH + H^+$$

SCHEME 63.

the reaction is generally believed to occur by an initial rate-determining proton transfer to the vinyl ether double bond to give an oxycarbonium ion which, in aqueous solution, decomposes rapidly to products. The oxycarbonium ion is equivalent to that found in acetal hydrolysis. Isotope effects, $k_{H_3O^+}/k_{D_3O^+}$, in the range 2·5–3·1[542,544] and Brönsted α values of 0·5–0·7[542,543] are consistent with a general acid catalysed process, with a rate-limiting proton transfer to the vinyl ether double bond. However, the kinetically equivalent specific acid–general base catalysed process (Scheme 64) cannot be ruled out[543]. In this mechanism deuterium exchange into unreacted material would be expected but has not been observed[539,545] indicating that the proton transfer is effectively irreversible.

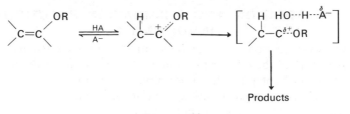

SCHEME 64.

b. *Electrostatic effects on vinyl ether hydrolysis.* The catalytic coefficients for the hydrolysis of ethyl vinyl ether catalysed by carboxylic acids containing dipolar substituents and by charged acids, show a distinct deviation from the Brönsted relation derived for neutral carboxylic acid catalysts[546,547]. An electrostatic interaction between the catalyst and the developing positive charge in the substrate is invoked to explain this observation. The interaction would be expected in the transition state for the rate-determining proton transfer, where catalyst and substrate are in close proximity. Positively charged catalysts would then be rate retarding whilst those bearing negative charges would cause a rate enhancement. Thus in the Brönsted relation, positively charged acids fall below the correlation line based on neutral acids, and negatively charged species lie above that line.

Hydrolyses of substituted α-methoxystyrenes yield the corresponding acetophenones except in the case where the ring contains an *ortho* carboxyl group, when the intermediate carbonium ion is trapped by the adjacent carboxyl group (Scheme 65) with formation of 3-methoxy-3-methylphthalide[543]. The hydronium ion and formic acid catalysed

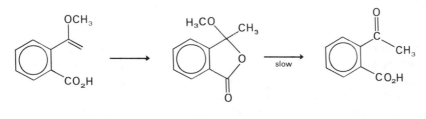

SCHEME 65.

reactions are accelerated by factors of 11·9 and 7·9, respectively, by ionization of the *o*-carboxyl group. These values are consistent with an electrostatic facilitation of the general acid catalysed oxycarbonium ion formation.

V. ADDITION AND ELIMINATION REACTIONS INVOLVING C=N

The cleavage reactions of the carbon–nitrogen double bond are discussed in Chapter 10 of the *Chemistry of the C=N bond*[548], and the methods of formation in Chapter 2 of the same volume[549]. In these sections the literature to 1967 is reviewed. Jencks's[550] book on *Catalysis in Chemistry and Enzymology* (1969) also discusses these topics.

A. C=N Forming Reactions

1. Mechanism

a. *General considerations*. The essential features of the formation of phenylhydrazones, semicarbazones, thiosemicarbazones, oximes and Schiff bases are well understood[549,550]. The reactions occur with a rate-determining addition of the nucleophile under slightly acidic conditions, but under neutral or basic conditions, the dehydration of the carbinol-amine intermediate becomes rate limiting. With strongly basic amines, for example hydroxylamine and aliphatic amines, the change in rate-determining step is normally observed between pH 2 and 5 (Scheme 66)[551–553]. The loss of hydroxide ion from the carbinolamine intermediate

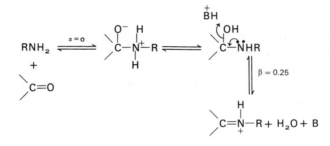

SCHEME 66.

is subject to general acid catalysis and values of the Brönsted coefficient α(0·75) indicate that the proton is close to oxygen in the transition state[552,553]. Electron-withdrawing groups on the carbonyl component inhibit the dehydration of the carbinolamine and favour the reverse reaction. Consequently the pH where the change in rate-determining step occurs decreases as increasingly electron-withdrawing substituents are introduced. A pH-independent process (equation 88) is also often observed in the dehydration of the carbinolamine formed from strongly basic amines. An exception is the reaction of hydroxylamine with aromatic carbonyl compounds where no pH-independent process has been

$$>\overset{|}{\underset{\curvearrowright}{N}}-\overset{|}{\underset{\curvearrowleft}{C}}-OH \;\rightleftharpoons\; >\overset{+}{N}=C< + \;^-OH \tag{88}$$

found[554,555]. Recently Amaral and coworkers[556] have demonstrated a pH-independent process in the dehydration of the carbinolamine formed from phenylhydrazine, a weakly basic amine, and substituted benzaldehydes.

For weakly basic amines (semicarbazide, phenylhydrazine, thiosemicarbazide and anilines) the change in rate-determining step with pH also occurs and is observed in the same, or slightly higher, pH region than for the more basic amines[556,557,558]. The mechanism of reaction is essentially the same as found with more basic amines but, owing to the presence of a less basic nitrogen, dehydration of the carbinolamine intermediate is more subject to acid catalysis and α approaches unity. Thus general acid catalysis of this step is less important. A base-catalysed pathway for dehydration is also possible as the amine becomes more acidic[554,557] and has been observed with hydroxylamines, semicarbazones and anilines. Transition states similar to the E2 and E1cB of olefin forming eliminations can be envisaged for the base-catalysed dehydration. These correspond to the concerted general base catalysis and specific base catalysis respectively, for the imine formation.

A second change in rate-determining step should occur when the base-catalysed dehydration becomes fast relative to the pH-independent addition. Until recently no examples of this had been observed as the addition step is also base catalysed, and generally cannot become rate determining. The attack of weakly basic amines on carbonyl compounds is subject to general acid catalysis ($\alpha = 0.25$). Several reports confirm that this is true general acid catalysis (**152**) rather than the kinetically equivalent specific acid–general base catalysed process (**153**)[559]. The addition is also subject to base catalysis.

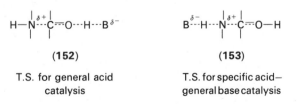

(152)	(153)
T.S. for general acid catalysis	T.S. for specific acid–general base catalysis

In recent years several authors have discussed whether general acid–base catalysis of complex reactions proceeds via stepwise proton transfers or by a concerted mechanism, and the subject has recently been reviewed by Jencks[559]. Evidence for discrete steps rather than a concerted process is based on changes in the rate-determining step not explained by bond forming and breaking involving carbon[560], breaks in Brönsted plots or in structure activity correlations[560,561], solvent isotope effects, and a dependence of rate on solution viscosity[562]. Jencks[559,563] has proposed a rule which states that concerted general acid–base catalysis of complex reactions in aqueous solution occurs only at sites where a large change in pK takes place during the reaction, and where the change in pK alters an unfavourable to a favourable proton transfer with respect to the catalyst.

Examples from the recent literature of Schiff base, hydrazone, semicarbazone, and oxime formation discussed in the next sections will illustrate some of the features of the introduction.

b. *Formation of aldimines, ketimines and Schiff bases.* Mixtures of a primary amine salt and a base may be used to catalyse the α-hydrogen exchange reactions of aldehydes and ketones[554,564,565]. An equilibrium mixture of imine and iminium ion is formed, the α-hydrogen being more acidic is then removed by the base. It has also been demonstrated that compounds with a primary amine group and another basic group in the same molecule may act as bifunctional catalysts (Scheme 67)[566,567].

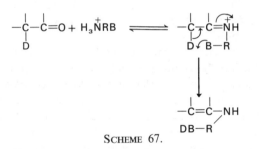

SCHEME 67.

In the de-deuteration of acetone-d_6 by this mechanism, deuterium transfer is rapid and imine formation is partly rate determining[567]. Equilibrium constants for the reaction of isobutyraldehyde with several amines have been determined[568–570]. The second-order rate constants for imine formation from isobutyraldehyde with n-propylamine, 3-methoxypropylamine, 2-methoxyethylamine, and 2,2,2-trifluoroethylamine in aqueous solutions are independent of pH above pH 10, and proportional to the hydrogen ion concentration below pH 9[570]. Electron-withdrawing groups decrease greatly the second-order rate constants for the uncatalysed reaction, whereas steric effects are less important. The equilibrium constant, K_{ca}, for carbinolamine formation is decreased by both electron withdrawing and bulky substituents. A plot of log K_{ca} against the pK_a values for the conjugate acids of the amines used, shows a deviation from linearity. Based on the linear free-energy relationship in equation (89), where P is a polar reaction

$$\log K_{ca} = PpK_a + SE_s + B \qquad (89)$$

constant, S is a steric reaction constant, E_s is the Taft steric substituent constant, and B is the intercept, a plot of $\log K_{ca}$ against $pK_a + (S/P)E_s$ gave a good linear correlation. Similar steric effects on equilibrium constants for additions to aldehydes have been reported by Jencks[571].

The second-order rate constants for the additions of primary amines containing a monoprotonated tertiary amino group in the same molecule $(H_2N-R-\overset{+}{N}HMe_2)$ to acetone are enhanced compared with rate constant for the addition of simple amines (primary n-alkylamines) and other bifunctional amines (ω-methoxy-, ω-dimethylamino- and 2-trimethyl-ammonio-n-alkylamines)[572].

The rate enhancement is rationalized in terms of catalysis of the dehydra-

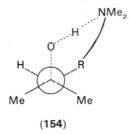

(154)

tion of the carbinol intermediate by the $\overset{+}{N}HMe_2$ substituent, via the transition state 154. Internal catalysis by phenolic groups has also been suggested[573]. Thus the rate constant for dehydration of the carbinol intermediate from 3-hydroxypyridine-4-carboxaldehyde and aniline is 60-times greater than that for pyridine-4-carboxaldehyde[573,574].

Page and Jencks[575] have examined reactions in which one amino group of a diamine acts as an internal basic catalyst for the other amino group.

The equilibrium constant for the formation of imidazolidines from acetone and isobutyraldehyde with ethylenediamines and N,N-dialkyl-ethylenediamines (Table 20) reflect the significance of steric factors[676], and decrease with increase in size of R″ (Scheme 68).

The corresponding values for the imidazolidinium ions formed from monoprotonated ethylenediamines and isobutyraldehyde were 658

TABLE 20. Equilibrium constants for imidazoline formation from isobutyraldehyde and ketones in Scheme 68[616]

Carbonyl component		Amine	K
R	R'	R"	(mol^{-1})
H	$CH(CH_3)_2$	H	2240
H	$CH(CH_3)_2$	CH_3	799
H	$CH(CH_3)_2$	C_2H_5	64·7
CH_3	CH_3	H	1·54
CH_3	CH_3	CH_3	0·63

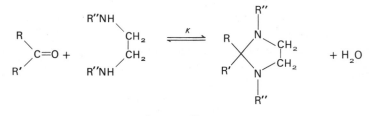

SCHEME 68.

(R" = H), 13·2 (R" = CH_3), and 2·27 mol^{-1} (R" = C_2H_5), and allow a calculation of the pK of the ions. These cannot be determined by standard titration owing to decomposition. The formation of imidazolidines may be important in the physiological action of folic acid derivatives[577].

The existence of the carbinolamine intermediate in the reaction of primary amines and carbonyl compounds has been confirmed using a stopped-flow technique, and the equilibrium constants for carbinolamine formation from the reaction of pyridine-4-aldehyde, p-chlorobenzaldehyde, and formaldehyde with primary and secondary amines have been measured[578]. The results are correlated by a free-energy relationship, $\log K_o = \Delta\gamma + A$, where Δ is a measure of the 'sensitivity' of the carbonyl compound to nucleophilic attack, γ is a measure of the ability of a compound to add to a carbonyl group, and A is a constant for a particular reaction series. The rates for carbinolamine formation are too fast to be studied by stopped-flow techniques[570,578], but recently Diebler and Thorneley[579] have determined rate constants for carbinolamine formation from piperazine and piperazine monocation with pyridine-4-carboxaldehyde using a temperature-jump study. The uncatalysed addition rate for piperazine (pK 9·97) is $2 \rightarrow 3 \times 10^5$ mol^{-1} sec^{-1}, and for piperazine monocation (pK 5·80) is 65 mol^{-1} sec^{-1}. The reaction of piperazine is subject to general base catalysis and the reaction of piperazine mono-cation to general acid catalysis.

A ketimine **155** (Scheme 69) has been detected as an intermediate in the decomposition of diacetone alcohol catalysed by *n*-propylamine[580], by spectral observation. The intermediate is formed rapidly and reversibly,

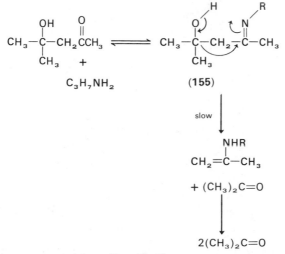

SCHEME 69.

decomposition to products being the slow step. A solvent isotope effect $(k_{H_2O}/k_{D_2O} = 1\cdot8)$ is in agreement with the accepted mechanism of imine formation from strongly basic amines, involving slow hydroxide ex-pulsion from a carbinolamine intermediate.

c. *Semicarbazone and hydrazone formation.* The attack of phenylhydra-zine and semicarbazide on carbonyl compounds is subject to both specific and general acid catalysis, and under mildly acidic conditions is the rate-determining step. Under neutral or basic conditions the rate-limiting process is dehydration of the carbinolamine intermediate. The dehydration step is subject to marked specific acid catalysis and weak general acid catalysis[550] $(\alpha \simeq 1\cdot0)$.

The addition of phenylhydrazine to benzaldehyde and its *p*-chloro-, *p*-methoxy, and *p*-hydroxy derivatives is subject to general acid catalysis by carboxylic acids and the conjugate acid of the nucleophile[581]. The reaction of *p*-nitrobenzaldehyde with phenylhydrazine, whilst catalysed by carboxylic acids, is inhibited by carboxylate ions. The inhibition may be due to complex formation between the carbonyl substrate and the carboxylate ion.

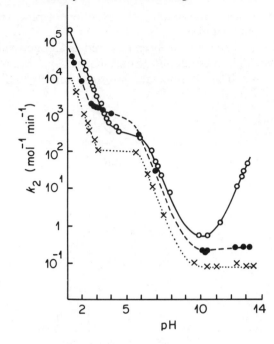

FIGURE 13. The logarithms of second-order rate coefficients for substituted benzaldehyde phenylhydrazone formation in 20% EtOH at 25 °C and ionic strength 0·5 M as a function of pH. (—○—, p-nitrobenzaldehyde; --●--, benzaldehyde; ··· × ···, p-methoxybenzaldehyde[581].

Rate constants for catalysis by the hydrated proton, chloroacetic acid, acetic acid, phenylhydrazinium ion, and water correlate with σ^+ substituent constants (Figure 13). Similar results have been obtained for semicarbazone formation[557]. The change to rate-determining carbinolamine dehydration occurs above pH 5. Between pH 5 and 9 the dehydration is specific acid catalysed. For benzaldehyde, p-chloro-, p-methoxy-, and p-nitro-benzaldehyde, a pH-independent process occurs above pH 9. For p-nitrobenzaldehyde a specific-base-catalysed carbinolamine dehydration, similar to that reported for semicarbazone and oxime formation[582], is observed. In the carbinolamine from p-nitrobenzaldehyde, assistance in the dehydration step from the nitrogen unshared electron pair is reduced by the electron-withdrawing nitro group and the base catalysed reaction becomes more important. The pH-independent processes are the first observed with weakly basic amines.

The kinetic α-deuterium isotope effects, k_D/k_H, for water-catalysed phenylhydrazone formation from p-methoxy- and p-nitro-benzaldehyde are 1·276 and 1·163, respectively[583,584]. The corresponding values for the reaction catalysed by the hydrated proton are 1·21 and 1·13. In the water-catalysed semicarbazone formation k_D/k_H was approximately 1·3, independent of the nature of the substituent. Values for the same reaction catalysed by hydrated proton and other acids are near 1·21. The values of the kinetic secondary isotope effects indicate that the degree of C—N bond formation in the transition states is decreased by increasing substrate reactivity, increasing reactivity of the nucleophile, and increasing acid strength of the catalyst. For semicarbazone formation from benzaldehyde, where the carbinolamine formation step is rate determining, k_D/k_H decreases from 1·31 to 1·20 as the substituent changes from hydrogen to p-nitro. The extent of C—O bond cleavage in the transition state for dehydration is therefore dependent on the substituent[584].

Ortho Cl, Br, NO_2, NH_2 and CH_3 substituents decrease the rate of reaction of acetophenone with semicarbazide but a methoxy substituent increases the rate by a factor of 3[585].

The second-order rate constants of pyruvic acid semicarbazone formations are highly dependent on pH[586]. In the pH ranges 0–2, 3–4, and 6–7 the rate constants show a linear dependence on the hydrated proton concentration. In the ranges 2–3 and 4–6 breaks occur in the pH-rate profile and the rate constants are insensitive to hydrated proton concentration. Rate-determining attack of semicarbazide on pyruvic acid below pH 2, and of semicarbazide on pyruvate between pH 3 and 6 are consistent with these results. Above pH 6 dehydration of the carbinolamine is rate limiting. The break in the pH-rate profile between pH 2–3 is not observed with pyruvic acid methyl ester semicarbazone formation (Figure 14). The reaction of semicarbazide with pyruvate is subject to general acid catalysis ($\alpha = 0·37$).

The reaction of 5-substituted furfurals with phenylhydrazine is subject to general acid catalysis by carboxylic acids ($\alpha = 0·35$)[587]. Acid-catalysed and pH-independent processes are observed in the dehydration of the carbinolamine. In addition, a base-catalysed dehydration is seen in the case of the carbinolamine from 5-nitrofurfural. The acid-catalysed dehydration is less sensitive to the nature of the polar substituent than is the pH-independent process. In contrast to previous observations[581,588] for phenylhydrazone formation, the rate constants for the water-catalysed attack are less sensitive to the nature of the polar substituent than those for the acid-catalysed reaction. No obvious explanation was offered for this observation.

The base-catalysed dehydration of carbinolamines from p-chlorobenzaldehyde and N-aminopyridinium chloride, p-toluenesulphonyl

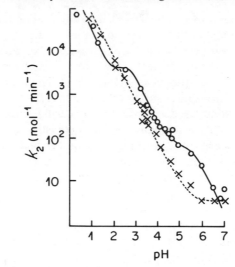

FIGURE 14. Logarithm of second-order rate coefficients for semicarbazone formation as a function of pH (—○—, pyruvic acid; ··· × ···, pyruvate)[586].

hydrazide, 2-methyl-3-thiosemicarbazide, thiosemicarbazide, acethydrazide, ethyl carbazate, semicarbazide, phenylhydrazine-*p*-sulphonic acid, and hydrazine have been reported[589]. pK_a values over this series vary from -7.2 for *N*-aminopyridinium chloride to 8.26 for hydrazine. Electron-withdrawing groups on nitrogen decrease the equilibrium constants for conversion of carbinolamine into hydrazone. A consistent value of the Brönsted coefficient β, for general base catalysis, of 0.7 is found for hydrazines varying in pK_a from -7.2 to 3.44. Of the two possible mechanisms of carbinolamine dehydration, equation (90) represents true general base analysis, whereas equation (91) corresponds to the kinetically

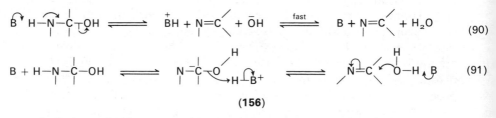

(**156**)

indistinguishable pre-equilibrium proton abstraction from nitrogen followed by a general acid catalysed hydroxide elimination. True general-base catalysis, in which hydroxide ion is lost essentially simultaneously, is preferred as calculations suggest that the nitrogen anion **156** is too unstable for its breakdown at a rate less than or equal to the diffusion-controlled limit to explain the observed reaction rate. It is, therefore, regarded as unnecessary to consider anion intermediates in the dehydration of moderately basic carbinolamines. A preassociation or spectator mechanism[559,590,591] in which the anion decomposition is faster than diffusion of BH$^+$ away from the ion pair 157, and without direct participation of the catalyst occurring in the transition state for the breakdown, is also possible. However, a value of β of 1·0 for general base catalysis would

$$\left[\begin{array}{c} R^1 \\ | \\ R-\bar{N}-C-H.\overset{+}{B}H \\ | \\ OH \end{array} \right]$$

(157)

be expected for this mechanism. The instability of the nitrogen anion also rules out the possibility of a 'one-encounter' mechanism[434] in which the catalyst carries a proton from nitrogen to oxygen. In a study of general acid catalysis of carbinolamine dehydration in this series[589], a small increase in α (hydrazine 0·62, thiosemicarbazide 0·73) with decreasing basicity of the hydrazine is found. Electron donation from nitrogen in the transition state is indicated by a decrease in rate caused by electron-withdrawing substituents on nitrogen. A more concerted than stepwise general-acid catalysis seems probable[559]. A pH-independent process is observed in the dehydration of carbinolamines derived from the more basic amines.

$\Delta H^{\neq}$ for attack of semicarbazide on p-hydroxybenzaldehyde catalysed by H_3O^+, HCO_2H and H_2O are all approximately 9 kcal/mol. Values of $\Delta S^{\neq}$ become more negative as the acidity of the catalyst decreases[592]. A tentative conclusion from this data is that the efficiency of the catalyst for this reaction is reflected mainly by changes in the entropy of activation.

d. *Thiosemicarbazone formation.* Thiosemicarbazide (pK = 1·88) is a weaker base than is semicarbazide (pK = 3·86) and although the essential mechanism of thiosemicarbazone formation would be expected to be the same as for its oxygen analogue, several interesting differences have been observed.

The change in rate-determining step for thiosemicarbazone formation from p-chlorobenzaldehyde occurs at pH 4[593]. The nucleophilic attack is subject to the general-acid catalysis by carboxylic acids (α = 0·25) and to basic catalysis. Above pH 8, concurrent with the usual dependence of carbinolamine dehydration on hydroxide ion concentration, a dependence on the basic component of the buffer is observed, that is, the reaction is general-base catalysed. Where the dehydration step is rate determining, a deuterium isotope effect k_{H_2O}/k_{D_2O}, for catalysis by tertiary amines is approximately 3, which is suggestive of loss of proton zero-point energy during its removal in the transition state.

A second change in rate-determining step with pH has been observed in the reaction of 2-methyl-3-thiosemicarbazide and p-chlorobenzalde-hyde[594,595].

Apart from the usual change from rate-determining addition to rate-determining carbinolamine dehydration occurring at pH 6, a second change at pH 10 is found where attack of the nucleophile again becomes rate limiting. The attack is subject to general-base catalysis. A non-linear Brönsted plot is indicative of a stepwise mechanism of proton transfer and C—N bond formation (Scheme 70). The postulated intermediate **158**

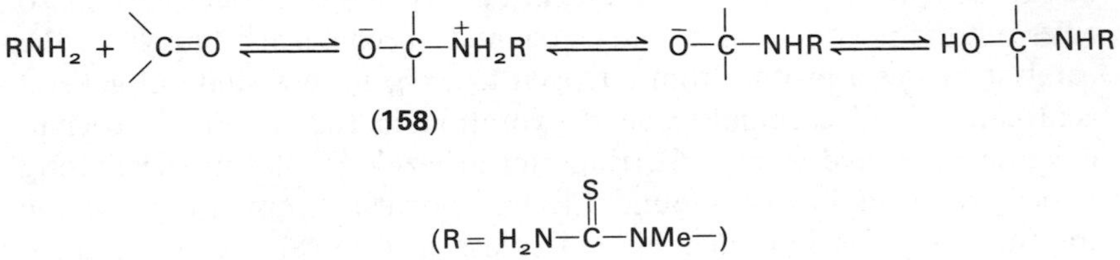

$$\text{RNH}_2 + \overset{\diagup}{\underset{\diagdown}{\text{C}}}=\text{O} \rightleftharpoons \bar{\text{O}}-\overset{|}{\underset{|}{\text{C}}}-\overset{+}{\text{N}}\text{H}_2\text{R} \rightleftharpoons \bar{\text{O}}-\overset{|}{\underset{|}{\text{C}}}-\text{NHR} \rightleftharpoons \text{HO}-\overset{|}{\underset{|}{\text{C}}}-\text{NHR}$$

(158)

$$(\text{R} = \text{H}_2\text{N}-\overset{\overset{\text{S}}{\|}}{\text{C}}-\text{NMe}-)$$

SCHEME 70.

would break down rapidly to starting material in the absence of a base-catalysed proton removal. If weaker nucleophiles than thiosemicarbazide are used it is likely that **158** would have too short a lifetime to exist as an intermediate and the reaction could be essentially concerted.

e. *Oxime formation.* Hydroxylamine is classed as a strongly basic amine and exhibits the expected rate-determining addition to carbonyl compounds at low pH, changing to a rate-limiting acid- and base-catalysed dehydration of the carbinolamine as the pH is increased. An inverse isotope effect ($k_{\bar{O}D}/k_{\bar{O}H}$) of 1·4, for hydroxide ion catalysed oxime formation is indicative of a specific base catalysed dehydration step[554].

The Hammett reaction constant ρ for the addition of hydroxylamine to a series of substituted acetophenones is 0·32[596]. This very small value is consistent with a reactant-like transition state for hydroxylamine addition.

Two breaks in the acid region of the pH-rate profile for the addition of O-methylhydroxylamine to benzaldehyde have been reported[597]. This observation is consistent with there being two changes in rate-determining step in this pH region and consequently necessitates the consideration of three kinetically significant steps in the reaction, one of which must be a proton transfer process. A dual mechanism for nucleophilic attack is suggested in which a concerted and stepwise process occur concurrently (Scheme 71). At a low pH the hydronium ion catalysed addition repre-

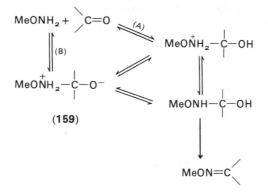

SCHEME 71.

sented by A is the predominant process. At a higher pH the mechanism proceeding via the dipolar intermediate **159** predominates. The change from path A to path B occurs because the rate-determining step for path B at low pH is the uncatalysed attack of the amine which becomes faster than the hydronium ion catalysed addition of path A as the pH is increased. The second break in the pH-rate profile occurs at pH 4–5 representing the usual transition from rate-limiting attack of the nucleophile to dehydration of the carbinolamine.

The α-deuterium isotope effect (k_H/k_D) for addition of hydroxylamine to ketones is generally less than unity[422,598]. Exceptions are 4-*t*-butyl-cyclohexanone and cyclohexanone where values slightly greater than unity are found (1·04 and 1·05 respectively). In these case steric effects are regarded as more important than electronic factors.

B. Addition Reactions of C=N

While several mechanistic studies on nucleophilic additions to C=O compounds have been reported, relatively few studies of equilibrium additions to the C=N bond have been made. Of these, most effort has been devoted to the investigation of the hydrolysis of Schiff bases, which will form the major part of this section. Additions of hydrogen cyanide, phenols, Grignard reagents, peracids and alcohols are mentioned only briefly.

1. Hydrolysis of C=N compounds

Since the formation of imines has been fully discussed it is unnecessary to deal in detail with the reverse reaction, as, by the principle of microscopic reversibility, the same transition state must be involved. Therefore, a reaction which is subject to general-acid catalysis in the forward direction should be subject to specific-acid catalysis and general-base catalysis in the reverse process (Scheme 72). General features of the hydrolysis mechanism

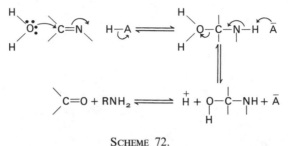

SCHEME 72.

are: (i) as the pH of the solution is lowered the rate-determining step for hydrolysis changes from nucleophilic attack to carbinolamine decomposition; (ii) decreasing the basicity of the component causes a decrease in the reactivity of the imine under basic conditions, but increases it under acidic conditions; (iii) protonated imines (i.e., the conjugate acids) are more reactive to nucleophilic attack than are the corresponding free bases.

In this section a few examples from the recent literature are discussed. An inverse secondary isotope effect ($k_D/k_H = 1.22$) is found for the attack of hydroxide ion on protonated N-benzylidene-1,1-dimethylethylamine and its 3-bromo- and 4-methoxy derivatives[599]. The isotope effect was independent of the nature of the substituent and its high value indicates that C—O bond formation in the transition state is almost complete. The first-order rate constants for the pH-independent hydrolysis of a series of substituted N-benzylideneanilines are insensitive to the nature of the polar substituent in either ring of the molecule[553,599], owing to the opposing effect of a polar substituent on substrate protonation and hydroxide ion attack. The second-order rate constants for the corresponding base-catalysed reaction increase as electron-withdrawing substituents are introduced. With the exception of Schiff bases derived from p-nitrobenzaldehyde ($\rho = 1.9$), values of ρ for alkaline hydrolysis of N-substituted benzylideneanilines are large (c. 2.7) and independent of the polar substituent on benzaldehyde, indicating adduct-like transition states for Schiff base hydrolysis.

In the hydrolysis of hydroxy and methoxy derivatives of N-benzylidene-2-aminopropane, it is important to take into account the tautomeric equilibrium (equation 92) when deriving an expression for the observed

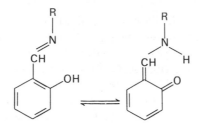

$$(92)$$

rate of hydrolysis of the *ortho* and *para* hydroxy derivatives[600]. A similar consideration is necessary in studies of the hydrolysis of salicylideneanilines[601]. No internal catalysis by the ionized hydroxy group in the hydrolysis of salicylideneanilines has been observed[602]. The *ortho* hydroxy group decreases the rate of hydrolysis between pH 8 and 9.

The hydrolysis of salicylideneanilines is faster in borate buffers than in phosphate buffers at the same pH[602,603]. A similar catalysis by borate is found in the hydrolysis of phenyl salicylate[604,605]. The structure **160** for the borate complex from salicylideneanilines is the most probable.

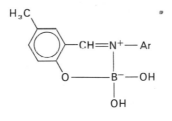

(160)

The nitrogen–boron bond formation then facilitates nucleophilic attack on the carbon–nitrogen double bond. Formation of the complex with boric acid is very rapid. In the alkaline region, the reaction rate is proportional to the concentration of the undissociated Schiff base, and in neutral solution the rate-determining step is the reaction of the complex with water or decomposition of the adduct formed.

Hoffmann has investigated the effect of –M substituents on the rate of hydrolysis of salicylideneanilines[606]. Deviations from a correlation of log k to σ or σ^-, in both acid and alkali, were found, indicating a change in rate-limiting step.

2. Addition of alcohols

In an investigation of the base-catalysed addition of alcohols to substituted benzylideneanilines, an inexplicable rate decrease in the order n-BuOH > EtOH > MeOH > s-BuOH > t-BuOH > i-PrOH is observed[607]. For the addition of methanol the equilibrium constant and rate constant are increased by electron-withdrawing substituents in either of the aromatic rings, showing correlation with σ^+ and σ, respectively. The observed correlation of the equilibrium constant with σ^+ for the substituent on the anilino ring has not been observed previously. A plausible mechanism for the addition of methanol (Scheme 73) involves a rate-limiting attack by methoxide ion.

$$MeO^- + \overset{\diagdown}{\underset{\diagup}{C}}=N- \underset{}{\overset{slow}{\rightleftharpoons}} MeO-\overset{|}{\underset{|}{C}}-\overset{-}{N}- \overset{MeOH}{\rightleftharpoons} MeO-\overset{|}{\underset{|}{C}}-NH-$$
$$+ MeO^-$$

SCHEME 73.

Unlike the acid-catalysed reaction, where a rapid solvolysis to aniline and a benzaldehyde dimethylacetal occurs, no solvolysis is observed in alkaline methanol.

3. Mechanism of addition of thiols to C=N

Hydrogen sulphide adds to ketimines in ether at low temperatures to yield *gem*-dithiols (equation 93).

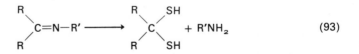

$$
\overset{R}{\underset{R}{\diagup}}\!\!\!\diagdown\!\! C=N-R' \longrightarrow \overset{R}{\underset{R}{\diagup}}\!\!\!\diagdown\!\! C\overset{\diagup SH}{\diagdown_{SH}} + R'NH_2 \qquad (93)
$$

Schiff bases react with thiols to give aminothioethers (Scheme 74), which may react further to give a disulphide and the reduced Schiff base[608].

ArCH=NAr $\xrightleftharpoons{\text{RSH}}$ ArCH(SR)NHAr $\xrightarrow{\text{RSH}}$ ArCH$_2$NHAr + RSSR

SCHEME 74.

A synthetically useful olefin synthesis is based on the addition of hydrogen sulphide to a hydrazone[609]. A synthesis of bicyclobutylidene (Scheme 75) has been developed using this method[610]. No detailed mechanistic studies have been made.

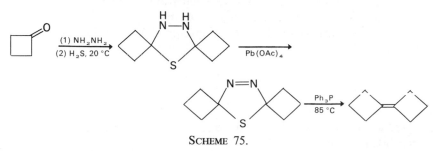

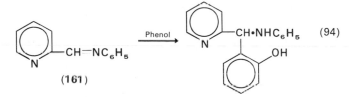

SCHEME 75.

4. Addition of phenols to Schiff bases

Whereas phenol does not react with *N*-benzylideneanilines, a facile addition to *N*-(2-pyridylmethylene)-aniline (**161**) is observed[611] (equation 94). Since *ortho*- and *para*-nitrobenzylidene anilines are unreactive under

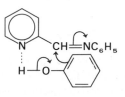

(94)

(161)

these conditions, an effect other than the electron-withdrawing ability of the pyridine nucleus is indicated. The mechanism proposed suggests the formation of a hydrogen bond between the phenol and the pyridine nitrogen atom (**162**). Support for this concept is provided by the failure of methyl salicylate and *ortho*-hydroxyacetophenone to react with the Schiff base.

(162)

5. Addition of HCN to C=N compounds

HCN adds to imines, Schiff bases, hydrazones, oximes and other similar compounds (equation 95). Hydrolysis of the product from imines and Schiff bases offers a synthetic approach to amino acids. More recently Schiff bases from optically active benzylamines and aliphatic aldehydes have been treated with HCN. Hydrolysis and hydrogenolysis of the resulting α-cyanoamines gives optically active amino acids[612–614]. Optical yields vary from 22–90%, although it seems probable that the higher yields[614] may have arisen by a fractionation during purification of the product.

$$\begin{array}{c}\diagdown \\ \diagup \end{array}C=N- \ + \ HCN \ \longleftrightarrow \ \begin{array}{c}\diagdown \\ \diagup \end{array}\underset{CN}{\overset{|}{C}}-NH- \qquad\qquad (95)$$

Ogata and Kawasaki[615,616] have reported the addition of HCN to benzylideneanilines.

6. Addition of Grignard reagents to oximes

Arylalkyl oximes react with Grignard reagents to yield aziridines (Scheme 76) The reaction is similar in many respects to the Neber

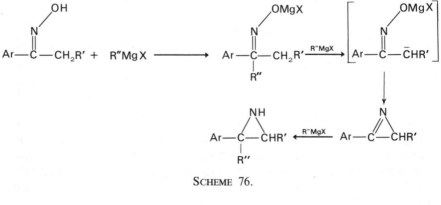

SCHEME 76.

$$\underset{NH}{\overset{\displaystyle PhCH\text{———}CH_2}{\diagdown\;\diagup}}$$

(163)

rearrangement[617]. Support for the azirine intermediate has been provided by the isolation of styrenimine **(163)**, when lithium aluminium hydride is added to the reaction mixture of ethylmagnesium bromide and acetophenone oxime[618]. Azirine formation is stereospecific in toluene, with ring formation occurring with the group *syn* to the hydroxy group, but in tetrahydrofuran a mixture of two possible products are found[619]. Addition of the Grignard reagent to the azirine takes place from the least hindered face of the molecule[619].

7. Reduction of C=N compounds

Imines, hydrazones and Schiff bases are reduced by $LiAlH_4$, $NaBH_4$ or by hydrogenation. Iminium salts are also reduced by $LiAlH_4$. Oximes are reduced to amines though recently the $LiAlH_4$ reduction of oximes has been developed as a useful synthesis of aziridines. A review on this subject has appeared recently[620].

8. Peracid oxidation of imines

Two mechanisms have been proposed for oxazirane formation from imines (equations 96, 97). Negative ρ values for substituted imines and a

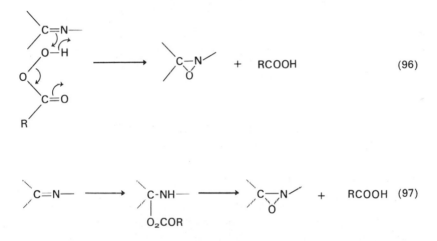

significant acid catalysis have been cited as evidence for the one-step mechanism (equation 96)[621]. The two-step mechanism (equation 97) is analogous to the Baeyer–Villiger oxidation of a carbonyl group, while the one-step process is equivalent to epoxide formation from alkenes. The two-step mechanism is favoured by Ogata and his coworkers for the oxidation of benzylidene *t*-butylamines with perbenzoic acid[622]. The reaction is accelerated by carboxylic acids and by alcohols, but is retarded by the addition of an excess of some alcohols. No explanation is obvious. Substituent effects on the imine varied with the reaction conditions and were negligible in the presence of ethanol. Nitrones are also formed in the addition of peracids to cyclic and acyclic imines[622,623]. Normally, oxazirane formation is predominant but nitrone formation (equation 98) is significant in aprotic solvents, and for imines bearing electron-donating substituents. This reaction represents the attack of the nitrogen lone-pair electrons on the peracid oxygen.

$$\ce{\overset{\diagdown}{\underset{\diagup}{C}}=N- + PhCOOOH -> \overset{\diagdown}{\underset{\diagup}{C}}=\overset{+}{\underset{\underset{O^-}{|}}{N}}- + PhCOOH}$$
(98)

The oxidation of 2-substituted cyclohexanone oximes by peroxytrifluoroacetic acid gives predominantly the *cis*-2-substituted-nitrocyclohexanes[624]. Protonation of a nitronate intermediate (**164**) from the less hindered face of the molecule can be used to rationalize this result (equation 99).

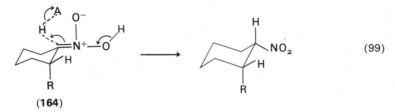

(99)

(**164**)

VI. THIOCARBONYL ADDITIONS AND ELIMINATIONS

A. Eliminations giving C=S

Although the term β-elimination is normally associated with eliminations giving rise to alkenes, the same mechanism in certain cases is applicable to reactions yielding carbon–heteroatom double bonds. In this section, eliminations leading to thiocarbonyls are briefly discussed. A review of routes to thiocarbonyl compounds is available[625].

The disulphide linkage may be cleaved by three main pathways involving nucleophilic reagents (e.g., metal alkoxides). A direct nucleophilic attack on sulphur (equation 100) leads to formation of a sulphenic acid derivative, while base promoted removal of a proton beta to the disulphide linkage favours olefin formation (equation 101). Thiocarbonyl compounds may arise if removal of the α-proton is preferred (equation 102). The nucleophilic attack on sulphur is dominant for reagents such as $\overline{C}N$, $\overline{S}CN$ and RS^-, whereas the elimination mechanisms are promoted by more basic nucleophiles such as hydroxide and alkoxide ions. The formation of a thiocarbonyl compound will obviously be encouraged by preventing alkene formation (i.e., substrates lacking β-hydrogens) and where the acidity of the α-hydrogen is high. Thus, when diphenylmethyl disulphides are treated with sodium isopropoxide (equation 103), the yield of the corresponding thione varied from 86–52% at 20°C, depending on the nature of the substituents X and Z[626,627]. Increased yields were obtained

$$RSSR + R'O^- \longrightarrow RS^- + RSOR' \tag{100}$$

$$RSS(CO_2)_2R' + R''O^- \rightleftharpoons RSSCH_2\bar{C}HR' + R''OH$$

$$\downarrow$$

$$RSS^- + R'CH=CH_2 \tag{101}$$

$$RSSCH_2R' + R''O^- \rightleftharpoons RSS\bar{C}HR' + R''OH$$

$$\downarrow$$

$$RS^- + R'CH=S \tag{102}$$

$$\tag{103}$$

with increasing electron-withdrawing ability of both X and Z. The main competing reaction was S_N2 displacement at sulphur (cf. equation 100). Contrary to expectations of $E2/S_N2$ behaviour, the yield of the thiobenzophenone decreased with increasing reaction temperature. A kinetic deuterium isotope effect, k_H/k_D, for the decomposition of α-biphenyl-4-yl-benzyl-4-tolyl disulphide and its α-deutero analogue at 30°C was 6·1, and is near the maximum of 6·6 predicted for intermediate proton transfer between carbon and oxygen. This high value is not consistent with a reversible carbanion mechanism and this is additionally ruled out by the lack of isotopic exchange with labelled solvent[627]. The Hammett reaction constants ρ_z and ρ_x for the thiocarbonyl elimination from $ZC_6H_4(Ph)$-$CHSSC_6H_4X$, for a limited number of substrates of 4 and 2, respectively, are indicative of a transition state with considerable carbanion character. Despite the extended cleavage of the C—H and S—S bonds, the authors proposed a transition state with little double-bond character (**165**). Delocalization of the developing negative charge on the central carbon atom would normally be expected but may not occur in this case due to the poor effectiveness of the sulphur 3p orbitals in forming π-bonds.

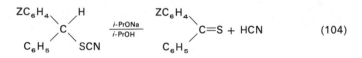

$$i\text{-PrO}\cdots\overset{\delta^-}{\text{H}}\cdots\overset{\overset{\displaystyle \text{Ar}}{|}}{\underset{|}{\overset{\delta^-}{\text{C}}}}\overset{\delta^+}{=\!\!=}\overset{}{\text{S}}\cdots\overset{\delta^-}{\text{S}}\!-\!\text{Ar}'$$
$$\underset{\text{Ar}}{}$$

(165)

In the thiocarbonyl elimination from diphenylmethylthiocyanates (equation 104), a Hammett reaction constant of 3·5 at 20 °C and a primary deuterium isotope effect of 3·0 (Z = p-Ph) have been determined[628]. These data indicate a concerted elimination involving a transition state with considerable carbanion character, similar to the E2 eliminations of 2-arylethyl derivatives with bases. Elimination from the unsubstituted thiocyanate occurs 300-times faster than the elimination from the corresponding disulphide. This difference may arise because of the greater acidity of the α-hydrogen in the thiocyanate as CN is more electron withdrawing than the SPh group.

$$\underset{C_6H_5}{\overset{ZC_6H_4}{\diagdown}}\!\!C\!\!\underset{SCN}{\overset{H}{\diagup}}\quad\xrightarrow[\;i\text{-PrOH}\;]{\;i\text{-PrONa}\;}\quad\underset{C_6H_5}{\overset{ZC_6H_4}{\diagdown}}\!\!C\!\!=\!\!S\;+\;HCN \qquad (104)$$

Block[629] has devised a method for generating sulphenic acids and a thiocarbonyl compound is formed simultaneously. The unstable sulphenic acid can be trapped by methyl acetylenecarboxylate (Scheme 77, cf. sulphoxide pyrolysis, Section II.H.2).

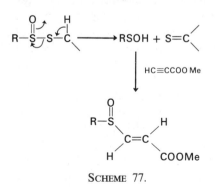

Scheme 77.

The flash thermolysis of allyl sulphides has been used to prepare thio-carbonyls[630] (equation 105). The resulting thiones are labile under the thermolysis conditions.

$$\begin{array}{c}\text{(structure)} \end{array} \longrightarrow \quad >C{=}S + H_2C{=}CHCH_3 \qquad (105)$$

B. Addition Reactions of C=S

Relatively few mechanistic studies of this type of addition have been reported. The oxidation of thiobenzophenones with perbenzoic acid has been examined[631] and also the 1,3-cyclo-additions of benzonitrile N-oxides with thiocarbonyls[632]. Organolithium additions to thiobenzo-phenones[633–636] occur by thiophilic addition to give an organolithium compound rather than a thiolate. However, in the reaction of prenyl lithium (166) with adamantanethione[637], the initial product (167) under-goes a [2,3]-sigmatropic rearrangement leading to formation of a C—C rather than a C—S bond (Scheme 78). The substrate was chosen because it is non-enolizable and stable. Reduction is generally a competing process in the reaction of alkyl lithiums with thiocarbonyl compounds.

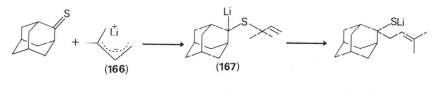

(166)   (167)

SCHEME 78.

The higher polarizability of the C=S bond than the C=O group is often invoked to explain the difference in reactivity, but the exact mecha-nism is unclear.

Diazomethane adds to C=S compounds to yield thiadiazolines[638–640] by a 1,3-dipolar addition (equations 106, 107). Whilst 168 and 169 give regiospecifically the Δ^3-1,3,4-thiadiazolines (170, 171), the cycloaddition of diazomethane with adamantanethione gives a mixture of the Δ^3-1,3,4-thiadiazoline (172) and the Δ^2-1,2,3-thiadiazoline (173). The product ratio is very dependent on the solvent[640], the proportion of 173 increasing with solvent polarity (e.g., ether 20%, methanol 78%).

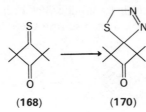

(168)   (170)

(106)

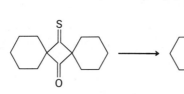

(169)   (171)

(107)

(172)   (173)

Allyl Grignard reagents **(174)** add to thioketones to give β,γ-unsaturated thiols**(175)**[641] (equation 108). The product may be cyclized to give substituted thiolanes **(176)**.

$$R_2C{=}S + R'CH{=}CHCH_2MgBr \longrightarrow R_2C(SH)CH_2CH{=}CHR' \qquad (108)$$

         (174)   **(175)**

(176)

VII. ACKNOWLEDGEMENT

We thank Dr. D. C. Horwell for his comments on the manuscript.

VIII. REFERENCES

1. A. Maccoll, in *The Chemistry of Alkenes* (Ed. S. Patai) Interscience, London, 1964, p. 203–240.
2. W. H. Saunders, Jr., in *The Chemistry of Alkenes* (Ed. S. Patai) Interscience, London, 1964, p. 149–202.
3. J. F. Bunnett, *Angew. Chem. Int. Ed.* **1**, 225 (1962).
4. E. H. White and D. J. Woodcock, in *The Chemistry of the Amino Group* (Ed. S. Patai), Interscience, London, 1968.
5. D. V. Banthorpe, in *The Chemistry of the Amino Group* (Ed. S. Patai), Interscience, London, 1968.
6. H. Knozinger, in *The Chemistry of the Hydroxyl Group* (Ed. S. Patai), Interscience, London, 1971, Part 2.
7. D. V. Banthorpe, *Elimination Reactions,* Elsevier, Amsterdam, 1963.
8. W. H. Saunders, Jr. and A. F. Cockerill, *Mechanisms of Elimination Reactions,* Wiley-Interscience, New York, 1973.
9. A. F. Cockerill, in *Comprehensive Chemical Kinetics* (Ed. C. H. Bamford and C. F. H. Tipper), Elsevier, Amsterdam, 1973, Volume 9, Chapter 3.
10. J. F. Bunnett, Olefin-forming Elimination Reactions, *Survey Prog. Chem.,* **5**, 53 (1969).
11. J. Sicher, *Angew, Chem. Int. Ed.,* **11**, 200 (1972).
12. D. J. McLennan, *Quart. Rev.* (*London*), **21**, 490 (1967).
13. F. G. Bordwell, *Accounts Chem. Research,* **3**, 281 (1970).
14. F. G. Bordwell, *Accounts Chem. Research,* **5**, 374 (1972).
15. A. Maccoll, *Chem. Rev.,* **69**, 33 (1969).
16. G. G. Smith and F. W. Kelly, *Prog. Phys. Org. Chem.,* **8**, 75 (1971).
17. H. Noller, P. Andréu, and M. Hunger, *Angew. Chem. Int. Ed.,* **10**, 172 (1971).
18. S. I. Miller, *Adv. Phys. Org. Chem.,* **6**, 185 (1968).
19a. W. Hanhart and C. K. Ingold, *J. Chem. Soc.,* 997 (1927).
19b. E. D. Hughes, *J. Amer. Chem. Soc.,* **57**, 708 (1935).
19c. E. D. Hughes, C. K. Ingold, and C. S. Patel, *J. Chem. Soc.,* 526 (1933).
20. P. S. Skell and W. L. Hall, *J. Amer. Chem. Soc.,* **85**, 2851 (1963).
21a. G. Wittig and R. Polster, *Ann. Chem.,* **599**, 13 (1956).
21b. A. C. Cope, N. A. LeBel, P. T. Moore, and W. R. Moore, *J. Amer. Chem. Soc.,* **83**, 3861 (1961).
22. F. G. Bordwell, A. C. Knipe, and K. C. Yee, *J. Amer. Chem. Soc.,* **92**, 5945 (1970).
23. Z. Rappoport, *Tetrahedron Letters,* 3601 (1968).
24. Z. Rappoport and E. Shohamy, *J. Chem. Soc.* (*B*), 2060 (1971).
25. R. Breslow, *Tetrahedron Letters,* 399 (1964).
26. D. J. McLennan and R. J. Wong, *Tetrahedron Letters,* 2891 (1972).
27. W. K. Kwok, W. G. Lee, and S. I. Miller, *J. Amer. Chem. Soc.,* **91**, 468 (1969).
28. W. H. Saunders, Jr., *Survey Prog. Chem.,* **3**, 109 (1966).
29. A. J. Parker, *Quart. Rev.,* (*London*), **16**, 163 (1962); *Adv. Org. Chem.,* **5**, 1 (1965).
30. J. Hine, *Physical Organic Chemistry,* McGraw-Hill Book Co., 2nd Ed., New York, 1962, Chapter 8.
31. F. C. Whitmore and M. G. Whitmore, *Org. Syntheses, Coll. Vol. 1,* 401 (1941).
32. A. Berndt, *Angew. Chem. Int. Ed.,* **8**, 613 (1969).
33. F. H. Westheimer, *Chem. Rev.,* **61**, 265 (1961).

34. F. G. Bordwell, M. M. Vestling, and K. C. Yee, *J. Amer. Chem. Soc.,* **92**, 5950 (1970).
35. P. F. Cann and C. J. M. Stirling, *J. Chem. Soc. Perkin II,* 817, 820 (1974).
36. J. Crosby and C. J. M. Stirling, *J. Amer. Chem. Soc.,* **90**, 6869 (1968); *J. Chem. Soc.* (*B*), 671, 679 (1970).
37. C. A. Bunton and V. J. Shiner, *J. Amer. Chem. Soc.,* **83**, 42 (1961).
38. Z. Rappoport, *J. Chem. Soc.* (*B*), 171 (1971).
39. W. von E. Doering and A. K. Hoffmann, *J. Amer. Chem. Soc.,* **77**, 521 (1955).
40. S. Ašperger, N. Ilakovac, and D. Pavlović, *J. Amer. Chem. Soc.,* **83**, 5032 (1961).
41. R. Bird and C. J. M. Stirling, *J. Chem. Soc.* (*B*), 111 (1968).
42. A. I. Biggs and R. A. Robinson, *J. Chem. Soc.,* 388 (1961).
43. A. Streitwieser, Jr., A. P. Marchand, and A. H. Pudjaatmaka, *J. Amer. Chem. Soc.,* **89**, 693 (1967).
44. R. A. More O'Ferrall and S. Slae, *J. Chem. Soc.* (*B*), 260 (1970).
45. R. A. More O'Ferrall, *J. Chem. Soc.* (*B*), 268 (1970).
46. L. R. Fedor, *J. Amer. Chem. Soc.,* **91**, 908 (1969).
47. L. R. Fedor, *J. Amer. Chem. Soc.,* **91**, 913 (1969).
48. R. C. Cavestri and L. R. Fedor, *J. Amer. Chem. Soc.,* **92**, 4610 (1970).
49. L. R. Fedor and W. R. Glave, *J. Amer. Chem. Soc.,* **93**, 985 (1971).
50. A. K. Colter and R. D. Johnson, *J. Amer. Chem. Soc.,* **84**, 3289 (1962).
51. J. Banger, A. F. Cockerill, and G. L. O. Davies, *J. Chem. Soc.,* (*B*), 498 (1971).
52. T. I. Crowell, R. T. Kemp, R. E. Lutz, and A. A. Wall, *J. Amer. Chem. Soc.,* **90**, 4638 (1968).
53. Reference 8, page 16.
54. K. N. Barlow, D. R. Marshall, and C. J. M. Stirling, *J. Chem. Soc. Chem. Comm.,* 175 (1973).
55a. S. J. Cristol, *J. Amer. Chem. Soc.,* **67**, 1494 (1945).
55b. S. J. Cristol *et al., J. Amer. Chem. Soc.,* **74**, 3333 (1952).
56. B. D. England and D. J. McLennan, *J. Chem. Soc.* (*B*), 696 (1966).
57. R. J. Anderson, P. Aug, B. D. England, V. H. McCann, and D. J. McLennan, *Aust. J. Chem.,* **22**, 1427 (1969).
58. D. J. McLennan and R. J. Wong, *J. Chem. Soc. Perkin II,* 279 (1972).
59. D. J. McLennan, *J. Chem. Soc.* (*B*), 705, 709 (1966).
60. D. J. McLennan and R. J. Wong, *Tetrahedron Letters,* 2887 (1972).
61. D. J. McLennan and R. J. Wong, *J. Chem. Soc. Perkin II,* 526 (1974).
62. A. F. Cockerill, *J. Chem. Soc.* (*B*), 964 (1967).
63. A. F. Cockerill and W. H. Saunders, Jr., *J. Amer. Chem. Soc.,* **89**, 4985 (1967).
64. A. V. Willi and M. Wolfsberg, *Chemistry and Industry,* (*London*), 2097 (1964).
65. F. G. Bordwell and W. J. Boyle, *J. Amer. Chem. Soc.,* **93**, 512 (1971).
66. see Reference 8, pp. 82, 514.
67a. D. J. McLennan and R. J. Wong, *Tetrahedron Letters,* 881 (1970);
67b. O. R. Jackson, D. J. McLennan, S. A. Short, and R. J. Wong, *J. Chem. Soc. Perkin II,* 2308 (1972).
68. J. Weinstock, R. G. Pearson, and F. G. Bordwell, *J. Amer. Chem. Soc.,* **78**, 3473 (1956).
69a. S. J. Cristol and N. L. Hause, *J. Amer. Chem. Soc.,* **74**, 2193 (1952);
69b. S. J. Cristol and E. F. Hoegger, *J. Amer. Chem. Soc.,* **79**, 3438 (1957);
69c. S. J. Cristol and R. P. Arganbright, *J. Amer. Chem. Soc.,* **79**, 3441 (1957).

70. J. Weinstock, J. L. Bernadi, and R. G. Pearson, *J. Amer. Chem. Sos.*, **80**, 4961 (1958).
71. J. Hine and O. B. Ramsay, *J. Amer. Chem. Soc.*, **84**, 973 (1962).
72. F. G. Bordwell, J. Weinstock, and T. F. Sullivan, *J. Amer. Chem. Soc.*, **93**, 4728 (1971).
73. A. V. Willi, *Helv. Chim. Acta*, **49**, 1735 (1966).
74. L. A. Kaplan and N. E. Burlinson, *J. Org. Chem.*, **37**, 3932 (1972).
75. J. A. Orvik and J. F. Bunnett, *J. Amer. Chem. Soc.*, **92**, 2417 (1970).
76. Z. Rappoport and R. Ta-Shma, *J. Chem. Soc.*, (*B*), 1461 (1971).
77a. E. Lord, M. P. Naan, and C. D. Hall, *J. Chem. Soc.* (*B*), 220 (1971);
77b. M. P. Naan, A. P. Bell, and C. D. Hall, *J. Chem. Soc. Perkin II*, 1821 (1973).
78. D. J. Cram and A. S. Wingrove, *J. Amer. Chem. Soc.*, **86**, 5490 (1964).
79. V. Fiandanese, G. Marchese, and F. Naso, *J. Chem. Soc. Chem. Comm.*, 250 (1972).
79b. V. Fiandanese, G. Marchese, and F. Naso, *J. Chem. Soc. Perkin II*, 1536 (1973).
80. S. Patai, S. Weinstein, and Z. Rappoport, *J. Chem. Soc.*, 1741 (1962).
81. J. Hine and L. A. Kaplan, *J. Amer. Chem. Soc.*, **82**, 2915 (1960).
82. R. F. Pratt and T. C. Bruice, *J. Amer. Chem. Soc.*, **92**, 5956 (1970).
83. A. F. Hegarty and L. N. Frost, *J. Chem. Soc. Perkin II*, 1719 (1973).
84. J. Zemlicka, R. Gasser, and J. P. Horwitz, *J. Amer. Chem. Soc.*, **92**, 4744 (1970).
85. F. G. Bordwell, D. A. R. Happer, and G. D. Cooper, *Tetrahedron Letters*, 2759 (1972).
86. W. H. Saunders, Jr., *Tetrahedron Letters*, 5129 (1972).
87. A. C. Knipe, *Organic Reaction Mechanisms* (Ed. B. Capon and C. W. Rees), Wiley, London, 1972, Chapter 4, p. 144.
88. R. A. More O'Ferrall, in *The Chemistry of the Carbon–Halogen Bond* (Ed. S. Patai), Interscience, London, 1973, Chapter 9.
89. R. A. More O'Ferrall, *J. Chem. Soc.* (*B*), 274 (1970).
90. E. R. Thornton, *J. Amer. Chem. Soc.*, **89**, 2915 (1967).
91. W. H. Saunders, Jr., and D. H. Edison, *J. Amer. Chem. Soc.*, **82**, 138 (1960).
92. W. H. Saunders, Jr., D. Bushmann, and A. F. Cockerill, *J. Amer. Chem. Soc.*, **90**, 1775 (1968).
93. C. H. DePuy and D. H. Froemsdorf, *J. Amer. Chem. Soc.*, **79**, 3710 (1957).
94. C. H. DePuy and C. A. Bishop, *J. Amer. Chem. Soc.*, **82**, 2532 (1960).
95. C. H. DePuy and C. A. Bishop, *J. Amer. Chem. Soc.*, **82**, 2535 (1960).
96. A. F. Cockerill and W. J. Kendall, *J. Chem. Soc. Perkin II*, 1352 (1973).
97. R. Baker and M. J. Spillett, *J. Chem. Soc.* (*B*), 481 (1969).
98. G. Ayrey, A. N. Bourns and V. A. Vyas, *Canad. J. Chem.*, **41**, 1759 (1963).
99a. P. J. Smith, *Ph.D. Thesis*, McMaster University, 1965;
99b. P. J. Smith and A. N. Bourns, *Canad. J. Chem.*, **52**, 749 (1974).
100. A. N. Bourns and P. J. Smith, *Proc. Chem. Soc.*, 366 (1964).
101. P. J. Smith and A. N. Bourns, *Canad. J. Chem.*, **48**, 125 (1970).
102a. A. Streitwieser, Jr. and H. F. Koch, *J. Amer. Chem. Soc.*, **86**, 404 (1964).
102b. M. Shima, D. N. Bhattacharyya, J. Smid, and M. Szwarc, *J. Amer. Chem. Soc.*, **85**, 1306 (1963);
102c. K. Bowden, A. F. Cockerill, and J. R. Gilbert, *J. Chem. Soc.* (*B*), 179 (1970).
103. L. J. Steffa and E. R. Thornton, *J. Amer. Chem. Soc.*, **89**, 6149 (1967).
104. C. H. DePuy, C. G. Naylor, and J. A. Beckman, *J. Org. Chem.*, **35**, 2750 (1970).

105. A. C. Frosst, *Ph.D. Thesis,* McMaster University, 1968.
106. E. L. Eliel, N. L. Allinger, S. J. Angyal, and G. A. Morrison, *Conformational Analysis,* Wiley-Interscience, New York, 1965, p. 482.
107. K. Bowden, *Chem. Rev., 66,* 119 (1966).
108. C. Reichardt, *Angew. Chem. Int. Ed., 4,* 29 (1965).
109. H. H. Jaffé, *Chem. Rev., 53,* 191 (1953).
110. D. J. Cram, C. H. DePuy, and F. D. Greene, *J. Amer. Chem. Soc., 78,* 790 (1956).
111a. H. M. R. Hoffmann, *J. Chem. Soc.,* 6753, 6762 (1965);
111b. G. M. Fraser and H. M. R. Hoffmann, *J. Chem. Soc.,* 265 (1967).
112. A. F. Cockerill, *Tetrahedron Letters,* 4913 (1969).
113. A. F. Cockerill, S. Rottschaefer, and W. H. Saunders, Jr., *J. Amer. Chem. Soc., 89,* 901 (1967).
114a. J. Banger, *M.Sc. Thesis,* University of East Anglia, Norwich, England, 1969;
114b. J. Banger, A. F. Cockerill, and G. L. O. Davies, unpublished results.
115. W. H. Saunders, Jr. and K. C. Brown, personal communication of unpublished results.
116. R. P. Bell and B. G. Cox, *J. Chem. Soc. (B),* 652, 783 (1971).
117. P. J. Smith and S. K. Tsui, *Tetrahedron Letters,* 917 (1972).
118. G. S. Hammond, *J. Amer. Chem. Soc., 77,* 334 (1955).
119a. A. M. Katz and W. H. Saunders, Jr., *J. Amer. Chem. Soc., 91,* 4469 (1969).
119b. W. H. Saunders, Jr., unpublished results and abstracts of papers p. 232, *14th Nordic Chemical Meeting,* Umea, Sweden, June 18–22, 1971 (quoted in Reference 8, p. 76).
120. R. A. More O'Ferrall and J. Kouba, *J. Chem. Soc. (B),* 985 (1967).
121. R. P. Bell and D. M. Goodall, *Proc. Roy. Soc. (London), A, 294,* 273 (1966).
122a. F. A. Long and J. L. Longridge, *J. Amer. Chem. Soc., 89,* 1292 (1967);
122b. L. F. Blackwell, P. D. Buckley, K. W. Jolly, and A. K. H. MacGibbon, *J. Chem. Soc.,* 169 (1973);
122c. S. Alunni and E. Baciocchi, *Tetrahedron Letters,* 4665 (1973);
122d. S. Alunni, E. Baciocchi, and V. Mancini, *J. Chem. Soc. Perkin II,* 1866 (1974).
123. H. C. Brown and G. J. McDonald, *J. Amer. Chem. Soc., 88,* 2514 (1966).
124. E. A. Halevi, M. Nussim, and A. Ron, *J. Chem. Soc.,* 866 (1963).
125. V. J. Shiner, Jr., *Tetrahedron, 5,* 243 (1959).
126. A. Streitwieser, Jr., R. H. Jagow, R. C. Fahey, and S. Suzuki, *J. Amer. Chem. Soc., 80,* 2326 (1958).
127. R. F. Hudson and G. Klopman, *J. Chem. Soc.,* 5 (1964).
128. W. Hückel, W. Tappe and G. Legutke, *Ann. Chem., 543,* 191 (1940).
129. S. J. Cristol, N. L. Hause and J. S. Meek, *J. Amer. Chem. Soc., 73,* 674 (1951).
130. J. Hine, *J. Amer. Chem. Soc., 88,* 5525 (1966).
131. O. S. Tee, *J. Amer. Chem. Soc., 91,* 7144 (1969).
132. K. Fukui and H. Fujimoto, *Tetrahedron Letters,* 4303 (1965).
133. W. T. Dixon, *Chem. Commun.,* 402 (1967).
134. J. P. Lowe, *J. Amer. Chem. Soc., 94,* 3718 (1972).
135. C. K. Ingold, *Proc. Chem. Soc.,* 265 (1962).
136. J. Sicher and J. Závada, *Coll. Czech. Chem. Commun., 33,* 1278 (1968).
137. J. Sicher, J. Závada, and J. Krupička, *Tetrahedron Letters,* 1619 (1966).
138. J. Závada, M. Svoboda and J. Sicher, *Tetrahedron Letters,* 1627 (1966).
139. J. Závada, M. Svoboda and J. Sicher, *Coll. Czech. Chem. Commun., 33,* 4027 (1968).

140. M. Svoboda, J. Závada and J. Sicher, *Coll. Czech. Chem. Commun.*, **33**, 1415 (1968).
141. J. Závada and J. Sicher, *Coll. Czech. Chem. Commun.*, **32**, 3701 (1967).
142. L. Schotsman, P. J. C. Fierens, and T. Verlie, *Bull. Soc. Chim. Belges*, **68**, 580 (1959).
143. J. Závada, J. Krupička, and J. Sicher, *Coll. Czech. Chem. Commun.*, **33**, 1393 (1968).
144. D. Bethell and A. F. Cockerill, *J. Chem. Soc.*, 913 (1966).
145. J. G. Traynham, W. B. Stone, and J. L. Couvillon, *J. Org. Chem.*, **32**, 510 (1967).
146. J. Sicher, *Prog. Stereochemistry*, **3**, 210 (1962).
147. J. L. Coke, M. P. Cooke, and M. C. Mourning, *Tetrahedron Letters*, 2247 (1968).
148. M. Svoboda, J. Závada, and J. Sicher, *Coll. Czech. Chem. Commun.*, **32**, 2104 (1967).
149. D. V. Banthorpe, A. Louden, and F. D. Waller, *J. Chem. Soc. (B)*, 509 (1967).
150. R. A. Bartsch, *Tetrahedron Letters*, 297 (1970); *J. Amer. Chem. Soc.*, **93**, 3683 (1971).
151. D. H. Froemsdorf, W. Dowd, W. A. Gifford, and S. Meyerson, *Chem. Commun.*, 449 (1968).
152. D. S. Bailey, F. C. Montgomery, G. W. Chodak, and W. H. Saunders, Jr., *J. Amer. Chem. Soc.*, **92**, 6911 (1970).
153. D. H. Froemsdorf, H. R. Pinnick, Jr., and S. Meyerson, *Chem. Commun.*, 1600 (1968).
154. W. F. Bayne and E. I. Snyder, *Tetrahedron Letters*, 571 (1971).
155. C. H. DePuy, G. F. Morris, J. S. Smith, and R. J. Smat, *J. Amer. Chem. Soc.*, **87**, 2421 (1965).
156. R. A. Bartsch and K. E. Wiegers, *Tetrahedron Letters*, 3819 (1972); see also V. Fiandanese, G. Marchese, F. Naso, and O. Sciacovelli, *J. Chem. Soc. Perkin II*, 1336 (1973).
157. R. A. Bartsch, G. M. Pruss, R. L. Buswell, and B. A. Bushaw, *Tetrahedron Letters*, 2621 (1972).
158. J. Závada, M. Pánková, and J. Sicher, *Coll. Czech. Chem. Commun.*, **37**, 2414 (1972).
159. J. Sicher, J. Závada, and M. Pánková, *Coll. Czech. Chem. Commun.*, **36**, 3140 (1971).
160. D. S. Bailey and W. H. Saunders, Jr., *J. Amer. Chem. Soc.*, **92**, 6904 (1970).
161. M. Pánková, M. Svoboda, and J. Závada, *Tetrahedron Letters*, 2465 (1972).
162. J. Závada, M. Pánková, and M. Svoboda, *J. Chem. Soc. Chem. Commun.*, 168 (1973).
163. J. Závada and M. Svoboda, *Tetrahedron Letters*, 23 (1972).
164. M. Svoboda, J. Hapala, and J. Závada, *Tetrahedron Letters*, 265 (1972).
165. H. C. Brown and O. H. Wheeler, *J. Amer. Chem. Soc.*, **78**, 2199 (1956).
166. W. H. Saunders, Jr., S. R. Fahrenholtz. E. A. Caress, J. P. Lowe, and M. R. Schreiber, *J. Amer. Chem. Soc.*, **87**, 3401 (1965).
167. R. A. Bartsch and J. F. Bunnett, *J. Amer. Chem. Soc.*, **90**, 408 (1968).
168. D. V. Banthorpe, E. D. Hughes, and C. K. Ingold, *J. Chem. Soc.*, 4054 (1960).
169. S. Alunni and E. Baciocchi, *Tetrahedron Letters*, 205 (1973).
170. R. A. Bartsch and J. F. Bunnett, *J. Amer. Chem. Soc.*, **91**, 1376 (1969).
171. I. N. Feit and W. H. Saunders, Jr., *J. Amer. Chem. Soc.*, **92**, 5615 (1970).

172. H. C. Brown, I. Moritani, and Y. Okamoto, *J. Amer. Chem. Soc.*, **78**, 2193 (1956).
173a. D. H. Froemsdorf and M. E. McCain, *J. Amer. Chem. Soc.*, **87**, 3983 (1965).
173b. D. H. Froemsdorf and M. D. Robbins, *J. Amer. Chem. Soc.*, **89**, 1737 (1967).
174. D. H. Froemsdorf, M. E. McCain, and W. Wilkinson, *J. Amer. Chem. Soc.*, **87**, 3984 (1965).
175. H. C. Brown, D. L. Griffith, and D. L. Meges, *Chem. Commun.*, 90 (1968).
176. H. C. Brown and R. L. Klimisch, *J. Amer. Chem. Soc.*, **87**, 5517 (1965).
177. D. H. Froemsdorf, W. Dowd, and K. E. Leimer, *J. Amer. Chem. Soc.*, **88**, 2345 (1966).
178. I. N. Feit, F. Schadt, J. Lubinkowski, and W. H. Saunders, Jr., *J. Amer. Chem. Soc.*, **93**, 6606 (1971).
179. J. K. Borchardt, R. Hargreaves, and W. H. Saunders, Jr., *Tetrahedron Letters*, 2307 (1972).
180. H. C. Brown and M. Nakagawa, *J. Amer. Chem. Soc.*, **78**, 2197 (1956).
180a. R. A. Bartsch, K. E. Wiegers, and D. M. Guritz, *J. Amer. Chem. Soc.*, **96**, 430 (1974).
181. S. P. Acharya and H. C. Brown, *Chem. Commun.*, 305 (1968).
182. H. C. Brown and I. Moritani, *J. Amer. Chem. Soc.*, **78**, 2203 (1956).
183. M. Pánková and J. Závada, *Tetrahedron Letters*, 2237 (1973).
184. J. M. A. Al-Rawi, J. A. Elvidge, D. K. Jaiswal, J. R. Jones, and R. Thomas, *J. Chem. Soc. Chem. Commun.*, 220 (1974).
185. S. Winstein, D. Darwish, and N. J. Holness, *J. Amer. Chem. Soc.*, **78**, 2915 (1956).
186. A. J. Parker, *Quart. Rev.* (*London*), **16**, 163 (1962): *Adv. Phys. Org. Chem.*, **5**, 173 (1967).
187. G. Biale, A. J. Parker, S. G. Smith, I. D. R. Stevens, and S. Winstein, *J. Amer. Chem. Soc.*, **92**, 115 (1970).
188. D. J. Lloyd, and A. J. Parker, *Tetrahedron Letters*, 637 (1971).
189. J. Hayami, N. Ono, and A. Kaji, *Tetrahedron Letters*, 1385 (1968).
190. W. T. Ford, *Accounts Chem. Research*, **6**, 410 (1973).
191a. E. L. Eliel and R. S. Ro, *Tetrahedron*, **2**, 353 (1958).
191b. E. L. Eliel and R. G. Haber, *J. Amer. Chem. Soc.*, **81**, 1249 (1959).
192. J. Csapilla, *Chimia*, **18**, 37 (1964).
193. A. J. Parker, M. Ruane, G. Biale, and S. Winstein, *Tetrahedron Letters*, 2113 (1968).
194. A. J. Parker, M. Ruane, D. A. Palmer, and S. Winstein, *J. Amer. Chem. Soc.*, **94**, 228 (1972).
195. J. F. Bunnett and E. Baciocchi, *J. Org. Chem.*, **35**, 76 (1970).
196. D. N. Kevill, G. A. Coppens, and N. H. Cromwell, *J. Amer. Chem. Soc.*, **86**, 1553 (1964).
197. D. J. Lloyd and A. J. Parker, *Tetrahedron Letters*, 5183 (1968).
198. P. Beltrame, G. Biale, D. J. Lloyd, A. J. Parker, M. Ruane, and S. Winstein, *J. Amer. Chem. Soc.*, **94**, 2240 (1972).
199a. D. N. Kevill and J. E. Dorsey, *J. Org. Chem.*, **34**, 1985 (1969);
199b. G. Biale, A. J. Parker, I. D. R. Stevens, J. Takahashi, and S. Winstein *J. Amer. Chem. Soc.*, **94**, 2235 (1972).
200a. R. A. More O'Ferrall, *J. Chem. Soc.* (*B*), 785 (1970);
200b. E. S. Lewis and M. C. R. Symons, *Quart, Rev.* (*London*), **12**, 230 (1958).

201. D. J. Lloyd and A. J. Parker, *Tetrahedron Letters,* 5029 (1970).
202. D. J. McLennan, *J. Chem. Soc. (B),* 705 (1966).
203. D. N. Kevill, E. D. Weiler, and N. H. Cromwell, *J. Amer. Chem. Soc.,* **88,** 4489 (1966).
204. P. B. D. de la Mare and C. A. Vernon, *J. Chem. Soc.,* 41 (1956).
205. J. F. Bunnett and E. Baciocchi, *J. Org. Chem.,* **32,** 11 (1967).
206. J. F. Bunnett and D. Eck, *J. Amer. Chem. Soc.,* **95,** 1897, 1900, 4473 (1973).
207. E. C. F. Ko and A. J. Parker, *J. Amer. Chem. Soc.,* **90,** 6447 (1968).
207a. *Carbonium Ions,* Volumes 1–4 (Ed. G. A. Olah and P. von R. Schleyer), Wiley-Interscience, New York, 1968–1974.
208a. R. A. Sneen and H. M. Robbins, *J. Amer. Chem. Soc.,* **91,** 3100 (1969);
208b. R. A. Sneen, *Accounts Chem. Research,* **6,** 46 (1973).
209. D. J. McLennan, *Tetrahedron Letters,* 2317 (1971).
210. A. Maccoll, *Chem. Rev.,* **69,** 33 (1969).
211. A. Maccoll, *Adv. Phys. Org. Chem.,* **3,** 91 (1965).
212a. G. G. Smith and F. W. Kelly, *Prog. Phys. Org. Chem.* **8,** 75 (1971);
212b. Reference 8, Chapter 8.
213a. S. W. Benson and G. Haugen, *J. Phys. Chem.,* **69,** 3898 (1965).
213b. J. T. Bryant and G. O. Pritchard, *J. Phys. Chem.,* **71,** 3439 (1967).
214a. N. Barroeta, A. Maccoll, M. Cavazza, L. Congiu, and A. Fava, *J. Chem. Soc. (B),* 1264, 1267 (1971).
214b. N. Barroeta, A. Maccoll, and A. Fava, *J. Chem. Soc. (B),* 347 (1969).
214c. N. Barroeta, V. De Sautis and R. Mazzali, *J. Chem. Soc. Perkin II,* 769 (1972).
215. H. M. R. Hoffmann and A. Maccoll, *J. Amer. Chem. Soc.,* **87,** 3774 (1965).
216. C. A. Kingsbury and D. J. Cram, *J. Amer. Chem. Soc.,* **82,** 1810 (1960).
217. J. R. Shelton and K. E. Davis, *J. Amer. Chem. Soc.,* **89,** 718 (1967).
218. T. C. Bruice and P. T. Markiw, *J. Amer. Chem. Soc.,* **79,** 3150 (1957).
219. T. S. Chou, J. R. Burgtorf, A. L. Ellis, S. R. Lambert, and S. P. Kukolja, *J. Amer. Chem. Soc.,* **96,** 1609 (1974).
220. D. N. Jones, D. Mundy, and R. D. Whitehouse, *J. Chem. Soc. Chem. Commun.,* 86 (1970).
221. R. E. Gilman *et al., Canad. J. Chem.,* **48,** 970 (1970).
222. E. E. Smissman, J. P. Li, and M. W. Creese, *J. Org. Chem.,* **35,** 1352 (1970).
223. C. Djerassi and W. S. Briggs, *J. Org. Chem.,* **33,** 1625 (1968).
224. K. G. Rutherford, R. M. Ottenbrite, and B. K. Tang, *J. Chem. Soc. (C),* 582 (1971).
225. W. E. Barnett and W. H. John, *J. Chem. Soc. Chem. Commun.,* 1002 (1971).
226. D. B. Bigley and C. M. Wren, *J. Chem. Soc. Perkin II,* 926, 1744, 2359 (1972).
227. S. Oae, K. Tsujihara, and N. Furukawa, *Tetrahedron Letters,* 2663 (1970).
228. D. G. Morris, B. W. Smith, and R. J. Wood, *Chem. Commun.,* 1134 (1968).
229. K. D. Berlin, J. G. Morgan, M. E. Peterson, and W. C. Pivonka, *J. Org. Chem.,* **34,** 1266 (1969).
230. L. C. Roach and W. H. Daly, *J. Chem. Soc. Chem. Commun.,* 66 (1970).
231. W. T. Borden, P. W. Concannon, and D. I. Phillips, *Tetrahedron Letters,* 3161 (1973).
232a. Reference 8, Chapter 6.
232b. H. Noller, P. Andréu, and M. Hunger, *Angew. Chem. Int. Ed.,* **10,** 172 (1971).
232c. G. Mochida, Y. Agu, A. Kato, and T. Seiyaina, *Bull. Chem. Soc. Japan,* **43,** 2245 (1970).

233a. D. S. Noyce, D. R. Hartter, and F. B. Miles, *J. Amer. Chem. Soc.*, **90**, 3794 (1968).

233b. D. S. Noyce, D. R. Hartter, and P. M. Pollock, *J. Amer. Chem. Soc.*, **90**, 3791 (1968).

234. A. P. G. Kieboom and H. van Bekkum, *Rec. Trav. Chim.*, **88**, 1424 (1969).

235a. R. S. Monson and L. N. Priest, *J. Org. Chem.*, **36**, 3826 (1971).

235b. J. S. Lomas, D. S. Sagatys, and J. E. Dubois, *Tetrahedron Letters*, 165 (1972).

236. R. E. A. Dear, E. E. Gilbert, and J. J. Murray, *Tetrahedron*, **27**, 3345 (1971).

237. D. N. Kirk and P. M. Shaw, *J. Chem. Soc.* (*C*), 182 (1970).

238a. G. M. Atkins, Jr. and A. M. Burgess, *J. Amer. Chem. Soc.*, **90**, 4744 (1968).

238b. P. Crabbé and C. Léon, *J. Org. Chem.* **35**, 2594 (1970).

239. R. O. Hutchins, M. G. Hutchins, and C. A. Milewski, *J. Org. Chem.*, **37**, 4190 (1972).

240. J. C. Martin and R. J. Arhart, *J. Amer. Chem. Soc.*, **93**, 4327 (1971).

241. I. M. Mathai, K. Schug, and S. I. Miller, *J. Org. Chem.*, **35**, 1733 (1970).

242a. S. I. Miller, *Adv. Phys. Org. Chem.*, **6**, 185 (1968).

242b. Reference 8, Chapter 7.

243a. C. S. Tsai Lee, I. M. Mathai, and S. I. Miller, *J. Amer. Chem. Soc.* **92**, 4602 (1970);

243b. I. M. Mathai and S. I. Miller, *J. Org. Chem.*, **35**, 3416 (1970).

244. E. Baciocchi and C. Lillocci, *J. Chem. Soc. Perkin II*, 38 (1973).

245a. J. Hine and W. H. Brader, *J. Amer. Chem. Soc.*, **77**, 361 (1955);

245b. J. Csapilla, *Chimia*, **18**, 37 (1964).

246a. E. Vogel, H. Kiefer, and W. R. Roth, *Angew. Chem. Int. Ed.*, **3**, 442 (1964).

246b. C. G. Scouten, *et al.*, *J. Chem. Soc. Chem. Commun.*, 78 (1969).

247a. J. K. Kochi and D. M. Singleton, *J. Amer. Chem. Soc.*, **90**, 1582 (1968).

247b. J. K. Kochi, D. M. Singleton, and L. J. Andrews, *Tetrahedron*, **24**, 3503 (1968).

248. S. Patai and Z. Rappoport, *The Chemistry of Alkenes* (Ed. S. Patai), Interscience, London, 1964, Chapter 8.

249a. Z. Rappoport, *Adv. Phys. Org. Chem.*, **7**, 1 (1969).

249b. G. Modena, *Accounts Chem. Research*, 4, 73 (1971).

249c. Reference 8, Chapter 9.

249d. G. Kobrich, *Angew. Chem. Int. Ed.*, **4**, 49 (1965); **6**, 41 (1967).

250. G. Marchese, F. Naso, N. Tangari, and G. Modena, *J. Chem. Soc.* (*B*), 1196 (1970).

251. E. Winterfeldt, *Angew. Chem. Int. Ed.*, **6**, 423 (1967).

252. W. K. Kwok, W. G. Lee, and S. I. Miller, *J. Amer. Chem. Soc.*, **91**, 468 (1969).

253. R. Huisgen, B. Giese, and H. Huber, *Tetrahedron Letters*, 1883 (1967).

254a. E. Lord, M. P. Naan, and C. D. Hall, *J. Chem. Soc.* (*B*), 1401 (1970); 213 (1971).

254b. M. P. Naän, A. P. Bell, and C. D. Hall, *J. Chem. Soc. Perkin II*, 1821 (1973).

255. Z. Rappoport and P. Peled, *J. Chem. Soc. Perkin II*, 616 (1973).

256a. D. Landini, F. Montanari, G. Modena, and F. Naso, *J. Chem. Soc.* (*B*), 243 (1969).

256b. G. Marchese, F. Naso, N. Tangari, and G. Modena, *J. Chem. Soc.* (*B*), 1196 (1970).

257. S. J. Cristol and C. A. Whittemore, *J. Org. Chem.*, **34**, 705 (1969).

258. M. Schlosser and V. Ladenberger, *Chem. Ber.,* **100**, 3877, 3893, 3901 (1967).
259a. E. D. Hughes, *Trans. Faraday Soc.,* **34**, 185 (1938); **37**, 603 (1941).
259b. L. L. Miller and D. A. Kaufman, *J. Amer. Chem. Soc.,* **90**, 7282 (1968).
260a. P. J. Stang and R. Summerville, *J. Amer. Chem. Soc.,* **91**, 4600 (1969).
260b. P. J. Stang, R. J. Hargrove, and T. E. Dueber, *J. Chem. Soc. Perkin II,* 843 (1974).
261a. S. A. Sherrod and R. G. Bergmann, *J. Amer. Chem. Soc.,* **91**, 2115 (1969).
261b. M. Hanack and T. Bassler, *J. Amer. Chem. Soc.,* **91**, 2117 (1969).
262. Z. Rappoport and Y. Apeloig, *J. Amer. Chem. Soc.,* **91**, 6734 (1969).
263. M. Hanack, *Accounts Chem. Res.,* **3**, 209 (1970).
264. A. T. Nielsen and W. J. Houlihan, *Organic Reactions,* **16**, 1 (1968).
265. G. Jones, *Organic Reactions,* **15**, 204 (1967).
266. A. Maercker, *Organic Reactions,* **14**, 270 (1965).
267. J. Reucroft and P. G. Sammes, *Quart. Rev. (London),* **25**, 135 (1971).
268. H. O. House, *Modern Synthetic Reactions,* 2nd ed., Benjamin Inc., New York 1972, Chapter 10.
269. J. Hine, J. G. Houston, J. H. Jenson, and J. Mulders, *J. Amer. Chem. Soc.,* **87**, 5050 (1965).
270. D. C. Gutsche, R. S. Buriks, K. Nowotny, and H. Grassner, *J. Amer. Chem. Soc.,* **84**, 3775 (1962).
271a. S. Cabani and N. Ceccanti, *J. Chem. Soc. (B),* 77 (1966).
271b. M. Stiles and A. Longroy, *Tetrahedron Letters,* 337, (1961).
271c. D. S. Noyce and W. L. Reed, *J. Amer. Chem. Soc.,* **80**, 5539 (1958).
272. R. L. Reeves in *The Chemistry of the Carbonyl Group* (Ed. S. Patai), Interscience, London, 1966, Vol. 1, p. 589.
273. J. E. Dubois and M. Dubois, *Chem. Commun.,* 1567 (1968).
274. J. E. Dubois and M. Dubois, *Tetrahedron Letters,* 4215 (1967).
275. B. Miller, H. Margulies, T. Drabb, Jr., and R. Wayne, *Tetrahedron Letters,* 3801, 3805 (1970).
276. J. E. Dubois and J.-F. Fort, *Tetrahedron,* **28**, 1653 (1972).
277. A. Schriesheim and C. A. Rowe, Jr., *Tetrahedron Letters,* 405 (1962).
278. J. E. Dubois and J. F. Fort, *Tetrahedron,* **28**, 1665 (1972).
279. H. O. House, D. S. Crumine, A. Y. Teranishi, and H. D. Olmstead, *J. Amer. Chem. Soc.,* **95**, 3310 (1973).
280a. G. Wittig, *Rec. Chem. Prog.,* **28**, 45 (1967).
280b. G. Wittig and A. Hesse, *Organic Syntheses,* **50**, 66 (1970).
280c. G. Wittig and H. Reiff, *Angew. Chem. Int. Ed.,* **7**, 7 (1968).
281. V. M. Belikov, Yu. N. Belokon, N. S. Martinkova, and N. G. Faleev, *Ser. Acad. Nauk S.S.S.R. Ser. Chim.,* **10**, 2136 (1969).
282. V. M. Belikov, Yu, N. Belokon, N. G. Faleev, and C. B. Korchemnaya, *J. Amer. Chem. Soc.,* **90**, 1477 (1968).
283. V. M. Belikov, Yu. N. Belokon, N. G. Faleev, and V. A. Maksakor, *Tetrahedron,* **28**, 3789 (1972).
284. R. G. Pearson and R. Dillon, *J. Amer. Chem. Soc.,* **75**, 2439 (1953).
285. F. S. Prout, V. D. Beauclaire, G. R. Dyrkacz, W. M. Koppes, R. E. Kuznicki, T. A. Marlewski, J. J. Pienkowski, and J. M. Puda, *J. Org. Chem.,* **38**, 1512 (1973).
286. R. A. Hann, *J. Chem. Soc. Perkin I,* 1379 (1974).

287. C. C. Tung, A. J. Speziale, and H. W. Frazier, *J. Org. Chem.*, **28**, 1514 (1963).
288. F. W. Bachelor and R. K. Bansal, *J. Org. Chem.*, **34**, 3600 (1969).
289. J. Seyden-Penne, M. C. Row-Schmitt, and A. Rowe, *Tetrahedron*, **26**, 2649, 2657 (1970).
290. J. A. Deyrup, *J. Org. Chem.*, **34**, 2724 (1969).
291. B. Deschamps and J. Seyden-Penne, *C. R. Acad. Sci. Paris, Ser. C*, **271**, 1097 (1970).
292. B. Deschamps and J. Seyden-Penne, *Tetrahedron*, **27**, 3959 (1971).
293. A. W. Johnson, *Ylide Chemistry*, Academic Press, New York, 1966, Chapter 4.
294. *The Chemistry of the Carbonyl Group* (Ed. S. Patai), Interscience, London, 1966, p. 570.
295. M. Schlosser and K. F. Christmann, *Ann. Chem.*, **708**, 1 (1967).
296. S. Trippett, *Quart. Rev. (London)*, **17**, 406 (1963).
297. G. Wittig and A. Haag, *Chem. Ber.*, **96**, 1535 (1963).
298. E. Vedejs and K. A. J. Snoble, *J. Amer. Chem. Soc.*, **95**, 5778 (1973).
299. F. Ramirez, C. P. Smith, and J. F. Pilot, *J. Amer. Chem. Soc.*, **90**, 6726 (1968).
300. M.-U. Hague, C. N. Caughlan, F. Ramirez, J. F. Pilot, and C. P. Smith, *J. Amer. Chem. Soc.*, **93**, 5229 (1971).
301. E. E. Schweizer, D. M. Crouse, T. Minami, and A. T. Wehman, *J. Chem. Soc. Chem. Commun.*, 1000, (1971).
302. D. J. H. Smith and S. Trippett, *J. Chem. Soc. Chem. Commun.*, 191 (1972).
303. N. J. De'ath and S. Trippett, *J. Chem. Soc. Chem. Commun.*, 172 (1969).
304a. D. J. Petersen, *J. Org. Chem.*, **33**, 780 (1968).
304b. T. H. Chan, E. Chang, and E. Vinokur, *Tetrahedron Letters*, 1137 (1970).
304c. F. A. Carey and A. S. Court, *J. Org. Chem.*, **37**, 939, 1926 (1972).
305. C. Trindle, J.-T. Hwang, and F. A. Carey, *J. Org. Chem.*, **38**, 2664 (1973).
306. C. Rüchardt, P. Panse, and S. Eichler, *Chem. Ber.*, **100**, 1144 (1967).
307. P. Frøyen, *Acta Chem. Scand.*, **26**, 2163 (1972).
308. W. P. Schneider, U. Axen, F. H. Lincoln, J. E. Pike, and J. L. Thompson, *J. Amer. Chem. Soc.*, **90**, 5895 (1968).
309. L. D. Bergelson and M. M. Shemyakin, *Tetrahedron*, **19**, 149 (1963); *Angew. Chem.*, **76**, 113 (1964); *Pure and Applied Chem.*, **9**, 271 (1964).
310. L. D. Bergelson, L. I. Barsukov, and M. M. Shemyakin, *Tetrahedron*, **23**, 2709 (1967).
311. W. P. Schneider, *J. Chem. Soc. Chem. Commun.*, 785 (1969).
312. U. Axen, F. H. Lincoln, and J. L. Thompson, *J. Chem. Soc., Chem. Commun.*, 303 (1969).
313. J. Boutagy and R. Thomas, *Chem. Rev.*, **74**, 87 (1974).
314. G. Lavielle and D. Reisdorf, *C. R. Acad. Sci. Paris*, **272**, 100 (1971).
315. E. J. Corey and D. E. Cane, *J. Org. Chem.*, **34**, 3053 (1969).
316. E. J. Corey and G. T. Kwaitkowski, *J. Amer. Chem. Soc.*, **88**, 5652, 5653, 5654 (1966).
317. F. Jung, N. K. Sharma, and T. Durst, *J. Amer. Chem. Soc.*, **95**, 3420 (1973).
318. S. C. K. Wong and A. W. Johnson, *Canad. J. Chem.*, **44**, 2793 (1966); *J. Org. Chem.*, **37**, 1850 (1972).
319. A. W. Johnson and H. Schubert, *J. Org. Chem.*, **35**, 2678 (1970).
320. Y. T. Huang, W. Y. Ting, and H. S. Cheng, *Acta Chim. Sinica.*, **31**, 37 (1965).
321. A. W. Johnson, *J. Org. Chem.*, **25**, 183 (1960).
322. D. Lloyd and M. I. C. Singer, *Chemistry and Industry (London)*, 510 (1967).
323. M. C. Henry and G. Wittig, *J. Amer. Chem. Soc.*, **82**, 563 (1960).

324. A. Maccioni and M. Secci, *Rend. Seminario, Fac. Sci. Univ. Cagliari,* **34**, 328 (1964); *Chem. Abstr.,* **63**, 5674e (1965).
325. T. Eicher in *The Chemistry of the Carbonyl Group* (Ed. S. Patai), Interscience, London, 1966, Vol. 1, p. 638.
326. E. C. Ashby, *Quart. Rev. (London),* **21**, 259 (1967); E. C. Ashby, J. Laemmle, and H. M. Neumann, *Accounts Chem. Res.,* **7**, 272 (1974).
327a. W. Schlenk and W. Schlenk, Jr., *Chem. Ber.,* **62**, 920 (1929).
327b. W. Schlenk, *Chem. Ber.,* **64**, 734 (1931).
328a. E. C. Ashby and W. E. Becker, *J. Amer. Chem. Soc.,* **85**, 118 (1963);
328b. E. C. Ashby, *J. Amer. Chem. Soc.,* **87**, 2509 (1965).
329. E. C. Ashby and W. B. Smith, *J. Amer. Chem. Soc.,* **86**, 4363 (1964).
330a. G. M. Whitesides and J. D. Roberts, *J. Amer. Chem. Soc.,* **87**, 4878 (1965).
330b. G. Fraenkel and D. T. Dix, *J. Amer. Chem. Soc.,* **88**, 979 (1966).
331. J. Toney and G. Stucky, *Chem. Comm.,* 1168 (1967).
332a. A. D. Vreugdenhil and C. Blomberg, *Recueil,* **39**, 84 (1965).
332b. E. C. Ashby and F. Walker, *J. Organometall. Chem.,* **7**, 17 (1967).
333. F. Walker and E. C. Ashby, *J. Amer. Chem. Soc.,* **91**, 3845 (1969).
334. S. E. Rudolph and L. F. Charbonneau, and S. G. Smith, *J. Amer. Chem. Soc.,* **95**, 7083 (1973).
335a. J. Billet and S. G. Smith, *J. Amer. Chem. Soc.,* **90**, 4108 (1968).
335b. H. O. House and J. E. Oliver, *J. Org. Chem.,* **33**, 929 (1968).
335c. T. Holm, *Acta. Chem. Scand.,* **21**, 2753 (1967).
336a. T. Holm, *Acta. Chem. Scand.,* **20**, 1139 (1966).
336b. J. Billet and S. G. Smith, *Tetrahedron Letters,* 4467 (1969).
336c. M. S. Singer, R. M. Salinger, and H. S. Mosher, *J. Org. Chem.,* **32**, 3821 (1967).
337. E. C. Ashby and S. H. Yu, *J. Org. Chem.,* **36**, 2123 (1971); *J. Organometall. Chem.,* **29**, 339 (1971).
338a. D. Bryce-Smith, *Bull. Soc. Chim. France,* 1418 (1963).
338b. D. Bryce-Smith and E. T. Blues, *Organic Syntheses,* **47**, 113 (1967).
339. E. C. Ashby and R. Reed, *J. Org. Chem.,* **31**, 985 (1966).
340a. H. F. Ebel and R. Schneider, *Angew. Chem.,* **77**, 914 (1965).
340b. T. Cuvigny and H. Normant, *Bull. Soc. Chim. France,* 2000 (1964).
341. R. Waack and M. A. Doran, *J. Amer. Chem. Soc.,* **91**, 2456 (1969).
342. E. C. Ashby, J. Laemmle, and H. M. Neumann, *J. Amer. Chem. Soc.,* **90**, 5179 (1968).
343. A. Tuulmets, *Reakts. Sposobnost. Org. Soedin,* **6**, 854 (1969).
344. J. Koppel, L. Margua, and A. Tuulmets, *Reakts. Sposobnost. Org. Soedin.,* **5**, 1041 (1968).
345. E. C. Ashby, R. Duke, and H. M. Neumann, *J. Amer. Chem. Soc.,* **89**, 1964 (1967).
346. T. Holm, *Acta Chem. Scand.,* **23**, 579 (1969).
347. S. G. Smith and J. Billet, *J. Amer. Chem. Soc.,* **89**, 6948 (1967).
348. E. C. Ashby, J. Laemmle, and H. M. Neumann, *J. Amer. Chem. Soc.,* **93**, 4601 (1971).
349. T. Holm and I. Crosland, *Acta Chem. Scand.,* **25**, 59 (1971).
350. H. Felkin and G. Roussi, *Tetrahedron Letters,* 4153 (1965).
351. H. Felkin and M. Chérest, *Tetrahedron Letters,* 2205 (1968).
352. E. C. Ashby, S. H. Yu, and P. V. Roling, *J. Organometall. Chem.,* **37**, 1918 (1972).

353. H. Felkin and C. Frajermann, *Tetrahedron Letters,* 1045 (1970).
354a. H. Felkin, Y. Gault, and G. Roussi, *Tetrahedron,* **26**, 3761 (1970).
354b. H. Felkin, C. Frajermann, and G. Roussi, *Ann. Chim.,* **6**, 17 (1971).
355. K. D. Berlin, R. D. Shupe, and R. D. Grigsby, *J. Org. Chem.,* **34**, 2500 (1969).
356. T. Holm, *J. Organometall. Chem.,* **29**, 45 (1971).
357. M. Chérest, H. Felkin, and C. Frajermann, *Tetrahedron Letters,* 379 (1971).
358a. E. Ghera and S. Shoua, *J. Chem. Soc. Chem. Commun.,* 398 (1971).
358b. S. Shoua and E. Ghera, *J. Org. Chem.,* **37**, 1292 (1972).
359. G. E. Dunn and J. Warkentin, *Canad. J. Chem.,* **34**, 75 (1956).
360. H. S. Mosher and J. D. Morrison, *Asymmetric Organic Reactions,* Prentice-Hall, Englewood Cliffs, N.J., 1971, Chapter 5.
361. J. Fauvarque and J.-F. Fauvarque, *C.R. Acad. Sci. Paris, Ser. C,* **263**, 488 (1966).
362. J. D. Buhler, *J. Org. Chem.,* **38**, 904 (1973).
363. P. J. Pearce, D. H. Richards, and N. F. Scilly, *J. Chem. Soc. Chem. Commun.,* 1160 (1970).
364. K. S. Rao, B. P. Stoicheff, and R. Turner, *Canad. J. Phys.,* **38**, 1516 (1960).
365. P. R. Jones, E. J. Goller, and W. J. Kauffman, *J. Org. Chem.,* **36**, 3311 (1971).
366. P. R. Jones, W. J. Kauffman, and E. J. Goller, *J. Org. Chem.,* **36**, 186 (1971).
367. E. C. Ashby and J. Laemmle, *J. Org. Chem.,* **33**, 3389 (1968).
368. J. Laemmle, E. C. Ashby, and P. V. Roling, *J. Org. Chem.,* **38**, 2526 (1973).
369. L. Lardicci and G. Giacomelli, *J. Chem. Soc. Perkin I,* 337 (1974).
370. M. Gaudemar, *Organometall. Chem. Rev.,* **A8**, 183 (1972).
371. M. W. Rathke and A. Lindert, *J. Org. Chem.,* **35**, 3966 (1970).
372. J. Canceill, J. Gabard, and J. Jacques, *Bull. Soc. Chim. France,* 231 (1968).
373. Y. Beziat and M. Mousseron-Canet, *Bull. Soc. Chim. France,* 1187 (1968).
374. F. Dardoize, J.-L. Moreau, and M. Gaudemar, *C.R. Acad. Sci., Paris, Ser. C,* **270**, 233 (1970).
375. B. Kurtev, M. Mladenova, and B. Blagoev, *C.R. Acad. Sci. Paris, Ser. C,* **271**, 871 (1970).
376. J. L. Luche and H. B. Kagan, *Bull. Soc. Chim. France,* 2260 (1971).
377. M. Bellassoued, R. Couffignal, and M. Gaudemar, *C.R. Acad. Sci. Paris, Ser. C,* **272**, 1686 (1971).
378. A. Balsamo, P. L. Barili, P. Crotti, B. Macchia, and F. Macchia, *Tetrahedron Letters,* 1005 (1974).
379. F. Dardoize, J.-L. Moreau, and M. Gaudmar, *Bull. Soc. Chim. France,* 3841 (1972); *C.R. Acad. Sci. Paris, Ser. C,* **268**, 2228 (1969).
380. *The Chemistry of the Carbonyl Group* (Ed. S. Patai), Interscience, London, Vol. 1, Chapter 11, p. 507.
381. H. O. House, *Modern Synthetic Reactions,* 2nd ed., Benjamin, New York 1972, p. 45.
382. H. C. Brown and S. Krishnamurthy, *J. Amer. Chem. Soc.,* **94**, 7159 (1972).
383. C. A. Brown, S. Krishnamurthy, and S. C. Kim, *J. Chem. Soc. Chem. Commun.,* 391 (1973).
384. J. Hooz, S. Ayiyama, F. J. Cedar, M. J. Bennett, and R. M. Tiggle, *J. Amer. Chem. Soc.,* **96**, 274 (1974).
385. R. O. Hutchins and D. Kandasamy, *J. Amer. Chem. Soc.,* **95**, 6131 (1973).
386. H. C. Brown and W. C. Dickason, *J. Amer. Chem. Soc.,* **92**, 709 (1970).
387. E. J. Corey, S. M. Albonico, U. Koelliker, T. K. Schaaf, and R. K. Varma, *J. Amer. Chem. Soc.,* **93**, 1491 (1971).

388. E. J. Corey and R. K. Varma, *J. Amer. Chem. Soc.*, **93**, 7319 (1971).
389. H. C. Brown and H. R. Deck, *J. Amer. Chem. Soc.*, **87**, 5620 (1965).
390. O. Červinka, *Chimia (Aarau)*, **13**, 332 (1959).
391. O. Červinka and O. Bělovoký, *Coll. Czech. Chem. Commun.*, **30**, 2487 (1965); **32**, 3897 (1967).
392. O. Červinka, *Coll. Czech. Chem. Commun.*, **30**, 1684 (1965).
393. H. Christol, D. Duval, and G. Solladié, *Bull. Soc. Chim. France*, 4151 (1968).
394. S. R. Landor, B. J. Miller, and A. R. Tatchell, *J. Chem. Soc., (C)*, 197 (1967).
395. S. R. Landor and A. R. Tatchell, *J. Chem. Soc. (C)*, 2280 (1966).
396. O. Červinka and A. Fabryová, *Tetrahedron Letters*, 1179 (1967).
397. G. M. Giongo, F. D. Gregorio, N. Palladino, and W. Marconi, *Tetrahedron Letters*, 3195 (1973).
398. W. G. Dauben, G. J. Fonken, and D. S. Noyce, *J. Amer. Chem. Soc.*, **78**, 2573 (1956).
399. E. L. Eliel and R. S. Ro, *J. Amer. Chem. Soc.*, **79**, 5992 (1957).
400. G. Chiurdoglu, H. Gonze and W. Masschelein, *Bull. Soc. Chim. Belges*, **71**, 484 (1962).
401. E. L. Eliel and S. H. Schroeter, *J. Amer. Chem. Soc.*, **87**, 5031 (1965).
402. J. Richer, *J. Org. Chem.*, **30**, 324 (1965).
403. J. A. Marshall and R. D. Caroll, *J. Org. Chem.*, **30**, 2748 (1965).
404. J. Klein, E. Dunkelblum, E. L. Eliel, and Y. Senda, *Tetrahedron Letters*, 6127 (1968).
405. E. L. Eliel and Y. Senda, *Tetrahedron*, **26**, 2411 (1970).
406. G. Chauvière, Z. Welvart, D. Eugène, and J.-C. Richer, *Canad. J. Chem.*, **47**, 3285 (1969).
407. J. M. Cense, *Tetrahedron Letters*, 2153 (1972).
408. A. Calvet and J. Levisalles, *Tetrahedron Letters*, 2157 (1972).
409. J. Klein and D. Lichtenberg, *J. Org. Chem.*, **35**, 2654 (1970).
410. J. Klein, *Tetrahedron Letters*, 4307 (1973).
411. P. Geneste, G. Lamaty, and J.-P. Roque, *Tetrahedron Letters*, 5007 (1970).
412. P. Geneste, G. Lamaty, C. Moreau, and J.-P. Roque, *Tetrahedron Letters*, 5011 (1970).
413a. D. J. Cram and F. A. Abd Elhafez, *J. Amer. Chem. Soc.*, **74**, 5828 (1952).
413b. T. J. Leitereg and D. J. Cram, *J. Amer. Chem. Soc.*, **90**, 4011, 4019 (1968).
413c. D. J. Cram and K. R. Kopecky, *J. Amer. Chem. Soc.*, **81**, 2748 (1959).
413d. D. J. Cram and D. R. Wilson, *J. Amer. Chem. Soc.*, **85**, 1245 (1963).
414. G. J. Karabatsos, *J. Amer. Chem. Soc.*, **89**, 1367 (1967).
415. M. Chérest, H. Felkin, and N. Prudent, *Tetrahedron Letters*, 2199 (1968).
416. P. Geneste, G. Lamaty, and B. Vidal, *Bull. Soc. Chim. France*, 2027 (1969).
417. E. C. Ashby, J. R. Boone, and J. P. Oliver, *J. Amer. Chem. Soc.*, **95**, 5427 (1973).
418. B. Rickborn and M. T. Wuesthoff, *J. Amer. Chem. Soc.*, **92**, 6894 (1970).
419. D. C. Wigfield and D. J. Phelps, *J. Chem. Soc., Perkin II*, 680 (1972).
420. G. Lamaty and J.-P. Roque, *Tetrahedron Letters*, 5293 (1967).
421. P. Geneste and G. Lamaty, *Bull. Soc. Chim. France*, 669 (1968).
422. P. Geneste, G. Lamaty and J.-P. Roque, *Tetrahedron Letters*, 5015 (1970).
423. V. J. Shiner and D. Whittaker, *J. Amer. Chem. Soc.*, **85**, 2337 (1963).
424. A. Streitwieser, Jr., *J. Amer. Chem. Soc.*, **75**, 5014 (1953).
425. V. J. Shiner, D. Whittaker, and V. P. Fernandez, *J. Amer. Chem. Soc.*, **85**, 2318 (1963).

426. W. Fieggen, H. Gerdind, and N. M. M. Nibbering, *Rec. Trav. Chem.*, **87**, 377 (1968).
427. V. J. Shiner, Jr. and D. Whittaker, *J. Amer. Chem. Soc.*, **91**, 394 (1969).
428. W. N. Moulton, R. E. von Atta, and R. R. Ruch, *J. Org. Chem.*, **26**, 290 (1961).
429. V. Hach, *J. Org. Chem.*, **38**, 293 (1973).
430. S. Yamashita, *J. Organometall. Chem.*, **11**, 377 (1968).
431. R. P. Bell, *Adv. Phys. Org. Chem.*, **4**, 1 (1966).
432. Y. Ogata and A. Kawasaki, in *The Chemistry of the Carbonyl Group* (Ed. J. Zabicky), Interscience, London, 1970, Vol. 2, p. 3.
433. P. Le Hénaff, *Bull. Soc. Chim. France*, 4687 (1968).
434. M. Eigen, *Discuss. Faraday Soc.*, **39**, 7 (1965).
435. J. L. Kurz and J. I. Coburn, *J. Amer. Chem. Soc.*, **89**, 3528 (1967).
436. R. P. Bell, J. P. Millington, and J. M. Pink, *Proc. Roy. Soc. A*, **303**, 1 (1968).
437. H. Dahn and J.-D. Aubort, *Helv. Chim. Acta*, **51**, 1348 (1968).
438. R. P. Bell and J. E. Critchlow, *Proc. Roy. Soc. A*, **325**, 35 (1971).
439. R. P. Bell and P. E. Sörensen, *J. Chem. Soc. Perkin II*, 1740 (1972).
440. R. Stewart and J. D. van Dyke, *Canad. J. Chem.*, **50**, 1992 (1972).
441. R. Stewart and J. D. van Dyke, *Canad. J. Chem.*, **48**, 3961 (1970).
442. A. Zanella and H. Taube, *J. Amer. Chem. Soc.*, **93**, 7166 (1971).
443. H. Dahn, H. P. Schlunke, and J. Temler, *Helv. Chim. Acta*, **55**, 907 (1972).
444. Y. Ogata and A. Kawasaki, in *The Chemistry of the Carbonyl Group* (Ed. J. Zabicky), Interscience Publishers, London, 1970, Vol. 2, p. 17.
445. B. Capon, *Chem. Rev.*, **69**, 407 (1969).
446. Y. Ogata and A. Kawasaki, in *The Chemistry of the Carbonyl Group* (Ed. J. Zabicky), Interscience, London, 1970, Vol. 2, p. 38.
447. G. E. Lienhard and W. P. Jencks, *J. Amer. Chem. Soc.*, **88**, 3982 (1966).
448a. R. Barnett and W. P. Jencks, *J. Amer. Chem. Soc.*, **89**, 5963 (1967).
448b. R. G. Kallen, *J. Amer. Chem. Soc.*, **93**, 6236 (1971).
449. C. Hassall, *Organic Reactions*, **9**, 73 (1957).
450. B. W. Palmer and A. Fry, *J. Amer. Chem. Soc.*, **92**, 2580 (1970).
451. Y. Ogata and Y. Sawaki, *J. Amer. Chem. Soc.*, **94**, 4189 (1972).
452. M. A. Winnik, V. Stoute, and P. Fitzgerald, *J. Amer. Chem. Soc.*, **96**, 1977 (1974).
453. *The Chemistry of the Carbonyl Goup* (Ed. S. Patai), Interscience, London, 1966, Vol. 1, p. 129.
454. *Oxidation in Organic Chemistry* (Ed. W. S. Trahanovsky), Academic Press, New York, 1973.
455a. E. J. Corey and C. U. Kim, *J. Amer. Chem. Soc.*, **94**, 7586 (1972).
455b. J. P. McCormick, *Tetrahedron Letters*, 1701 (1974).
456. K. E. Pfitzner and J. G. Moffatt, *J. Amer. Chem. Soc.*, **87**, 5661, 5670 (1965).
457. J. G. Moffatt in *Oxidation* (Ed. R. L. Augustine and D. J. Trecker), Marcel Decker, New York, 1971, p. 1–64.
458. J. G. Moffatt, *J. Org. Chem.*, **36**, 1909 (1971).
459. A. H. Fenselau and J. G. Moffatt, *J. Amer. Chem. Soc.*, **88**, 1762 (1966).
460. N. M. Weinshenker and C.-M. Shen, *Tetrahedron Letters*, 3281, 3285 (1972).
461. R. E. Harmon, C. V. Zenarosa, and S. K. Gupta, *J. Org. Chem.*, **35**, 1936 (1970); *Tetrahedron Letters*, 3781 (1969).
462. E. J. Corey and G. W. J. Fleet, *Tetrahedron Letters*, 4499 (1973).
463. J. C. Collins, W. W. Hess, and F. J. Frank, *Tetrahedron Letters*, 3363 (1968).

464. R. Ratcliffe and R. Rodehorst, *J. Org. Chem.*, **35**, 4000 (1970).
465. E. Buncel and A. N. Bourns, *Canad. J. Chem.*, **38**, 2457 (1960).
466. J. W. Baker and T. G. Heggs, *J. Chem. Soc.*, 616 (1955) and references therein.
467. A. N. Bourns and P. J. Smith, *Canad. J. Chem.*, **44**, 2553 (1966).
468. J. Wicha, A Zarecki, and M. Kocór, *Tetrahedron Letters*, 3635 (1973).
469. D. H. R. Barton, P. D. Magnus, G. Smith, G. Stretchert, and D. Zurr, *J. Chem. Soc. Perkin I*, 542 (1972).
470. C. F. Cullis and A. Fish in *The Chemistry of the Carbonyl Group* (Ed. S. Patai), Interscience, London, 1966, Vol. 1, p. 142.
471. K. B. Wiberg, *Oxidation in Organic Chemistry*, Academic Press, New York, 1965.
472. K. B. Wiberg and H. Schäfer, *J. Amer. Chem. Soc.*, **91**, 927 (1969); *J. Amer. Chem. Soc.*, **91**, 933 (1969).
473. J. Roček and A. E. Radkowsky, *J. Amer. Chem. Soc.*, **95**, 7123 (1973).
474. K. B. Wiberg and S. K. Mukherjee, *J. Amer. Chem. Soc.*, **93**, 2543 (1971).
475. P. M. Nave and W. S. Trahanovsky, *J. Amer. Chem. Soc.*, **92**, 1120 (1970).
476. J. Roček and A. E. Radkowsky, *J. Amer. Chem. Soc.*, **90**, 2986 (1968).
477. K. B. Wiberg and S. K. Mukherjee, *J. Amer. Chem. Soc.*, **96**, 1884 (1974).
478. K. Meyer and J. Roček, *J. Amer. Chem. Soc.*, **94**, 1209 (1972).
479. J. Roček and A. E. Radkowsky, *J. Org. Chem.*, **38**, 89 (1973).
480. J. Roček and A. E. Radkowsky, *Tetrahedron Letters*, 2835 (1968).
481. M. Rahman and J. Roček, *J. Amer. Chem. Soc.*, **93**, 5455, 5462 (1971).
482. W. S. Trahanovsky and J. Cramer, *J. Org. Chem.*, **36**, 1890 (1971).
483. J. Hampton, A. Leo, and F. H. Westheimer, *J. Amer. Chem. Soc.*, **78**, 306 (1956).
484. M. P. Doyle, R. J. Swedo, and J. Roček, *J. Amer. Chem. Soc.*, **92**, 7599 (1970); **95**, 8352 (1973).
485. J. Y.-P. Tong and E. L. King, *J. Amer. Chem. Soc.*, **82**, 3805 (1960).
486. J. Roček, A. M. Martinez, and G. E. Cushmac, *J. Amer. Chem. Soc.*, **95**, 5425 (1973).
487. F. Hasan and J. Roček, *J. Amer. Chem. Soc.*, **94**, 3181 (1972).
488. F. Hasan and J. Roček, *J. Org. Chem.*, **38**, 3812 (1973).
489. F. J. Kakis, *J. Org. Chem.*, **38**, 2536 (1973).
490. M. Fetizon and M. Golfier, *C.R. Acad. Sci. Paris Ser. C*, **276**, 900 (1968).
491. M. Echert-Maksie, Lj Tusek, and D. E. Sunko, *Croat. Chim. Acta*, **43**, 79 (1971).
492. F. J. Kakis, M. Fetizon, N. Douchkine, M. Golfier, P. Mourgues, and T. Prange, *J. Org. Chem.*, **39**, 523 (1974).
493. C. W. Rees and R. C. Storr, *J. Chem. Soc.* (*C*), 1474 (1969).
494. R. Filler, *Chem. Rev.*, **63**, 21 (1963).
495. V. Thiagarajan and N. Venkatasubramanian, *Canad. J. Chem.*, **47**, 694 (1969); *Tetrahedron Letters*, 3349 (1967).
496. N. S. Srinivasan and N. Venkatasubramanian, *Tetrahedron*, **30**, 419 (1974).
497. K. B. Wiberg and W. G. Nigh, *J. Amer. Chem. Soc.*, **87**, 3849 (1965).
498. H. A. Connon *et al.*, *J. Org. Chem.*, **38**, 2020 (1973).
499. T.-L. Ho, *Synthesis*, 347 (1973).
500. P. Salomaa in *The Chemistry of the Carbonyl Group* (Ed. S. Patai), Interscience, London, 1966, Vol. 1, p. 188.

501. E. H. Cordes, *Prog. Phys. Org. Chem.*, **4**, 1 (1967); E. H. Cordes and H. G. Bull, *Chem. Revs*, **74**, 581 (1974).
502. T. H. Fife, *Accounts Chem. Research*, **5**, 264 (1972).
503. W. P. Jencks, *Catalysis in Chemistry and Enzymology*, McGraw-Hill Book Co., London, 1969, p. 226.
504. T. H. Fife, *J. Amer. Chem. Soc.*, **89**, 3228 (1967).
505. M. S. Newman and R. E. Dickson, *J. Amer. Chem. Soc.*, **92**, 6880 (1970).
506. T. H. Fife and L. H. Brod, *J. Org. Chem.*, **33**, 4136 (1968).
507. B. Capon and M. I. Page, *J. Chem. Soc. Chem. Comm.*, 1443 (1970).
508. T. H. Fife and L. K. Jao, *J. Amer. Chem. Soc.*, **90**, 4081 (1968).
509. T. H. Fife and L. H. Brod, *J. Amer. Chem. Soc.*, **92**, 1681 (1970).
510. B. Capon and M. I. Page, *J. Chem. Soc. Perkin II*, 522 (1972).
511. B. Capon, M. I. Page, and G. H. Sankey, *J. Chem. Soc. Perkin II*, 529 (1972).
512. E. Anderson and B. Capon, *J. Chem. Soc. Chem. Comm.*, 390 (1969).
513. E. Anderson and B. Capon, *J. Chem. Soc. (B)*, 1033 (1969).
514. E. Anderson and T. H. Fife, *J. Amer. Chem. Soc.*, **91**, 7163 (1969).
515. A. Kankaanperä and M. Lahti, *Acta Chem. Scand.*, **23**, 2465, 3266 (1969).
516. E. Anderson and T. H. Fife, *J. Amer. Chem. Soc.*, **93**, 1701 (1971).
517. R. H. De Wolfe, K. M. Ivanetich, and N. F. Perry, *J. Org. Chem.*, **34**, 848 (1969).
518. T. H. Fife and E. Anderson, *J. Org. Chem.*, **36**, 2357 (1971).
519. T. H. Fife and E. Anderson, *J. Amer. Chem. Soc.*, **92**, 5464 (1970).
520. A. Kankaanperä and M. Merilahti, *Acta Chem. Scand.*, **26**, 685 (1972).
521. M. Lahti and A. Kankaanperä, *Acta Chem. Scand.*, **26**, 2130 (1972).
522. L. N. Johnson and D. C. Phillips, *Nature*, 761 (1965).
523. R. F. Atkinson and T. C. Bruice, *J. Amer. Chem. Soc.*, **96**, 819 (1974).
524. B. N. Dunn and T. C. Bruice, *Adv. Enzymology*, **37**, 1 (1973).
525. T. H. Fife and E. Anderson, *J. Amer. Chem. Soc.*, **93**, 6610 (1971).
526. A. Anderson and T. H. Fife, *J. Chem. Soc. Chem. Comm.*, 1470 (1971).
527. B. Capon, M. C. Smith, E. Anderson, R. H. Dahm, and G. H. Sankey, *J. Chem. Soc. (B)*, 1938 (1969).
528. B. M. Dunn and T. C. Bruice, *J. Amer. Chem. Soc.*, **92**, 6589 (1970).
529. G.-A. Kraze and A. J. Kirby, *J. Chem. Soc. Perkin II*, 61 (1974).
530. J. N. Brönsted and W. F. K. Wynne-Jones, *Trans. Faraday Soc.*, **25**, 59 (1929).
531. R. H. De Wolfe and R. M. Roberts, *J. Amer. Chem. Soc.*, **76**, 4379 (1954).
532. P. Salomaa, A. Kankaanperä, and M. Lahti, *J. Amer. Chem. Soc.*, **93**, 2084 (1971).
533. M. Lahti and A. Kankaanperä, *Acta Chem. Scand.*, **24**, 706 (1970).
534. C. A. Bunton and R. H. De Wolfe, *J. Org. Chem.*, **30**, 1371 (1965).
535. A. J. Kresge and R. J. Preto, *J. Amer. Chem. Soc.*, **87**, 4593 (1965).
536. H. Kwart and M. B. Price, *J. Amer. Chem. Soc.*, **82**, 5123 (1960).
537. M. Price, J. Adams, C. Lagenaur, and E. H. Cordes, *J. Org. Chem.*, **34**, 22 (1969).
538. E. Anderson and T. H. Fife, *J. Org. Chem.*, **37**, 1993 (1972).
539. A. J. Kresge and Y. Chiang, *J. Chem. Soc. (B)*, 53 (1967).
540. M. M. Kreevoy and R. Eliason, *J. Phys. Chem.*, **72**, 1313 (1968).
541. R. D. Frampton, T. T. Tidwell, and V. A. Young, *J. Amer. Chem. Soc.*, **94**, 1271 (1972).
542. A. J. Kresge and H. I. Chen, *J. Amer. Chem. Soc.*, **94**, 2819 (1972).

543. G. M. Louden, C. K. Smith, and S. E. Zimmerman, *J. Amer. Chem. Soc.*, **96**, 465 (1974).
544. D. S. Noyce and P. M. Pollack, *J. Amer. Chem. Soc.*, **91**, 119 (1969).
545. P. Salomaa, A. Kankaanperä, and M. Lajunen, *Acta Chem. Scand.*, **20**, 1790 (1966).
546. A. J. Kresge and Y. Chiang, *J. Amer. Chem. Soc.*, **95**, 803 (1973).
547. A. J. Kresge, *et al.*, *J. Amer. Chem. Soc.*, **93**, 413 (1971).
548. A. Bruylants and E. F. de Medicis, in *The Chemistry of the Carbon Nitrogen Double Bond* (Ed. S. Patai), Interscience, London, 1970, Chapter 10, p. 465.
549. S. Dayagi and Y. Degani in *The Chemistry of the Carbon–Nitrogen Double Bond* (Ed. S Patai), Interscience, London, 1970, Chapter 2, p. 61.
550. W. P. Jencks, *Catalysis in Chemistry and Enzymology*, McGraw-Hill Book Co., 1969, p. 490.
551. E. H. Cordes and W. P. Jencks, *J. Amer. Chem. Soc.*, **85**, 2843 (1963).
552. J. E. Reimann and W. P. Jencks, *J. Amer. Chem. Soc.*, **88**, 3973 (1966).
553. L. do Amaral, W. A. Sandstrom, and E. H. Cordes, *J. Amer. Chem. Soc.*, **88**, 2225 (1966).
554. A. Williams and M. L. Bender, *J. Amer. Chem. Soc.*, **88**, 2508 (1966).
555. M. Masui and C. Yijima, *J. Chem. Soc.* (*B*), 56 (1966).
556. L. do Amaral and M. P. Bastos, *J. Org. Chem.*, **36**, 3412 (1971).
557. E. H. Cordes and W. P. Jencks, *J. Amer. Chem. Soc.*, **84**, 832 (1962).
558. R. L. Reeves, *J. Amer. Chem. Soc.*, **84**, 3332 (1962).
559. W. P. Jencks, *Chem. Rev.*, **72**, 705 (1972).
560a. R. E. Barnett and W. P. Jencks, *J. Amer. Chem. Soc.*, **91**, 2358 (1969).
560b. S. L. Johnson and D. L. Morrison, *J. Amer. Chem. Soc.*, **94**, 1323 (1972).
560c. G. M. Blackburn, *J. Chem. Soc. Chem. Commun.*, 249 (1970).
561a. M. I. Page and W. P. Jencks, *J. Amer. Chem. Soc.*, **94**, 8828 (1972).
561b. D. Drake, R. L. Schowen, and H. Jayaraman, *J. Amer. Chem. Soc.*, **95**, 454 (1973).
561c. M. Caplow, *J. Amer. Chem. Soc.*, **90**, 6795 (1968).
562. C. Cerjan and R. E. Barnett, *J. Phys. Chem.*, **76**, 1192 (1972).
563. W. P. Jencks, *J. Amer. Chem. Soc.*, **94**, 4731 (1972).
564. J. Hine, B. C. Menon, J. H. Jensen, and J. Mulders, *J. Amer. Chem. Soc.*, **88**, 3367 (1966).
565. J. Hine, K. G. Hampton, and B. C. Menon, *J. Amer. Chem. Soc.*, **89**, 2664 (1967).
566. J. Hine, F. E. Rogers, and R. E. Notari, *J. Amer. Chem. Soc.*, **90**, 3279 (1968).
567. J. Hine, M. S. Cholod, and J. H. Jensen, *J. Amer. Chem. Soc.*, **93**, 2321 (1971).
568. J. Hine and C. Y. Yeh, *J. Amer. Chem. Soc.*, **89**, 2669 (1967).
569. J. Hine, C. Y. Yeh, and F. C. Schmalstieg, *J. Org. Chem.*, **35**, 340 (1970).
570. J. Hine and F. A. Via, *J. Amer. Chem. Soc.*, **94**, 190 (1972).
571. R. G. Kallen and W. P. Jencks, *J. Biol. Chem.*, **241**, 5864 (1966).
572. J. Hine, M. S. Cholod, and W. K. Chess, Jr., *J. Amer. Chem. Soc.*, **95**, 4270 (1973).
573a. T. C. French, D. S. Auld, and T. C. Bruice, *Biochemistry*, **4**, 77 (1965).
573b. R. L. Reeves, *J. Org. Chem.*, **30**, 3129 (1965).
574. D. S. Auld and T. C. Bruice, *J. Amer. Chem. Soc.*, **89**, 2083 (1967).
575. M. I. Page and W. P. Jencks, *J. Amer. Chem. Soc.*, **94**, 8818 (1972).

576. J. Hine and K. W. Narducy, *J. Amer. Chem. Soc.*, **95**, 3362 (1973).
577a. R. G. Kallen, *Methods Enzymol.*, **18**, 705 (1971).
577b. S. J. Benkovic, P. H. Benkovic, and R. Chrzanowski, *J. Amer. Chem. Soc.*, **92**, 523 (1970).
578. E. G. Sander and W. P. Jencks, *J. Amer. Chem. Soc.*, **90**, 6154 (1968).
579. H. Diebler and R. N. F. Thorneley, *J. Amer. Chem. Soc.*, **95**, 896 (1973).
580. R. M. Pollack and S. Ritterstein, *J. Amer. Chem. Soc.*, **94**, 5064 (1972).
581. L. do Amaral and M. P. Bastos, *J. Org. Chem.*, **36**, 3412 (1971).
582. R. Wolfenden and W. P. Jencks, *J. Amer. Chem. Soc.*, **83**, 2763 (1961).
583. L. do Amaral, H. G. Bull, and E. H. Cordes, *J. Amer. Chem. Soc.*, **94**, 7579 (1972).
584. L. do Amaral, M. P. Bastos, H. G. Bull, and E. H. Cordes, *J. Amer. Chem. Soc.*, **95**, 7369 (1973).
585. V. Baliah, V. N. V. Desikan, and V. N. Vedanta, *Indian J. Chem.*, **9**, 1088 (1971).
586. T. Pino and E. H. Cordes, *J. Org. Chem.*, **36**, 1668 (1971).
587. L. do Amaral, *J. Org. Chem.*, **37**, 1433 (1972).
588. E. H. Cordes and W. P. Jencks, *J. Amer. Chem. Soc.*, **84**, 4319 (1962).
589. J. M. Sayer, M. Peskin, and W. P. Jencks, *J. Amer. Chem. Soc.*, **95**, 4277 (1973).
590. W. P. Jencks and K. Salvesen, *J. Amer. Chem. Soc.*, **93**, 1419 (1971).
591. L. D. Kershner and K. Salvesen, *J. Amer. Chem Soc.*, **93**, 2014 (1971).
592. R. K. Chaturvedi and E. H. Cordes, *J. Amer. Chem. Soc.*, **89**, 4631 (1967).
593. J. M. Sayer and W. P. Jencks, *J. Amer. Chem. Soc.*, **91**, 6353 (1969).
594. J. M. Sayer and W. P. Jencks, *J. Amer. Chem. Soc.*, **94**, 3262 (1972).
595. J. M. Sayer and W. P. Jencks, *J. Amer. Chem. Soc.*, **95**, 5637 (1973).
596. P. Geneste, G. Lamaty, and J.-P. Roque, *Recueil*, **91**, 188 (1972).
597. S. M. Silver and J. M. Sayer, *J. Amer. Chem. Soc.*, **95**, 5073 (1973).
598. P. Geneste, G. Lamaty, and J.-P. Roque, *Tetrahedron*, **27**, 5561 (1971).
599. J. Archila, H. Bull, C. Lagenaur, and W. P. Jencks, *J. Org. Chem.*, **36**, 1345 (1971).
600. W. Bruyneel, J. J. Charette, and E. de Hoffmann, *J. Amer. Chem. Soc.*, **88**, 3808 (1966).
601. E. de Hoffmann, *Coll. Czech. Chem. Commun.*, **36**, 4115 (1971).
602. J. Hoffmann, J. Klionar, V. Štěrba, and M. Večeřa, *Coll. Czech. Chem. Commun.*, **35**, 1387 (1970).
603. J. Hoffmann and V. Štěrba, *Coll. Czech. Chem. Commun.*, **37**, 2043 (1972).
604. B. Capon and C. Ghosh, *J. Chem. Soc. (B)*, 472 (1966).
605. W. Tanner and T. C. Bruice, *J. Amer. Chem. Soc.*, **89**, 6954 (1974).
606. J. Hoffmann, J. Klienar, and V. Štěrba, *Coll. Czech. Chem. Commun.*, **36**, 4057 (1971).
607. Y. Ogata and A. Kawasaki, *J. Org. Chem.*, **39**, 1058 (1974).
608. H. Harada in *The Chemistry of the Carbon–Nitrogen Double Bond* (Ed. S. Patai), Interscience, London, 1970, p. 260.
609. D. H. R. Barton and B. J. Willis, *J. Chem. Soc. Perkin I*, 305 (1972).
610. J. W. Everett and P. J. Garratt, *J. Chem. Soc. Chem. Commun.*, 642 (1972).
611. S. Miyano and N. Abe, *Tetrahedron Letters*, 1909 (1970).
612. J. C. Fiaud and A. Horeau, *Tetrahedron Letters*, 2565 (1971).
613. K. Harada and T. Okawara, *J. Org. Chem.*, **38**, 707 (1973).
614. M. S. Patel and M. Worsley, *Canad. J. Chem.*, **48**, 1881 (1970).
615. Y. Ogata and A. Kawasaki, *J. Chem. Soc. (B)*, 325 (1971).

616. Y. Ogata and A. Kawasaki, *J. Chem. Soc. Perkin II*, 1792 (1972).
617. C. O'Brien, *Chem. Rev.*, **64**, 81 (1964).
618. S. Eguchi and Y. Ishii, *Bull. Chem. Soc. Japan*, **36**, 1434 (1963).
619. G. Alvernhe and A. Laurent, *Bull. Soc. Chim. France*, 3304 (1970).
620. K. Kotera and K. Kitahonoki, *Org. Prep. Proc.*, **1**, 305 (1969).
621. V. Madan and L. B. Clapp, *J. Amer. Chem. Soc.*, **91**, 6078 (1969); **92**, 4902 (1970).
622. Y. Ogata and Y. Sawaki, *J. Amer. Chem. Soc.*, **95**, 4687 (1973).
623. Y. Ogata and Y. Sawaki, *J. Amer. Chem. Soc.*, **95**, 4692 (1973).
624. R. J. Sundberg and P. A. Bukowick, *J. Org. Chem.*, **33**, 4098 (1968).
625. D. Paquer, *Int. J. Sulphur Chem.* (*B*), **7**, 269 (1972).
626. U. Miotti, A. Sinico, and A. Ceccon, *Chem. Commun.*, 724 (1968).
627. U. Miotti, U. Tonellato, and A. Ceccon, *J. Chem. Soc.* (*B*), 325 (1970).
628. A. Ceccon, U. Miotti, U. Tonellato, and M. Padovan, *J. Chem. Soc.* (*B*), 1084 (1969).
629. E. Block, *J. Amer. Chem. Soc.*, **94**, 642 (1972).
630. H. G. Giles, R. A. Marty, and P. de Mayo, *J. Chem. Soc. Chem. Commun.*, 409 (1974).
631. A. Battaglia, A. Dondoni, P. Giorgianni, G. Maccagnani, and G. Mazzanti, *J. Chem. Soc.* (*B*), 1547 (1971).
632. A. Battaglia, A. Dondoni, G. Maccagnani, and G. Mazzanti, *J. Chem. Soc.* (*B*), 2096 (1971).
633. P. Beak and J. W. Worley, *J. Amer. Chem. Soc.*, **94**, 597 (1972).
634. M. Dagonneau, D. Paquer, and J. Vialle, *Bull. Soc. Chim. France*, 1699 (1973).
635. M. Dagonneau and J. Vialle, *Tetrahedron Letters*, 3017 (1973).
636. M. Dagonneau, P. Metzner, and J. Vialle, *Tetrahedron Letters*, 3675 (1973).
637. V. Rautenstrauch, *Helv. Chim. Acta*, **57**, 496 (1974).
638. A. P. Krapcho, D. R. Rao, M. P. Silvan, and B. Abegaz, *J. Org. Chem.*, **36**, 3885 (1971).
639. C. E. Diebert, *J. Org. Chem.*, **35**, 1501 (1970).
640. A. P. Krapcho, M. P. Silvan, I. Goldberg, and E. G. E. Jahngen, Jr., *J. Org. Chem.*, **39**, 860 (1974).
641. M. Dagonneau and J. Vialle, *Tetrahedron*, **30**, 415 (1974).

CHAPTER **5**

The electrochemistry of X=Y groups

ALBERT J. FRY

Wesleyan University, Middletown, Connecticut, U.S.A.

and

ROBERTA GABLE REED

The Mary Imogene Bassett Hospital, Cooperstown, New York, U.S.A.

I. INTRODUCTION

This review includes a discussion of the anodic and cathodic chemistry
of the X=Y bond, i.e. **1**, where **1** can include the various combinations
shown. The emphasis of the chapter is upon the electrochemical behaviour

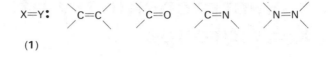

(1)

of the isolated functional group as opposed to other chemistry it may
exhibit when conjugated to other functional groups. Hence, for example,
the electrochemical reduction and oxidation of the carbon–carbon
double bond is discussed, including effects upon this chemistry of attached
functional groups, but the electrolytic hydrodimerization of activated
olefins[1] is not reviewed since this reaction may more properly be considered as a reaction of the extended conjugated π-system of the activated
olefin, and at any rate has been adequately reviewed elsewhere[2]. The
distinction is somewhat difficult to apply rigorously: benzophenone is
easier to reduce than acetophenone, which in turn is easier to reduce than
acetone[3], thus demonstrating that a conjugative effect of an aromatic
nucleus does operate upon the electrochemical behaviour of the carbonyl
group. Nevertheless, all three ketones undergo electrochemical reaction
at the carbonyl group, hence all three are discussed herein.

II. GENERAL CONSIDERATIONS

The initial step in a true electrochemical reduction of any double bond
consists of addition of an electron to the lowest unfilled molecular orbital
(LUMO) of the double bond. This definition excludes, therefore, such
processes as the electrochemical reduction of acetylenes to *cis* olefins at a
nickel cathode in acid[4]; such processes are probably more accurately
represented as catalytic hydrogenations using electrochemically generated
hydrogen than as direct electrochemical reductions. Conversely, the
first step in a direct electrochemical oxidation involves abstraction of an
electron from a non-bonding molecular orbital, or the highest occupied
molecular orbital (HOMO) of the substrate, though again a number of
anodic oxidations actually involve electrochemical generation of a
reagent, e.g. $ClO_2{}^{.5}$, $NO_3{}^{.6}$, $PbO_2{}^7$, etc., and subsequent chemistry
involving this reagent. The primary emphasis of this review will be upon
direct electrochemical reactions initiated by electron transfer to or from
the electroactive species.

Much of the electrochemical behaviour of simple, doubly-bonded systems follows directly from the fact that reduction and oxidation involve electron injection into the LUMO and electron removal from the HOMO, respectively. Thus olefins become more readily reduced when electron-withdrawing, and more readily oxidized when electron-supplying, groups are attached to the double bond.

Because the electronegativity of the atom attached to carbon increases as one proceeds across the following series, one would expect that reduction would become easier as one proceeds from **2** to **4**. It appears that

$$R_2C{=}C\diagup\diagdown \qquad\qquad R_2C{=}N\diagup \qquad\qquad R_2C{=}O$$

$$\textbf{(2)} \qquad\qquad\qquad \textbf{(3)} \qquad\qquad\qquad \textbf{(4)}$$

this is indeed the case: olefins are not electrochemically reducible[8] under conditions where imines (**3**)[9] and carbonyl compounds[10] are, e.g. in ethanol or water containing a tetra-alkylammonium salt as supporting electrolyte. (Actually, another factor is also operating in such solvents: trialkylimines and dialkyl ketones are not electrochemically reducible in dipolar aprotic solvents[11], suggesting that their reducibility in protic solvents is associated with the hydrogen-bonding ability of the solvent.)

The initial product of electron transfer to a substance of structure such as **1**–**4** is a radical anion, which may be represented, for **1**, as

$$X\bar{Y}{\cdot} \equiv \dot{X}{-}\bar{Y} \longleftrightarrow \bar{X}{-}\dot{Y}$$

and, for **2**–**4** as

$$R_2C{=}\bar{X}{\cdot} \equiv R_2\dot{C}{-}\bar{X} \longleftrightarrow R_2\bar{C}{-}X{\cdot}$$

$$\textbf{(5)} \qquad\quad \textbf{(5a)} \qquad\qquad \textbf{(5b)}$$

$$X = CR_2,\ NR,\ O$$

Since electronegativity increases in the order C, N, O, structure **5a** should become a relatively more important contributor to the resonance hybrid as one proceeds through radical anions in the series **2**–**4**. With olefins, of course **5a** and **5b** are equivalent; the point is moot, since radical anions of olefins unsubstituted by activating groups such as carbonyl, nitro, cyano, etc. cannot be generated at potentials accessible in organic solvents. The unshared electron of the radical anion of α,β-unsaturated ketones appears, from e.s.r. spectroscopy, to be located far more on the β than the α-carbon[12].

The radical anions of carbonyl compounds (ketyls) are more accurately described by structure **5a** than **5b**, and much of their chemistry follows directly from this feature. Hence reactions of ketyls with electrophiles should take place at oxygen, and radical reactions should occur at carbon. These expectations are fulfilled in practice: protonation, for example, occurs at oxygen, and ketyls dimerize at carbon to produce, after subsequent protonation *vic*-diols (pinacols) (**6**)[13] (equation 1). Structure **5b**

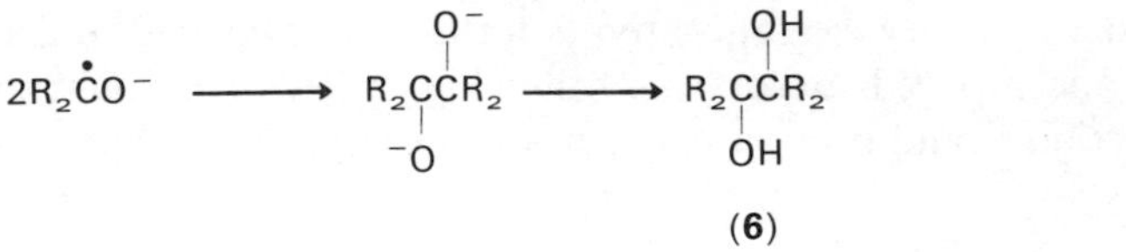

$$2R_2\overset{\cdot}{C}O^- \longrightarrow \underset{\underset{-O}{|}}{R_2\overset{\overset{O^-}{|}}{C}CR_2} \longrightarrow \underset{\underset{OH}{|}}{R_2\overset{\overset{OH}{|}}{C}CR_2} \qquad (1)$$

$$(6)$$

should be a more important contributor (though not the major one) in imine radical anions, and indeed reaction with electrophiles can apparently take place at either nitrogen or carbon. Protonation appears to occur on nitrogen[11], but reaction with carbon dioxide takes place at carbon (equation 2).[14]

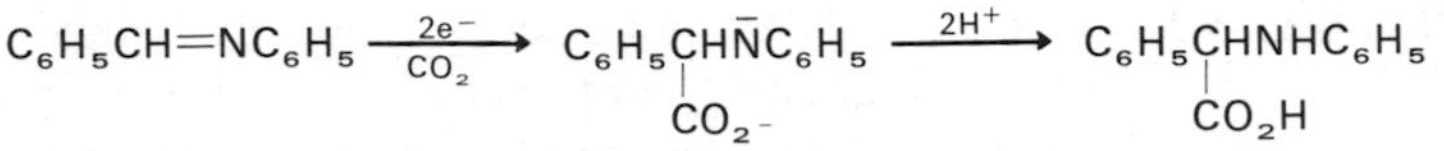

$$C_6H_5CH{=}NC_6H_5 \xrightarrow[CO_2]{2e^-} \underset{\underset{CO_2^-}{|}}{C_6H_5\overset{}{C}H\bar{N}C_6H_5} \xrightarrow{2H^+} \underset{\underset{CO_2H}{|}}{C_6H_5\overset{}{C}HNHC_6H_5} \qquad (2)$$

Compounds of type X=Y exhibit well-defined changes in electrochemical behaviour as the proton donor characteristics of the medium are changed. In alkaline media or aprotic organic solvents radical anions (**7**) are moderately stable, though slow dimerization may occur. In such solvents it is generally possible to introduce a second electron, generating a dianion (**8**). Because of electron repulsion in the dianion it is generally produced only at a potential E_2 negative of that (E_1) necessary to produce **7** (Scheme 1)[11]. Under conditions of moderate proton availability (or

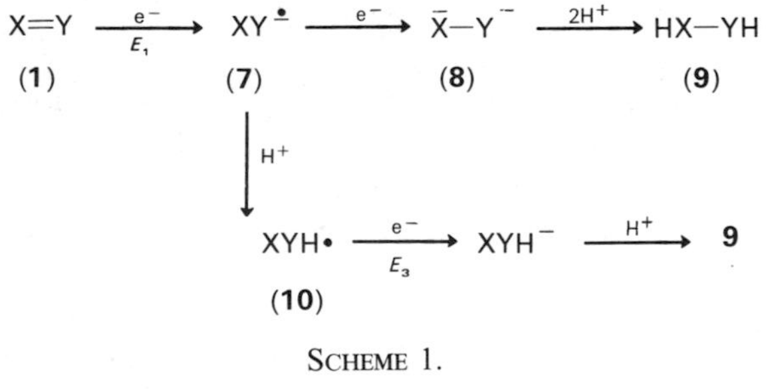

$$X{=}Y \xrightarrow[E_1]{e^-} XY^{\cdot-} \xrightarrow{e^-} \bar{X}{-}Y^- \xrightarrow{2H^+} HX{-}YH$$

$$(1) \qquad\qquad (7) \qquad\qquad (8) \qquad\qquad (9)$$

$$\Big\downarrow H^+$$

$$XYH\cdot \xrightarrow[E_3]{e^-} XYH^- \xrightarrow{H^+} 9$$

$$(10)$$

SCHEME 1.

where **7** is very basic[11]), e.g. neutral protic media or aprotic organic solvents containing added proton donors, **7** can be protonated to afford a neutral radical (**10**), whose reduction potential E_3 is generally *positive* of E_1, and hence **10** is usually reduced immediately upon formation. It has generally been believed that **10** receives its electron from the electrode, but it now appears that **10** is reduced in homogeneous solution by another molecule of **7**[18]. In media of high proton availability (low pH) or where **1** is moderately basic, the initial electrochemical step is reduction of the conjugate acid of **1**. This reduction also generates **10** (equation 3), but

$$1 \xrightarrow{\text{H}^+} \overset{+}{\text{X}}{=}\text{YH} \xrightarrow[E_4]{e^-} \text{XYH}\bullet \longrightarrow \text{Dimer(s)} \qquad (3)$$
$$(\mathbf{10})$$

since it occurs at a potential (E_4) which can be positive of E_3, **10** is usually sufficiently long-lived to undergo reactions, e.g. dimerization, not observed when it is generated in media of low proton availability. One is able to exert a great deal of control over the reduction by judicious control of potential and proton availability: dimers can be generated by reduction at the potential of the first reduction step in acid or base, and dihydro compounds can be generated in neutral media or by reduction at the potential of the second reduction step in acid or base. The exact pH range over which these transitions occur will depend upon the structure of **1**, of course, as will be seen below in the discussion of individual functional groups.

The reader unfamiliar with electrochemical nomenclature and practice may refer to various introductory reviews and textbooks on the subject[19-25], several of which[19-21] are intended for the organic chemist with no previous exposure to electrochemistry.

III. THE OLEFINIC LINKAGE

A. Reduction

Isolated double bonds are not reducible electrochemically; the anti-bonding orbital of isolated olefinic bonds lies too high to be accessible via a direct electron-transfer process. Isolated double bonds are, however, slowly reduced by solvated electrons generated by electrolysis of a solution of lithium chloride in methylamine[26]. Alkynes are reduced to alkenes at porous platinum black[27] or bright platinum[28] electrodes, but these reactions are probably regarded as catalytic hydrogenations using electrochemically generated hydrogen. Alkynes can be reduced to alkenes in similar fashion at spongy nickel[4] or silver–palladium[29] electrodes.

The double bond can be reduced by direct electron transfer from the electrode when it is part of an extended π-system, as, for example, in butadiene[30], 1,3,5-hexatriene, styrene[32], stilbene[33] and some styryl-heterocyclic compounds[34]. Likewise, direct electron transfer to olefinic compounds bearing electron-withdrawing substituents, e.g. α,β-unsaturated esters, nitriles, and ketones is facilitated because the LUMO in such compounds lies considerably lower than that of an isolated olefin, but such activated olefins exhibit electrochemical properties which differ in some respects from those of simpler olefins, and has already been discussed elsewhere[2]. For this reason, the electrochemistry of activated olefins is not discussed in the present review. In the present context, it is however interesting to note that the LUMO of the highly activated olefin tetracyanoethylene (TCNE, **11**) lies so low as to be weakly binding[35].

(11)

One and even two electrons may be added to TCNE, forming the radical anion and dianion, respectively, at quite positive potentials.

B. Oxidation

Electrochemical oxidation of simple olefins is difficult but can be made to occur at very positive potentials, which can be reached in acetonitrile containing a tetraalkylammonium tetrafluoroborate or hexafluorophosphate[36]. Sample oxidation potentials are: ethylene [+2·90 V vs. 0·01 M-Ag$^+$(Ag)], propylene (+2·84 V), 2-butene (stereochemistry unknown) (+2·26 V)[36]. Fleischmann and Pletcher observed in fact a reasonably good correlation between the ionization potential of olefins (and also alkanes) and their oxidation potential in this solvent[36]. Oxidation of olefins at a gold anode occurs at a lower potential[37]; presumably some sort of chemical interaction of the olefin with the gold surface facilitates oxidation. Similarly, it has been found that mercury(II)–olefin complexes are readily oxidized to carboxylic acids[38].

Allylic substitution, presumably via an intermediate allyl cation, appears to be a common pathway in the electrochemical oxidation of alkenes[39,40] (equation 4). Intramolecular analogues of this reaction have

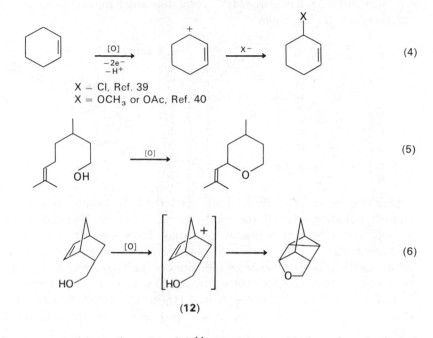

been reported (equations 5 and 6)[41]. The latter oxidation of a substituted norbornene to a homoallylic cation (12) may be contrasted with the anodic conversion of the parent compound, norbornene, to a norbornyl cation[42] (equation 7).

The oxidation of olefins becomes easier as they are substituted with electron-supplying groups (cf. the oxidation potentials of ethylene, propylene and 2-butene above). Hence phenyl substitution also lowers the oxidation potential of the double bond[43,48]. The products of oxidation of arylethylenes also depend upon the number of aryl groups attached to the double bond. In styrenes[43,44] and stilbene[45] the products are derived from a radical cation formed by removal of one electron from the double bond. With polarylethylenes, particularly when the aromatic rings bear strong electron-supplying groups, e.g. p-Me$_2$N or p-CH$_3$O, removal of a second electron to afford the dication can compete[45,48]. Tetra-p-anisyl-ethylene dication (13) undergoes an interesting cyclization (equation 8)

338   Albert J. Fry and Roberta Gable Reed

to a phenanthrene derivative (14)[49], a reaction which may be governed by orbital symmetry principles[50].

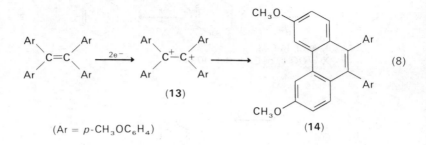

$$(Ar = p\text{-}CH_3OC_6H_4)$$

Oxidation of an olefinic bond is also rendered facile by substitution with electron-donating heteroatoms, e.g. N, S, and O. Fritsch, Weingarten and Wilson have tabulated oxidation potentials for a number of nitrogen-containing olefins[51]. A reaction of potential synthetic interest reported by them, which may be regarded as indicative of the type of electrochemistry generally to be expected for electron-rich olefins, is the oxidative dimerization of 1,1-bis(dimethylamino)-ethylene (15) to the butadiene derivative 16 (equation 9). The reaction presumably proceeds via nucleophilic attack

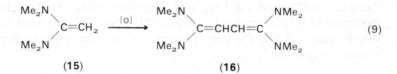

upon the radical cation of 15, i.e. 17, by another molecule of 15 (equation 10).

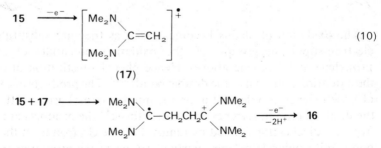

Carbon–carbon double bonds bearing four electron-supplying heteroatoms are oxidized extremely easily. A striking example is provided by *tetrakis*-(dimethylamino)-ethylene (TDME) (**18**), which is as good a reducing agent as metallic zinc[52]! Reaction between **18** and TCNE (**11**)

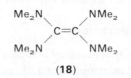

(**18**)

results[53] in two-electron transfer to produce the salt $TDME^{+2}\ TCNE^{-2}$. It may be shown[51] that the spacing between the first and second oxidation potentials (ΔE), corresponding to stepwise formation of the radical cation and dication of a species, may be equated to the equilibrium constant K_{disp} for disproportionation of the radical cation into dication and neutral starting material, viz.,

$$X \underset{E_1}{\overset{-e^-}{\rightleftharpoons}} X^+ \underset{E_2}{\overset{-e^-}{\rightleftharpoons}} X^{+2} \tag{10}$$

$$X^{+2} + X \overset{K_{disp}}{\rightleftharpoons} 2X^+ \tag{11}$$

$$\Delta G^\circ = -RT \ln K_{disp} = -n\mathscr{F}\,(E_1 - E_2) \tag{12}$$

For **18**, $\Delta E = -0.14$ V in acetonitrile at 25 °C, i.e. $K_{disp} = 230$[54] (Fritsch, Weingarten, and Wilson[51] report $\Delta E = 0.12$ V and $K_{disp} = 106$). Hunig and coworkers[55] have tabulated values of ΔE and K_{disp} for other electron-rich olefins. Chambers and coworkers[56] carried out an extensive study of the electrochemical behaviour of tetrathioethylenes (**19**), whose voltammetric behaviour closely resembles that of compounds such as **18**.

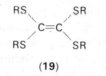

(**19**)

IV. THE CARBONYL GROUP

A. Reduction

Electrochemically the carbonyl group represents the most thoroughly investigated and best characterized of the double-bonded functional groups. Reduction occurs in both aqueous and non-aqueous solvents. The nature of the reduction process is largely determined by the presence of proton donors. Changes in electrochemical behaviour with increasing pH (or added proton donors in the case of non-aqueous solvents) occur in a regular pattern common to most carbonyl compounds.

1. Mechanistic considerations

a. *Diaryl ketones.* Benzophenone, as a typical diaryl ketone, produces stable radical and anionic intermediates and serves as an excellent model for the entire spectrum of electrochemical behaviour observed for carbonyl compounds. In the absence of proton donors electron transfer produces a stable radical anion and dianion. Increasing acidity leads to protonation of the dianion, radical anion, and finally benzophenone itself. Less fully arylated carbonyl compounds are reduced to more reactive intermediates and consequently, their electrochemical behaviour may not be as clearly defined as for benzophenone.

Reduction of benzophenone (**20**) in liquid ammonia at $-50\,°C$ occurs in two one-electron steps to produce a stable radical anion (**21**) and dianion (**22**)[57] (equations 13 and 14). Both steps are reversible by cyclic

$$(C_6H_5)_2C{=}O + e^- \rightleftharpoons (C_6H_5)_2CO{-} \qquad (13)$$
$$\textbf{(20)} \qquad\qquad\qquad \textbf{(21)}$$

$$21 + e^- \rightleftharpoons (C_6H_5)_2CO^{2-} \qquad (14)$$
$$\textbf{(22)}$$

voltammetry: a reoxidation peak is seen corresponding to each reduction peak. Addition of ethanol shifts the reduction wave for the second step to a more positive potential with loss of the corresponding reoxidation peak. Such behaviour is typical of electron transfer followed by a rapid, irreversible chemical reaction—in this case protonation of the dianion.

Electrochemical behaviour of benzophenone in anhydrous dimethylformamide[16], acetonitrile[58], or pyridine[58] is similar to that in ammonia containing ethanol. Although the former solvents are often regarded as aprotic, cyclic voltammetric and polarographic data support a mechanism including rapid protonation of **22**. Recent work by Jensen and Parker[59] suggests that the irreversible nature of the second step is a result of reaction of the dianion with electrophilic impurities in the solvent, and that reductions are indeed reversible when a small amount of alumina is added to the solvent to scavenge these impurities. If true, this discovery will be of considerable significance, but this preliminary report will require further corroboration.

Addition of phenol to benzophenone in pyridine increases the first polarographic wave at the expense of the second[16]. Protonation of the radical anion by phenol produces an electroactive radical which is responsible for the increased current at the first wave.

Addition of a strong acid, e.g. benzoic acid, to a dimethylformamide solution of benzophenone produces a new polarographic wave anodic of benzophenone reduction, representing reduction of the conjugate acid of benzophenone[60].

In aqueous solvents benzophenone exhibits electrochemical behaviour similar to that outlined for non-aqueous systems with pH reflecting relative proton availability. In acid (pH $\leq$ 4) benzophenone exhibits two diffusion controlled one-electron polarographic waves. The first is reduction of the protonated carbonyl to a radical (23); the second, reduction of the radical to a carbanion[16]. This is analogous to behaviour in dimethylformamide containing benzoic acid. Controlled potential electrolysis at the first wave gave quantitative yield of benzpinacolone (24) from radical coupling and subsequent acid-catalysed rearrangement of the initially formed benzpinacol (equation 15).

$$(C_6H_5)_2C{=}OH^+ \xrightarrow{\ e^-\ } (C_6H_5)_2\,{}^{\bullet}COH \xrightarrow{\ 23\ } \underset{\overset{|}{OH}}{\overset{\overset{OH}{|}}{(C_6H_5)_2C{-}C(C_6H_5)_2}}$$

(23)

$$\xrightarrow[H_2O]{H^+} \underset{\overset{\|}{O}}{C_6H_5CC(C_6H_5)_3} \qquad (15)$$

(24)

In neutral or weakly alkaline solutions ($4.9 < pH < 11.3$), benzophenone is reduced to a radical anion which rapidly abstracts a proton from the solvent. The resulting radical is reduced to an anion which is also protonated. Large scale reduction of benzophenone at pH 8·6 produced benzhydrol[16] (equation 16). This is analogous to behaviour in pyridine containing phenol.

$$(C_6H_5)_2C{=}O \xrightarrow[2H^+]{2e^-} (C_6H_5)_2CHOH \qquad (16)$$

In strongly alkaline media benzophenone exhibits two one-electron waves. Under these conditions, the radical anion is not protonated and reduction to the dianion occurs at a more negative potential[16,61]. This behaviour is analogous to that in dimethylformamide or pyridine.

b. *Monoaryl carbonyls*. Reduction of aryl alkyl ketones and aryl aldehydes (ArCOR, where R = alkyl or H) is similar to that of diaryl ketones. In acid two one-electron waves correspond to reduction of the protonated compound and subsequently the neutral radical[61,62]. Preparative reduction at the first wave gives pinacol, usually in high yield[2a,63,64]. Reduction at the second wave would be expected to produce alcohol but such results have not been reported.

In the pH range 7·2 to 11·3, one two-electron wave is observed for reduction of aryl alkyl ketones. Elving and Leone reported consumption of 1·9 electrons/mole during controlled potential reduction of acetophenone but did not attempt to identify the product—presumably the alcohol[62]. In strongly alkaline solutions (pH > 12) and in aprotic media two one-electron waves are observed for ArCOR. Pinacols are formed upon reduction at the first wave. For example, reduction of benzaldehyde at the first wave in dimethylformamide affords benzoin. In the presence of phenol, benzyl alcohol is the product[65].

c. *Alkyl carbonyls*. Reduction of dialkyl ketones and alkyl aldehydes occurs only at extremely negative potentials. Frequently reduction is not observed at all since the cation of the electrolyte (particularly if it is a metal ion) is often easier to reduce than an alkyl carbonyl. In tetrabutylammonium salts in 80–90% ethanol or methanol a single two-electron wave has been reported for a number of aliphatic ketones[10a,c] and alcohols have been isolated from electrochemical reduction in high yields[66]. Absence of dimeric products indicates the reduction of the neutral ketyl radical must be much faster than dimerization presumably because of the extremely negative potentials involved.

The aliphatic ketones camphor and norcamphor are reducible in alcoholic solvents but have been shown to be electrochemically inactive in dimethylformamide–tetraethylammonium bromide, even though more negative potentials are accessible in dimethylformamide[11]. This suggests that reduction of ketones is facilitated through hydrogen bonding between protic solvents and the carbonyl. Support for this argument comes from a polarographic investigation of reduction of saturated aldehydes in dimethylformamide by Kirrman and coworkers[67]. Addition of phenol shifted the half-wave potential 0·1 V anodic and increased the wave height to correspond to a one-electron process. The small level of electroactivity (30%) in the absence of phenol was attributed to reduction of a complex of the carbonyl with residual water in the solvent.

d. *Non-conjugated olefinic aldehydes and ketones*. Polarographic reductions of cyclic aliphatic aldehydes **25–28** and their semicarbazones indicate that the presence of a non-conjugated double bond facilitates reduction of aldehydes **27** and **28**. Half wave potentials of **27** and **28**

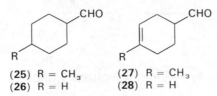

(**25**) R = CH$_3$ (**27**) R = CH$_3$
(**26**) R = H   (**28**) R = H

are 0·03 and 0·02 V anodic of their saturated counterparts. The semi-carbazones were less affected by the presence of a non-conjugated double bond[68].

Electron transfer to the carbonyl group of a non-conjugated olefinic ketone (**29**) produced a cyclic tertiary alcohol in a stereoselective intra-molecular cycloaddition[69] (equation 17). However, cyclization was found

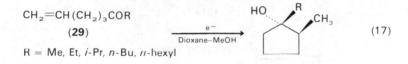

$$CH_2=CH(CH_2)_3COR$$
(**29**)

$$\xrightarrow[\text{Dioxane–MeOH}]{e^-}$$

R = Me, Et, i-Pr, n-Bu, n-hexyl   (17)

to occur only for olefinic ketones which might form five- or six-membered rings (equations 18–21).* Reduction of **30** produced cyclic alcohols when n was 3 or 4, but the corresponding olefinic alcohol when n was 2 or 5[69].

$$CH_2=CH(CH_2)_n\overset{\overset{\displaystyle O}{\|}}{C}CH_3$$

(**30**)

$$30\ (n=2)\ \xrightarrow{e^-}\ CH_2=CH(CH_2)_2\overset{\overset{\displaystyle OH}{|}}{C}HCH_3 \qquad (18)$$

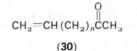

$$30\ (n=3)\ \xrightarrow{e^-} \qquad\qquad\qquad\qquad (19)$$

---

* Electroreductive coupling of ketones with non-activated olefins (e.g. acetone and 1-octene) has been reported by T. Shono and M. Mitani, *Nippon Kagaku Kaishi*, 975 (1973).

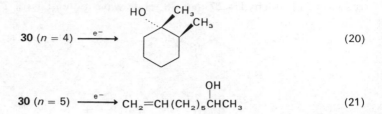

$$30 \ (n = 4) \xrightarrow{\ e^- \ } \qquad\qquad\qquad\qquad (20)$$

$$30 \ (n = 5) \xrightarrow{\ e^- \ } CH_2{=}CH(CH_2)_5\overset{\displaystyle OH}{\underset{\displaystyle |}{C}}HCH_3 \qquad (21)$$

e. *Diketones*. Evans and Woodbury have reported intramolecular coupling of ketyl radicals generated from reduction of **32** which is itself a coupling product from reduction of dibenzoylmethane at the first polarographic wave at pH 4·2[70] (Scheme 2). Reduction of dibenzoyl-

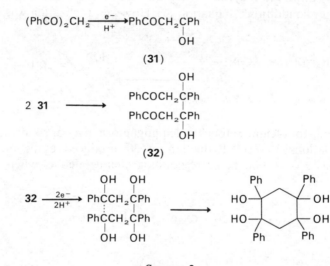

$$(PhCO)_2CH_2 \xrightarrow[H^+]{e^-} PhCOCH_2\overset{\displaystyle |}{\underset{\displaystyle OH}{C}}Ph$$

$$\textbf{(31)}$$

$$2 \ \textbf{31} \longrightarrow \begin{array}{c} OH \\ PhCOCH_2\overset{\displaystyle |}{C}Ph \\ PhCOCH_2\overset{\displaystyle |}{C}Ph \\ OH \end{array}$$

$$\textbf{(32)}$$

$$\textbf{32} \xrightarrow[2H^+]{2e^-} \begin{array}{c} OH \quad OH \\ Ph\overset{\displaystyle |}{C}CH_2\overset{\displaystyle |}{C}Ph \\ Ph\overset{\displaystyle |}{C}CH_2\overset{\displaystyle |}{C}Ph \\ OH \quad OH \end{array} \longrightarrow$$

SCHEME 2.

methane at more negative potentials produces monomeric product, 1,3-diphenyl-1,3-propanediol (**33**) since the rate of reduction of the radical (**31**) is much higher than the rate of dimerization at very negative potentials (equation 22).

$$\textbf{31} \xrightarrow{e^-} PhCOCH_2\overset{\displaystyle OH}{\underset{\displaystyle |}{\underline{C}}}Ph \xrightarrow[3H^+]{2e^-} PhCHCH_2CHPh \qquad (22)$$
$$\qquad\qquad\qquad\qquad\qquad\qquad \overset{\displaystyle |}{OH} \quad \overset{\displaystyle |}{OH}$$

$$\textbf{(33)}$$

Reduction of **34** by controlled potential electrolysis gave products **37–40**. The mechanism outlined below suggests formation of pinacol intermediates by both intramolecular (**35**) and intermolecular (**36**) processes [71] (Scheme 3).

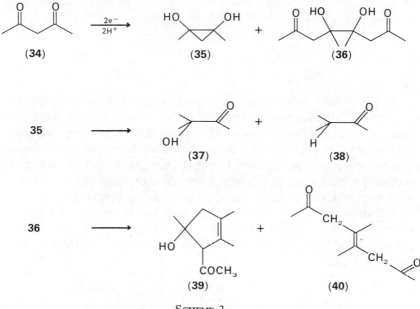

(34)   (35)   (36)

35 ⟶   (37)   +   (38)

36 ⟶   (39)   +   (40)

SCHEME 3.

Cyclopropanediol was reported to be formed from reduction of 2-methyl-2-acetylcyclohexanone at mercury in tetrahydrofuran with tri-butylethylammonium tetrafluoroborate as the supporting electrolyte. Acetic anhydride was used to trap the labile cyclopropanediol and the diacetate **41** was isolated[72].

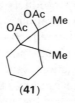

(41)

Reduction of **42** gave a mixture of monomeric alcohol and pinacol with the relative amount of monomeric product increasing with increasing reduction potential[73] (equation 23). This illustrates a case where dimerization rates are competitive with electron transfer rates. As the electrode potential is made more negative, the rate of electron transfer to the radical increases while the rate of dimerization is unchanged.

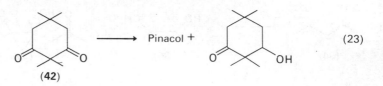

$$\text{(42)} \xrightarrow{\hspace{2cm}} \text{Pinacol} + \hspace{2cm} \tag{23}$$

(42)

Benzil exhibits two one-electron polarographic waves in dimethylformamide[74]. Reduction at the first wave affords a blue material—assumed to be the radical anion (**43**) (equation 24). Further reduction to the dianion occurs at a more negative potential. Subsequent protonation may produce *cis*- or *trans*-stilbenediol (Section IV.A.1.b). When the supporting electrolyte is changed from tetrabutylammonium iodide to lithium perchlorate a single two-electron wave is observed[75]. The e.s.r. signal observed with tetrabutylammonium iodide present, is no longer observed in the presence of lithium ion. Philp, Layloff and Adams[75] suggested that reduction of benzil in the presence of lithium salts produces **44** (equation 25).

$$\underset{\text{PhC}-\text{CPh}}{\overset{\text{O} \quad \text{O}}{\overset{\|\quad\|}{}}} \xrightarrow{e^-} \left[\underset{\text{PhC}-\text{CPh}}{\overset{\text{O} \quad \text{O}}{\overset{\|\quad\|}{}}} \right]^{\overline{}} \xrightarrow{e^-} \underset{\text{PhC}-\text{CPh}}{\overset{\text{O}^- \;\; \text{O}^-}{\overset{|\quad|}{}}} \tag{24}$$

(43)

$$\textbf{43} \xrightarrow{\text{Li}^+} \underset{\text{PhC}-\text{CPh}}{\overset{\text{O} \quad \text{OLi}}{\overset{\|\quad|}{}}} \xrightarrow{e^-,\text{Li}^+} \underset{\text{PhC}=\text{CPh}}{\overset{\text{OLi} \;\; \text{OLi}}{\overset{|\quad|}{}}} \tag{25}$$

(44)

Alkali metal reduction of *o*-, *m*-, and *p*-dibenzoylbenzene was shown to produce the same radical anion products as electrochemical reduction at the potential of the first polarographic wave. Products were identified by e.s.r. and visible spectra. Further reduction of the *para* isomer produced a blue diamagnetic dianion. The blue material was converted to starting material upon oxidation[76]. Reduction of the *para* isomer by lithium metal was thought to form a dilithium product (**45**) is a process similar to the electrochemical reduction of benzil in the presence of Li$^+$. Electrochemical behaviour of *p*-dibenzoylbenzene in the presence of Li$^+$ was not investigated.

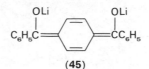

(45)

f. *Reduction to hydrocarbons.* Aldehydes and ketones may be reduced to hydrocarbons in strongly acidic media at cathodes with high over-voltages[77]. Literally nothing is known about the mechanism, however, since attention has focused on empirical determination of experimental conditions to effect reduction. It is known that an alcohol is not an intermediate since alcohols may be recovered unchanged when exposed to these reduction conditions[77a]. It is probable that the cathode itself is intimately involved in the reduction process with the possible partici-pation of organometallic intermediates.

g. *Carboxylic acids.* Aromatic carboxylic acids and amides are reducible at very negative potentials[78]. Reduction of benzoic acid was reported to occur at mercury and lead cathodes by electron transfer to an adsorbed ion pair formed from the benzoate and the cation of the electro-lyte[79]. This reduction occurred at potentials cathodic of hydrogen evolution and it is not clear that this represents a true electrochemical reduction involving electron transfer to benzoate.

In electrochemical reduction in the absence of hydrogen evolution, the initial reduction product is thought to be an aldehyde which may or may not be further reduced, depending on experimental conditions. If the aldehyde is hydrated, as would be the case for heterocyclic compounds no further reduction occurs[80a] (equation 26). In contrast, aromatic acids may be reduced to alcohols since the intermediate aldehyde is not pro-tected by hydration[81]. Wagenknecht[81b] reported reduction of a number

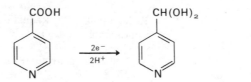

$$(26)$$

of aromatic carboxylic acids to aldehydes. It was suggested that the aldehyde was not further reduced because the initially formed hydrate lost water very slowly in solutions buffered to pH 6.

Reduction of phthalic acid resulted in reduction of the aromatic ring rather than the carboxyl group. The principal product was *trans*-1,2-dihydrophthalic acid[82] (equation 27).

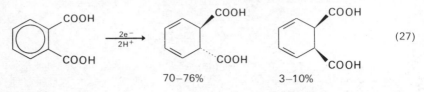

$$70-76\% \qquad 3-10\% \qquad (27)$$

2. Stereochemistry

For a series of steroidal ketones in which the carbonyl is located at different positions on the steroid nucleus, the half-wave potential increases as accessibility of the carbonyl decreases[83]. Reduction of the more accessible carbonyl is favoured implicating orientation of the steroid at the electrode surface as an important feature of the reduction process. This steric selectivity can be used to effect selective reduction of a difunctional molecule. For example, in large scale reductions of 3,12-diketosteroids, the more accessible 3-keto group was reduced selectively and 3-hydroxy-12-cholanic acid was the product[84].

Products from the reduction of a series of cyclic ketones in isopropanol and in $H_2SO_4-H_2O-MeOH$ are compared in Table 1. In isopropanol the product distributions reflect the thermodynamic stabilities of the product isomers. In acid medium the thermodynamically unstable epimer predominates[66a]. In the poorer proton donating medium, isopropanol, protonation takes place after the active species has diffused from the electrode surface into bulk solution. Thermodynamic stability of the products controls the stereochemical outcome. In the better proton donating medium, the anionic species is protonated in the vicinity of the

TABLE 1. Products from electrochemical reduction of some cyclic ketones[66a]

	Reduction in isopropanol	Reduction in $H_2SO_4-H_2O-MeOH$	Relative stabilities
	Epimer ratio (*trans/cis*)		
3-Methylcyclohexanone	22/78	52/48	22/78
4-Methylcyclohexanone	78/22	38/62	70/30
4-*t*-Butylcyclohexanone	85/15	48/52	79/31
3,3,5-Trimethylcyclohexanone	26/74	40/60	
2-Methylcyclohexanone	59/42	42/58	
	Epimer ratio (*endo/exo*)		
Norcamphor	16/84	98/2	20/80
Camphor	76/24		71/29

electrode while still oriented or absorbed at the surface and its least hindered side is protected by the bulk of the electrode.

Similar results for reduction of 4-*t*-butylcyclohexanone and 3,3,5-trimethylcyclohexanone were reported by Coleman, Kobylecki, and Utley[66b]. In acidic media at a lead cathode, roughly equal quantities of axial and equatorial alcohols were formed. In a poorer proton-donating medium, hexamethylphosphoramide, the more stable equatorial alcohols predominated. Organolead intermediates were suggested to be responsible for the non-thermodynamic product distribution in aqueous acid. This suggestion is supported by reported formation of dialkylmercury products from reductions at a mercury cathode[85]. Presumably similar trapping of an intermediate species could occur at lead. Significantly, such organo-mercurials are known to be more stable to protolysis than their lead counterparts[86]. On the other hand, nearly identical non-thermodynamic product distributions were reported by Shoni and Mitani[66a] for reduction of 4-*t*-butylcyclohexanone and 3,3,5-trimethylcyclohexanone in aqueous acid at a carbon rod electrode where organometallic intermediates obviously could not play any role. Formation of thermodynamically less favoured alcohols apparently may occur by several different pathways.

Reduction of an unsymmetrical ketone such as ArCOR may lead to two diasteromeric pinacols, *meso* and *dl*. Stocker and Jenevein found that the *dl/meso* product ratio from reduction of acetophenone ranges from 1·0 to 1·4 in acid, 2·5 to 3·2 in alkaline media[87], and 5 to 9 in aprotic media[88]. In acid, where pinacols arise from dimerization of neutral radicals, steric factors would be expected to favour formation of *meso* product. Predominance of *dl* isomer is thought to arise from favourable hydrogen bonding in the *dl* transition state (**46**). This postulate is supported by

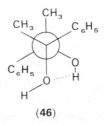

(**46**)

studies of the electrochemical pinacolization of 2-acetylpyridine. Strong intramolecular hydrogen bonding in the radical (**47**) results in the predominance of the *meso* isomer, also predicted on steric grounds[89]. Likewise electrochemical reduction of $C_6H_5COCONH_2$ in acid affords

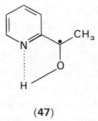

(47)

the *meso* pinacol exclusively, presumably due to strong intramolecular hydrogen bonding in the intermediate[90].

Grimshaw and coworkers[91] have reported that reduction of 2-acetylnaphthalene affords isomeric 2,3-di-(2-naphthyl)butane-2,3-diol with *dl* isomer predominating. Stereoselectivity is more apparent at pH 14 than at pH 3 but *dl* is favoured in both acid and alkali.

In alkali, it has been suggested that pinacols arise from dimerization of neutral radicals with radical anions. This suggestion accounts for the increased preference for formation of *dl* pinacols in alkali since the hydrogen bond in the transition state leading to **50** should be stronger than in **46** (Scheme 4). Propiophenone and benzaldehyde exhibit behaviour similar to that observed for acetophenone[92].

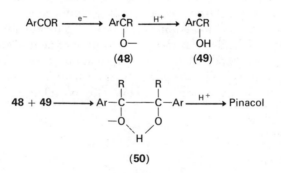

SCHEME 4.

An exception to the generalization that the *dl*-pinacol predominates in reductions carried out in alkali occurs when the aromatic ring bears a phenolic hydroxyl[93]. Reduction of **51** produces a mixture of *meso*- and *dl*-pinacol but reduction of **52** affords pure *meso*-pinacol[94].

It is interesting to note that complex formation with the electrode has not been implicated in explanations of stereochemical preferences of radical coupling even though many metals, especially mercury, are known to be excellent radical traps.

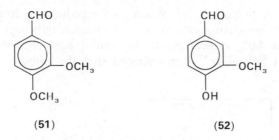

(51)   (52)

Vincenz-Chodkowska and Grabowski found that benzil (53) is reduced electrochemically to a mixture of *cis*- (55) and *trans*- (54) stilbenediols (Scheme 5). The relative proportions of stereoisomeric products are strongly dependent on experimental parameters which affect the strength of the electric field at the electrode surface such as electrode potential,

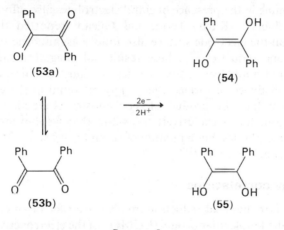

SCHEME 5.

ionic strength, temperature, and nature of the supporting electrolyte. The ratio of *cis* to *trans* could be varied from 1/50 to 2/1 with *cis* pre-dominating at high field strengths[95]. It is likely that field strength controls product stereochemistry by affecting the relative population of rotamers in the starting material.

In an interesting experiment, Horner and Degner[96] effected asymmetric induction through the use of an optically active supporting electrolyte. From reduction of acetophenone in methanol containing (−)-ephedrine hydrochloride, they isolated R-(+)-α-methylbenzyl alcohol of 4.2%

optical purity (equation 28). When (+)-ephedrine hydrochloride was used as the supporting electrolyte, S-(−)-α-methylbenzyl alcohol was produced in 4·6% optical purity. In similar acetophenone reductions Horner and Schneider[97] demonstrated that the optical purity of the

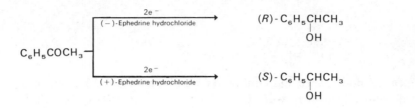

$$C_6H_5COCH_3$$

$$\xrightarrow[\;(-)\text{-Ephedrine hydrochloride}\;]{2e^-} (R)\text{-}C_6H_5\underset{\overset{|}{OH}}{C}HCH_3$$

$$\xrightarrow[\;(+)\text{-Ephedrine hydrochloride}\;]{2e^-} (S)\text{-}C_6H_5\underset{\overset{|}{OH}}{C}HCH_3$$

(28)

product alcohol was significantly affected by temperature and electrode material and solvent and to a lesser extent by acidity and reduction potential. Reduction of a series of aryl alkyl ketones and acetophenone-N-benzylimine in the presence of chiral electrolytes also afforded optically active products[98]. Kariv, Terni, and Gileadi reported that alkaloids absorbed on the electrode surface also induce asymmetric products from acetophenone reduction[99]. These results all suggest a high degree of orientation of both electroactive species and inactive chiral entity.

Further evidence of surface effects may be found in the effect cathode material may have on product stereochemistry. Reduction of 2-methyl-cyclohexanone forms exclusively $trans$-2-methylcyclohexanol at mercury or platinum cathodes, but a mixture of cis and $trans$ alcohols at nickel and pure cis alcohol at copper[100].

3. Energy correlations

Since electrochemical reduction involves transfer of an electron to the lowest unfilled molecular orbital (LUMO), of the electroactive compound, reduction potentials might reasonably be expected to correlate with LUMO values. Compounds chosen for such correlations should exhibit electrochemically reversible behaviour. Irreversibility leads to distortion of polarographic waves and measured half-wave potentials will reflect not only molecular energy levels but additional kinetic and thermo-dynamic parameters as well. Although this limitation to reversible systems is frequently neglected, successful correlations have been reported for a wide variety of organic compounds[101].

The success of correlations between reduction potentials of carbonyls and LUMO values led to use of reduction potential as a criterion for evaluating the validity of Coulomb and resonance integrals for sulphur.

A linear relationship was observed between reduction potentials and LUMO values for a series of aromatic carbonyls and for their sulphur analogues. Consequently the resonance and Coulomb integrals for sulphur which had been calculated from electron densities derived from e.p.r. data were confirmed as reasonable values[102].

Loutfy and Loutfy[103] have demonstrated a linear correlation between half-wave potentials and $n \rightarrow \pi^*$ triplet energies for a number of carbonyl and thiocarbonyl compounds. For a series of benzophenones and thio-benzophenones of structure **56**, this relationship was used to estimate the energy of n, π^* transitions for compounds whose n, π^* triplets could not

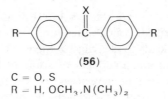

(**56**)

C = O, S
R = H, OCH$_3$, N(CH$_3$)$_2$

be observed spectroscopically but whose half-wave potentials could be measured[103a] (Table 2).

In a similar manner a linear correlation between $E_{\frac{1}{2}}$ and n, π^* triplet energies for aliphatic and aromatic ketones was used to estimate the n, π^* triplet energy for acetone to be 76.6 kcal/mol[103b].

TABLE 2. Reduction potentials and spectroscopic energies for a series of benzophenones and thiobenzophenones[103a]

X	R	$-E_{\frac{1}{2}}$ (V)	$E_T, n \longrightarrow \pi^*$ (kcal/mol)
O	H	1·84	68·6
O	OCH$_3$	2·01	69·4
O	N(CH$_3$)$_2$	2·16	70·1
S	H	1·17	40·6
S	OCH$_3$	1·35	42·1
S	N(CH$_3$)$_2$	1·51	43·4

The half-wave potentials of compounds of structure **56** were also corre-lated with Hammett substituent constants. The reaction parameter ρ

was found to be identical for the carbonyl and the thiocarbonyl series (0.384)[103a].

B. Oxidation

Electrochemical oxidation of C=O (and C=N) is not commonly observed, presumably because most solvent–electrode systems are more easily oxidized than the functional group. Manousek and Volke[104] reported on the anodic oxidation of aromatic aldehydes at mercury. The oxidation process was found to be an indirect oxidation of the aldehyde by HgO or the corresponding hydroxide which formed when the electrode itself was oxidized.

Miller and coworkers[105] reported an interesting fragmentation upon electrooxidation of benzylic aldehydes and ketones at platinum in acetonitrile. Oxidation leads to cleavage of the C—C bond, which mimics mass spectral fragmentation. Products of some of these oxidations are shown in Table 3. The oxidative process for benzoin is represented in Scheme 6. Product yield from this relatively simple oxidative process

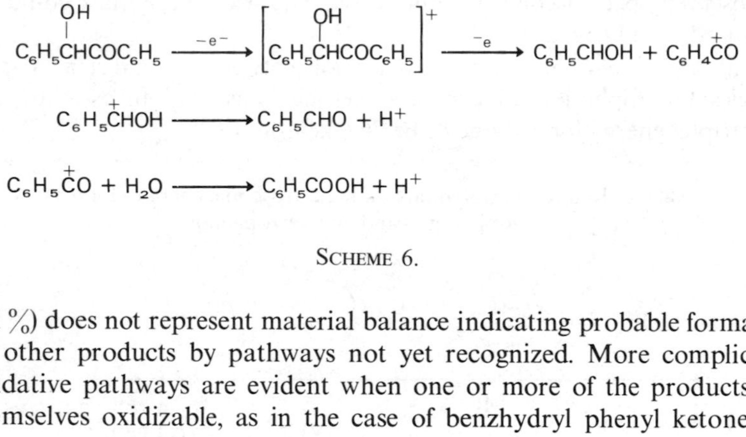

$$
\underset{\substack{\text{OH}\\|}}{C_6H_5CHCOC_6H_5} \xrightarrow{-e^-} \left[\underset{\substack{\text{OH}\\|}}{C_6H_5\overset{\cdot}{C}HCOC_6H_5}\right]^{+} \xrightarrow{-e} C_6H_5\overset{+}{C}HOH + C_6H_4\overset{+}{C}O
$$

$$
C_6H_5\overset{+}{C}HOH \longrightarrow C_6H_5CHO + H^+
$$

$$
C_6H_5\overset{+}{C}O + H_2O \longrightarrow C_6H_5COOH + H^+
$$

Scheme 6.

(61%) does not represent material balance indicating probable formation of other products by pathways not yet recognized. More complicated oxidative pathways are evident when one or more of the products are themselves oxidizable, as in the case of benzhydryl phenyl ketone. By

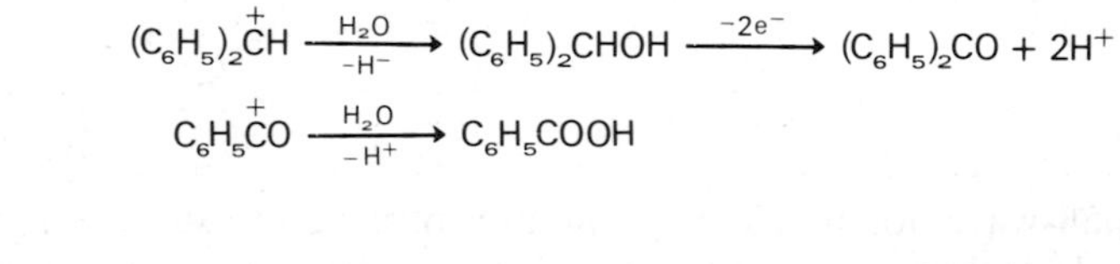

$$
(C_6H_5)_2CHCOC_6H_4 \xrightarrow{-2e^-} (C_6H_5)_2\overset{+}{C}H + C_6H_5\overset{+}{C}O
$$

$$
(C_6H_5)_2\overset{+}{C}H \xrightarrow[-H^-]{H_2O} (C_6H_5)_2CHOH \xrightarrow{-2e^-} (C_6H_5)_2CO + 2H^+
$$

$$
C_6H_5\overset{+}{C}O \xrightarrow[-H^+]{H_2O} C_6H_5COOH
$$

Scheme 7.

analogy to the oxidation of benzoin in Scheme 6, oxidation of this ketone would be expected to result in formation of benzhydrol and benzoic acid. However, oxidation of benzhydrol occurs at 1·7 V while oxidation of the starting material does not occur until 1·8 V. Consequently any benzhydrol formed will be rapidly oxidized. Scheme 7 summarizes the oxidation of benzhydryl phenyl ketone. The amides generated in certain oxidations (Table 3) are presumably formed by a Ritter reaction between intermediate carbonium ions and the solvent, acetonitrile.

TABLE 3. Products from electrochemical oxidation of benzylic aldehydes and ketones in acetonitrile[105]

Compound	V^a	Products
$(C_6H_5)_2CHCOC_6H_5$	1·82	$(C_6H_5)_2CO$ C_6H_5COOH
$(C_6H_5)_2CHCHO$	1·71	$(C_6H_5)_2CO$ $(C_6H_5)_2CHNHCOCH_3$
$C_6H_5CH_2COC_6H_5$	1·80	C_6H_5COOH $C_6H_5COCOC_6H_5$ C_6H_5CHO
$C_6H_5CH(CH_3)CHO$	1·80	$C_6H_5CH(CH_3)NHCOCH_3$ $C_6H_5COCH_3$
$C_6H_5CH(OH)COC_6H_5$	1·89	C_6H_5CHO C_6H_5COOH $C_6H_5COCOC_6H_5$
$(C_6H_5)_2CHOH$	1·70	$(C_6H_5)_2CO$ $(C_6H_5)_2CHOCH(C_6H_5)_2$

^{a} Potential vs. $Ag/AgNO_3$

V. THE AZOMETHINE LINKAGE

The electrochemical behaviour of compounds containing an azomethine group is quite similar to that of corresponding carbonyls although azomethines are generally more easily reduced than carbonyl compounds. As early as 1902 it was observed that solutions containing a primary amine and a carbonyl compound could be reduced under conditions where neither could be reduced alone[78]. Electroactivity was correctly attributed to the condensation product—an azomethine. Condensation has proved a useful method for conversion of a polarographically inactive carbonyl to a polarographically active species[106].

The electrochemical behaviour of imines R₂R=NR depends on the degree of arylation. Completely arylated imines such as benzophenone anil (57) are easily reduced to relatively stable intermediates. As aryl groups are replaced by hydrogen or alkyl groups, ease of reduction and product stability both decrease.

$$(C_6H_5)_2C{=}NC_6H_5$$

(57)

The electrochemical behaviour of imines R₂R=NR depends on the ketones in electrochemical behaviour. Below pH 9 two polarographic waves are observed for reduction of the protonated anil and of the resulting radical[9]. Reduction of neutral anil to *N*-benzhydrylaniline takes place in a single two-electron process above pH 9[9]. In contrast the transition from reduction of protonated to reduction of neutral benzophenone takes place at pH 5, thus reflecting the greater basicity of imines over carbonyl compounds. Above pH 13 and in aprotic solvents two one-electron waves are observed for reduction of 57 to a stable radical anion and subsequently to a dianion. Cyclic voltammetry confirms the first wave as reversible and the e.s.r. spectrum of the radical anion may be observed[107].

Diaryl imines Ar₂C=NR and ArRC=NAr undergo transition from reduction of protonated to reduction of neutral imine around pH 13 in contrast to pH 9 for triarylated imines reflecting increased basicity over their triarylated counterparts[108]. Polarographic behaviour of diaryl imines in aprotic media depends on the nature of the two aryl groups. For example two one-electron waves are observed for imines with an extended aromatic system such as 58[17] while a single one-electron wave is observed for 59[109].

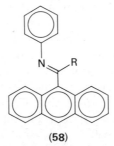

(58)

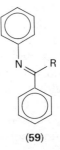

(59)

Cyclic voltammetry and e.s.r. spectroscopy show that the lifetime of the radical anion formed in the initial electron transfer step is highly dependent on the nature of the aryl substituents[109]. This is reasonable since the basicity of the radical anion and the rate of proton abstraction are influenced by the nature of the substituents. For highly resonance-stabilized structures, proton abstraction is slow on a polarographic time scale and two one-electron waves are seen. In contrast, rapid protonation of a less stabilized radical anion, e.g. that derived from **59**, produces an electroactive radical which accepts a second electron and a single two-electron wave is seen. In some cases behaviour is intermediate between these extremes indicating partial protonation of the intermediate radical anion.

Monoaryl and trialkyl imines have not been investigated extensively. A single two-electron process is observed in protic media. Controlled potential electrolysis yields a secondary amine. In aprotic media, trialkyl imines are not reducible, suggesting that as for alkyl ketones (Section IV.A.1.c) reduction in protic solvents is enhanced by hydrogen bonding to the solvent. Several monoaryl imines (**64a** and **b** have been investigated in dimethylformamide[11]. A single two-electron polarographic wave was observed and cyclic voltammetry at high sweep rates (100 V/sec) showed no anodic peak on scan reversal. Controlled potential coulometry confirmed consumption of two electrons per molecule and the corresponding secondary amines were isolated in high yields[11].

Compounds of the type $R_2C=NX$ are reduced in four-electron processes when X is a group bound through N or O. The N—X bond is cleaved first, followed by reduction of the intermediate imine to an amine.

Diaryl, arylalkyl and dialkyl oximes exhibit a single four-electron polarographic wave in acid solutions[9,110]. Controlled potential coulometry confirms a four electron reduction and the product amine may be isolated in good yields. The reduction has been shown to involve the conjugate acid of the oxime and to proceed via initial cleavage of the N—O bond to form an imine which is immediately protonated and reduced (equation 29). The fact that hydroxylamines are not reducible

$$R_2C=NOH_2^+ + 4e^- + 4H^+ \longrightarrow R_2CHNH_3^+ + H_2O \qquad (29)$$

at potentials where oximes are reduced to amines rules out reduction of the C=N bond first[9]. As added evidence, several oximes exhibit two two-electron polarographic waves when reduction occurs in two discrete steps[111]. In these cases controlled potential electrolysis at the first wave produced the imine in good yield. Above pH 6 the wave observed in acid media disappears and a new four-electron wave for reduction of the neutral oxime appears at more cathodic potential. In strongly alkaline solution the oxime conjugate base is not reducible.

Phenylhydrazones and semicarbazones are reduced in a single four-electron step in acid as demonstrated by polarography, coulometry, and isolation of amine from controlled potential reduction. Reduction has been shown to proceed via initial cleavage of the N—N bond of the conjugate acid[9,108] (equation 30).

$$R_2C{=}N\overset{+}{N}H_2X + 4e^- + 4H^+ \longrightarrow R_2CHNH_3^+ + XNH_2$$

(30)

$$X = C_6H_5, \overset{\overset{O}{\|}}{C}NH_2$$

Reduction of a number of immonium salts in dimethylformamide and acetonitrile has been reported by Andrieux and Saveant[112]. Immonium salts **60** and **61** exhibited irreversible behaviour while immonium salts

$$(CH_3)_2C{=}\overset{+}{N}\diagdown$$

(60)

$$(CH_3)_2C{=}\overset{+}{N}(C_6H_5)_2$$

(61)

$$(C_6H_5)_2C{=}\overset{+}{N}(CH_3)_2$$

(62)

$$(p{-}CH_3OC_6H_4)_2C{=}\overset{+}{N}(CH_3)_2$$

(63)

62 and **63** showed reversible behaviour and e.s.r. spectra of the product radicals were obtained.

Reduction of camphor anil (**64a**) and norcamphor anil (**64b**) may produce *endo* (**65**) or *exo* (**66**) amines while reduction of **67** results in a new chiral centre[11] (equations 31 and 32). Table 4 compares the stereo-

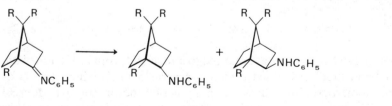

(31)

(64) a: R = CH₃   **(65)**   **(66)**
       b: R = H

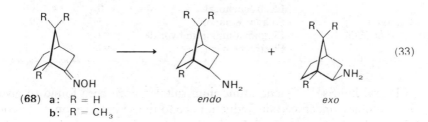

$$
\begin{array}{c}
\text{CH}_3 \\
\underset{\text{C}_6\text{H}_5}{\text{C}_6\text{H}_5} \diagdown \quad \overset{\text{CHC}_6\text{H}_5}{\diagup} \\
\text{C}=\text{N} \\
\underset{\text{CH}_3}{\diagup}
\end{array}
\quad\longrightarrow\quad
\begin{array}{c}
\underset{\text{C}_6\text{H}_5}{} \quad \underset{\text{C}_6\text{H}_5}{} \\
\text{H}-\text{C}-\text{NH}-\text{C}-\text{H} \\
\underset{\text{CH}_3}{} \quad \underset{\text{CH}_3}{}
\end{array}
\tag{32}
$$

(67)

chemical consequences of treatment of these imines with a variety of reducing agents. Based on product stereochemistry, electrochemical reduction of these imines at a mercury cathode resembles dissolving metal reduction more closely than catalytic hydrogenation.

The electrochemical reduction of norcamphor oxime (**68a**) and camphor oxime (**68b**) to the corresponding amines (equation 33) follows a very

$$
\tag{33}
$$

(**68**) a: R = H
b: R = CH$_3$

endo   exo

TABLE 4. Stereochemistry of reduction of some imines[11]

Compound	Mode of reduction	Relative products (%)	
	Bicyclic imines		
		exo Amine	*endo* Amine
Norcamphor anil (**64b**)	Sodium–ethanol	33	67
	Catalytic hydrogenation	0	100
	Electrochemical	20	80
Camphor anil (**64a**)	Sodium–ethanol	0	100
	Catalytic hydrogenation	100	0
	Electrochemical	0	100 ⁻
N-α-Methylbenzylidine-α-methylbenzylamine (**67**)			
		dl Amine	*meso* Amine
Racemic **67**	Sodium–ethanol	46	54
	Catalytic hydrogenation	93	7
	Electrochemical	39	61
		Optically active (−)-amine	*meso* Amine
Optically active (−)-(*R*) **67**	Sodium–ethanol	42	58
	Catalytic hydrogenation	90	10
	Electrochemical	43	57

different stereochemical course from the anils[113] (Table 5) and products from electrochemical reduction of the oximes resemble those from catalytic hydrogenation.

The product differences observed here bear a striking similarity to the cyclic ketone reductions reported in Table 1. The electrochemical reductions of the anils were carried out in dimethylformamide while oxime reductions were performed in ethanol.

TABLE 5. Stereochemistry of reduction of bicyclic oximes[113]

		Relative products (%)	
Compound	Mode of reduction	*exo* Amine	*endo* Amine
Norcamphor	Sodium–ethanol	75	25
oxime (**68a**)	Lithium aluminium hydride	0	100
	Electrochemical	0	100
Camphor	Sodium–ethanol	4	96
oxime (**68b**)	Lithium aluminium hydride	99	1
	Electrochemical	99	1

Diaryl imines[114,98b] and immonium salts[112] have been found to give *vic*-diamines under certain reduction conditions, but the only stereochemical information available is the report by Law that benzal-*p*-toluidine (**69**) affords a mixture of the *dl* and *meso* isomers of **70** upon reduction in neutral ethanol at a copper electrode (equation 34)[115].

$$C_6H_5CH{=}NAr \xrightarrow{\ e^-\ } \begin{array}{c} C_6H_5CHNHAr \\ | \\ C_6H_5CHNHAr \end{array} \qquad (34)$$

(**69**)   (**70**)

$$(Ar = p\text{-}CH_3C_6H_4)$$

Scott and Jura measured the half-wave potentials of a series of 24 Schiff bases of the type ArCH=NAr′ where Ar and Ar′ represent unsubstituted aryl groups[17]. The effect of variation of half-wave potential with structural change is reflected very well by a linear combination of atomic orbitals[116]. Since half-wave potentials may be viewed as free energies for reversible reductions, the data were recast in a Hammett σ–ρ form with σ_r defined as the shift in half-wave potential with respect to reduction of benzanil. Values of σ were found to correlate well with chemical reactivities[17]. Bezuglyi and coworkers reported a similar correlation between σ

substituent constants and half-wave potentials for another series of Schiff bases[117].

VI. THE AZO LINKAGE

The electrochemical reduction of aromatic azo compounds has been the subject of several reviews[21,118]. Azobenzene is reduced to hydrazobenzene in the pH range 2–6 (equation 35). At pH 2 and below further reduction of

$$C_6H_5N{=}NC_6H_5 \xrightarrow[2H^+]{2e^-} C_6H_5NHNHC_6H_5 \qquad (35)$$

hydrazobenzene to aniline (equation 36) takes place, but at a much more

$$C_6H_5NHNHC_6H_5 \xrightarrow[2H^+]{2e^-} 2C_6H_5NN_2 \qquad (36)$$

negative potential than necessary for formation of hydrazobenzene. If electrolysis is carried out at low pH at a potential sufficient to reduce azobenzene but not hydrazobenzene, the product is benzidine, formed by acid-catalysed rearrangement of hydrazobenzene (equation 37). Hence it is possible to convert azobenzene into hydrazobenzene, benzidine, or

$$C_6H_5NHNHC_6H_5 \xrightarrow{H^+} H_2N{-}\langle\bigcirc\rangle{-}\langle\bigcirc\rangle{-}NH_2 \qquad (37)$$

aniline by appropriate choice of pH and potential.

The *cis* and *trans* isomers of azobenzene are reduced at approximately the same potential (*c.* −0·6 V vs. S.C.E. in aqueous ethanol containing 0·1 M-tetraethylammonium perchlorate), but the reduction of the *cis* isomer appears to be irreversible, judging from the drawn-out appearance of the polarogram[119]. The reduction of both isomers is pH dependent, becoming easier as the pH is lowered[119,120]. In acid, the electroactive species may be the conjugate acid of the azo compound (71), but this is not clear, since only small amounts of 71 will be present in all but the

$$C_6H_5N{=}\overset{+}{N}HC_6H_5$$

(71)

most acidic media (the pK_a of 71 is −2·90 in 20% aqueous ethanol[121]).

362   Albert J. Fry and Roberta Gable Reed

The electrochemical behaviour of most substituted azobenzenes parallels that of azobenzene itself. When one of the rings bears an electron-supplying heteroatom (usually —OH or —NH$_2$), more complex behaviour may be observed[21,122]. The reduction of *p*-aminoazobenzene (72) provided an illustrative example (Scheme 8)[123]. The azo linkage undergoes

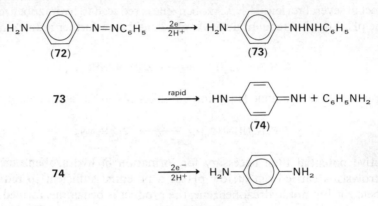

SCHEME 8.

the usual two-electron reduction, producing hydrazo compound 73, can then fragment to aniline and the *bis*-quinone imine 74, which is immediately reduced to *p*-phenylenediamine. The reduction of *p*-hydroxy-azobenzene (75) proceeds in similar fashion (Scheme 9)[122].

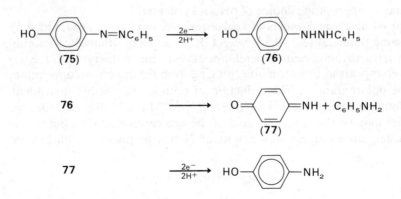

SCHEME 9.

In aprotic solvents aromatic azo compounds exhibit two one-electron reduction steps, corresponding to stepwise reduction to the radical anion **(78)**, and dianion, respectively (equation 38)[124]. In the presence of lithium

$$ArN=NAr \xrightarrow{\ e^- \ } ArN=NAr^- \bullet \xrightarrow{\ e^- \ } Ar\bar{N}-\bar{N}Ar \qquad (38)$$

(78)

ion azobenzene exhibits a single two-electron wave[125]. Likewise, addition of a lithium salt to a solution of the azobenzenide ion **(78**, with tetrabutyl-ammonium ion as counter ion) results in instantaneous disproportionation of the radical anion into azobenzene and the dilithio derivative of hydrazo-benzene **(79)**. The driving force for the disproportionation of the radical

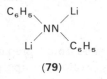

(79)

anion appears to be the formation of two highly covalent nitrogen–lithium bonds in **79**[125]. Similar behaviour has been observed, it will be recalled (Section IV.A), with benzil.

In an interesting application of electrochemical principles, Iversen has used **79** as an electrochemically generated base for synthetic pur-poses[126,127]. Iversen and Lund electrolysed a mixture of azobenzene, benzyltriphenylphosphonium bromide, and benzaldehyde at a potential where only azobenzene is reduced[126]. Stilbene **(81)** was formed in 98% yield (Scheme 10). The key reaction here is the deprotonation of the

$$C_6H_5N=NC_6H_5 \xrightarrow[2Li^+]{2e^-} 79$$

$$79 + 2C_6H_5CH_2P^+(C_6H_5)_3 \longrightarrow C_6H_5NHNHC_6H_5 + 2C_6H_5CH=P(C_6H_5)_3$$

(80)

$$80 + C_6H_5CHO \longrightarrow (C_6H_5)_3PO + C_6H_5CH=CHC_6H_5$$

(81)

SCHEME 10.

phosphonium salt by **79**; a Wittig reaction completes the sequence. As Lund and Iversen pointed out, this 'electrochemical generation of strong base' has several features which recommend it for use in other base-promoted reactions: by controlling the current passed, one can carefully control both the total amount of base produced and also its rate of production, and one avoids the use of highly nucleophilic reagents such as phenyl lithium. Iversen has also carried out a Stevens rearrangement of a quaternary ammonium salt using electrochemically generated base[127].

Lund examined the electrochemical behaviour of 3,3-pentamethyleneazirine (82)[128]. In alkali (pH = 8) 82 exhibits a single polarographic wave corresponding to the uptake of two electrons per molecule of 82

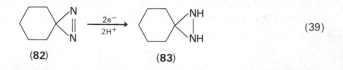

$$(39)$$

(82) (83)

and 83 was isolated in 80% yield from a preparative electrolysis (equation 39). In acid (pH = 6), 83 is protonated upon formation, weakening the N—N bond, and hence undergoes further reduction to a hydrolytically unstable *gem*-diamine (equation 40).

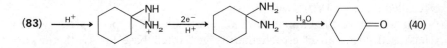

$$(40)$$

The electrochemical interconversion of diazenium salts and hydrazyl radicals (84 ⇌ 85) has been accomplished in either direction[129,130].

$$R_2\overset{+}{N}=NR^1 \underset{-e^-}{\overset{e^-}{\rightleftarrows}} R_2N-\overset{\bullet}{N}R^1$$

(84) (85)

Nelsen and Landis[129] reported the electrochemical reduction of several bicyclic diazenium ions (84) while Solon and Bard[130] showed that the stable free radical diphenylpicrylhydrazyl (85, R = phenyl, R^1 = 2,4,6-trinitrophenyl) is reversibly oxidized to the corresponding diazenium ion.

VII. REFERENCES

1. M. M. Baizer, J. D. Anderson, J. H. Wagenknecht, M. R. Ort, and J. P. Petrovich, *Electrochim. Acta*, **12**, 1377 (1967).

2a. M. M. Baizer and J. P. Petrovich, *Progr. Phys. Org. Chem.*, **7**, 189 (1970).
2b. J. D. Anderson, J. P. Petrovich, and M. M. Baizer, *Advan. Org. Chem.*, **6**, 257 (1969).
3. L. Meites, *Polarographic Techniques*, 2nd Ed., Wiley, New York, 1965, pp. 671–711.
4. K. N. Campbell and E. E. Young, *J. Amer. Chem. Soc.*, **65**, 965 (1943).
5. G. Cauquis and D. Serve, *J. Electroanal. Chem.*, **27**, App. 2 (1970).
6. S. D. Ross, M. Finkelstein, and R. C. Peterson, *J. Amer. Chem. Soc.*, **89**, 4088 (1967).
7. A. Nilsson, A. Ronlan, and V. D. Parker, *J. Chem. Soc., Perkin I*, 2337 (1973).
8. L. Horner and H. Röder, *Ann. Chem.*, **723**, 11 (1969).
9. H. Lund, *Acta Chem. Scand.*, **13**, 249 (1959).
10a. R. M. Powers and R. A. Day, Jr., *J. Org. Chem.*, **24**, 722 (1959).
10b. W. Wiesener and K. Schwabe, *J. Electroanal. Chem.*, **15**, 73 (1967).
10c. P. Kabasakalian, J. McGlotten, A. Basch, and M. D. Yudis, *J. Org. Chem.*, **26**, 1738 (1961).
11. A. J. Fry and R. G. Reed, *J. Amer. Chem. Soc.*, **91**, 6448 (1969).
12. K. W. Bowers, R. W. Giese, J. Grimshaw, H. O. House, N. H. Kolodny, K. Kronberger, and D. K. Roe, *J. Amer. Chem. Soc.*, **92**, 2783 (1970).
13. S. Wawzonek and A. Gundersen, *J. Electrochem. Soc.*, **107**, 537 (1960).
14. N. Weinberg, A. K. Hoffman, and T. Reddy, *Tetrahedron Letters*, 2271 (1971).
15. G. A. Russell, E. J. Geels, F. J. Smentowski, K.-Y. Chang, J. Reynolds, and G. Kaupp, *J. Amer. Chem. Soc.*, **89**, 3821 (1967).
16. R. F. Michielli and P. J. Elving, *J. Amer. Chem. Soc.*, **90**, 1989 (1968).
17. J. M. W. Scott and W. H. Jura, *Can. J. Chem.*, **45**, 2375 (1967).
18. M. Fujihira, H. Suzuki, and S. Hayano, *J. Electroanal. Chem.*, **33**, 393 (1971).
19. A. J. Fry, *Synthetic Organic Electrochemistry*, Harper and Row, New York, 1972.
20. M. M. Baizer, ed., *Organic Electrochemistry*, Marcel Dekker, New York, 1973.
21. C. L. Perrin, *Progr. Phys. Org. Chem.*, **3**, 165 (1965).
22. P. Zuman, *Progr. Phys. Org. Chem.*, **5**, 81 (1967).
23. R. N. Adams, *Electrochemistry at Solid Electrodes*, Marcel Dekker, New York, 1969.
24. J. Heyrovsky and J. Kuta, *Principles of Polarography*, Academic Press, New York, 1966.
25. L. Meites, *Polarographic Techniques*, 2nd Ed., Wiley, New York, 1965.
26. R. A. Benkeser and S. J. Mels, *J. Org. Chem.*, **34**, 3970 (1969).
27. H. J. Barger, *J. Org. Chem.*, **34**, 1489 (1969).
28. A. T. Kuhn, *Electrochim. Acta*, **13**, 477 (1968).
29. J. B. Lee and P. Cashmore, *Chem. Ind.*, 1758 (1966).
30. J. W. Loveland, *U.S. Patent* 3,032,489 (May 1, 1962); *Chem. Abstr.*, **57**, 4470 (1962).
31. G. J. Hoijtink, *Rec. Trav. Chim.*, **73**, 895 (1954).
32. H. A. Laitinen and S. Wawzonek, *J. Amer. Chem. Soc.*, **64**, 1765 (1942).
33. R. Dietz and M. E. Peover, *Discuss. Faraday Soc.*, **45**, 154 (1968).
34. J. Grimshaw and J. Trocha-Grimshaw, *J. Chem. Soc., Perkin I*, 1275 (1973).
35. T. L. Cairns and B. C. McKusick, *Angew. Chem.*, **73**, 520 (1961).
36a. M. Fleischmann and D. Pletcher, *Tetrahedron Letters*, 6255 (1968).
36b. J. Bertram, M. Fleischmann, and D. Pletcher, *Tetrahedron Letters*, 349 (1971).

37. For example, C. Cwiklinski and J. Perichon, *Electrochim. Acta*, **19**, 297 (1974).
38. M. Fleischmann, D. Pletcher, and G. M. Race, *J. Chem. Soc., B*, 1746 (1970).
39. M. Fleischmann, G. Faita, and D. Pletcher, *J. Electroanal. Chem.*, **25**, 455 (1969).
40a. T. Shono and A. Ikeda, *J. Amer. Chem. Soc.*, **94**, 7892 (1972).
40b. T. Shono and T. Koska, *Tetrahedron Letters*, 6207 (1968).
41. T. Shono, A. Ikeda, and Y. Kimura, *Tetrahedron Letters*, 3599 (1971).
42. T. Inone, K. Koyama, and S. Tsutsumi, *Bull. Chem. Soc. Japan*, **40**, 162 (1967).
43. J. J. O'Connor and I. A. Pearl, *J. Electrochem. Soc.*, **111**, 335 (1964).
44. Reference 19, p. 306.
45. V. D. Parker and L. Eberson, *Chem. Commun.*, 340 (1969).
46. L. Eberson and V. D. Parker, *Acta Chem. Scand.*, **24**, 3553 (1970).
47. A. J. Bard and J. Phelps, *J. Electroanal. Chem.*, **25**, App. 2 (1970).
48. V. D. Parker, K. Nyberg, and L. Eberson, *J. Electroanal. Chem.*, **22**, 150 (1969).
49a. J. D. Stuart and W. E. Ohnesorge, *J. Amer. Chem. Soc.*, **93**, 4531 (1971).
49b. J. D. Stuart and W. E. Ohnesorge, *J. Electroanal. Chem.*, **30**, App. 11 (1971).
50. R. Reynolds, L. L. Line, and R. F. Nelson, *J. Amer. Chem. Soc.*, **96**, 1087 (1974).
51. J. M. Fritsch, H. Weingarten, and J. D. Wilson, *J. Amer. Chem. Soc.*, **92**, 4038 (1970).
52. N. Wiberg, *Angew. Chem. Int. Ed.*, **7**, 766 (1968).
53. N. Wiberg and J. W. Buchler, *Chem. Ber.*, **96**, 3223 (1963).
54. K. Kuwata and D. H. Geske, *J. Amer. Chem. Soc.*, **86**, 2101 (1964).
55. S. Hunig, H. Schlaf, G. Kiesshich, and D. Scheutzow, *Tetrahedron Letters*, 2271 (1969).
56. D. L. Coffen, J. Q. Chambers, D. R. Williams, P. F. Garrett, and N. D. Canfield, *J. Amer. Chem. Soc.*, **93**, 2258 (1971).
57. A. Demortier and A. J. Bard, *J. Amer. Chem. Soc.*, **95**, 3495 (1973).
58. P. H. Given, M. E. Peover, and N. M. Schoen, *J. Chem. Soc.*, 2764 (1958).
59. B. S. Jensen and V. D. Parker, *Chem. Commun.*, 367 (1974).
60. C. K. Mann and K. K. Barnes, *Electrochemical Reactions in Nonaqueous Systems*, Marcel Dekker, New York, 1970.
61. R. Pasternack, *Helv. Chim. Acta.*, **31**, 753 (1948).
62. P. J. Elving and J. T. Leone, *J. Amer. Chem. Soc.*, **80**, 1021 (1958).
63. A. H. Maki, *J. Chem. Phys.*, **35**, 761 (1961).
64. J. H. Stocker and R. M. Jenevein, *J. Org. Chem.*, **33**, 294 (1968).
65. S. Wawzonek and A. Gundersen, *J. Electrochem. Soc.*, **107**, 537 (1960).
66a. T. Shono and M. Mitani, *Tetrahedron*, **28**, 4747 (1972).
66b. J. P. Coleman, R. J. Kobylecki, and J. H. P. Utley, *Chem. Commun.*, 104 (1971).
67. A. Kirrman, J. M. Saveant, and N. Moe, *C.R. Acad. Sci. Paris*, **253**, 1106 (1961).
68. G. P. Kugatova-Shemyakina, G. M. Nikolaev, S. G. Mairanovskii, and N. V. Surikova, *Tetrahedron Lett.*, 1725 (1974).
69. T. Shono and M. Mitani, *J. Amer. Chem. Soc.*, **93**, 5284 (1971).
70. D. H. Evans and E. C. Woodbury, *J. Org. Chem.*, **32**, 2155 (1967).
71. A. D. Thomsen and H. Lund, *Acta Chem. Scand.*, **25**, 1576 (1971).
72. T. J. Curphey, C. W. Amelotti, T. P. Layloff, R. L. McCartney, and J. H. Williams, *J. Amer. Chem. Soc.*, **91**, 2817 (1969).

73. E. Kariv and E. Gileadi, *Coll. Czech, Chem. Commun.*, **36**, 476 (1971).
74. R. H. Philp, Jr., R. L. Flurry, and R. A. Day, Jr., *J. Electrochem. Soc.*, **111**, 328 (1964).
75. R. H. Philp, Jr., T. Layloff, and R. N. Adams, *J. Electrochem. Soc.*, **111**, 1189 (1964).
76. J. A. Campbell, R. W. Koch, J. V. Hay, M. A. Ogliaruso, and J. F. Wolfe, *J. Org. Chem.*, **39**, 146 (1974).
77a. S. Swann, Jr., *Trans. Electrochem. Soc.*, **62**, 177 (1932).
77b W. R. J. Simpson, D. Bobbe, J. A. Edwards, and J. H. Freed, *Tetrahedron Lett.*, 3209 (1967).
77c. L. Throop and L. Tokes, *J. Amer. Chem. Soc.*, **89**, 4789 (1967).
78. M. J. Allen, *Organic Electrode Processes*, London, Chapman and Hall (1958).
79. R. G. Barradas, O. Kutowy, and D. W. Shoesmith, *Electrochimica Acta*, **19**, 49 (1974).
80a. H. Lund, *Acta Chem. Scand.*, **17**, 972 (1963).
80b. P. E. Iversen and H. Lund, *Acta Chem. Scand.*, **21**, 279 (1967).
80c. P. E. Iversen, *Acta Chem. Scand.*, **24**, 2459 (1970).
81a. G. H. Coleman and H. L. Johnson, *Org. Syn.*, **21**, 10 (1941).
81b. J. H. Wagenknecht, *J. Org. Chem.*, **37**, 1513 (1972).
82. S. A. Cerefice and E. F. Fields, *J. Org. Chem.*, **39**, 971 (1974).
83. P. Kabasakalian and J. McGlotten, *Anal. Chem.*, **31**, 1091 (1959).
84. M. Schenck and H. Kirchhof, *Z. Physiol. Chem.*, **163**, 125 (1927).
85. C. Schall and W. Kirst, *Z. Elektrochim.*, **29**, 537 (1923).
86. W. P. Newman and K. Kuhlein, *Adv. Organometallic Chem.*, **7**, 241 (1968).
87. J. H. Stocker and R. M. Jenevein, *J. Org. Chem.*, **33**, 2807 (1968).
88a. J. H. Stocker and R. M. Jenevein, *Coll. Czech. Chem. Commun.*, **36**, 925 (1971).
88b. A. Bewick and H. P. Cleghorn, *J. Chem. Soc. Perkin II*, 1410 (1973).
89. J. H. Stocker and R. M. Jenevein, *J. Org. Chem.*, **34**, 2807 (1969).
90. J. Armand, G. Dane, and F. Valentini, *Bull. Soc. Chim. France*, 4581 (1968).
91. J. Grimshaw, J. T. Grimshaw, and E. J. F. Rea, *J. Chem. Soc. (C)*, 683 (1971).
92. J. H. Stocker and R. M. Jenevein, *J. Org. Chem.*, **33**, 2145 (1968).
93. J. H. Stocker, R. M. Jenevein, and D. H. Kern, *J. Org. Chem.*, **34**, 2810 (1969).
94. J. Grimshaw and J. S. Ramsey, *J. Chem. Soc.*, 254 (1952).
95. A. Vincenz-Chodkowska and Z. R. Grabowski, *Electrochim. Acta*, **9**, 789 (1964).
96. L. Horner and D. Degner, *Tetrahedron Lett.*, 5889 (1968).
97. L. Horner and R. Schneider, *Tetrahedron Lett.*, 3133 (1973).
98a. L. Horner and D. Degner, *Tetrahedron Lett.*, 1241, 1245 (1971).
98b L. Horner and D. Skaletz, *Tetrahedron Lett.*, 3679 (1970).
99. E. Kariv, H. A. Terni, and E. Gileadi, *J. Electrochem. Soc.*, **120**, 639 (1973).
100. P. Anziani, A. Anbry, and R. Cornubert, *C.R. Acad. Sci. Paris*, **225**, 878 (1947).
101. A. J. Streitwieser, Jr., *Molecular Orbital Theory for Organic Chemists*, Wiley, New York, 1961.
102. L. Lunazzi, G. Maccagnani, G. Mazzanti, and G. Placucci, *J. Chem. Soc. (B)*, 162 (1971).
103a. R. O. Loutfy and R. O. Loutfy, *J. Phys. Chem.*, **76**, 1650 (1972).
103b. R. O. Loutfy and R. O. Loutfy, *J. Phys. Chem.*, **77**, 336 (1973).
103c. R. O. Loutfy and R. O. Loutfy, *Can. J. Chem.*, **51**, 1169 (1973).

103d. R. O. Loutfy and R. O. Loutfy, *Tetrahedron*, 2251 (1973).
104. O. Manousek and J. Volke, *Electroanal. Chem.*, **43**, 365 (1973).
105. L. L. Miller, V. R. Koch, M. E. Larscheid, and J. F. Wolf, *Tetrahedron Lett.*, 1389 (1971).
106. P. J. Elving and R. E. VanAtta, *J. Electrochem. Soc.*, **103**, 676 (1956).
107. A. J. Bard, private communication to A. J. Fry.
108. P. Zuman and O. Exner, *Coll. Czech. Chem. Commun.*, **30**, 1832 (1965).
109. V. D. Bezuglyi, L. V. Kononenko, A. F. Korunova, V. N. Dmitrieva, and B. L. Timan, *J. Gen. Chem. U.S.S.R.*, **39**, 1647 (1969).
110. P. Souchay and S. Ser, *J. Chim. Phys.*, **49C**, 172 (1952).
111. H. Lund, *Acta Chem. Scand.*, **18**, 563 (1964).
112. C. P. Andrieux and J. M. Saveant, *Bull. Soc. Chim. France*, **35**, 4671 (1968).
113. A. J. Fry and J. H. Newberg, *J. Amer. Chem. Soc.*, **89**, 6374 (1967).
114. L. Horner and D. H. Skaletz, *Tetrahedron Lett.*, 1103 (1970).
115. H. D. Law, *J. Chem. Soc.*, **101**, 154 (1912).
116. H. G. Benson and J. M. W. Scott, *Can. J. Chem.*, **46**, 2895 (1968).
117. N. F. Levchenko, L. Sh. Afanasiadi, and V. D. Bezuglyi, *J. Gen. Chem. U.S.S.R.*, **37**, 624 (1967).
118. Reference 19, p. 234–236.
119. R. G. Reed and R. W. Murray, unpublished research.
120. S. Wawzonek and J. D. Frederickson, *J. Amer. Chem. Soc.*, **77**, 3985 (1955).
121. S. J. Yeh and H. H. Jaffe, *J. Amer. Chem. Soc.*, **81**, 3279 (1969).
122. T. M. Florence, *J. Electroanal. Chem.*, **52**, 115 (1974), and many references therein.
123a. G. A. Gilbert and E. K. Rideal, *Trans. Faraday Soc.*, **47**, 396 (1951).
123b. H. A. Laitinen and T. J. Kneip, *J. Amer. Chem. Soc.*, **78**, 736 (1956).
124. J. L. Sadler and A. J. Bard, *J. Amer. Chem. Soc.*, **90**, 1979 (1968).
125. G. H. Aylward, J. L. Garnett, and J. H. Sharp, *Chem. Commun.*, **137** (1966).
126. P. E. Iversen and H. Lund, *Tetrahedron Letters*, 3523 (1969).
127. P. E. Iversen, *Tetrahedron Letters*, 55 (1971).
128. H. Lund, *Coll. Czech. Chem. Commun.*, **31**, 4175 (1966).
129. S. F. Nelsen and R. T. Landis, II, *J. Amer. Chem. Soc.*, **95**, 6454 (1973).
130. E. Solon and A. J. Bard, *J. Amer. Chem. Soc.*, **86**, 1926 (1964).

CHAPTER **6**

1,3-Dipolar cycloadditions involving X=Y groups

GIORGIO BIANCHI, CARLO DE MICHELI and
REMO GANDOLFI

Institute of Organic Chemistry, University of Pavia, Italy

I. INTRODUCTION

Huisgen in the early 1960's fully recognized the general concept of 1,3-dipolar cycloadditions. Since that time, there has been a steadily increasing output of papers on the subject. In order to confine the treatment of the reaction mechanism and chemistry of these important cycloadditions to a reasonable length, it is only possible to consider selected examples from the literature. General principles and typical reactions have been presented reporting references which provide a broad survey of the specific topic.

The literature covering the reactions of hetero double bonds with 1,3-dipoles and 1,3-dipolar cycloadditions involving carbon–carbon double bonds has been considered up to 1974 following a previous comprehensive survey appearing in the first volume of this series[1]. The present survey also considers the 1,3-dipolar retrocycloadditions which have been reported in some instances.

Owing to the theoretical importance of the allyl anion, the parent of all 1,3-dipoles, some space has been dedicated to the anionic cycloadditions.

1,3-Dipoles without octet stabilization, and other systems such as nitrous oxide, nitro compounds, azimines, azoxy compounds, nitrile sulphides and sulphur diimides have not been considered in this review, because they have been too little studied or studied in 1,3-dipolar cycloadditions with triple bonds only.

II. GENERAL

A. 1,3-Dipoles

Although 1,3-dipolar cycloadditions have been known for a long time, the reaction was not rationalized until the early sixties[1].

The 1,3-dipole $a\!\!=\!\!\overset{+}{b}\!\!-\!\!\overset{=}{c}$ or $a\!\!\equiv\!\!\overset{+}{b}\!\!-\!\!\overset{=}{c}$ is a system isoelectronic with the allyl or propargyl anion which has four π electrons distributed on three adjacent atoms. This can react with a multiple bond system d=e, called dipolarophile, to give five-membered cyclic compounds the reaction being symbolized $3 + 2 \rightarrow 5^2$.

The ground state of a 1,3-dipole in valence bond theory can be described by resonance structures, e.g. **1–5** and **6–9** for formonitrile oxide and ozone respectively. The term 1,3-dipole derives from structures **4, 5, 8, 9** which, together with carbenic structure **3**, should hardly contribute to the stabilization of the compounds.

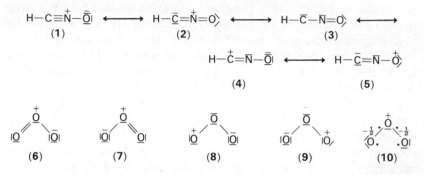

The non-classical structure **10**, derived from Linnett's bond theory, has been considered to be the representation in unique form of the electronic structure of the molecule[3]. A singlet biradical structure was recently proposed for diazomethane, ozone and methylene peroxide[3a].

The a, b, c components of the dipole are generally C, N, and O; less common are 1,3-dipoles containing second row elements (see Table 1).

TABLE 1. 1,3-Dipoles with octet stabilization

Structure	Name
1,3-Dipoles with an orthogonal double bond (propargyl–allenyl anion type)	
−C≡N̄−C̄⟨ ⟷ −C̄=N̄=C⟨	Nitrile ylides
−C≡N̄−N̄− ⟷ −C̄=N̄=N−	Nitrile imines
−C≡N̄−Ō⟩ ⟷ −C̄=N̄=O	Nitrile oxides
−C≡N̄−S̄I ⟷ −C̄=N=SI	Nitrile sulphides
IN≡N̄−C̄⟨ ⟷ IN̄=N̄=C⟨	Diazo compounds
IN≡N̄−N̄− ⟷ N̄=N̄=N−	Azides
IN≡N̄−Ō I ⟷ N̄=N̄=O⟩	Nitrous oxide

TABLE 1. 1,3-Dipoles with octet stabilization (*continued*)

Structure	Name

1,3-Dipoles without an orthogonal double bond (allyl anion type)

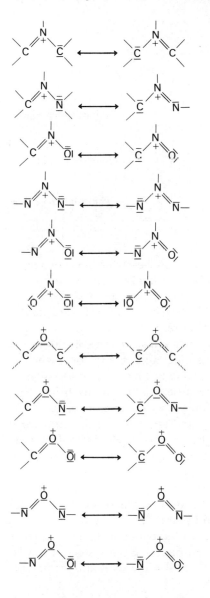

Azomethine ylides

Azomethine imines

Nitrones (azomethine oxides)

Azimines

Azoxy compounds

Nitro compounds

Carbonyl ylides

Carbonyl imines[a]

Carbonyl oxides

Nitroso imines[a]

Nitroso oxides[a]

TABLE 1. 1,3-Dipoles with octet stabilization (*continued*)

Structure	Name
1,3-Dipoles without an orthogonal double bond	

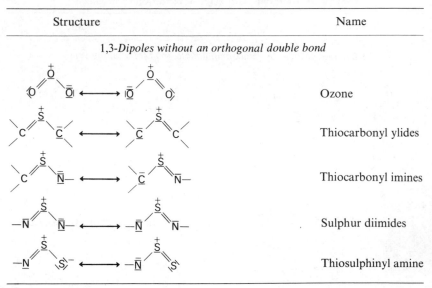

Ozone

Thiocarbonyl ylides

Thiocarbonyl imines

Sulphur diimides

Thiosulphinyl amine

[a] 1,3-Dipoles not yet discovered.

For the sake of classification, 1,3-dipoles may be divided into two main categories: 1,3-*dipoles with* and 1,3-*dipoles without octet stabilization*[1] (the latter not discussed in this chapter). The former dipoles are further divisible into two classes: 1,3-*dipoles with and without an orthogonal double bond to the allyl anion system*, these two corresponding to 1,3-*dipoles with and without a double bond in the sextet form* in Huisgen's classification. The geometry of the two classes of dipoles is different, since the orthogonal π bond causes the 1,3-dipoles of the first class to be linear (e.g., diazoalkanes, nitrile oxides), whilst the other dipoles are characterized by being bent, even in the ground state (e.g., azomethine ylides, nitrones, ozone).

The uses of molecular orbital theory, and in particular the knowledge of energy values of frontier orbitals [highest occupied (HOMO) and lowest unoccupied (LUMO)], their coefficients and symmetry properties, are of great importance for the perturbation approach to reactivity and regio-chemistry of 1,3-dipolar cycloadditions[4-11].

The molecular orbital features of allyl anion (a), azomethine ylide (b) and formonitrile oxide (c) are shown in Figure 1[4].

By considering a plane perpendicular to the plane of the three carbon atoms of the allyl anion [Figure 1(a)] and bisecting the system, it can be seen that orbitals ψ_1 and ψ_3 are symmetric (S) while ψ_2 is anti-symmetric

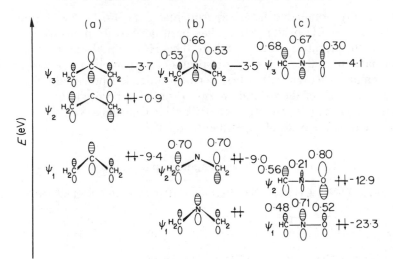

FIGURE 1. CNDO/2 π frontier orbital energies of (a) allyl anion, (b) azomethine ylide, and (c) formonitrile oxide (approximate scale).

(A). The same considerations hold for the azomethine ylide [Figure 1(b)] whereas in the case of formonitrile oxide [Figure 1(c)] there is only symmetry as far as sign of the wavefunction is concerned (consider sizes of the lobes).

The presence of heteroatoms (more electronegative than carbon) in the 1,3-dipoles causes a lowering of the orbital energies which are, in the case of the azomethine ylide, all lower than those of the allyl parent system. However, the geometry of a system can reverse this effect, e.g. formonitrile oxide which has its LUMO at a higher energy level than that of the allyl anion. The high energy value of the LUMO of the formonitrile oxide is mainly due to its shorter bonds and to the linearity of the molecule which cause a greater antibonding interaction between positions 1 and 2 and a weaker bonding interaction between positions 1 and 3 respectively.

An important point for linear 1,3-dipoles is that calculations show that bending causes only small changes in the energies and coefficients of HO and LU orbitals[4,12]. As far as the effect of substituents on the 1,3-dipoles is concerned, they affect the molecular orbital picture in a manner similar to that reported for the dipolarophiles (see Section II.B). The orbital energy levels and the sizes of lobes are modified according to coefficient values for the unsubstituted term.

Although the energy values for HO and LU orbitals and their coefficients can be calculated theoretically by a number of different methods, they are not readily obtainable from experimental data. For a few 1,3-dipoles, however, the LU orbital energy has been derived from electron affinity whose the negative value is taken. The HO orbital energy has been calculated from the photoelectron spectra of the compounds. According to Koopman's theorem[13], the first vertical ionization potential (IP_v) is the negative value of the orbital energy. Frontier orbital energies of representative 1,3-dipoles are reported in Table 2. Calculated values have been corrected on the basis of experimental data[4,5].

TABLE 2. Energy levels of frontier orbitals for some 1,3-dipoles

Energy (eV)	PhN_3	CH_2N_2	PhCHN(Me)O	PhCNO	PhCNNPh	$(NC)_2COC(CN)_2$
LUMO	−0·2	1·8[a]	−0·4	−1	−0·5	−1·1
HOMO	−9·5	−9	−8	−10	−7·5	−9

[a] Value for NLUMO. LUMO and NHMO are lying in the plane of the molecule.

Calculations carried out on several non-substituted 1,3-dipolar systems show that the size of coefficients on the two ends a (neutral) and c (anionic) are different for all dipoles with the exception of the bent ones having C_{2v} symmetry[4,7].

As indicated in the Table 3 the 'anionic' atom c has a larger coefficient in the HOMO whereas in the LUMO the reverse is found. However exceptions to this rule are known (e.g., CH_2N_2). It has also been reported that the calculated values for the nitrile ylides are in disagreement with experimental data (see also Section IV.A.2)[4,5]. Substituents on the 1,3-dipole affect the coefficient of the atoms but do not reverse the relative size (compare benzonitrile oxide with fulminic acid and phenylazide with azotidric acid) unless the parent 1,3-dipole has very similar coefficients on the two ends.

Figure 2 gives a more complete picture of π electrons of two characteristic 1,3-dipoles with orthogonal double bond, namely formonitrile oxide and nitrile ylide. As shown in the figure formonitrile oxide possesses two orthogonal heteroallyl anion systems. Frontier orbital energies of this class of 1,3-dipoles are straightforward: those possessing $C_{\infty v}$ (e.g., formonitrile oxide) and C_{3v} (e.g., acetonitrile oxide) symmetry will have two degenerate HOMO's and LUMO's each, while those of C_s (e.g., phenyl azide) and C_{2v} (e.g., diazomethane) symmetry will not present a couple of degenerate frontier orbitals but two single frontier orbitals each with very close orbitals called NHOMO (next to HOMO) and NLUMO (next to LUMO) respectively[4].

TABLE 3. Relative coefficient values of frontier orbitals of some 1,3-dipoles

$\overset{+}{a}\!\equiv\!\overset{-}{b}\!-\!c$	N_2CH_2[a]	HCNO	PhCNO	HCNNH	HCNCH$_2$	N$_3$H	N$_3$Ph
LUMO	$C_a < C_c$	$C_a > C_c$	$C_a > C_c$	$C_a > C_c$	$C_a > C_c$	$C_a > C_c$	$C_a > C_c$
HOMO	$C_a < C_c$	$C_a < C_c$	$C_a < C_c$	$C_a < C_c$	$C_a < C_c$	$C_a < C_c$	$C_a < C_c$

$\overset{+}{b}\diagdown_{a}\overset{c^-}{}$	CH$_2$N(H)CH$_2$	CH$_2$N(H)NH	CH$_2$N(H)O	CH$_2$OCH$_2$	CH$_2$ONH	CH$_2$OO
LUMO	$C_a = C_c$	$C_a > C_c$	$C_a > C_c$	$C_a = C_c$	$C_a > C_c$	$C_a > C_c$
HOMO	$C_a = C_c$	$C_a < C_c$	$C_a < C_c$	$C_a = C_c$	$C_a < C_c$	$C_a < C_c$

[a] Similar values $C_a(0\cdot628) \simeq C_b(0\cdot625)$ for the coefficients of the LUMO of diazomethane have been recently reported: T. Minata, S. Yamabe, S. Inagaki, H. Fujimoto, and K. Fukui, *Bull. Chem. Soc. Japan*, **47**, 1619 (1974).

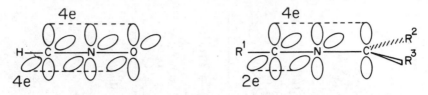

FIGURE 2. π-Orbitals of formonitrile oxide and a nitrile ylide.

B. Dipolarophiles

Double bonds acting as dipolarophiles are always composed of first-
and second-row elements. The most widely studied are: C=C, C=N,
C=O, C=S, C=P, N=N, N=S, N=O, N=P, and N=B.

The energy levels and the coefficients of the frontier orbitals of the
dipolarophile are properties of the whole molecule and depend principally
on the nature of the atoms involved and the substituents on the double
bond (see Figure 3)[4,7,14].

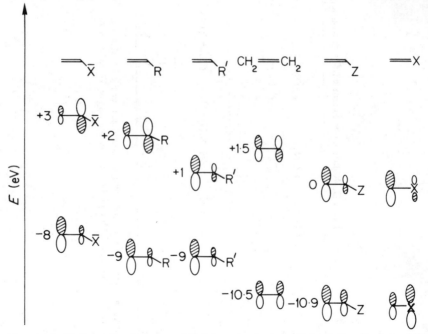

FIGURE 3. Estimated π frontier orbital energies and coefficients for dipolarophiles
($\overline{X}$ = OR, NR$_2$ etc.; R = alkyl; R′ = CH=CH$_2$, Ph, etc.; X = O, N etc., Z = CO$_2$R,
CHO, etc.).

If the frontier orbitals of ethylene are taken as a model for C=C systems it is possible to make two general points. Firstly the substituents on the double bond change the energy content of the orbitals, and secondly the coefficients on the two carbons of the double bond are modified. Introduction of an electron-donating group ($\overline{X}$, R) in the ethylene raises the energy levels of both LUMO and HOMO by nearly the same shift; the opposite is observed when the substituent is electron withdrawing by inductive effect only. Conjugating substituents (R') compress the frontier orbital separation by raising HOMO and lowering LUMO energies. A group, both electron-withdrawing and -conjugating (Z) lowers much less HOMO than LUMO energy. Electron-releasing groups (alkyl, OR, NR$_2$) give rise to a larger β-coefficient in HOMO, whereas the ratio of the coefficients is the reverse in LUMO ($\alpha > \beta$), as is to be expected on inductive grounds[15,16]. For alkyl groups the two coefficients are nearly identical in the LUMO[4,11,14]. Electron-withdrawing groups have exactly the opposite effect from that of donor groups if they interact only inductively (e.g., CCl$_3$). Conjugating (—CH=CH$_2$, Ph) and conjugating and electron-withdrawing substituents such as CHO, CO$_2$R, CN in the ethylenic dipolarophile give rise to larger coefficients in β than in α position in HO (actually quite small differences for the cases of conjugating electron-withdrawing group) as well as in LU orbitals. The coefficients of polysubstituted dipolarophiles represent the sum of the individual effects of the substituents.

When the double bond is made up of one carbon and one heteroatom, or of two heteroatoms, the levels of the frontier orbitals are lowered (see Figure 3) in comparison with those of ethylene. This is due to the higher electronegativity of heteroatoms. Furthermore the coefficients on the two atoms are different, the heteroatom having the larger coefficient in the HOMO and conversely the smaller in the LUMO.

The ratio of the coefficients is related to the difference in electronegativity of the two elements, therefore a larger ratio will result for ketones than for thioketones.

C. Stereospecificity

The cycloaddition of 1,3-dipoles with alkenes is strictly a stereospecific reaction: *cis* addition occurs with retention of the stereochemistry of the olefin in the final adduct. For kinetically controlled cycloadditions no exceptions to this rule are known at present, even though a scrupulous search for a mixture of adducts has been carried out for several 1,3-dipoles,

including diphenyl nitrile imine [17-18], benzonitrile oxide[19-21], diazomethane[22-23], 4-nitrophenyl azide[24], azomethine ylide **11**[25], azomethine
imines **12**[26] and **13**[27-28], 3,4-dihydroisoquinoline-N-oxide[29], tetracyano-
carbonyl ylide[30] and benzonitrile dimethylmethylide[31].

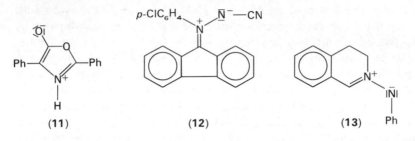

    (11)   (12)   (13)

It is worth pointing out the stereospecificity observed in the reactions
(1)[24] and (2)[31] in which the possibility of discrete intermediates, e.g. the
zwitterions **14** and **15**, can be seriously considered.

Those cases where stereospecificity has not been observed were due to
the primary adduct being transformed into the more stable stereoisomer
under the reaction conditions[17].

All these experimental results, together with the fact that zwitterionic
intermediates of the type **16** and spin-paired diradicals of the type **17** can
lose their stereochemistry, allowed Huisgen to propose a concerted process
for the 1,3-dipolar cycloadditions[32-33]. The two new σ bonds are formed
simultaneously (at the same time) but not necessarily in a synchronous
manner (at the same time and rate). The unequal progress of bond formation in the transition state together with the fact that one addend acts as

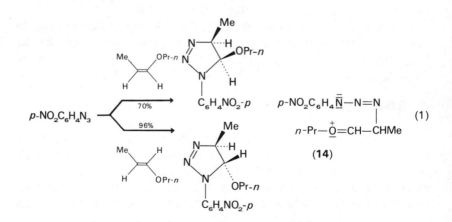

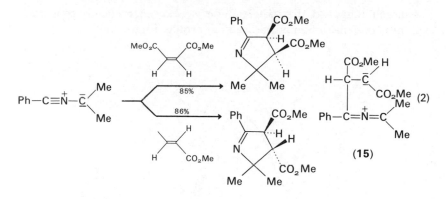

electron donor and the other as electron acceptor leads to partial charges, e.g. **18**. This picture corresponds to a 'bondingly-concerted' cyclo-

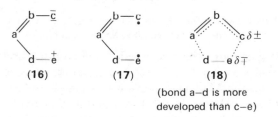

(bond a–d is more
developed than c–e)

addition[34]. The plot of bond indices versus reaction coordinates for synchronous-concerted and simultaneous-concerted cycloadditions are shown in Figure 4[34].

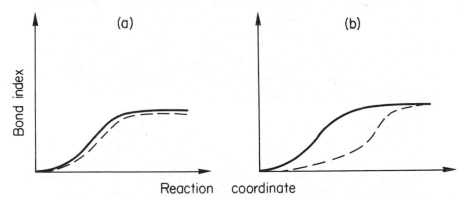

FIGURE 4. Plots of bond indices for two new bonds formed for (a) synchronous and (b) simultaneous concerted cycloadditions.

Huisgen suggested that the reaction is also 'energetically-concerted' with only one maximum in the reaction profile (Figure 5)[32,34].

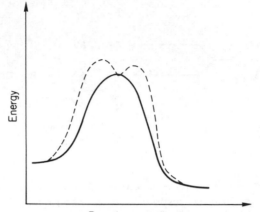

FIGURE 5. Plots of energies for energetically-concerted (heavy line) and energetically-non-concerted (dashed line) cycloadditions.

The 'simultaneous-concerted' model with unequal bond formation in the transition state for 1,3-dipolar cycloadditions has recently found a theoretical basis in M.O. perturbation treatment (see Section II.I.J).

MINDO/2 calculations suggest also that in the cycloadditions leading to products in which the two new single bonds are equivalent, these bonds will very likely be non-equivalent in the transition state which, therefore, is not symmetrical[35].

In conclusion we might consider the 1,3-dipolar cycloadditions as a continuum in the mechanism from quasi-synchronous to non-concerted (two steps). The latter mechanism may be at work for particular double bond systems such as enolates[35a]. It has been suggested in fact that cyclo-additions of azides with enolates (see Section IV.E.7) go to the final adducts through a discrete anionic intermediate. A similar mechanism is also plausible for cycloadditions of systems derived from allyl anion (see Section III).

Representation of 1,3-dipoles according to Linnett's notation is that of a diradical, e.g. **10**. On this basis, Firestone postulated the reversible formation of a discrete intermediate, a spin-paired diradical, for all 1,3-dipolar cycloadditions which would explain why both monosubstituted alkenes with electron-withdrawing and electron-donating groups exhibit the same orientational preference in the reaction with 1,3-dipoles (except azides)[36–38] (equation 3).

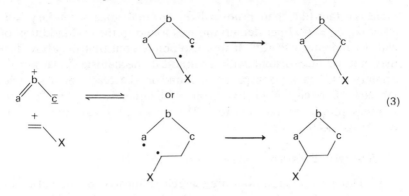

$$(3)$$

The complete stereospecificity of the reaction is hardly compatible with the diradical mechanism. Furthermore the rationalization of the regio-chemistry of 1,3-dipolar cycloadditions by M.O. perturbation approach (see Section II.J) makes Firestone's proposal unnecessary. All of the Firestone's arguments have been thoroughly refuted by Huisgen[33].

D. Isotope Effects

A suitable modern experimental technique to study concertedness in cycloadditions is the analysis of primary and secondary kinetic isotope effects.

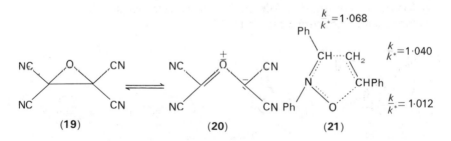

(19) (20) (21)

In the study of 1,3-dipolar cycloadditions of the carbonyl ylide **20** in thermal equilibrium with tetracyanoethylene oxide (**19**) to the three isomeric monodeuterated styrenes, specifically labelled at the olefinic positions, an identical secondary isotope effect k_H/k_D 0·96–0·97 was found[39]. This identity suggests that both the new bonds have been formed to a comparable extent in the transition state: a simultaneous-(*quasi synchronous-*)concerted process seems to be at work. This small inverse isotope effect was also interpreted as signifying that the transition state is more reactant-like than product-like. An analogous secondary isotope effect ($k_H/k_D = 0.97$ per deuterium) was found in the cycloaddition of **20** with 1,1-dideuteroallene[40]. It was also pointed out that the above results may not be inconsistent with a diradical mechanism[39]. However the primary ^{14}C-kinetic-isotope effect found in the reaction of specifically labelled ^{14}C-phenyl-N-phenylnitrone with either α-^{14}C or β-^{14}C labelled styrenes (displayed on structure **21**) are only in agreement with the concerted mechanism[41].

E. Solvent Effects

1,3-Dipolar cycloadditions are scarcely influenced by the nature of the solvent; the solvent effect on the reaction rate does not exceed the value of 10 (see Table 4). In general, no correlation between the reaction rates and empirical parameters of solvent polarity was found. Two examples in which a fair correlation was observed are represented by cycloadditions (d) and (g) which, however, are characterized by opposite trends (Table 4).

The above data do not mean that there is no solvent role in 1,3-dipolar cycloadditions. In fact while adducts were not isolated, or isolated in very low yields from the reaction of diazomethane with Schiff bases in anhydrous ether, good yields of $\varDelta^2$-1,2,3-triazolines were obtained when the cycloaddition was carried out in water–dioxane solution[50].

Furthermore, the influence of the solvent polarity on the ratio of the two possible regioisomers **22–23 24–25** has recently been reported for the cycloadditions of mesitonitrile oxide with benzalacetone[51] and of diazomethane with thioadamantane[52] (Table 5).

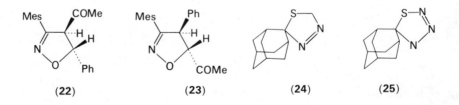

(22)   (23)   (24)   (25)

Solvent polarity was also found to affect the *syn/anti* ratio in the cycloaddition of nitrile oxides with *cis*-3,4-dichlorocyclobutene[53].

The small solvent effect on reaction rate is considered to be evidence against a zwitterionic intermediate and consistent with a concerted process[32].

TABLE 4. Solvent effect on the rate coefficients k_2 (l/mol sec) of some 1,3-dipolar cycloadditions

Solvent	E_T^a (25°C)	$10^3 k_2^b$ (25°C)	$10^2 k_2^c$ (25°C)	$10^4 k_2^d$ (85°C)	$10^2 k_2^e$ (25°C)	$10^3 k_2^f$ (24–27°C)	$10^{-5} k_2^g$ (25°C)
Ethanol	51·9			0·86	1·55	2·56	
Nitromethane	46·3			1·65		3·24	
Acetonitrile	46·0		2·70	1·63	1·12		14·0
Acetone	42·2		1·70	1·88			
Nitrobenzene	42·0	19·2					
Dichloroethane	41·9						
Dichloromethane	41·1				1·21	8·45	9·1
Pyridine	40·2	23·0		2·22			
Chloroform	39·1				0·93	13·0	6·0
Ethyl acetate	38·1	8·15		2·62		1·50	
Chlorobenzene	37·5					9·64	
Tetrahydrofuran	37·4		1·60				
Dimethylformamide	43·8	32·9		1·64			
Dioxane	36·0			2·77		2·66	
Benzene	34·5	11·5		4·24		5·59	
Toluene	33·9		2·00	4·80			
Carbon tetrachloride	32·5				1·66		3·8
Cyclohexane	31·2	5·42					
n-Hexane	30·9		0·92				

[a] E_T is an empirical solvent polarity parameter[49].
[b] Phenylazide + 1-pyrrolidino cyclopentene[42].
[c] Diazomethane + methyl methacrylate[43,44].
[d] C-Phenyl-N-methylnitrone + ethyl acrylate[45].
[e] p-Chloro benzonitrile oxide + styrene[46].
[f] Picryl azide + dicyclopentadiene[47].
[g] Ozone + trans-stilbene[48].

TABLE 5. Solvent effect on regioisomer ratio for cycloadditions of mesitonitrile oxide with benzalacetone (at
18 °C) and diazomethane with thioadamantane (at
0 °C)[a]

	Ratio (22):(23)	Ratio (24):(25)
Cyclohexane	13:87	87:13[b]
Benzene	21:79	76:24
Ether	20:80	80:20
Methylene chloride	47:53	58:42
Acetonitrile	55:45	32:68
Methanol[c]	32:68	30:70

[a] An ethereal solution of diazomethane was added
to a solution of thioadamantane in the appropriate
solvent.
[b] Petrol ether.
[c] $E_T = 55 \cdot 5$.

F. Activation Parameters

Cycloadditions of several 1,3-dipoles with either $C{=}C$, $C{=}X$ or $X{=}Y$ double bonds follow second-order kinetics characterized by moderate activation enthalpies from $\simeq 8 \cdot 0$ to $\simeq 17 \cdot 0$ kcal/mol, and large negative activation entropies from $\simeq -20$ to $\simeq -40$ e.u. (Table 6). It is noteworthy that in the reactions of ozone with alkenes, activation enthalpies near zero were found. Large negative entropies are in agreement with a concerted process requiring a highly-ordered transition state.

The reaction of azides with norbornene showed an increased rate in passing from phenyl azide to p-nitrophenyl azide and was accompanied by a diminution of $\Delta H^{\neq}$ which is responsible for such a variation (see Table 6). It is therefore correct to attribute this variation to electronic effects which influence the $\Delta H^{\neq}$.

Diazoethane is more reactive with methyl methacrylate than diazomethane in spite of a larger $\Delta H^{\neq}$. Its higher reactivity is therefore due to the entropy factor. The last example suggests that caution should be used when explaining reactivity on the basis of electronic effects only.

However, an almost completely satisfactory explanation for the changes in reaction rates based on both inductive and resonance effects of substituents and good Hammett or Taft relationships have been encountered for 1,3-dipolar cycloadditions (ozone) and cycloreversions (fragmentations of primary ozonides) characterized by $\Delta H^{\ddagger}$ values near to zero and therefore controlled by their $\Delta S^{\ddagger}$ (see Section E.3 and F).

TABLE 6. Activation parameters of typical 1,3-dipolar cycloadditions

Reaction (solvent)	$T(°C)$	$10^4 k_2$(l/mol sec)	$\Delta H^{\neq}$(kcal/mol)	$\Delta S^{\neq}$(e.u.)	Ref.
p-X—C_6H_4—N_3 + norbornene (AcOEt)					
X=H	25	0·171	14·7	−31·0	54
X=NO_2	25	1·05	12·5	−34·9	
Phenyl azide + morpholino cyclopentere (benzene)	25	2·58	11·7	−36	42
Diazomethane + methyl methacrylate (tetrahydrofurane)	25	160	8·0	−40	43
Diazoethane + methyl methacrylate (tetrahydrofurane)	25	1400	10·6	−27	43
p-Chlorobenzonitrile oxide + styrene (CCl_4)	25	166	11·8	−27	46
Ozone + $trans$-stilbene (CCl_4)	25	38×10^8	0	−33	48
C-Phenyl-N-methyl nitrone + ethyl crotonate (toluene)	85·5	0·37	17·0	−23·4	45
$(PhO)_2PO$—N_3 + EtO_2C—CH=CH—PPh_3 (benzene)	25	16·9	11·8	−31·5	55
3-Methyl-2,4-diphenyl oxazolium-5-oxide + methyl acrylate (benzonitrile)	50	26,000	12·3	−19	56
3-Methyl-2,4-diphenyl oxazolium-5-oxide + carbon disulphide (benzonitrile)	50	24·0	14·8	−25	56
3-Methyl-2,4-diphenyl oxazolium-5-oxide + benzaldehyde (benzonitrile)	50	4·75	10·0	−43	56
3,5-Dichloro-2,4,6-trimethyl benzonitrile oxide + p-$EtOC_6H_4$—N=S=O (CCl_4)	25	6·21	11·7	−34	57

G. The Geometry of the Transition State

According to Huisgen's proposal, 1,3-dipole and dipolarophile approach each other on two parallel planes[32]. The π orbitals begin to interact in a σ manner when the 1,3-dipole is still linear (for 1,3-dipoles with an orthogonal π bond), e.g. the *transoid*-oriented complex between benzonitrile benzylide and dimethyl maleate **26**. (We propose the use of the terms *cisoid* and *transoid* orientations for the stereochemical relationship between the substituent on a sp^2-hybridized end-centre of a 1,3-dipole and the substituent on the dipolarophile which are eventually found *cis* and *trans* respectively in the final adduct, instead of the *syn* and *anti* convention previously used[58].) M.O. perturbational calculations (see Section II.J) on the system are carried out at this stage of low interaction of the two addends[5]. A rehybridization of the two systems takes place successively with bending of the 1,3-dipole to give via an *endo*-**27** or *exo*-**28** orientation of the 1,3-dipole with respect to substituents on the dipolarophile, adduct **29** (30%). Similarly compound **30** (70%) was formed via a *cisoid*-oriented complex[31]. However, on bending of linear 1,3-dipoles, the allyl anion resonance energy is not lost. Hückel M.O. calculations indicate that for phenyl azide this bending process requires low energy ($\simeq 5$ kcal/mol)[59]. More recently, CNDO/2 calculations indicate the energy required for the in-plane bending of diazomethane is $> 20{\cdot}0$ kcal/mol[4,12] and, on this basis, the hypothesis was formulated that the bending begins after substantial

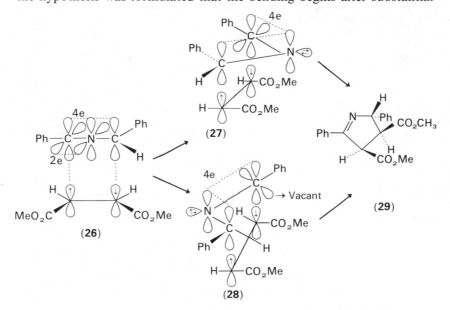

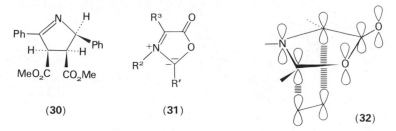

(30)   (31)   (32)

bonding takes place between 1,3-dipole and dipolarophile[4]. While *endo*
and *exo* orientations have no consequences on the stereochemistry of the
adducts from cycloadditions of 1,3-dipoles with orthogonal π-bonds, it
should be pointed out that bonding secondary orbital interactions
between the 1,3-dipole central atom and substituents on the dipolarophile
for *endo* transition states can *a priori* have relevant consequences on
enhancing reaction rates[60].

It should also be noticed that the double bond of the product **29**
originates from the lone pair present on the nitrogen atom in **27** or **28**[32].

The Huisgen proposal was supported by the experimental findings that
mesoionic oxazolones **31** and sydnones give cycloadditions with alkenes
which show all the typical features of the 1,3-dipolar type. For these
1,3-dipoles only an orientation complex such as **32** is possible.

Finally it should be noted that, as one of the two bonds is forming faster
than the other, the planes of the two interacting systems are not perfectly
parallel.

H. Allowedness of 1,3-Dipolar Cycloadditions

Woodward and Hoffmann proposed the principle of conservation of
orbital symmetry as a theoretical basis for all pericyclic reactions[61]. All
1,3-dipolar cycloadditions were shown to occur with the 1,3-dipoles and
alkenes approaching each other in a suprafacial–suprafacial manner (see
top of Figure 6). They are classified as $[\pi^4s + \pi^2s]$ cycloadditions in the
Woodward–Hoffmann notation.

The orbital correlation diagrams for allyl anion–ethylene *supra–supra*
thermal (ground state) cycloaddition (the prototype of 1,3-dipolar cycloadditions) are also shown in Figure 6.

In the diagram, the bonding occupied orbitals in the reactants are only
correlated with bonding occupied orbitals of the same symmetry in the
product. The reaction is therefore symmetry allowed in the ground state.

The correlation of the π-bond of the dipolarophile with the lone pair of
the adduct has a physical basis. As the reaction proceeds the olefin mixes

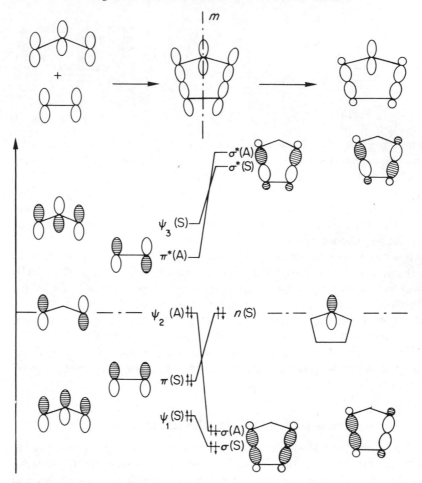

FIGURE 6. Correlation diagram for the $[\pi^4 s + \pi^2 s]$-cycloaddition of allyl anion with ethylene (symmetries, A = antisymmetric and S = symmetric, are with respect to the mirror plane m).

with ψ_1 allyl (which has a lower energy than ethylene π) in an antibonding way, e.g. **33** and ψ_3 allyl (with a higher energy than ethylene π) in a bonding way, e.g. **34**. ψ_1 Allyl, ψ_3 allyl and π ethylene all have the same symmetry with respect to the mirror plane m. The dipole (allyl anion) contribution cancels at C_1 and C_3 but reinforces at C_2. Thus in the transition state this orbital is essentially half in the dipole and the other half in the dipolarophile as shown by **35**[61].

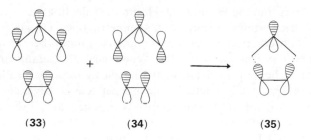

(33)   (34)   (35)

The orbital correlation diagram for allyl anion–butadiene *supra–supra* thermal cycloaddition, $[\pi^4s + \pi^4s]$, leads to the correlation of bonding orbitals of reactants with antibonding orbitals of products and therefore the reaction is symmetry forbidden.

Woodward and Hoffmann also assumed that removal of molecular symmetry (e.g., passing from allyl anion to benzonitrile oxide) does not exclude the possibility that the reaction is orbital-symmetry controlled.

In conclusion the Woodward–Hoffmann rules state that thermal *supra–supra* 1,3-dipolar cycloadditions are symmetry allowed with conjugated polyenes containing $4n + 2\pi$ electrons whereas they are symmetry forbidden with $4n$-electron systems. The same rules apply to *antara–antara* (unknown) cycloaddition whereas the rules have to be reversed (allowed with $4n\pi$ electron polyenes) for *supra–antara* reactions (unknown). These types of cycloadditions are greatly disfavoured by steric constraints.

1,3-Dipolar cycloadditions involving reactants in excited states have not yet been reported.

A brilliant verification of the Woodward–Hoffmann rules was the formation of compounds by *supra–supra* concerted 1,3-dipolar cycloadditions with trienes indicated as $[\pi^4s + \pi^6s]$ or $6 + 3 \rightarrow 9$.

Compounds of type **36** have been isolated from the reaction of arylnitrile oxides[62] or diphenylnitrile imine[63,64] with tropone, whilst compounds derived from adducts of type **37** have been formed as the sole adducts from the cycloadditions of diazomethane with 6,6-dimethylfulvene[65] and of benzonitrile oxide with 6-dimethylaminofulvene respectively[66].

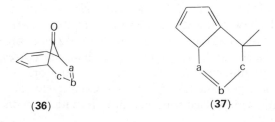

(36)   (37)

Considering that the Woodward–Hoffmann rules hold just as rigorously in asymmetric systems as they do in symmetric ones, Dewar pointed out that the 'conservation of orbital symmetry cannot be regarded as a satisfactory physical basis for the Woodward–Hoffmann rules'[67]. He proposed another approach for predicting allowedness (or disallowedness) of a concerted pericyclic reaction quite independent of reagent and product symmetry. This approach can readily be applied to 1,3-dipolar cyclo-additions which are often lacking in symmetry elements. According to Dewar's approach the formation of multicentre MO's by mutual inter-action of AO's depends on the latter overlapping in space but it does not depend on the geometry of the overlap. The properties of these multi-centre MO's are similar whether they are orbitals of stable molecules or of a transition state.

The properties of cyclic conjugated transition states of thermal peri-cyclic reactions can be derived from those well known for cyclic conjugated polyenes. There are two topologically-distinct types of conjugated polyene systems: normal Hückel systems in which all AO's of the basis set can be chosen so that they all overlap in phase, e.g. **38**; anti-Hückel systems where there is at least one phase dislocation, no matter how the phases of the AO's of the basis set are chosen. An example of an anti-Hückel system is a conjugated chain, twisted through 180° and then joined up into a ring: the resulting π system will have a phase dislocation, e.g. **39**.

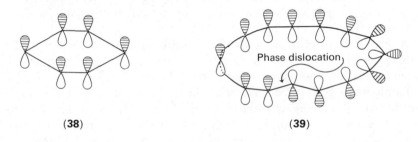

(**38**)   (**39**)

The latter twisted π-system is comparable to a Möbius strip[68]. Zimmer-mann, in an approach similar to that of Dewar, defines the anti-Hückel system as a Möbius system[69]. In general a ring is of the Hückel type if the number of phase dislocations is even, or of the anti-Hückel type if the number is odd. The Hückel systems with $4n + 2$ electrons have aromatic, or closed shells, and are more stable than the corresponding linear polyenes. Systems with $4n\pi$ electrons, are anti-aromatic and less stable than the corresponding linear polyenes. Conversely anti-Hückel systems are aromatic with $4n$ electrons and anti-aromatic with $4n + 2$ electrons.

The Dewar and Zimmermann approaches simply state that thermal pericyclic reactions preferentially take place via aromatic transition states. As a consequence a *supra–supra* concerted cycloaddition of a 1,3-dipole with a double bond or a triene will be favoured over a dipolar or diradical intermediate. The delocalized MO's of transition states **40** and **41**, in fact, have all overlaps in phase and are aromatic Hückel systems having six and ten electrons respectively.

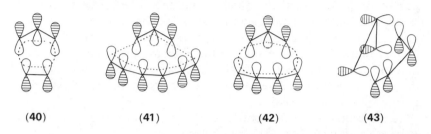

(40)   (41)   (42)   (43)

The Hückel system with eight electrons formed by a 1,3-dipole and a 1,3-diene, shown in the *supra–supra* orientation **42** is anti-aromatic and therefore not favoured. The other possible *supra–antara* orientation **43**, a $[\pi^4s + \pi^4a]$ anti-Hückel system, is aromatic but again disfavoured by steric constraints.

According to Dewar, it should be preferable to talk of aromatic and anti-aromatic instead of allowed and forbidden pericyclic processes. This formulation expresses that anti-aromatic reactions are just as possible as is the existence of anti-aromatic conjugated systems.

I. Reactivity

The dipolarophilic activity of double bonds depends to a large extent on the effect of substituents[1,32]. Conjugation with electron-withdrawing or electron-donating substituents has been shown to play an important role in reaction rates. It was observed, for instance, that additions of diazoalkanes and nitrile ylides were only enhanced by electron-withdrawing groups in the dipolarophile. On the contrary, cycloadditions with ozone and nitrous oxide were accelerated only by electron-donating substituents on the double bond. Other 1,3-dipoles, such as azides, azomethine imines, azomethine oxides, azomethine ylides, carbonyl ylides, nitrile imines and nitrile oxides showed low reactivity with common alkenes while the cycloadditions were accelerated both by electron-withdrawing and electron-donating substituents on the double bond. The original rationalization of these phenomena, which was proposed some years ago[32], took into account two electronic effects: (i) any type of conjugation enhances the polarizability of the double bond involved in the 1,3-dipolar cycloaddition, (ii) concerted formation of two new σ bonds does not need to be synchronous and therefore partial charges can be present in the transition state. The two effects are interrelated and the partial charges can be stabilized to some degree by the substituents on the two reactants. On the basis of the above explanation it was difficult to account for the low reactivity found, for instance, in the reaction of nitrile imines with β-dimethylaminoacrylic ester. The dimethylamino and ester groups, each strongly activating when present alone on a double bond, together resulted in no activation of the dipolarophile. Another property not explained by the partial charge model is the acceleration observed when a second group of the same nature as the first is introduced symmetrically onto the double bond.

More recently MO perturbation theory has been successfully used to explain the problem of the reactivity in the 1,3-dipolar cycloadditions[4—11]. Second-order MO perturbation theory provides equation (4)[10]. This equation was derived from the more general expression, proposed by Salem[70,71], for the energy gain in bond formation between centres a and c of the 1,3-dipole and centres d and e of the dipolarophile when they approach each other (see Section II.A).

$$\Delta E = \frac{(C_a C'_d \gamma_{ad} + C_c C'_e \gamma_{ce})^2}{E_{\psi_2} - E_{\psi_B}} + \frac{(C'_a C_d \gamma_{ad} + C'_c C_e \gamma_{ce})^2}{E_{\psi_A} - E_{\psi_3}} \qquad (4)$$

$\psi_A = $ HOMO(dipolarophile); $\quad \psi_B = $ LUMO(dipolarophile)

$\psi_2 = $ HOMO(dipole); $\qquad\quad \psi_3 = $ LUMO(dipole).

Equation (4) considers the energy change occurring when the frontier orbitals of the 1,3-dipole and dipolarophile are interacting. E represents orbital energy, C and C' are atomic orbital coefficients within the molecular orbitals of the HOMO's and LUMO's respectively. The resonance integral γ is a function of the distance between the reacting centres and of the nature of the atoms involved. The resonance integral is larger for the formation of C—C than C—O and C—N bonds (being $\gamma_{CC} > \gamma_{CN} > \gamma_{CO}$)[5,7].

Equation (4) only considers interactions between occupied and unoccupied MO's as interactions between occupied MO's do not yield a net energy change (neglecting overlap integrals).

A schematic and qualitative picture of the newly formed molecular orbitals arising from symmetry-allowed frontier-orbital interactions is shown in Figure 7. ΔE of equation (4) is equal to the summation of the two stabilizing energy terms ΔE_I and ΔE_{II}.

It can easily be deduced from Figure 7 and equation (4) that the stabilization energy ΔE is inversely proportional to the orbital energy differences
of the interacting frontier orbitals. Evaluation of the HOMO's and
LUMO's only in the equation (4) is therefore justified.

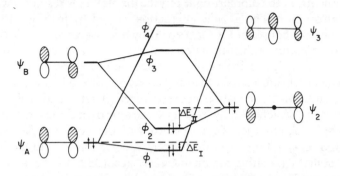

FIGURE 7. Schematic representation of the newly-formed molecular orbitals arising
from the interaction of frontier orbitals of a 1,3-dipole and a dipolarophile.

According to Sustmann[9,10], and other authors[5,72] 1,3-dipolar cycloadditions can be classified into three types (Figure 8): Type I, where the
interaction of the HOMO dipole with LUMO dipolarophile is greatest;
Type II, where both frontier orbital interactions must be taken into
account; Type III, where the LUMO dipole, HOMO dipolarophile
interaction is greatest.

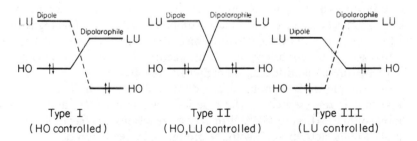

FIGURE 8. Classification of 1,3-dipolar cycloadditions. (In the original papers by
Sustmann[9,10], Type I and Type III were inverted.)

The above three types have also been called more concisely HOMO,
HOMO–LUMO and LUMO controlled 1,3-dipolar cycloadditions
respectively. It follows that the ΔE of equation (4) is dependent for Type I

mainly on the energy difference $\psi_2 - \psi_B$, for Type III on $\psi_3 - \psi_A$ and for Type II on both $\psi_2 - \psi_B$ and $\psi_3 - \psi_A$. The energy gap $\psi_2 - \psi_B(\psi_3 - \psi_A)$ in cycloadditions of Type I (Type III) is lowered by electron-donating substituents on the dipole (on the dipolarophile) and electron-withdrawing substituents on the dipolarophile (on the dipole). This results in a stronger F.O. interaction. The greater interaction caused by lowering the energy difference of a couple of F.O. in Type II by both electron-withdrawing and electron-donating substituents on any of the two addends, overcome the adverse effect due to the raising of the other F.O. energy difference. An enhanced F.O. interaction obviously corresponds to a higher reaction rate.

A quantitative correlation between reaction rates and frontier orbital energies can be achieved by making some approximations in equation (4). Holding constant the two numerators of the equation for the cyclo- additions of a 1,3-dipole with different dipolarophiles and assuming the same shifts x (positive for an increase, according to the nature of the substituent) in orbital energies for ψ_A and ψ_B, Sustmann and Trill derived equation (5) from equation (4)[10].

$$\Delta E = AB^2 \left[\frac{1}{D_I + x} + \frac{1}{D_{II} - x} \right] \tag{5}$$

$$D_I = E_{\psi_2} - E_{\psi_B} \qquad D_{II} = E_{\psi_A} - E_{\psi_3}$$

$$\ln k_2 = k\Delta E + k' \tag{6}$$

When one part of the sum in equation (5) is dominant, the other can be neglected in a first approximation and equation (5) will represent a hyper- bola. When D_I and D_{II} are of comparable importance, the superposition of the two hyperbolae gives rise to a parabola-shaped curve. The logarithms of the rate constants for cycloadditions, assuming $\Delta S^{\neq}$ as constant, are related to ΔE by equation (6)[6]. Consequently, the nearer in energy the frontier orbitals of addends, the higher the reaction rates.

An example of 1,3-dipolar cycloaddition where both D_I and D_{II} are important is represented by the reactions of phenyl azide and different dipolarophiles (Figure 9)[10]. For these reactions a parabola-shaped dipolarophilic activity scale with the minimum in the region of unactivated alkenes, has been found on plotting the logarithms of the rate constants for cycloadditions of phenyl azide against the ionization potentials of substituted olefins. The variations of the IP_v are taken as a measure of the variation either of the HOMO olefins–LUMO azide or HOMO azide– LUMO olefins energy differences, because the energy levels of the 1,3- dipole are constant and the HOMO and LUMO energies of the olefins shift in the same direction. The curve will therefore describe, from left to

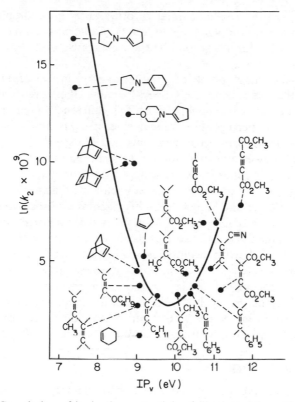

FIGURE 9. Correlation of ionization potentials of dipolarophiles with $\ln k_2$ values for cycloadditions of phenyl azide in CCl_4 or C_6H_6 at 25 °C. [Reproduced with permission from Sustmann and Trill, *Angew. Chem. Int. Ed.*, **11**, 838 (1972).]

right, reactions of Type III LUMO-azide controlled with electron rich olefins, of Type II with unactivated olefins and of Type I HOMO-azide controlled with electron-poor olefins.

A plot of $\ln k_2$ v.s. IP_v of olefins with 1,3-dipoles whose cycloadditions fall in Type I (e.g., diazomethane, nitrile ylides) or all in Type III (e.g., ozone), should be hyperbolae, with lower reaction rates respectively for electron-rich and for electron-poor olefins.

It should be appreciated how the perturbation model explains some peculiarities discussed at the beginning of this section, e.g. the different behaviour of 1,3-dipoles such as ozone, nitrile ylides and nitrile oxides. However a few other points have to be considered here. The energy levels of frontier orbitals are not always the determining factors for the reactivity in 1,3-dipolar cycloadditions. Benzene, for instance, with an ionization potential of 9·24 eV, a value similar to other fairly good dipolarophiles, enters only a few 1,3-dipolar cycloadditions. The loss of aromaticity upon undergoing cycloaddition should be the main cause of its low dipolarophilic activity.

Another important point for the reactivity in 1,3-dipolar cycloadditions is the steric element. *cis–trans*-Isomeric alkenes react at different rates with the *trans* isomers being faster. This difference in reactivity is not explained by the perturbation model since the *cis* and *trans* isomers possess very similar frontier orbital energies. Huisgen[1,32] attributed the different reactivity of the two isomers to steric compression of the two *cis*-substituents on going from reactants to transition state†. Strain relief in the dipolarophile is also important in enhancing reaction rates. To this element is attributed the higher reactivity of *trans*-cyclooctene compared with that of *cis*-cyclooctene, even though the two geometrical isomers have very close IP values. For instance, high ratios were found for the 1,3-dipolar cycloadditions of the two cyclooctenes with mesitonitrile oxide[73], with phenyl azide (10,460)[47,74—76] and with diazomethane[75]. Steric strain relief was also considered to play an important role, even though it was not the only cause, in the high dipolarophilic reactivity of norbornene with diphenylnitrile imine[77], benzonitrile oxide[72], diazomethane[78], phenyl azide[42] and 3,4-dihydroisoquinoline-N-oxide[74].

J. Regioselectivity

The 1,3-dipolar cycloaddition between two asymmetrical reagents can be either regiospecific or regioselective or non-regiospecific. When the reaction gives exclusively (within experimental error) one of the two possible orientational isomers, it is called regiospecific. In the case of the formation of both the orientational isomers in nearly equimolar amounts it is said to be non-regiospecific, whilst a preponderance of one isomer is the result of a regioselective cycloaddition[79].

In equation (7) the three cases are represented for the 1,3-dipole $a\overset{+}{=}b\overset{-}{-}\overset{-}{c}$ and dipolarophile $d=e$.

† Mock and coworkers [W. L. Mock, *Tetrahedron Letters*, 475 (1972) and L. Radom, J. A. Pople, and W. L. Mock, *Tetrahedron Letters*, 479 (1972)] advanced the proposal that a state of bond distortion at the unsaturated centres of *cis*-disubstituted alkenes and *trans*-cyclooctene are responsible for their reactivity. This state is probably responsible for the high reactivity shown by *trans*-cyclooctene in the reported cases and also explains the generally-low reactivity of *cis*-alkenes and cyclohexane.

Also, in Diels–Alder reactions *trans*-alkenes are generally more reactive than their *cis*-isomers. O. Einstein and N. T. Anh [*Bull. Soc. Chim. France*, 2721, (1973)] proposed that along with the steric effects, orbital interaction also favours the cycloadditions of *trans* over *cis*-alkenes.

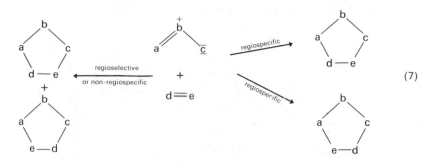

$$(7)$$

The phenomenon of orientation in 1,3-dipolar cycloadditions has represented an intricate problem for a long time. In recent years, perturbation theory provided a complete rationalization of regioselectivity of several 1,3-dipole cycloadditions[4-8,11].

According to Fukui[80], reactions take place in the direction which allows maximal frontier orbital overlapping. Figure 10 shows the two possible orientations (a) and (b) for asymmetrical addends.

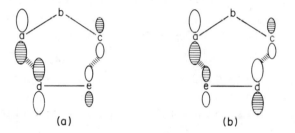

(a)   (b)

FIGURE 10. Regioisomeric transition states. Transition state (a) is more stable than (b) due to better interaction of terminal coefficients.

Orientation (a) is more stabilized than (b) and in the transition state there is unequal bond formation, bond a–d being more developed than bond c—e. As a rule, the favoured regioisomer will be the one formed through that transition state in which atoms with largest coefficients overlap (see equation 4).

When one or more heteroatoms are present in the reacting centres, the resonance integral $\gamma_{xy}(\gamma_{CC} > \gamma_{CN} > \gamma_{CO})$ should also be taken into account (see equation 4). However, when the addends are at a distance of 1·75 Å, γ values become very similar so that the above rules regarding coefficients only hold as a reliable qualitative guide to regioisomerism[5].

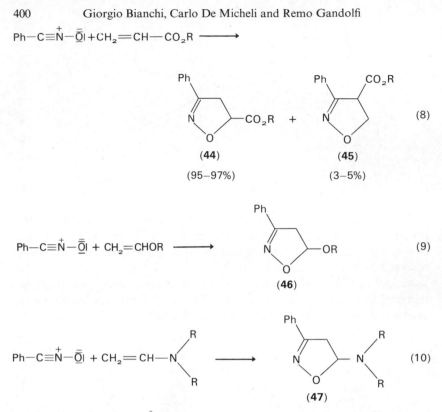

The distance of 1.75 Å chosen for the M.O. calculations by Houk and coworkers, was considered reasonable in order of taking into account the large negative entropy values and the high steric susceptibility of the reaction[5].

For the sake of brevity, only typical cycloadditions of benzonitrile oxide will be considered here, namely those with acrylic esters[21], vinyl ethers[20] and enamines[20] which would serve as general examples (equations 8, 9, and 10).

As shown in the above equations, mixtures of 3-phenyl-5- and 4-Δ^2-isoxazoline carboxylates (44) and (45) were obtained from reaction (8) whereas reactions (9) and (10) gave rise to the formation of only one regioisomer, the 3-phenyl-5-substituted-Δ^2-isoxazolines 46 and 47 respectively. Why is the reaction of benzonitrile oxide regiospecific with vinyl ethers and enamines and regioselective with acrylic esters? 1,3-Dipolar cycloadditions of benzonitrile oxide are placed in between Types II and III in Sustmann's classification. As shown in Figure 11, benzonitrile oxide is characterized by a set of atomic orbitals in which carbon possesses a

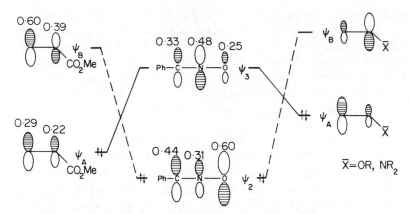

FIGURE 11. HO, LU orbitals of methyl acrylate, benzonitrile oxide and electron-rich polarophiles.

larger coefficient in ψ_3 and a smaller in ψ_2 compared with that of the oxygen atom.

The electron-deficient double bond of acrylic esters has the larger coefficients at the unsubstituted carbon in both ψ_A and ψ_B orbitals. The dominant interaction is between the dipole ψ_3 and the dipolarophile ψ_A.

Union of the largest coefficients on each frontier orbital leads to the 3-phenyl-5-substituted-Δ^2-isoxazoline **44** which is the major regioisomer found. However, as the coefficient s at C and O in ψ_3 are comparable and the difference between the α and β orbital coefficients in ψ_A of acrylic esters is not large, the formation of a small quantity of isoxazoline **45** is not surprising. Furthermore the interaction $\psi_2 - \psi_B$ [first term of equation (4)] cannot be ignored. Despite the larger $E_I = \psi_2 - \psi_B$ in the denominator, the larger coefficients at the dipole ends and at the α- and β-carbons of acrylic ester increase the numerator considerably. The latter interaction undoubtedly favours the formation of the minor regioisomer **45**. For electron rich dipolarophiles, e.g., vinyl ethers and enamines, the $\psi_3 - \psi_A$ interaction is controlling and the orentation is unambiguous (see Figure 11). In the latter interaction, the carbons of the benzonitrile oxide and β-position of the alkenes give rise to the largest overlapping. The large coefficient at the alkene α-position and oxygen of benzonitrile oxide in the other interaction $\psi_2 - \psi_B$ also favours the same cycloaddition direction. Hence, clearly both electron-rich and moderately electron-poor olefins present the same orientation preference.

The perturbation approach applied to 1,3-dipolar cycloaddition involving heterodipolarophiles fully explains also the observed regiochemistry. Moreover, it turned out to be the best approach in explaining the regiochemistry of several cycloadditions (e.g., dimerization of nitrile oxides and the reaction of diazomethane with Schiff bases) where Huisgen's principle of 'maximum gain of σ bonding energy'[32] failed.

The perturbational treatment can similarly be applied to all other cases of regioselectivity in 1,3-dipolar cycloadditions, provided steric and electrostatic interactions (e.g., dipole–dipole interaction) are similar in the two regioisomeric transition states.

As a corollary of this discussion, another generally accepted point about the electronic nature of 1,3-dipole should be revised. The idea that 'it is not meaningful to assign an electrophilic end and a nucleophilic end to a 1,3-dipole'[33] needs to be tempered. In frontier-orbital controlled cycloadditions (allowing for the importance of γ, see above) the end of the 1,3-dipole with the largest coefficient in the HOMO should be considered more nucleophilic, while the end with the largest coefficient in the LUMO will be more electrophilic. Phenylazide, for instance, has the nucleophilic attributes on the $Ph-\overline{\overline{N}}-$ end while the other end $-\overset{+}{N}\equiv N|$ behaves like an electrophilic centre[8].

K. Periselectivity

Periselectivity is a term introduced by Houk and coworkers[82] which recognizes the possibility of the selective formation of one of the thermally-allowed pericyclic reaction products.

Tropone and fulvenes are triene systems for which the thermally allowed $[\pi^4 s + \pi^2 s]$ and $[\pi^4 s + \pi^6 s]$ cycloadditions are geometrically possible. Figure 12 shows the π-orbital picture of 6,6-dimethylfulvene[83]. The reaction of the latter compound with diazomethane, a HOMO-controlled, 1,3-dipolar cycloaddition, gives as the only adduct **48**[65]. This result shows that the interaction of the HOMO dipole with the 2 and 6 centres of the LUMO in 6,6-dimethylfulvene is more important than that with the 2 and 3 centres of the same dipolarophile. The adduct **48** shows also the correct and expected regiochemistry.

The nearly non-existent lobe at position 6 of HOMO in fulvene makes the interaction LUMO dipole-2,6-fulvene centres much smaller than LUMO dipole-2,3-fulvene centres. In fact reaction of benzonitrile oxide, a LUMO-controlled 1,3-dipolar cycloaddition, results in a 3 + 2 cycloaddition with the formation of the regioisomer **49a** (prevalent) and its regioisomer **49b**[66]. The prevalent formation of **49a** seems in contrast with frontier orbital prediction of a sole LUMO dipole–HOMO dipolarophile interaction. However, it should be pointed out that, due to the quite high

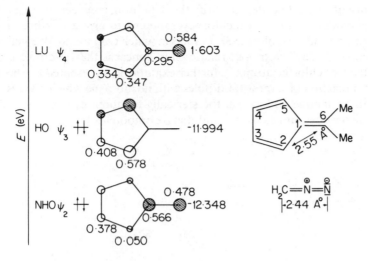

FIGURE 12. CNDO/2 frontier orbitals of dimethylfulvene.

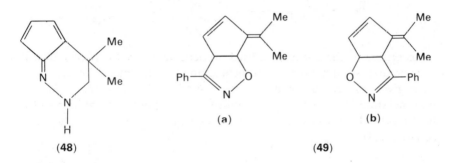

(48)   (49)

energy level of NHOMO ψ_2, the interaction LUMO benzonitrile oxide–
NHOMO 6,6-dimethylfulvene can be in competition and therefore favour
regioisomer **49a**. Moreover in this cycloaddition both the transition states
leading to a 3 + 6 → 9 cycloaddition and to **49b** are sterically disfavoured
with respect to that giving rise to **49a**.

L. syn–anti *and* endo–exo *Isomerism*

syn–anti and *endo–exo* Isomerism represent important aspects of the
stereochemistry in 1,3-dipolar cycloadditions which have received little
attention up to now.

For substituted dipolarophilic systems which have diastereotopic faces
there are two manners of approaching the 1,3-dipole definable as *syn* and

anti orientations. In order to clarify the definition, consider the reaction of
nitrile oxides with *cis*-3,4-dichlorocyclobutene to give a mixture of *anti*-
chlorine **50** and *syn*-chlorine **51** adducts in which compound **51** is generally
dominant[53]. The *syn* or *anti* relationship concerns the 1,3-dipole moiety
and the two chlorine atoms. A further example is represented by the well-
known reaction of several 1,3-dipoles with norbornene which always gives
rise to an unique adduct, i.e. the sterically favoured cycloadduct **52**[1,28].
(See also azides, nitrile oxides and diazo compounds.)

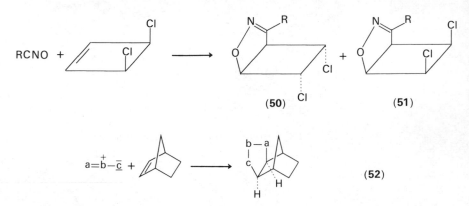

According to the above-proposed nomenclature, cycloadducts **52** will
be indicated as *syn*-methylene adducts although in the literature they have
been given the confusing name of *exo*-adducts. In fact the *endo–exo*
nomenclature is more properly applied to the relative orientations of a
bent 1,3-dipole in respect to substituents on the double bond as shown in
equation (11).

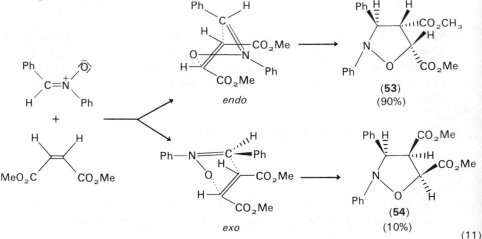

(11)

For instance, in consequence of *endo* and *exo* dispositions in the transition state, cycloaddition at room temperature of *C*-phenyl-*N*-phenyl nitrone (a ground-state bent 1,3-dipole) with a substituted dipolarophile such as dimethyl maleate, gives rise to the formation of a mixture of the two diastereoisomeric isoxazolidines **53** and **54**[84].

The kind of stereoisomerism (*syn–anti, exo–endo*) reported above is governed by secondary orbital interactions and by other elements such as steric effects, dipole–dipole and van der Waals–London interactions (see in particular diazo compounds, nitrones and nitrile oxides).

M. 1,1-Cycloadditions of 1,3-β-Dipoles

Among the resonance structures of nitrilium and diazonium betaines carbene (**55**) or azene (**56**) structures occur respectively. This suggests the possibility of a 1,1-cycloaddition as the first step in the 1,3-dipolar cycloaddition, followed by a vinyl cyclopropane–cyclopentene rearrangement (equation 12).

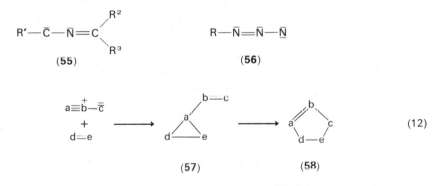

However, in the cases of nitrile ylides, diazoalkanes and azides[85], it has been demonstrated that a three-membered ring **57** is not the precursor of **58**. Thermolysis of 1-(4-nitrophenylazo)-2-phenylaziridine (**59**) in fact gives *p*-nitrophenyl azide which has been isolated or trapped with norbornene. The additional product, styrene, has also been trapped with *C*-benzoyl-*N*-phenylnitrone.

$$\underset{\textbf{(59)}}{\text{aziridine with } N=NC_6H_4NO_2\text{-}p} \quad \xrightarrow[50—80\,°C]{\Delta} \quad p\text{-}NO_2C_6H_4\text{-}N_3 + Ph\text{-}CH=CH_2$$

It has been reported recently[86] that benzonitrile oxide reacts with arylidene isoxazolones **60** to give compounds **62** along with the two usual regioisomeric adducts to the exocyclic double bond. The formation of **62** has been rationalized through a mechanism involving a primary-carbene addition of benzonitrile oxide giving an unstable, non-isolable compound, **61**. The yield of **62** was found to be temperature dependent, the best yields being obtained in boiling benzene. The intramolecular cycloaddition of benzonitrile allyl methyl methylide (**62a**) represents a recent example of the same type of reaction[86a].

Ph—C̄—N̄=Ō

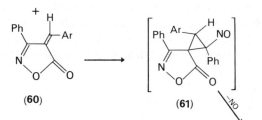

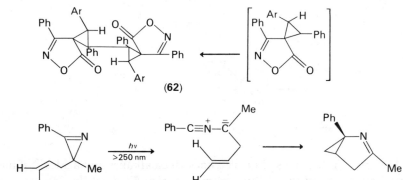

III. CYCLOADDITIONS OF ALLYL AND HETERO-ALLYL ANION SYSTEMS[86b]

Systems derived from the allyl anion have recently been reported as reacting with double bonds to give a $3^- + 2 \rightarrow 5^-$ polar cycloaddition. Treatment of the cyanocyclopropane **63** with lithium diisopropyl amide in tetrahydrofurane at $-30\,°C$ gives the cyclopropyl anion **64** isomerizing to the allyl anion **65** on electrocyclic conrotatory ring opening, and **65** reacts with alkenes to give (upon protonation) cyclopentane derivatives (equation 13)[87].

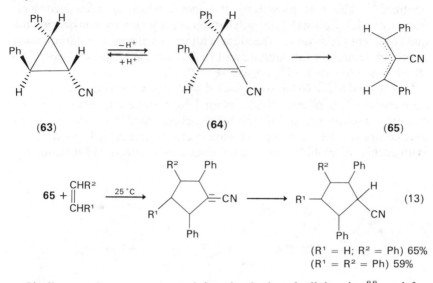

(63)   (64)   (65)

Similar results were reported for the 2-phenyl allyl anion[88] and for 2-acyl allyl anion derivatives[88a]. These polar anionic cycloadditions are energetically favoured when an electron-withdrawing group is cross-conjugated with the allyl system as in the case of 2-aza-allyl anions. In all these cases the negative charge is more stabilized in the final adducts, where it is localized at the central atom, than in the allyl anions where it is largely localized at the two end atoms.

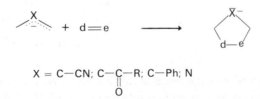

$X = C—CN; C—\underset{\underset{O}{\|}}{C}—R; C—Ph; N$

Kauffmann and coworkers have extensively studied the cycloadditions of 2-aza-allyl anions[86a,89—94]. *trans,trans*-1,3-Diphenyl-2-aza-allyl lithium (**66**), prepared by reaction of *N*-benzylidenbenzylamine with lithium diisopropylamide in tetrahydrofuran at −30 to −70 °C, does not react with cyclohexene while it reacts readily (at 0–20°C) with strained (norbornadiene) and conjugated (styrene, stilbene, dienes) double bonds to give stereospecific cycloadducts (equation 14)[86b,89,91—93]. With acenaphthylene, in a kinetically controlled reaction, an equimolar mixture of *endo* and *exo* adducts (the latter being thermodynamically more stable) was obtained[92]. The role played by the metal which is associated with *trans, trans*-1,3-diphenyl-2-aza-allyl anion is evident when considering the quite different yields for the reactions with *trans*-stilbene. For cations such as lithium (equation 14), sodium, and potassium the yields were 80–85%, 60–65% and less than 10% respectively[86b].

Hetero-allyl anion **66** has been reacted with several other dipolarophiles: azomethynes[90], azoderivatives[90], phenyl isocyanate and phenylisothiocyanate (equation 15), and dicyclohexylcarbodiimide[94]. All the reactions were completely stereospecific and gave cycloadducts in fairly good yields. With carbon disulphide a bis cycloadduct was obtained whilst with CO_2

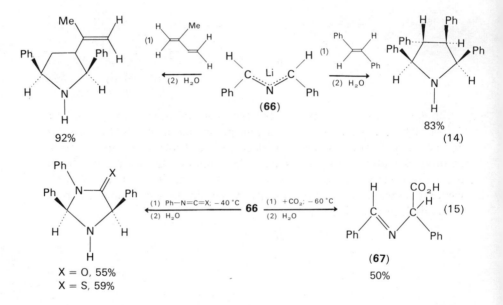

the open adduct **67** was the result (equation 15)[94]. As far as regiochemistry is concerned, the results of cycloadditions (at 20 °C) of 1,1-diphenyl-2-aza-allyl lithium **68** with styrene[89] and dienes[92] to give only the sterically less favoured adducts are quite relevant. **68** shows an even more marked tendency than **66** to give open adducts as exemplified by its reaction with benzylidenaniline[86b].

Again **68**[94a] and 1,2-diazaallyl **69**[95] react with aldehydes and ketones to give open adducts in fairly good yields. It has not yet been ascertained

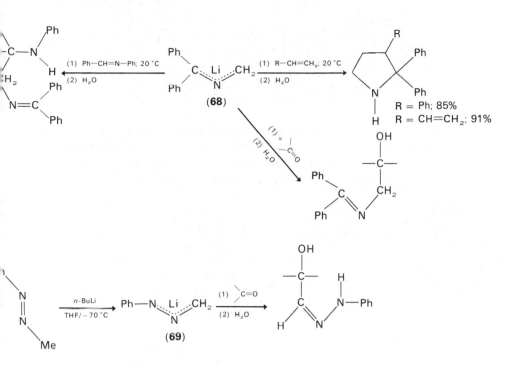

whether the final products arise from intermediate cycloadducts or not.

69 turned out to be less reactive than **66** and **68** towards alkenes and whereas it reacts with acenaphthylene (32 %)[95] it was found to be unreactive towards styrene[86b]. This can be accounted for by the presence of a peripheral nitrogen atom which stabilizes the negative charge on the anion making it less reactive.

Two reaction mechanisms have been proposed for the cycloadditions of allyl anions and hetero-allyl anions in order to account for the observed stereospecificity: (i) a one-step, 1,3-dipolar cycloaddition as depicted in **70** or a two-step process via an intermediate of type **71**. In the latter case, rotation around axes a and b is blocked by a strong chelate bridge either to the carbon–nitrogen double bond or to the nitrogen atom[92].

Very recently a two-step anionic cycloaddition was reported[96].

Cycloadditions of **66** (with alkenes and azoderivatives) and of **69** (with alkenes) have recently been reported to be reversible[97].

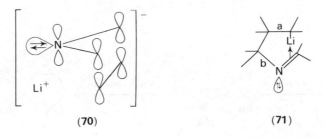

(70)   (71)

Several examples of retro-cycloadditions reported in the literature can be considered as anionic retro-cycloadditions[98—102], e.g. equation (16). Evidence for a concerted anionic retro-cycloaddition has been given for the transformation of **72** into **73** and nitrogen (equation 17)[103].

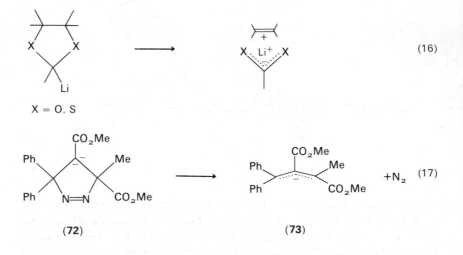

(16)

X = O, S

+N$_2$ (17)

(72)   (73)

IV. CYCLOADDITIONS OF 1,3-DIPOLES WITH AN ORTHOGONAL DOUBLE BOND

A. Nitrile Ylides

1. Synthetic approaches

Nitrile ylides cannot be isolated but they can be generated in several ways in low concentrations and allowed to react with suitable dipolarophiles. The treatment of imidoyl chlorides **74** and **75** with triethylamine in benzene produces small amounts of benzonitrile 4-nitrobenzylide (**76**) and 4-nitrobenzonitrile benzylide (**77**) respectively, which can be trapped in a regiospecific reaction by benzaldehyde, a very reactive dipolarophile towards nitrile ylides. In both cases a mixture of *cis*- and *trans*-Δ^3-oxazolines was obtained in which the *cis* isomer predominated[104].

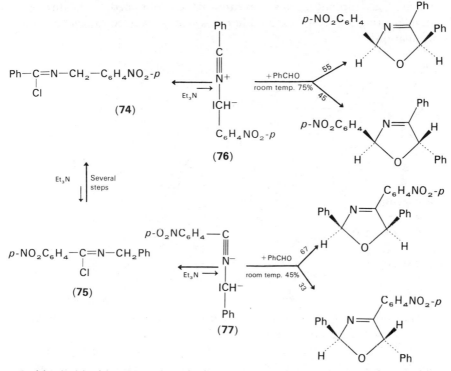

Imidoyl chlorides **74** and **75** in benzene containing triethylamine equilibrated with nitrile ylides **76** and **77** and with themselves (ratio **74**:**75** = 92:8)[105]. With the less-reactive dipolarophile norbornene, an identical mixture of stereoisomeric adducts **78** and **79** (ratio **79**:**78** = 1·5:1) was obtained either from **74** (74% yield) or from **75** (59% yield) through the more reactive nitrile ylide **76**[106].

Nitrile ylides **80**, which have recently been prepared by dehydrohalogenation of the corresponding imidoyl chlorides[107], or by thermolysis[107—109] or photolysis[107,109,110] of Δ^4-1,4,2 λ^5 oxazaphospholines **81**,

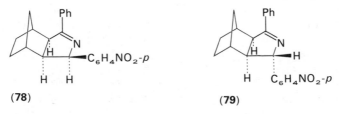

were similarly trapped by benzaldehyde to give Δ^2-oxazolines **82** in a regiospecific reaction [111]. The regiochemistry of this cycloaddition is the opposite to that found for nitrile ylides **76**, **77** and **85**. When isonitriles are used as trapping agents 1-azetines **83** are obtained[112,113] through an interesting $3 + 1 \rightarrow 4$ cycloaddition, which is reversed to reactants on irradiation[114].

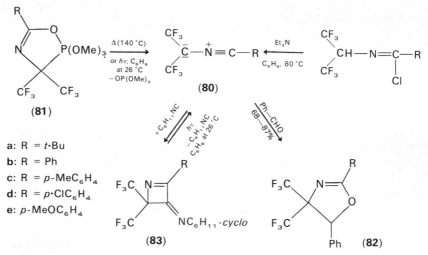

a: R = *t*-Bu
b: R = Ph
c: R = *p*-MeC$_6$H$_4$
d: R = *p*-ClC$_6$H$_4$
e: *p*-MeOC$_6$H$_4$

The photolysis of 3-substituted 2*H*-azirines (both aromatic and aliphatic substituents) is a more versatile method to generate nitrile ylides[115–118].

Irradiation of 3-phenyl-2*H*-azirines **84** in benzene in the presence of aliphatic and aromatic aldehydes give Δ^3-oxazolines in good yields through the intermediate nitrile ylides **85**[119,120]

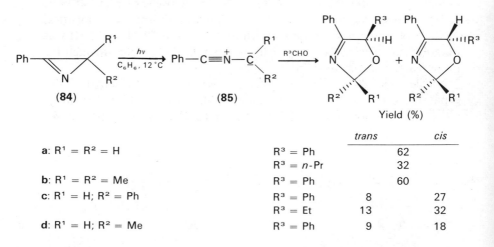

		Yield (%)	
		trans	*cis*
a: R^1 = R^2 = H	R^3 = Ph		62
	R^3 = *n*-Pr		32
b: R^1 = R^2 = Me	R^3 = Ph		60
c: R^1 = H; R^2 = Ph	R^3 = Ph	8	27
	R^3 = Et	13	32
d: R^1 = H; R^2 = Me	R^3 = Ph	9	18

An explanation has been proposed for the prevailing formation of cis- Δ^3-oxazolines, namely that in the transition state the linear nitrile ylide bends so as to minimize the steric interactions between all the substituents, Ph, R^2 and R^3 [119].

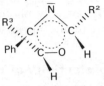

Azirines with aldehydes and acylazirines are intermediates in the photoisomerization of Δ^2-isoxazolines to Δ^3-oxazolines[121] and of isoxazoles to oxazoles respectively[122-124]. As far as the mechanistic aspect is concerned photocycloaddition of 2H-azirines proceeds through a n-π* excited singlet state (A^{*1}): the reaction in fact is neither quenched by standard triplet quenchers nor sensitized by standard triplet donors. Moreover, the similarity of the reactivities (see Table 7 and the reactions of **76** or of **85** with aldehydes) found in the reaction of nitrile ylides generated photochemically and thermally suggest that the intermediate nitrile ylides (NY) are electronically and vibrationally relaxed. The proposed reaction mechanism for 3-phenyl-2H-azirines is shown below[31,125,139]. Decay $A^{*1} \rightarrow A_0$ was shown to be radiationless and no fluorescence emission was detected.

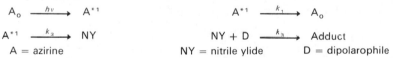

It was recently noted that irradiation with polychromatic light of the 2,2,3-triphenyl-2H-azirine (**84e**) in a matrix at $-185\,°C$ or of the Δ^3-oxazolin-5-one **86** at $-190\,°C$ allowed the benzonitrile diphenyl methylide

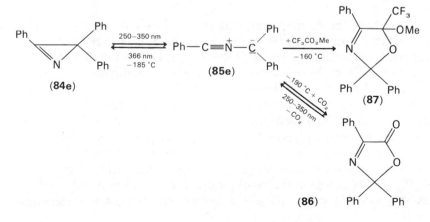

(**85e**) to be detected by u.v. spectroscopy. This was shown by e.s.r. to be in a singlet state. The nitrile ylide, **85e**, in the presence of methyl trifluoroacetate at $-160\,°C$ gives Δ^3-oxazoline **87** regiospecifically and even at $-190\,°C$ reacts again with the CO_2 trapped in the matrix to give **86**[126]. Δ^3-oxazolin-5-ones lose CO_2 not only photochemically[126,139] but also thermally at high temperature to give nitrile ylides[127].

2. Reactivity and regioselectivity of the reaction with carbonyl compounds

The nitrile ylide **85b** with two electron-releasing methyl groups was found to be more reactive than **85a** or **85c** towards acetone and acetophenone[116,128], which are less reactive dipolarophiles than aldehydes[120]. Furthermore, the carbonyl reactivity with nitrile ylides (**85**) is enhanced by electron-withdrawing groups as found for phenyl (or methyl) trifluoromethyl ketone, diethyl mesoxalate ($R^3=R^4=O_2Et$), methyl haloacetate, acid chlorides, acetic anhydride, acyl cyanides and α-ketoesters[117,126,128]. As a general trend the ketonic group of α-ketoesters was found the sole reactive part of the molecule[128]. However, the reaction of **85b** with ethylpyruvate gave 21% of the adduct to ketonic double bond and 6% of adduct to ester group[128a].

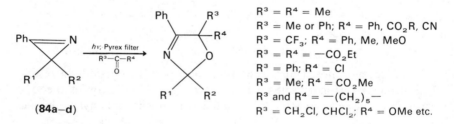

$R^3 = R^4 = Me$
$R^3 = Me$ or Ph; $R^4 = Ph$, CO_2R, CN
$R^3 = CF_3$; $R^4 = Ph$, Me, MeO
$R^3 = R^4 = -CO_2Et$
$R^3 = Ph$; $R^4 = Cl$
$R^3 = Me$; $R^4 = CO_2Me$
R^3 and $R^4 = -(CH_2)_5-$
$R^3 = CH_2Cl$, $CHCl_2$; $R^4 = OMe$ etc.

(**84a–d**)

These trends in reactivity and a Taft $\rho_x^* = 2.06$ found for the reactions of **85b** with α-halosubstituted acetates have their rationale in the fact that the reported cycloadditions belong to the Type I, HOMO-dipole controlled reaction. All the above reactions with carbonyl compounds are regiospecific and give Δ^3-oxazolines.

As far as regiochemistry is concerned, the M.O. calculations of the values of the HOMO coefficients on the two carbon atoms of the nitrile ylides are not reliable. This is because they vary greatly depending on the values of the two C—N bond lengths[4]. Experimentally, the problem has been solved by generating, photochemically, nitrile ylides **80** in $C_6H_6/MeOH$ and **85** in methanol (or other protic solvent) and obtaining **88**[114] and **89**[129], which possess a reverse regiochemistry. Compounds

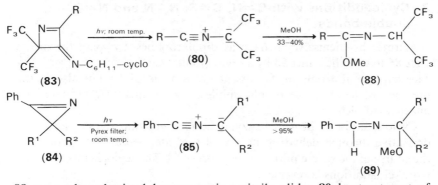

88 were also obtained by generating nitrile ylides **80** by treatment of imidoyl chlorides with tertiary base in the presence of methanol[107].

If we assume that the addition of methanol is initiated by frontier-controlled protonation, this means that the greater coefficient of the HOMO of **80** is on the 'anionic' trisubstituted carbon, whereas it is on the 'neutral' disubstituted carbon of **85**.

The regiospecificity found for carbonyl compounds (aldehydes, ketones and esters) with the 'neutral' carbon of **85** and the anionic carbon of **80** attacking the carbon of the carbonyl, which generally possesses larger coefficient than that of oxygen in the LUMO, is in full agreement with the simple F.O. perturbation approach.

Few examples of regioselective reactions of nitrile ylides with carbonyl compounds are known: the reaction of **76** and **77** with diethyl mesoxalate and benzoyl chloride[104] and of nitrile ylides **80** with diethyl pyruvate[111].

Finally an example in which the two reactants add in the opposite direction (probably because of steric hindrance) was found in the intramolecular nitrile ylide–aldehyde cycloaddition reported below[128].

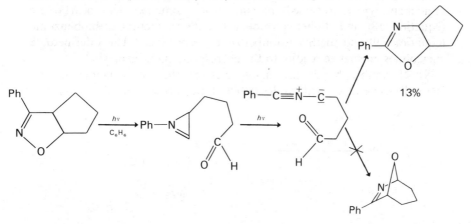

3. Cycloadditions with C=N, C=S, N=N and N=O double bonds

Azirines by themselves behave as dipolarophiles, trapping the nitrile ylide **85** to give the dimer **90** by a regiospecific reaction[115,116,120,130–132]. This happens not only in the absence of dipolarophiles but also in the presence of less-reactive dipolarophiles, e.g. *cis*-methyl crotonate and unactivated olefins[31,118].

Cycloaddition of **76** with benzylidene methylamine gives a imidazole derivative through dehydrogenation of the intermediate imidazoline in the course of the severe purification process[133]. The regiochemistry of the two cycloadditions is reversed.

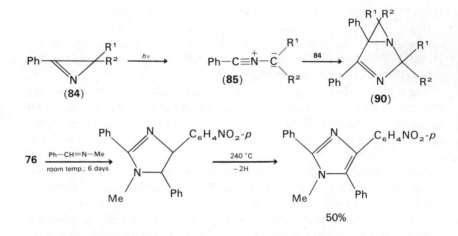

Cycloadditions with C=S double bonds were more studied[120,133]. The regiochemistry is reversed passing from dimethyl trithiocarbonate (equation 18) and diphenyl thioncarbonate to methyl dithiobenzoate (equation 19) and methyl thionbenzoate respectively. The intermediate thiazolines aromatize readily to the corresponding thiazoles[133].

N=N double bonds are also reactive: the photochemically- or thermally-generated nitrile ylides (**80**) and the photochemically-generated

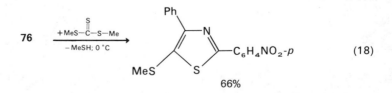

(18)

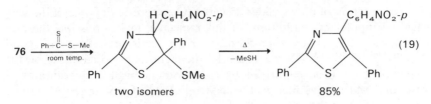

(19)

85 react smoothly with dimethyl or diethyl azodicarboxylate to give 1,2,4-Δ^3-triazolines in both cases[111,134].

Reaction of **76** with nitroxobenzene gave benzonitrile and C-p-nitro-phenyl-N-phenyl nitrone through a retro-1,3-dipolar cycloaddition of the non-isolable 1,2,4-oxadiazoline intermediate **91**[133].

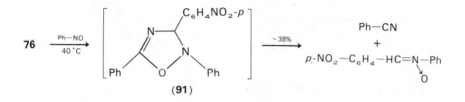

(91)

4. Cycloadditions with cumulated double bonds

3-Phenyl-2H-azirines **84b** and **84c** in benzene on irradiation react smoothly with ketenes[135], isocyanates, isothiocyanates and carbodi-imides[136]. The bonds C=S and C=O were found more reactive than the cumulated C=N and C=C bonds. The reaction is regiospecific with the disubstituted carbon atom of the nitrile ylide attacking the central atom of the cumulated system which possesses largest coefficients in the LUMO. The higher reactivity of the C=O than the C=C bond for ketene is consistent with the fact that the coefficient at oxygen in the LUMO is greater than that at the terminal carbon[137].

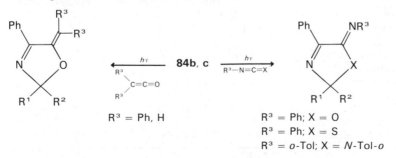

R³ = Ph, H

R³ = Ph; X = O
R³ = Ph; X = S
R³ = o-Tol; X = N-Tol-o

418   Giorgio Bianchi, Carlo De Micheli and Remo Gandolfi

The reaction of nitrile ylides **85** with CO_2, representing one of the very few 1,3-dipolar cycloadditions of this molecule, makes possible the preparation of Δ^3-oxazolin-5-ones from azirines **84** in good yields[116,117,136,138]. More recently the synthetic possibilities of the reaction have been confirmed by the photochemical cycloadditions of **84b** and **84d** to give high yields ($\simeq 90\%$) of Δ^3-oxazolinones. Moreover in the case of **84c** an equimolar mixture of the two regioisomeric oxazolinones, Δ^2- and Δ^3-oxazolinone respectively, was obtained[139].

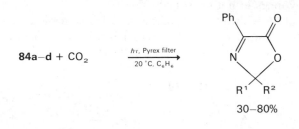

$$84a\text{--}d + CO_2 \xrightarrow[20\,^\circ C.\ C_6H_6]{hv,\ \text{Pyrex filter}}$$

30–80%

5. Cycloadditions with olefins

The most reactive olefins were found to be those substituted with electron-withdrawing groups, as would be expected for HOMO-dipole-controlled reactions. Table 7 shows some reactivity data typical of nitrile ylides generated either thermally or photochemically, and demonstrates the low reactivity of *trans*-methyl crotonate, the additive effect of two substituents on the double bond, and the higher dipolarophilic activity of *trans*- compared to the *cis*-1,2-disubstituted olefins[125].

TABLE 7. Relative reactivities of a series of olefins towards nitrile ylides **85c** and **76** generated from 2,3-diphenyl-2H-azirine (**84c**) and from N-(p-nitrobenzyl)-benzimidoyl chloride (**74**) respectively

Olefin	2,3-Diphenyl-2H-azirine	Imidoyl chloride
Methyl crotonate	1	1
Diphenyl azirine	2·5	
Methyl methacrylate	9	10
Methyl acrylate	160	
Diethyl maleate	135	51
Dimethyl maleate	166	61
Dimethyl fumarate	84,000	
cis-Dicyanoethylene	2300	
trans-Dicyanoethylene	189,000	

trans-Dicyanoethylene is more reactive than *cis*-dicyanoethylene, although according to Epiotis *cis*-dicyanoethylene should be more reactive than the *trans*-isomer in cycloaddition reactions as a consequence of attractive interaction between the two cyano groups which should increase along the reaction coordinate[140].

Unactivated alkenes (1-octene, cyclohexene) do not react photochemically with azirine **84c**[31], whereas norbornene and styrene were reactive[120].

Interestingly enough the nitrile ylide **85c** reacts with 1,4-benzo- and 1,4-naphtho-quinones only at the carbon–carbon double bond to give, in fairly good yields, 1,3-diphenyl-2*H*-isoindole-4,7-diones and 1,3-diphenyl-2*H*-benzo[*f*]isoindole-4,9-diones derived from initial adducts on oxidation by atmospheric oxygen[140a].

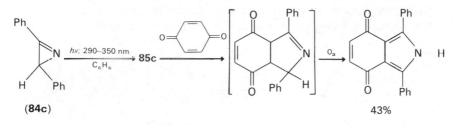

(84c)   43%

Formation of the less-sterically-favoured Δ^1-pyrrolines from azirine **84b** with methacrylonitrile and methyl methacrylate respectively (equation 20) illustrate the electronic control of the regiochemistry[31].

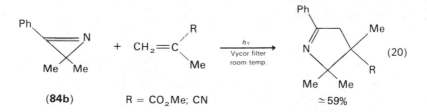

(84b)   R = CO$_2$Me; CN $\simeq$ 59%

A mixture of *cis*- and *trans*-Δ^1-pyrrolines was obtained in the regiospecific reaction of **84c** with methyl acrylate and acrylonitrile (equation 21)[31].

As a general rule, nitrile ylides **76** and **85** react with electron-poor conjugated olefins (methyl acrylate, acrylonitrile, methyl crotonate, methacrylonitrile and methyl methacrylate)[31,106,118,139] and with styrene[85,120] to give only one regioisomer. The 'neutral' carbon atom of the

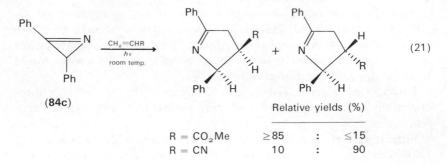

(21)

	Relative yields (%)		
R = CO$_2$Me	≥ 85	:	≤ 15
R = CN	10	:	90

1,3-dipole interacts with the β-carbon (relative to the electron-withdrawing or phenyl group) of the olefin possessing the greater LUMO coefficient (see equations 20 and 21). However, the reactions of phenyl azirine (**84a**) with methyl methacrylate and methacrylonitrile (equation 22)[31] and of **84b** with diethyl vinylphosphonate[128a] represent exceptions to this rule.

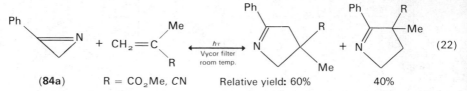

(22)

In sharp contrast with the above results, the cycloadditions of the nitrile ylides **80** with methyl acrylate give always a mixture of the two regioisomers **92a** and **93a** in good yields, and the ratio of the two products is very similar (**92a**:**93a** ≃ 60:40) irrespective of whether **80** is obtained thermally or photochemically[107].

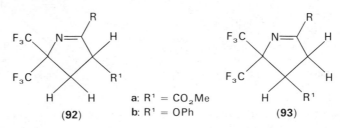

A mixture of two regioisomers was also obtained in the reaction of **80**, thermally generated from **81**, with phenyl vinyl ether. In this case the Δ^1-pyrroline **93** was dominant (for R = Ph, ratio **93b**:**92b** = 73:27; for R = t-Bu, ratio **93b**:**92b** = 88:12). **80a** and n-butyl vinyl ether gave only **93c** (R = t-Bu)[141].

Two electron-withdrawing groups, such as CF_3, present in the nitrile ylides **80** produce a lowering of frontier-orbital energies: HOMO dipole–LUMO dipolarophile and LUMO dipole–HOMO dipolarophile interactions for cycloadditions with both moderately electron-poor and electron-rich olefins are therefore important (Type II cycloadditions). As the two frontier-orbital interactions might favour two different orientations, the formation of two regioisomers as reported above, has here a theoretical explanation.

B. Nitrile Imines

The earliest report on a 1,3-dipolar cycloaddition involving the nitrile imine system is probably the reaction of benzphenylhydrazidoyl chloride with benzamidine[142], although the formation of the intermediate 1,3-dipole was not formulated. The most widely studied disubstituted nitrile imines (**96**) are commonly prepared either by thermolysis of 2,5-disubstituted tetrazoles **94**[1] (a retro-1,3-dipolar cycloaddition)[143] or by base-catalysed elimination of HX from hydrazidoyl halide **95a** and α-nitrohydrazone **95b**[1], or by lead tetra-acetate oxidation of aldehyde hydrazones[144].

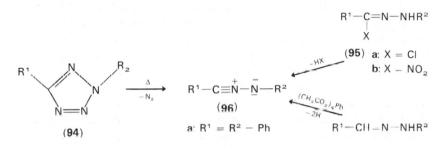

Although nitrile imines are not isolable, their existence has been proved by chemical reactions[145]. The exchange of chlorine in **95a** with that in $Et_3NH^{36}Cl$ in the presence of Et_3N is considered evidence for their existence.

1. Cycloadditions with carbon–carbon double bonds

Cycloaddition of **96** with alkenes is a general method for the synthesis of Δ^2-pyrazolines. With monosubstituted aryl[146], alkyl-[18], alkoxy-[147,148], carbalkoxy-[147], and amino-alkenes[147,149] the reaction gives 5-substituted Δ^2-pyrazolines **97** in fairly good yields.

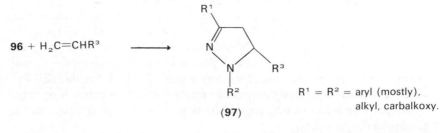

96 + H₂C=CHR³

R¹ = R² = aryl (mostly),
alkyl, carbalkoxy.

(97)

The same regiochemistry was found for cycloadditions with piperilene, isoprene, 2-chloro-1,3-butadiene, and styrene all of which gave 5-mono-substituted Δ^2-pyrazolines[150].

Perturbation approach explains the regiochemistry found in 1,3-dipolar cycloadditions of diphenylnitrile imine[5]. As with the parent formonitrile imine (see Section II.A), in diphenylnitrile imine the coefficients should be: C anionic > C neutral in the HOMO and C neutral ≫ C anionic in the LUMO. Consequently in 1,3-dipolar cycloadditions with electron-rich alkenes, LUMO nitrile imine-controlled cycloadditions, the most favoured overlap is between the carbon of the 1,3-dipole and the carbon of dipolarophile remote from the substituent, to give 5-substituted pyrazolines. For conjugated alkenes the ρ value + 0·88 found for the reaction of **96a** with styrenes[77] seems, however, to indicate a slight prevalence of HOMO dipole–LUMO dipolarophile interaction, both HOMO and LUMO inter-actions should be at work and a mixture of the two regioisomers is expected. The formation of only 5-substituted pyrazolines is probably due to the greater differences in the LUMO dipole coefficient magnitudes and steric effects. For electron-poor alkenes, on the basis of calculated frontier orbital energy and coefficient values, a dominant HOMO diphenylnitrile imine–LUMO dipolarophile interaction should be at work and consequently a prevalence of 4-substituted pyrazolines due to the largest, more-favoured overlap between the anionic end of nitrile imine and β-carbon of the dipolarophile is to be expected. A possible explanation for the formation of 5-substituted pyrazolines in the latter type of 1,3-dipolar cycloadditions, is that steric elements overcome electronic effects. It is in fact well documented that the C atom of nitrile imines is more sensitive than N to the steric requirements of the dipolarophile[18,146,77].

Equations (23)[145], (24)[147], (25)[151,144], (26)[146] and (27)[152] represent examples of regioselective 1,3-dipolar cycloadditions of nitrile imines.

A logical explanation of this regioselectivity is that the coefficients on the two double-bond carbon atoms (with the exceptions of that in equation 25) have similar values[5].

Aromatic dipolarophiles such as furan[153], benzofuran[154] and N-methyl-indole[155] react with nitrile imines to give the adducts shown opposite

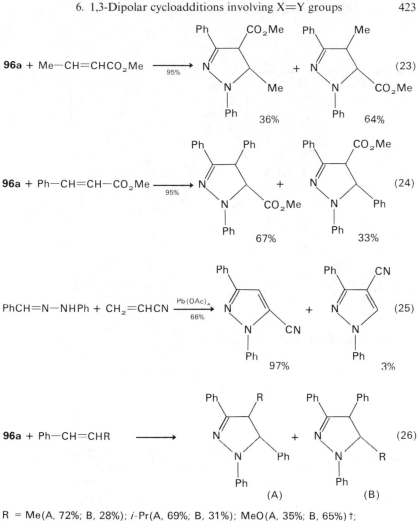

R = Me(A, 72%; B, 28%); *i*-Pr(A, 69%; B, 31%); MeO(A, 35%; B, 65%)†;
   Br(A, 85%; B, 15%)*; NO$_2$(A, 31%; B, 69%);†
   p-MeOC$_6$H$_4$(A ≃ B); *p*-O$_2$NC$_6$H$_4$(A, 65%; B, 35%)

in moderate yields. The high-lying HOMO and the largest HOMO coefficient at position 2 of furan leads to a dominant LUMO dipole–HOMO dipolarophile interaction with the formation of **98**[156].

---

† These compounds are isolated as pyrazoles owing to the easy loss of CH$_3$OH, HBr and HNO$_2$ from the corresponding pyrazolines.

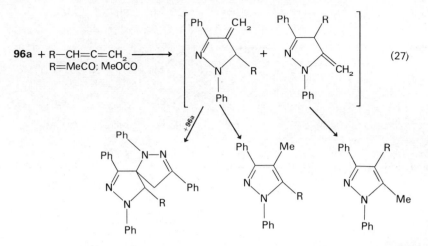

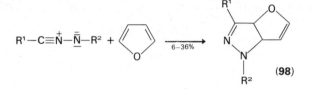

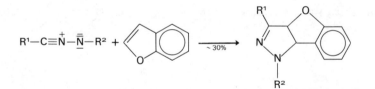

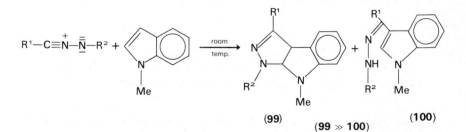

N-Methylindole reacts with nitrile imines to give **99** and the hydrazonic derivatives, **100**. The latter is probably formed by a mechanism analogous to that reported for the reaction of terminal acetylenes with nitrile oxides and nitrile imines which gave a mixture of cyclic compounds, acetylenic oximes and arylhydrazones[157,158].

Diphenylnitrile imine has been reacted with several enynes[159–161]; in every case the double bond was seen to be more reactive than the triple bond and consequently 5-alkynyl-Δ^2-pyrazolines **101** were formed.

In contrast with the latter results is the finding that the *cis-* and *trans-*methoxybutenynes react with diphenyl nitrile imine to give the two regioisomers **102** and **103**[162]. In both reactions it is the terminal multiple bond only which enters the 1,3-dipolar cycloaddition.

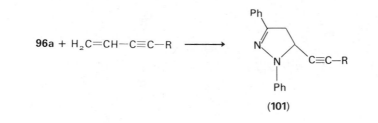

$$\text{96a} + H_2C=CH-C\equiv C-R \longrightarrow$$

(**101**)

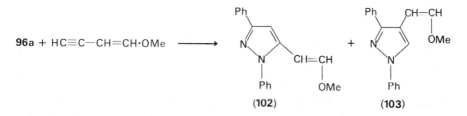

$$\text{96a} + HC\equiv C-CH=CH\cdot OMe \longrightarrow$$

(**102**)   (**103**)

An essential non-selectivity was shown in the reaction of diphenyl-nitrile imine with vinyl alkynyl ketones. Both double and triple bonds were reactive and so pyrazoles as well as pyrazolines were obtained[163].

Interesting results have been found for the reactions of aryl substituted nitrile imines with cyclopropenes[164].

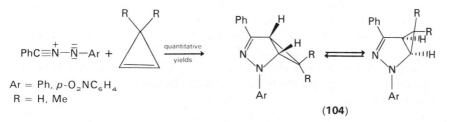

$$\text{PhC}\equiv\overset{+}{N}-\overset{-}{N}-\text{Ar} +$$

$$\text{Ar} = \text{Ph, } p\text{-O}_2\text{NC}_6\text{H}_4$$
$$\text{R} = \text{H, Me}$$

quantitative yields

(**104**)

The adducts **104** display easy ring inversion, manifested by line broadening and coalescence of the R absorption in their n.m.r. spectra at increased temperatures.

Diarylnitrile imines easily react at room temperature with cyclooctatetraene perispecifically to give the tricyclopyrazoline **106** probably through an electrocyclic ring-closure of the primary adduct **105**; with compound **107** they give a mixture of three isomers **108**, **109** and **110**. The formation of the latter three isomers showed the non-selectivity of the cycloaddition of the 1,3-dipole towards three different bonds in the same molecule. Thermal behaviour of adducts **106** and **108–110**, was studied; for instance, adduct **106** gives benzene and 1-phenyl-3-arylpyrazole on heating[165].

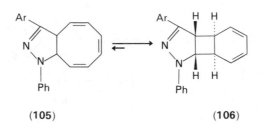

(**105**)   (**106**)

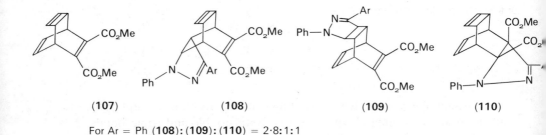

(**107**)   (**108**)   (**109**)   (**110**)

For Ar = Ph (**108**):(**109**):(**110**) = 2·8:1:1

The reactions reported so far for nitrile imines are all examples of $[\pi^4s + \pi^2s]$ cycloadditions. An example of 1,3-dipolar cycloadditions involving the $[\pi^4s + \pi^6s]$ electronic system is the $3 + 6 \rightarrow 9$ reaction of diphenylnitrile imine with 2,4,6-cycloheptatrienone (tropone) (**111**, a planar molecule with a $C_{(2)}$–$C_{(7)}$ distance of about 2·55 Å) to give **112** along with the $3 + 2 \rightarrow 5$ adducts shown below[63,64]. The latter products were formed on dehydrogenation and internal hydrogen transposition of primary adducts.

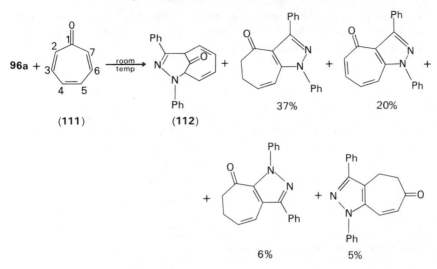

Another example of addition which does not belong to the general classification $3 + 2 \to 5$ has been reported for the reaction of nitrile imines with vinyl derivatives of trivalent phosphorus[166,167]. The dipolarophile is represented by the P—C—C system (see equation 28) which reacts with the nitrile imine according to a $3 + 3 \to 6$ classification. A dipolar intermediate, **113**, has been proposed, which is transformed into the phosphonium salt **114** in the presence of triethylamine hydrochloride.

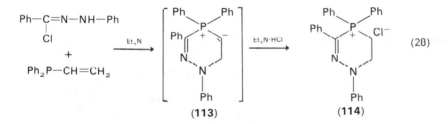

Vinyl derivatives of pentavalent phosphorus, however, undergo 1,3-dipolar cycloadditions with nitrile imines according to the classification $3 + 2 \to 5$ to give 5-substituted-Δ^2-pyrazolines[168].

2. Cycloadditions with hetero double bonds

The reactions of carbon–nitrogen double bonds with 1,3-dipoles have been reviewed by Anselme in a previous volume of this series[169]. Cyclo-additions of nitrile imines with this dipolarophilic system have been shown to be always regiospecific to give 4,5-dihydro-1,2,4-triazoles (115) and 1,2,4-triazoles (116)[142,170-3]. With diphenylcarbodiimide and 96a bisadduct 117 was formed in good yield[174].

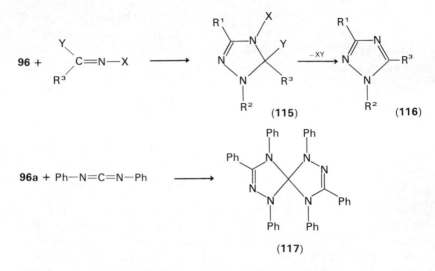

The cycloaddition of diphenylnitrile imine with diazines gave 1,2,4-triazoles and the corresponding 4,5-dihydroderivatives[171].

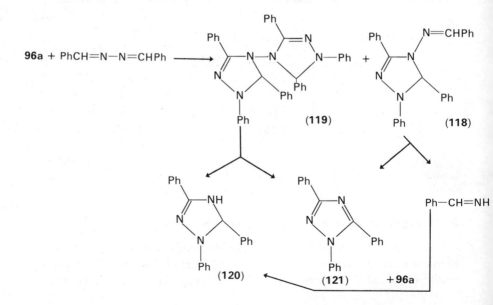

It was proposed that the mono-adduct **118** from diphenylnitrile imine and benzalazine, undergoes a second addition of the same 1,3-dipole to give **119**. Compound **119** gives **120** and **121** through an elimination–disproportion reaction. Alternatively, the adduct **118** could give **121** by elimination of benzaldimine which is then trapped by the nitrile imine to give **120**.

Isocyanates can react with diphenylnitrile imine to give cycloaddition either across the carbon–nitrogen or the carbon–oxygen double bond to give **122** and **123** respectively[170].

Similarly with phenyl isothiocyanate 2,3-dihydro-5-phenylimino-1,3,4-thiadiazole (**124**) and 2,3-dihydro-1,2,4-triazol-5-thione (**125**) have been obtained in 58% and 29% yields respectively[175].

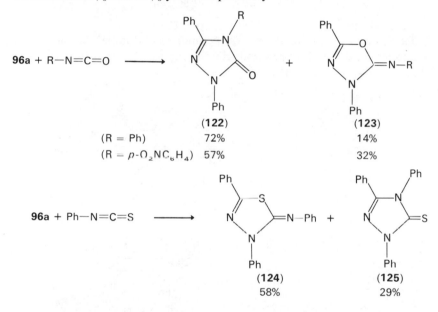

Carbonyl groups of aldehydes and ketones activated by electron-withdrawing groups, are capable of reacting with diphenylnitrile imine to give 2,3-dihydro-1,3,4-oxadiazole derivatives **127**[175].

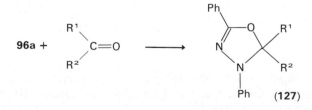

The reactions between diphenylnitrile imine and thioketones[175], thiocarboxylic esters[175], thioamides[175], thiouretanes[175] and carbon disulphide[175,176] are particularly facile and lead to the formation of several 2,3-dihydro-1,3,4-thiadiazole derivatives.

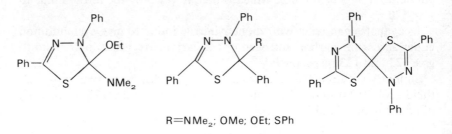

R=NMe₂; OMe; OEt; SPh

Sulphines, which contain two cumulated double bonds, react regiospecifically with diphenylnitrile imine to give 2,3-dihydro-1,3,4-thiadiazolin-S-oxides **128**[177,178].

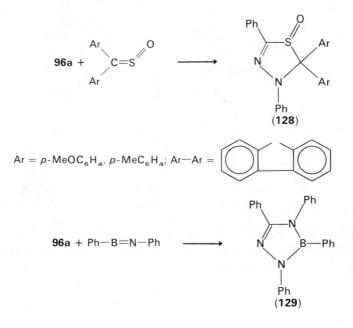

Phenyl-(phenylimino)-borane, an unusual dipolarophile, was found to react with diphenylnitrile imine to give 1,2,4,5-tetraphenyl-1-bora-2,3,5-triaza-Δ^3-cyclopentene (**129**)[179].

Hammerum and Wolkoff[180] investigated the adducts from diphenyl nitrile imine and compounds incorporating C=C, C=O and C=S double bonds by mass spectrometry. The five-membered ring compounds decomposed, upon electron impact, by way of a retro-1,3-dipolar cyclo-addition. In all the studied cases, the ion corresponding to the parent 1,3-dipole was observed.

3. 1,3-Dipolar reactivity of nitrile imines[1]

As shown by the above reported examples, nitrile imines are quite reactive 1,3-dipoles towards numerous dipolarophilic systems. The few kinetic data reported in the literature for heterodipolarophiles are those of benzylidene methylamine, benzaldoxime and benzaldehyde (k relative to styrene: 3·2, 0·060 and 0·032, respectively)[77]. On the other hand, there are many kinetic data on 1,3-dipolar cycloadditions of nitrile imines with differently substituted alkenes. Figure 13 shows the parabola obtained by plotting log k_2 of various reactions against the IP's of alkenes[72].

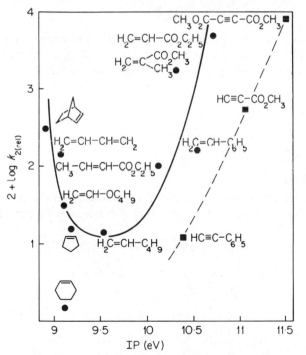

FIGURE 13. Logarithms of the rate constants for cycloadditions of diphenyl nitrile imines (in benzene, 80 °C) with ionization potentials of substituted olefins. [Reproduced with permission from Bast and coworkers, *Chem. Ber.*, **106**, 3312 (1973).]

Common alkenes show low rate constants, while electron-donating as well as electron-withdrawing substituents increase the reactivity of the dipolarophile, electron-withdrawing substituents being more active.

The position of cyclohexene is remarkable: it is quite far from the curve and its low reactivity has also been observed in other 1,3-dipolar cyclo-additions, e.g. with nitrile oxides and azides.

C. Nitrile Oxides [81]

Nitrile oxides **130** are often crystalline isolable compounds, which are available, e.g. by dehydrogenation of aldoximines with a variety of oxidizing agents, by dehydrohalogenation of hydroximic acid halides and from nitroparaffins.

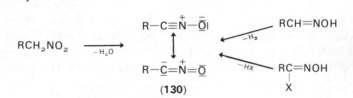

Nitrile oxides, on heating, either rearrange to isocyanates or dimerize to furoxane **131**. The latter reaction is a $3 + 2 \rightarrow 5$ cycloaddition of the nitrile oxides with the carbon–nitrogen multiple bond of the same 1,3-dipole.

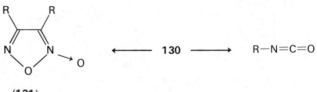

The regiochemistry found in furoxanes is contrary to the principle of maximum gain in σ bond energy[32], but consistent with perturbation theory[5].

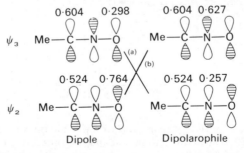

As shown above for acetonitrile oxide frontier orbitals ψ_2 and ψ_3, interactions (a) and (b), in which the greatest overlaps are between C—C and N—O respectively, favour the formation of regioisomer **131**.

It has been found that the rate of dimerization of nitrile oxides is increased by electron-withdrawing substituents and decreased by electron-releasing groups[181,182]. A Hammett-type relationship holding with $\rho = +0.86$ was found for the dimerization of aromatic nitrile oxides. *Ortho*-disubstituted nitrile oxides dimerize slowly (Cl) or not at all (CH$_3$) at room temperature due to steric hindrance.[81].

1. Cycloadditions with C=C

The 1,3-dipolar cycloadditions of nitrile oxides with olefins have been widely studied. Figure 14 and Table 8 show the effect of substituents on the

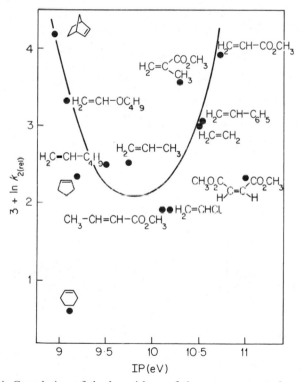

FIGURE 14. Correlation of the logarithms of the rate constants for cycloadditions of nitrile oxides with the ionization potentials of substituted olefins. [Reproduced with permission from Bast and coworkers, *Chem. Ber.*, **106**, 3312 (1973).]

TABLE 8. Rate constants k (l/mol sec) for the cycloadditions of substituted aryl nitrile oxides with styrenes at 25°C in carbon tetrachloride[183]

$X{-}C_6H_4C{\overset{+}{\equiv}}N{-}\overset{=}{\underset{}{O}}|$	$p\text{-}O_2NC_6H_4CH{=}CH_2$ $k \times 10^2$	$C_6H_5CH{=}CH_2$ $k \times 10^2$	$p\text{-}CH_3OC_6H_4CH{=}CH_2$ $k \times 10^2$
X = p-MeO	2·25	0·731	0·856
p-Me	2·42	0·855	0·975
H	2·45	1·10	1·23
p-Cl	2·91	1·66	2·17
m-Cl	3·67	2·20	2·94
m-NO$_2$	4·67	4·38	6·25
	$\rho_X = +0·33$	$\rho_X = +0·79$	$\rho_X = +0·90$

reactivity of nitrile oxides with alkenes[72,183]. The reactivity of alkenes is enhanced to a similar extent both by electron-withdrawing and electron-donating groups.

The positive Hammett ρ values found for substituted nitrile oxides (see Table 8) seem to be evidence for a role of acceptor on the part of nitrile oxide in these 1,3-dipolar cycloadditions[184]†. It is also worth mentioning that while the left side of the parabola corresponds to Type III, LUMO-dipole-controlled cycloadditions, the right side is not formed by Type-I cycloadditions but, as indicated by the regiochemistry of the products (see Section II), represents Type-II dipolar cycloadditions where LUMO dipole–HOMO dipolarophile interaction is still prevalent. Reactivity of nitrile oxides (and other 1,3-dipoles) with electron-poor olefins is greater than expected from frontier-orbital interactions. This was attributed to Coulombic effect, i.e. to electrostatic interaction between 1,3-dipoles possessing excess negative charge on both terminal atoms and the centres of dipolarophiles lacking in electrons, owing to the presence of electron-withdrawing groups.

The general trend in the reactions of mono- and 1,1-disubstituted olefins with nitrile oxides to give the 5-substituted-Δ^2-isoxazolines is again observed (equation 29)[81].

$$\mathbf{130} + H_2C\!=\!C\!\begin{smallmatrix}R^1\\[2pt]R^2\end{smallmatrix} \longrightarrow \qquad (29)$$

Among monosubstituted alkenes only acrylic acid and its esters have been reported to give the two regioisomers with nitrile oxides[186] (cf. Section II). Regioisomers have been more commonly found in cyclo-additions of nitrile oxides with asymmetrically 1,2-disubstituted ethylenes such as α,β-unsaturated ketones[51], esters[185–188], styrenes[20], indenes[189–190], dihydronaphthalene[191] and with furan and benzofuran[192] (equation 30). According to the perturbational approach, two regio-isomers are formed since the two double-bond carbons have similar

---

† The values expressing reactivity of each nitrile oxide with the three styrenes in Table 8 [see also A. Battaglia and A. Dondoni, *Ric. Sci.*, **38**, 201 (1968)] all define a V shaped Hammett relationship. Hence, it would follow a prevalent interaction HOMO dipole–LUMO dipolarophile for *p*-nitrostyrene (positive ρ) and a LUMO dipole–HOMO dipolarophile for *p*-methoxystyrene (negative ρ). This will be clearly in contrast with what is stated above and with the regiochemistry found for the reactions considered. In conclusion, caution should be used when choosing the prevalent frontier orbital interaction on the basis of ρ values only.

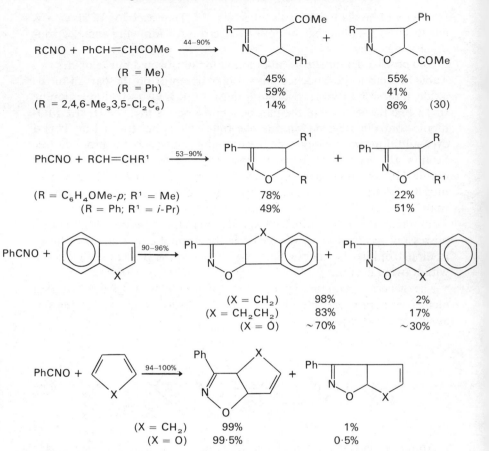

RCNO + PhCH=CHCOMe $\xrightarrow{44-90\%}$

(R = Me)
(R = Ph)
(R = 2,4,6-Me$_3$3,5-Cl$_2$C$_6$)

45%	55%
59%	41%
14%	86% (30)

PhCNO + RCH=CHR^1 $\xrightarrow{53-90\%}$

(R = C$_6$H$_4$OMe-p; R^1 = Me)
(R = Ph; R^1 = i-Pr)

78%	22%
49%	51%

PhCNO +

(X = CH$_2$)	98%	2%
(X = CH$_2$CH$_2$)	83%	17%
(X = O)	~70%	~30%

PhCNO + $\xrightarrow{94-100\%}$

(X = CH$_2$)	99%	1%
(X = O)	99·5%	0·5%

coefficients. Alternatively, two different operating interactions, favouring either of the two possible orientations, can be at work.

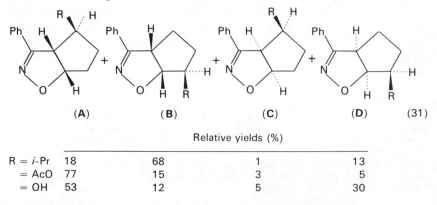

(A)   (B)   (C)   (D)   (31)

Relative yields (%)

R = i-Pr	18	68	1	13
= AcO	77	15	3	5
= OH	53	12	5	30

Caramella and Cellerino[193] tried to evaluate and separate electronic from steric effects by studying the results of the cycloaddition of benzonitrile oxide with 3-substituted cyclopentenes. The reaction gave the two pairs of diastereoisomers **A, C** and **B, D**.

The formation of the two *anti*-regioisomers **A, B** is clearly controlled only by electronic effects. As equation (31) shows, the regiochemistry trend is reversed in passing from electron-donating to electron-withdrawing groups following the inversion of size of the frontier-orbital coefficients of the double bond. The preferred formation of *anti* over *syn* adducts is attributable to steric requirements. The quite large amount of isomer **D**(R = OH) is attributed to the formation of a hydrogen bond between the hydroxyl and the benzonitrile oxide oxygen in the transition state.

Several nitrile oxides react with *cis*-3,4-dichlorocyclobutene giving always a mixture of the *syn* and *anti* adducts **51** and **50** respectively (see Table 9)[53].

TABLE 9. Product ratios in the reaction of various nitrile oxides with *cis*-3,4-dichlorocyclobutene[53]

Nitrile oxide	Total yield (%)[a]	(**50**) (rel. %)	(**51**) (rel. %)
MeCNO	69	31	69
p-MeOC$_6$H$_4$CNO	90 (79)[b]	59 (34)	41 (66)
PhCNO	82 (79)[b]	52 (29)	48 (71)
p-O$_2$NC$_6$H$_4$CNO	84 (78)[b]	31 (19)	69 (81)
2,4,6-Me$_3$C$_6$H$_2$CNO	80	72·5	27·5
2,4,6-Me$_3$-3,5-Cl$_2$C$_6$CNO	88	58	42

[a] For reactions in ether.
[b] Values in brackets for reaction in acetonitrile.

The following main points were derived: (i) steric effect (repulsive) favours the formation of the *anti* isomer although, in the cases of cyclo-additions of *ortho*-disubstituted nitrile oxides the percentages of *syn* isomers were fairly high; (ii) in the series of *para*-substituted benzonitrile oxides, the higher the dipole moment of the 1,3-dipole, the larger is the percentage of the *anti* isomer formed; (iii) increasing solvent polarity favours the *syn* isomer. Furthermore, a through-space bonding interaction shown in **132** between the non-bonding occupied orbitals of chlorine and the LUMO of the nitrile oxides was postulated to explain the high percentage of the *syn* isomer. Alternatively London–van der Waals forces between chlorine atoms and nitrile oxide moieties might account for the same results[53]†.

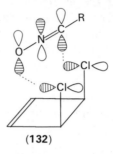

(**132**)

Benzonitrile oxide has been reacted with norbornene[78], apobornene[78] and norbornadiene[196]. The first two dipolarophiles are attacked by benzonitrile oxide only from the side of the methylene bridge of the molecules to give **133a** and **133b** as the sole adducts. These results have been attributed tentatively to the 'torsional effect' and steric hindrance by the 5,6-hydrogens of the two dipolarophiles. In the case of the reaction of benzonitrile oxide with excess norbornadiene, a mixture of the mono-adducts **134** and **135** is obtained in good yield. The *syn*-methylene adduct **134** is the major component in the mixture, being formed in 88–90% yield. Adduct **135**, however was not isolated as a pure compound.

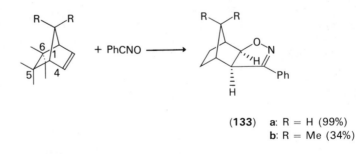

(**133**)   **a**: R = H (99%)
            b: R = Me (34%)

† For other possible explanations see diazoalkanes; for a discussion of the same problem in the Diels–Alder reaction, see References 194 and 195.

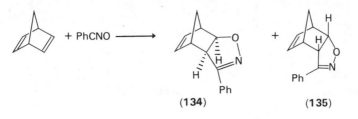

(134)   (135)

Benzonitrile oxide was reported to react with 6-dimethylaminofulvene to give only a compound derived from a [π^4s + π^6s] cycloaddition, and with 6,6-dimethyl- and 6,6-diphenylfulvene to give a mixture of regio-isomeric adducts to the $C_{(2)}$—$C_{(3)}$ double bond (see Section II)[66].

Benzonitrile oxide and mesitonitrile oxide react slowly with tropone at room temperature to give 136 and a complex mixture of compounds derived from the non-isolated adducts 137, 138 and 139[62]. Cycloocta-

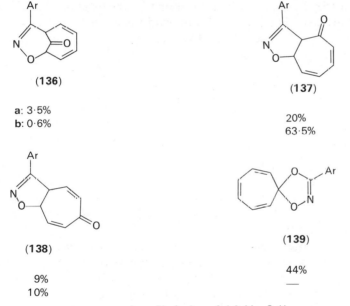

(136)

a: 3·5%
b: 0·6%

(137)

20%
63·5%

(138)

9%
10%

(139)

44%
—

a. Ar = Ph; b: Ar = 2,4,6-Me$_3$C$_6$H$_2$

tetraene readily reacts with nitrile oxides to give the bicyclic adduct 140 stable at low temperatures but rearranging to 141 at temperatures exceeding 5 °C[20,197]. Compound 141 reacts with dimethyl acetylendicarboxylate to give 142 which in turn is readily pyrolysed to 2,3-oxazabicyclo[3.2.0]-hepta-3,6-dienes (143) and dimethyl phthalate[197].

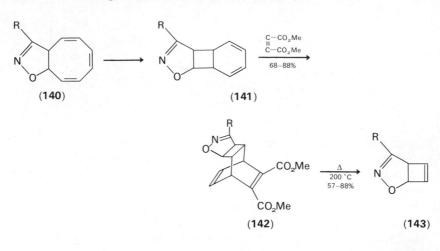

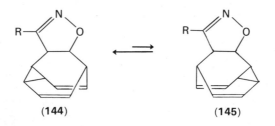

A perispecific cycloaddition was found for the reaction of nitrile oxides with bullvalene. The $3 + 2 \rightarrow 5$ monoadducts formed were shown to be a mixture of two valence isomers **144** and **145**[198].

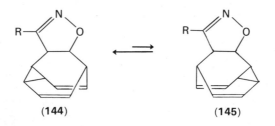

Interconversion between the two isomers is rapid at room temperature on the n.m.r. scale. In the equilibrium state isomer **144** is always prevalent, especially when R is a bulky group.

The mass spectra of several 3-aryl-Δ^2-isoxazolines have recently been studied. The isoxazolinic ring decomposed upon electron impact by way of competitive processes. Among these processes a retro-1,3-dipolar cycloaddition going through the nitrile oxide and ethylene radical ions, was observed[199,200].

2. Cycloadditions with C=O and C=S

The cycloaddition of aromatic nitrile oxides generated *in situ* from hydroximic acid chlorides and triethylamine in the presence of carbonyl compounds gives 1,3,4-dioxazole derivatives **146** which are otherwise difficult to prepare[81].

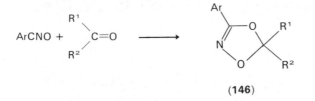

(146)

Aromatic aldehydes[201], chloral[201], glyoxylic esters[201], α-diketones[201], α-ketoesters[201] and quinones[202,203] were found to be very reactive carbonyl dipolarophiles. Although carbon–carbon double bonds are generally more reactive than carbon–oxygen double bonds, in the case of o-naphthoquinone the hetero double bond was found the most active.

Less reactive aldehydes and ketones have their reactivity enhanced by the formation of a complex with BF_3[204].

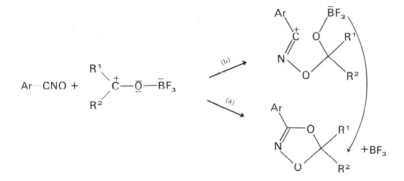

The higher reactivity of the carbonyl BF_3 complex is attributable to a lower LUMO carbonyl energy compared to that of the free carbonyl group, which gives rise to a stronger HOMO dipole–LUMO dipolarophile interaction. A dipolar intermediate such as that depicted above (path b) cannot be dismissed on the basis of present knowledge of the reaction.

The carbonyl groups of tropone[62] and diphenylcyclopropenone[204a] also react with benzonitrile oxide; in the case of the former, a non-concerted reaction mechanism has been postulated.

As previously reported for Δ^2-isoxazolines, also 1,3,4-dioxazoles undergo a retro-1,3-dipolar cycloaddition upon electron impact[204b].

An easy route to 1,4,2-oxathiazole derivatives is the cycloaddition of aromatic nitrile oxides with carbon–sulphur double bonds of thioketones, dithiocarboxylic and thioncarboxylic esters, thioamides, trithiocarbonates and thiocarbonic acids, to give 147[201a,205]. The latter compounds are fairly stable, but decompose on heating to give isothiocyanates and the carbonyl compounds corresponding to the starting thiocarbonyl dipolarophiles.

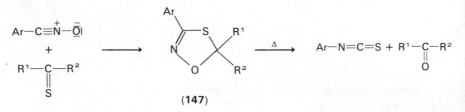

(147)

The two cumulated double bonds of carbon disulphide react with mesitonitrile oxide to give the spiro compound **148** which spontaneously decomposes to give 3-mesityl-1,4,2-oxathiazoline-5-one (**149**) and mesityl isothiocyanate[206]. Arylnitrile oxide reacts with thioketenes on their carbon–sulphur double bond to give **150**[207].

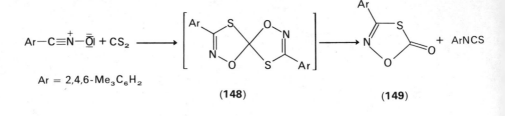

(148)   (149)

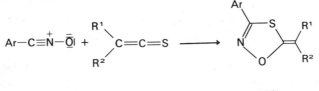

(150)

3. Cycloadditions with C=N, N=N, N=S, S=O, N=B, C=P and N=P

The carbon–nitrogen double bond of different classes of compounds reacts with nitrile oxides to give Δ^2-1,2,4-oxadiazoline derivatives[81]. Equations (32)[208,209], (33)[210], and (34)[211] serve as examples of such a reaction.

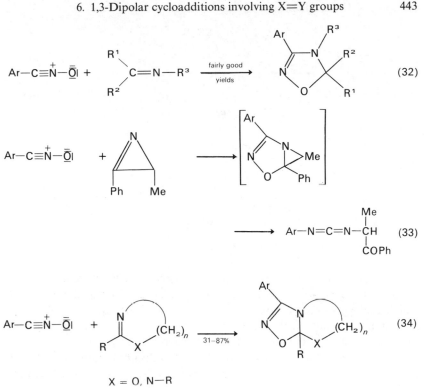

$$Ar-C\equiv\overset{+}{N}-\overline{\underline{O}}| + \underset{R^2}{\overset{R^1}{\diagdown}}C=N-R^3 \xrightarrow[\text{yields}]{\text{fairly good}} \tag{32}$$

$$Ar-C\equiv\overset{+}{N}-\overline{\underline{O}}| \quad + \quad \triangle \quad \longrightarrow \tag{}$$

$$\longrightarrow \quad Ar-N=C=N-\underset{\underset{COPh}{|}}{\overset{\overset{Me}{|}}{CH}} \tag{33}$$

$$Ar-C\equiv\overset{+}{N}-\overline{\underline{O}}| \quad + \quad \underset{R \quad X}{\diagup}(CH_2)_n \xrightarrow[31–87\%]{} \tag{34}$$

$$X = O, N-R$$

The less-reactive carbon–nitrogen double bond of oximes has enhanced reactivity when complexed with ethereal BF_3[212]. The reaction of the latter complexes of ketoximes and aldoximes with nitrile oxides represents the only way known for the synthesis of 4-hydroxy-1,2,4-oxadiazolines **151**.

The carbon–nitrogen double bond of Δ^2-isoxazolines[191] and Δ^2-pyrazolines[213] was found to be a reactive dipolarophile towards nitrile oxides yielding **152** and **153** respectively.

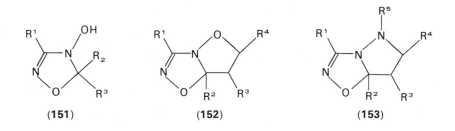

(151)   **(152)**   **(153)**

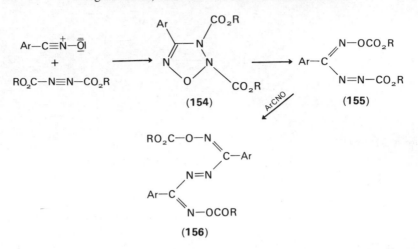

(154)   (155)

(156)

Dimethyl and diethyl azodicarboxylates react with aryl nitrile oxides to give the unstable intermediate 4-aryl-Δ^4-1,2,3,5-oxatriazoline (154) which undergoes ring cleavage followed by rearrangement to 155 which subsequently reacts with the nitrile oxide to give 156[214,215]. In some cases the primary cycloadducts 154 have been isolated[215].

The heterodipolarophilic system of the N-sulphinylamines reacts with nitrile oxides to give 2-oxo-1,2,3,5-oxathiadiazoles 157[57]. Furthermore, nitrile oxides add to the sulphur–oxygen double bond to give 1,3,2,4-dioxathiazole-S-oxides 158[216].

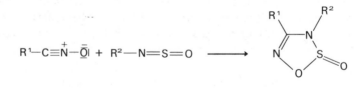

(157)

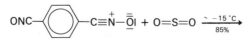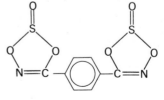

(158)

The unusual heterocycle 1,3,5,2-oxadiazaboroline **160** is readily obtained by cycloaddition at $-25\,°C$ of aryl nitrile oxides with boroimides **159**[217].

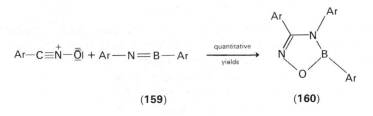

(159)   (160)

A concerted 1,3-dipolar cycloaddition has been proposed for the reaction of nitrile oxides with phosphorus ylides[218,219] and phosphorus imines[220]. The former dipolarophiles react with nitrile oxides to give the cycloadducts **161** which in some cases are isolable, the latter giving rise to the formation of triphenylphosphine oxide and carbodiimides, isolated as disubstituted ureas, through the unstable intermediate **162**.

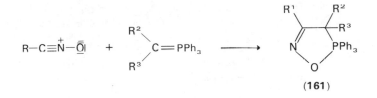

(161)

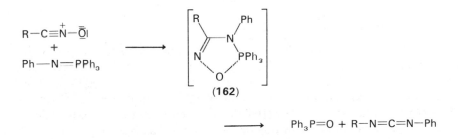

(162)

$$\longrightarrow \quad Ph_3P{=}O \;+\; R{-}N{=}C{=}N{-}Ph$$

D. Diazo Compounds

1. Structure

The molecular structure of diazomethane, which is the most studied diazoalkane, has been determined by microwave spectroscopy[221]. Bond angles and lengths are shown in formula **163**.

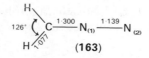

(163)

The geometrical data together with the low dipole moment[221] ($\mu = 1\cdot5$ D) of the molecule agree with the representation of the molecule by canonical structures with the charges distributed on the three positions of the 1,3-dipole. Theoretical calculations of charge distribution carried out for diazomethane and several other diazoalkanes (Table 10) show that the larger negative charge is on the carbon atom.

TABLE 10. Charge distribution on various diazo compounds:

$$\begin{array}{c} R^2 \\ \diagdown \\ \diagup \\ R^1 \end{array} C-N_{(1)}-N_{(2)}$$

R^2	R^1	qC^a	$qN_{(1)}{}^a$	$qN_{(2)}{}^a$	Ref.
H	H	$-0\cdot26$	$+0\cdot31$	$-0\cdot14$	4
Me	H	$-0\cdot4221$	$+0\cdot6825$	$-0\cdot2605$	222
Ph	H	$-0\cdot3613$	$+0\cdot7172$	$-0\cdot0980$	222
EtOCO	H	$-0\cdot4373$	$+0\cdot7347$	$-0\cdot0891$	222
Me	Me	$-0\cdot3648$	$+0\cdot6612$	$-0\cdot2964$	222
Ph	Ph	$-0\cdot3633$	$+0\cdot7170$	$-0\cdot0748$	222

$^a q$ = total charge.

2. Cycloadditions with C=C

a. *General features.* The charge distributions reported in Table 10 for some diazo compounds confirm an earlier opinion that diazoalkanes add to alkenes through an initial nucleophilic attack[223], e.g. the carbon of diazomethane adds to the β position of α,β-unsaturated esters to give the Δ^2-pyrazoline-3-carboxylate (**164**). More recently, however, the formation of the two possible regioisomeric pyrazolines has been reported[224,225] e.g. methyl *p*-nitrocinnamate reacts with diazomethane to give **165** and

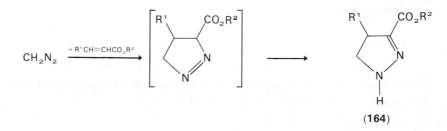

(**164**)

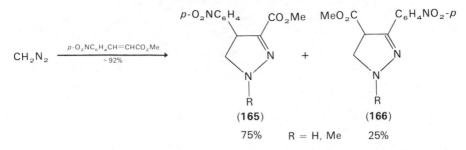

$$CH_2N_2 \xrightarrow[\sim 92\%]{p\text{-}O_2NC_6H_4CH=CHCO_2Me}$$

(165)   (166)
75%   R = H, Me   25%

166. The nucleophilic character of diazoalkanes does not mean that their cycloadditions proceed through a completely defined dipolar intermediate. Although a two-step reaction mechanism has been invoked in few cases[226,227], entropy and enthalpy values[43,228], solvent effect[229,32] and the stereospecificity found for all the reactions studied are in favour of a concerted mechanism.

b. *Reactivity.* Cycloadditions of diazoalkanes belong to Type I, where the most important interaction is HOMO dipole LUMO dipolarophile[4,5]. Diazoalkanes therefore behave as donors and the alkenes are the acceptors; the carbon atom of the 1,3-dipole represents the most nucleophilic centre. It follows that cycloadditions are favoured by electron-releasing groups on the 1,3-dipole and electron-withdrawing groups on the dipolarophile. In full agreement with this is the scale of reactivity found for several diazo compounds[5]: $(Me)_2CN_2 > MeCHN_2 > CH_2N_2 \gg Ph_2CN_2 > H_5C_2OCOCHN_2$.

The experimental reactivity scale has further support in the first values for ionization potential found for diazomethane (IPV = 9·0 eV) and dimethyldiazomethane (IPV = 7·9 eV)[230]. The smaller dipolarophilic activity of alkenes such as acenaphthylene and styrene compared with those of dimethyl fumarate and maleic anhydride for cycloadditions with diphenyldiazomethane[1] confirms the acceptor role of alkenes. Further support is found in cycloadditions of 2,2,2-trifluorodiazoethane which are slow with ethylene but moderately accelerated with electron-poor olefins and decelerated with alkyl-substituted olefins[231].

A significant feature of this reaction is the formation of two diastereoisomers such as **167** and **168**. These are formed through two different

$$R=H; Me; CF_3$$
$$CF_3-CHN_2 + R-CH=CH_2 \xrightarrow{83-100\%}$$

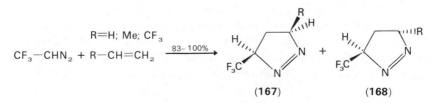

(167)   (168)

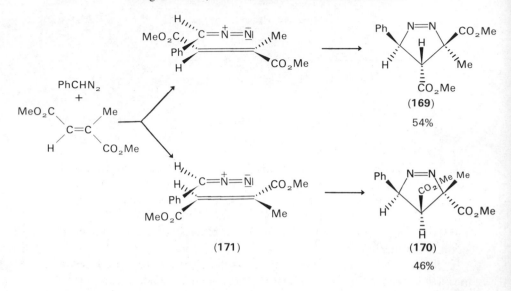

approaches of the diazoalkane to the olefin as also proposed for the reaction of phenyldiazomethane with methyl mesaconate to give the Δ^1-pyrazolines **169** and **170** in nearly equal yields[58]. This suggests that in the 'cisoid complex' **171**, π overlap on the one hand and steric effects between the phenyl group of the 1,3-dipole and the ester group of dipolarophile on the other are balanced.

A mixture of diastereoisomeric pyrazolines was also obtained in the reaction of *p*-methoxyphenyldiazomethane with *p*-methoxystyrene[226]. In sharp contrast with the most widely accepted mechanism for 1,3-dipolar cycloadditions (see Section II), this result was rationalized as being due to a zwitterionic intermediate leading to two epimeric final adducts.

A kinetic study in DMF for the cycloadditions of diazomethane with substituted styrenes gave a value of $\rho = +1\cdot31$[229] (increased reactivity with electron-withdrawing groups) which again should be direct evidence for an operating HOMO dipole–LUMO olefin interaction[184]. The alternative frontier-orbital interaction LUMO dipole–HOMO olefin is probably of importance in the cycloaddition of electron-poor diazo compounds and electron-rich dipolarophiles[5], e.g. in the reaction of acetyldiazomethane with pyrrolidinocyclohexene to give **172**[232].

A strong dependence of reactivity on pressure has recently been reported for reactions of diazomethane with various dipolarophiles. Stilbenes, benzylidenanilines and α-phenyl cinnamates have enhanced reaction rates when the reactions are carried out under high pressure (5000 atm)[232a].

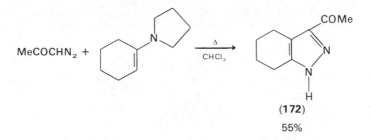

(**172**)

55%

c. *Regio-, peri- and* syn–anti *isomerism.* The regiochemistry found in reactions (35)[233], (36)[234], and (37)[234] is fully rationalized in the perturbational approach considering the data reported in Table 11.

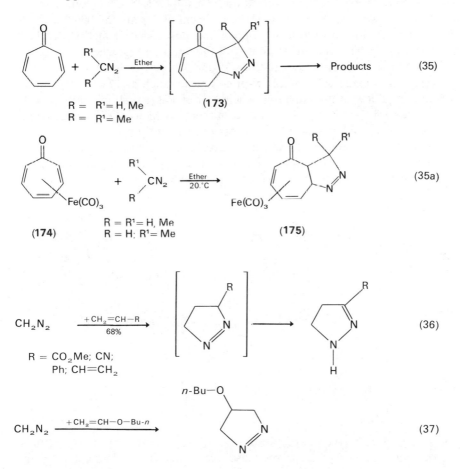

TABLE 11. LUMO coefficients (C) for some dipolarophiles

Dipolarophile	C_α	C_β	Ref.
Tropone	-0.52	$+0.23$	70
Methyl acrylate	-0.414	$+0.607$	11
Butadiene	-0.427	$+0.564$	11
Methoxyethylene[a]	-0.720	$+0.673$	11

[a] Coefficients of n-butoxyethylene are similar to those of methoxyethylene.

Table 11 shows that the two α,β-unsaturated carbonyl compounds, methyl acrylate and tropone, have the absolute values of LUMO coefficients reversed at the α,β positions. This gives rise to a stronger interaction between the largest HOMO coefficient of the 'anionic' centre of the 1,3-dipole with position α of tropone and position β of methyl acrylate respectively. This is in agreement with the opposite regiochemistry found for the two reactions (see equations 35 and 36). Also worthy of note is the reaction of diazo compounds with complexes such as **174** to give stable adducts **175** with the same regiochemistry of **173** (equation 35a)[235].

The perfect correspondence between theory (which considers electronic factors only) and results in the case of the cycloadditions of diazomethane is not always applicable to substituted diazoalkanes. Dimethyldiazomethane, in fact, reacts with methyl acrylate to give **176** whilst methyl-β-t-butylacrylate gives **177** with a reversed regiochemistry[236].

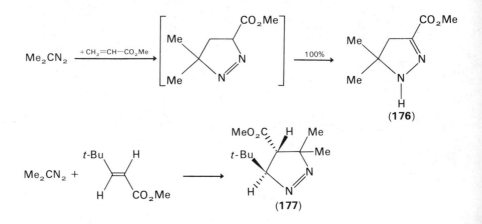

The normal direction of addition to α,β-unsaturated nitro compounds can also be reversed, in the case of disubstituted diazoalkanes, by encumbering β-substituents. The latter results have been attributed to steric effects[236].

The results from reactions of diazo compounds with thiete sulphone are difficult to account for. Both diazomethane and dimethyldiazomethane react with thiete sulphone to give mixtures of the two regioisomers. As expected the dominant pyrazolines were those resulting from attack by the anionic centre of the 1,3-dipole on the β-position of sulphone as was observed in the case of the cycloadditions of diazomethane with divinyl and phenyl vinyl sulphone[236a].

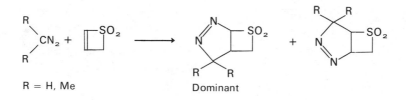

R = H, Me   Dominant

Surprisingly, an opposite regiochemistry was observed in the regio-specific cycloadditions of the below reported diazoalkanes with the same thiete sulphone. The 'wrong' isomer was also dominant in the cyclo-

R_1 = Ph, p-MeOC$_6$H$_4$, Me, Ph, Ph, Ph
R_2 = H, H H, D, Ph, Me

additions of diazomethane to methyl and phenyl-β-styryl sulphones[236a].

The reaction of diazomethane with dimethylfulvene gives the sole $[\pi^4s + \pi^6s]$ adduct (see Section II.K) while reaction between diazomethane and diphenylfulvene gives only a $[\pi^4s + \pi^2s]$ adduct[65].

The different perispecificity shown by the two latter dipolarophiles is probably due to a prevalence of electronic and steric effects respectively.

A further example of the interplay of electronic and steric effects is given by the reactions of diazoalkanes with cyclobutene and norbornadiene derivatives[237].

The results shown in equation (38) were rationalized by considering a secondary stabilizing interaction in the transition state between the empty p-orbital on the terminal nitrogen of the bent 1,3-dipole and the lone-pair of the chlorine atom[238]. The explanation illustrated by **179** is

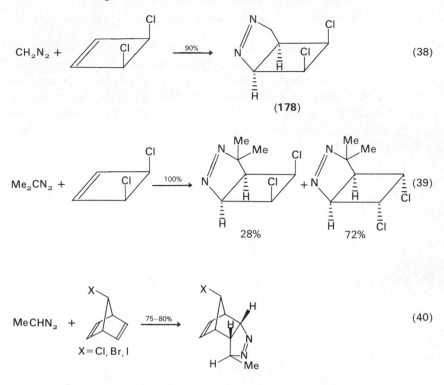

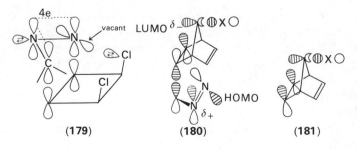

valid in this case, even though liable to criticism, but breaks down for cycloadditions of *cis*-3,4-dichlorocyclobutene with 1,3-dipoles not having orthogonal double bonds, e.g. nitrones (see Section V.C.).

The problem received further attention recently and a more sophisticated reaction mechanism was proposed implying a so-called 'π–σ*' effect'[239] which explains the results of equations (38) and (40). Formulae **180** and **181** serve to clarify the manner of operation of this effect. Formula

181 shows that interaction between σ^*_{C-Cl} and π orbital brings about an electron deficiency in the side of the $C_{(5)}-C_{(6)}$ double bond situated *anti* to methylene. This face of the molecule, therefore, will be more easily attacked by those 1,3-dipoles, such as diazoalkanes, the cycloadditions of which belong to Type I. In addition, interaction $\sigma^*_{C-Cl} - \pi^*_{C_5-C_6}$ shown in **180** allows a better dispersion of the partial negative charge arising on the dipolarophile in the oriented complex.

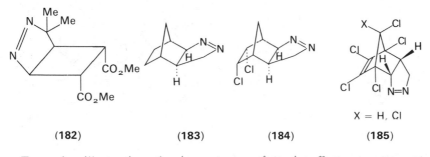

(182)   (183)   (184)   (185)

Examples illustrating the importance of steric effects on *syn–anti* isomerism in the reactions of diazoalkanes are given by the sole formation of adducts **182**[237], **183**[240], **184**[241] and **185**[242] and those in equation (39) from related addends.

3. Cycloadditions with C=N

The reaction of the hetero-double bond C=N with diazoalkanes gives 1,2,3-triazolines[243–245]. These can be formed alternatively and less easily by cycloaddition of azides with alkenes. Cycloadditions of diazomethane with several Schiff bases give the triazolines **186**. The regiochemistry of the latter compounds is in contrast with the 'maximum σ gain' principle postulated by Huisgen[32] but in agreement with the perturbational approach. In the interaction HOMO dipole–LUMO dipolarophile the greatest overlap is between the 1,3-dipole carbon and Schiff-base carbon[5].

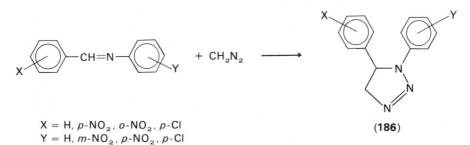

X = H, p-NO$_2$, o-NO$_2$, p-Cl
Y = H, m-NO$_2$, p-NO$_2$, p-Cl

(186)

Diazomethane reacts with 2-aza-1,3-butadiene **187** specifically to give the mono-adduct **188** and the bis-adduct **189**. There is no trace of the pyrazoline which would be expected to arise from an attack by diazomethane on the carbon–carbon double bond only[246].

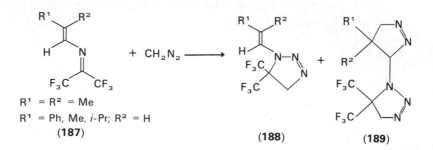

R^1 = R^2 = Me
R^1 = Ph, Me, *i*-Pr; R^2 = H
(**187**)   (**188**)   (**189**)

4. Cycloadditions with C=O

The most documented reaction of diazoalkanes with carbonyl compounds is that with aldehydes and ketones to give mixtures of epoxides and homologous carbonyl compounds[247]. The ester group was also found to be reactive[248].

The reaction mechanisms postulated are described below for the reaction of diazomethane with acetone.

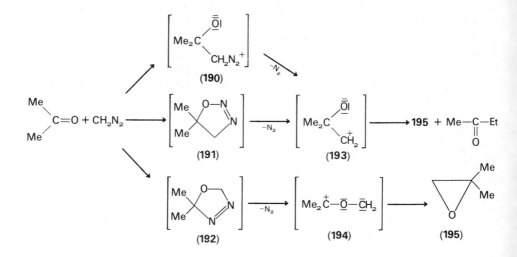

The most widely accepted mechanism is that implying a nucleophilic attack by the diazoalkane onto the carbonyl carbon to give epoxide **195** and 2-butanone through **190** and **193**[247]. Alternative pathways consider the formation of true cycloadducts such as **191** and **192** transforming into the products through the dipolar intermediates **193** and **194**, respectively[249].

Reaction of ethyldiazoacetate with pentafluoronitroacetone gave a stable Δ^3-1,3,4-oxadiazoline of the type **192**, not isolated in the other reported cases, which, surprisingly, is transformed into the corresponding epoxide only if heated above 170 °C[250].

p-Benzoquinones can react with diazoalkanes either at C=O or at C=C double bond.

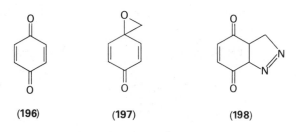

(196)   **(197)**   **(198)**

X = F, Cl, Br, I, OMe, OCOMe

Epoxides **197** were obtained when diazomethane was reacted with *p*-benzoquinones **196**[251]. The C=C double bond was found more reactive than the hetero double bond in the case of *p*-benzoquinone, tetramethyl benzoquinone[252a], 2-amino-*p*-benzoquinone and some 2,5- or 2,6-disubstituted benzoquinones with the formation of adducts of the type **198** which can successively enter other reactions[252]. The C=O double bond of 1,4-naphthoquinone was also found reactive with diazoalkanes to give epoxides[252].

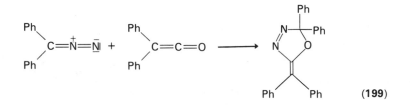

(199)

In agreement with the predicted higher reactivity of the carbon–oxygen compared with the carbon–carbon double bond of ketene for HOMO 1,3-dipole controlled cycloadditions[137] is the sole formation of the 1,3,4-oxadiazoline **199** from the reaction of diphenyldiazomethane with diphenylketene[253–254].

5. Cycloadditions with C=S

Diazoalkanes react with carbon–sulphur double bonds to give 1,2,3-thiadiazoline and 1,3,4-thiadiazoline derivatives[255–262] as shown by equations (41)[256], (42)[257], (43)[260], (44)[261]. Δ^3-1,3,4-Thiadiazolines lose nitrogen on heating to give thiiranes[255–257] (equation 41).

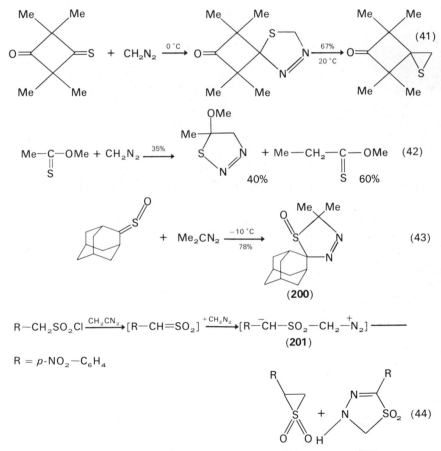

The coefficients of carbon and sulphur in the frontier orbitals of thio-ketones are similar and the formation of mixtures of regioisomers in their 1,3-dipolar cycloadditions with diazo compounds is not therefore sur-prising, e.g. the reaction of adamantanethione with diazomethane (see Section II.E). Sulphines are somewhat less reactive dipolarophiles than are thioketones. In contrast with the parent thioketone, adamantan-ethione S-oxide gives no reaction with diazomethane while it reacts with 2-diazopropane to give 1,3,4-thiadiazoline-1-oxide **200**[260].

Sulphenes were shown to be quite reactive dipolarophilic systems; they react with diazoalkanes to give 1,3,4-thiadiazoline-1,1-dioxide **202** and the unstable episulphide-1,1-dioxide **203**. The authors assumed the dipolar intermediate **201**.

Equations (45) and (46) are examples of 1,3-dipolar cycloadditions of diazoalkanes with cumulated hetero double bonds.

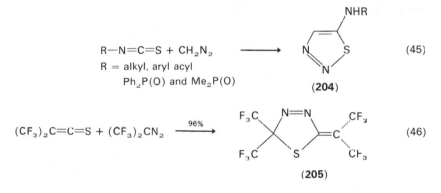

$$R-N=C=S + CH_2N_2 \longrightarrow \text{(204)} \tag{45}$$

R = alkyl, aryl acyl
Ph₂P(O) and Me₂P(O)

$$(CF_3)_2C=C=S + (CF_3)_2CN_2 \xrightarrow{96\%} \text{(205)} \tag{46}$$

As shown by the final adducts **204** and **205**, the carbon–sulphur double bond is more reactive than either the carbon–nitrogen[263-266] or the carbon–carbon double bond[262].

E. Azides

1. Cycloadditions with alkenes

The structural aspects and chemistry of azides **206** have recently been comprehensively reviewed[267-269].

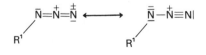

(206)

Azides readily undergo addition at room temperature to nor-bornene[54,270] and 7,7-dimethylene norbornene derivatives[271] on the less hindered *syn*-methylene side of the molecule. 7,7-Dimethyl[272] and 7,7-tetramethylene norbornene[271], which present steric hindrance against *syn*-methylene attack, do not react. Cycloaddition of phenyl azide with norbornadiene (at 65 °C) gives both *syn*- and *anti*-methylene adducts **207** and **208** respectively (ratio **207**:**208** = 11:1)[273]; with 7-*t*-butoxynorborna-diene **209** (at room temperature) a mixture of **211**, **212** and **213** was detected[274]. The high yield of **211** indicates that the HOMO in **209** is mainly localized on the *anti*-methylene side of the $C_{(2)}$—$C_{(3)}$ double bond. This orbital state may be due to two through-space operating interactions: the first between the oxygen sigma lone pair and the double bond $C_{(2)}$—$C_{(3)}$[275,276], the second between σ^*_{C-O} and $C_{(5)}$—$C_{(6)}$ double bond[239] (see **210** and **181**). The former interaction raises, and the latter lowers the energy levels of electrons of $C_{(2)}$—$C_{(3)}$ and $C_{(5)}$—$C_{(6)}$ double bonds respectively.

(207)

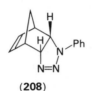

(208)

(209)

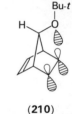

(210)

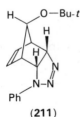

(211)

Relative yield: 55%

(212)

30%

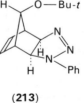

(213)

15%

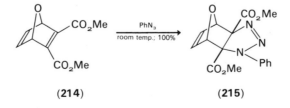

(214)   **(215)**

Interestingly, oxanorbornadiene **214** reacts at room temperature with phenyl azide only at the tetrasubstituted double bond to give the non-isolable adduct **215** which then undergoes a retro-Diels–Alder reaction to give furan and 1-phenyl-4,5-dicarbomethoxy-1,2,3-triazole[277].

Cyclobutenes are not very reactive towards azides[74] but hexafluoro[278] and hexamethyl-Dewar-benzene[279], on heating, add to aryl azides to give fairly good yields of *syn*-fluoro- or *syn*-methyl-Δ^2-1,2,3-triazolines **216**.

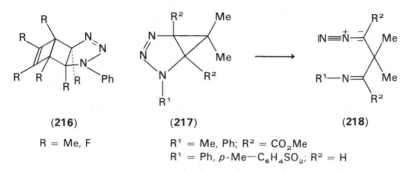

(216)   **(217)**   **(218)**

R = Me, F

R¹ = Me, Ph; R² = CO₂Me
R¹ = Ph, *p*-Me—C₆H₄SO₂; R² = H

3,3-Dimethylcyclopropenes were found to react smoothly with azides to give unstable adducts **217** which undergo a retro-1,3-dipolar cyclo-addition to diazoimino compounds **218**[280,281].

Very long reaction times are required for the reactions of aryl azides at room temperature with unactivated linear olefins (up to some months) and with dienes (several days)[282]. The reactions are regiospecific and give 5-substituted-Δ^2-1,2,3-triazolines, e.g. **219**, **220** and **221**. The less-hindered double bond was involved in the reaction with isoprene to give only **221**[282].

Again, regiospecificity was observed with aryl azides reacting with tetramethylallene to give triazolines **222**[283].

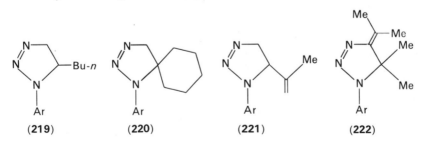

(219)   **(220)**   **(221)**   **(222)**

Phenyl azide reacts with styrene[28] as well as *cis*- and *trans*-β-methyl-styrenes[284] to give only 1,5-diphenyl-Δ^2-1,2,3-triazolines whereas *p*-nitrophenyl azide with styrene gives a mixture of the two regioisomers[85].

Mixtures of the two possible regioisomers have also been evidenced in the reaction of aryl azides with homodienes[285] and in the reaction, at high temperatures, of azide **223** with dihydronaphthalene and 1,1-diphenyl-ethylene[286].

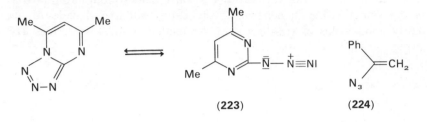

(223)   (224)

The low reactivity of unactivated double bonds make possible the synthesis of olefinic azides[287,288]. The most investigated have been vinyl azides[267,289], e.g. **224**, which were shown to enter 1,3-dipolar cyclo-additions with active methylene compounds[290,291], acetylenic derivatives[292,293] and α-ketophosphorus ylides[294] to give $N_{(1)}$-vinyltriazoles.

A disadvantage in the synthesis of Δ^2-1,2,3-triazolines is their thermal instability. On heating above 100 °C, most of the Δ^2-triazolines give aziridines and anils through the dipolar intermediate **225**[295].

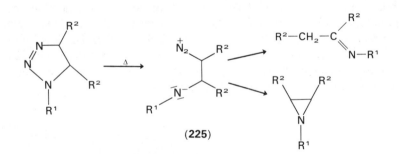

(225)

Δ^2-Triazolines are very unstable when R^1 is an electron-withdrawing group: primary cycloadducts from picryl azide[47,296] (for an exception see ref. 283) cyanogen azide[297,298] and sulphonyl azides[268,299,300] have never been isolated. In the latter instances, kinetic studies and the structure of decomposition products proved that the first reaction step is a true regiospecific 1,3-dipolar cycloaddition (with the exception of *quasi*-symmetrically substituted double bonds)[297] with the electrophilic azide end attacking the less-substituted carbon atom of the double bond.

2. Cycloadditions with enamines, vinyl ethers and azomethines

The electron-rich double bonds of enamines[268,301-305], vinyl ethers[24,268,306-308], and ketene dialkylketals[309] readily react with azides in a stereospecific and regiospecific manner to give 5-amino- and 5-alkoxy-Δ^2-1,2,3-triazolines (see equations 1 and 47). Enamines were found to be more reactive than vinyl ethers, and azides with electron-withdrawing substituents were more reactive than those with electron-releasing substituents (see Table 13)[42].

n-Butyl azide does not react with n-butyl vinyl ether even at 100 °C[310].

The two double bonds of 1-diethylaminobutadiene show quite different reactivity, the double bond bearing the amino group being more reactive[311,312].

The reaction of p-nitrophenyl azide with the mixture of equilibrating enamines **226** and **227** (equation 47) is of interest[304].

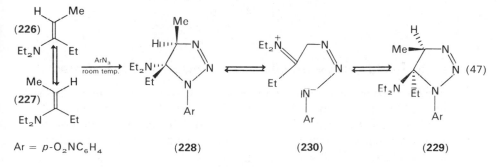

Ar = p-O$_2$NC$_6$H$_4$ (228)   (230)   (229)

At room temperature and in benzene the *trans*-enamine **226** was found to react faster than the *cis* isomer **227** and only the triazoline **228** was formed. The latter however, equilibrates quickly with epimer **229** through the dipolar intermediate **230** (ratio, **228:229** = 55:45 in CHCl$_3$).

The carbon–nitrogen double bond of azomethines is not reactive towards azides.† Azomethines with an α-methylene group, however, can equilibrate with the tautomeric enamine form, which readily reacts with aryl azides in anhydrous chloroform to give 5-amino-Δ^2-1,2,3-triazolines (equation 48)[313-315].

Triazolines **231** are generally unstable: when R^1 is an acyl, carboalkoxy, or aryl group, they spontaneously lose an aliphatic amine molecule to give 1-aryl-1,2,3-triazoles (**232**); when R^1 is H or alhyl, intermediate **231** gives

---

† Note that Δ^2-tetrazolines undergo a retro-1,3-dipolar cycloaddition to azides and azomethines on heating or on electron impact. See T. Isida, T. Akiyama, N. Mihara, S. Kozima, and K. Sisida, *Bull. Chem. Soc. Japan*, **46**, 1250 (1973).

rise (spontaneously or on heating) to a 1-alkyl-1,2,3-triazole **233** and an aromatic amine. This rearrangement is catalysed by TsOH[313,314].

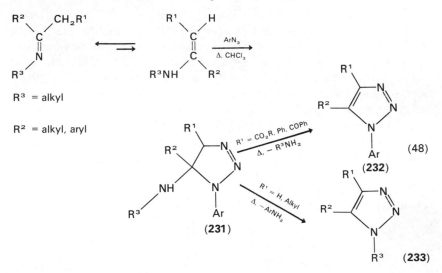

R³ = alkyl

R² = alkyl, aryl

(48)

(232)

(231)

(233)

3. Cycloadditions with conjugated, electron-poor olefins

N-Butyl azide reacts readily with excess ethyl acrylate at room temperature (24 h) in a regiospecific manner to give the 4-carboethoxy-Δ^2-triazoline **234**. This compound can be isolated in the pure state but, on standing, reaches an equilibrium with 3-*n*-butylamino-2-diazopropionate (**235**). If a mixture of *n*-butyl azide and excess ethyl acrylate is left standing for a longer time, a 95% yield of **236** is isolated[310].

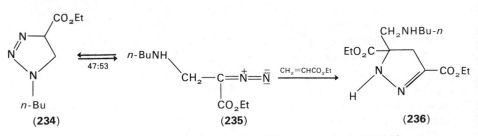

(234)   (235)   (236)

Similarly azides add to acrylamide[310], acrylonitrile[310,316], methyl acrylate[316–318], methyl crotonate and ethylidenacetone[316] regiospecifically to give adducts which show base-catalysed ring opening (e.g. Et₃N or triazoline itself) to diazo derivatives.

Reactions of aryl azides with ethylidene and arylidenemalonic esters and nitriles, α-acylacrylic esters, benzylidenacetone, methyl cinnamates and cinnamonitriles have been studied recently[319—321]. All these compounds showed a low dipolarophilic activity and the cycloadditions were stereospecific and regiospecific, with the anionic nitrogen of the azide attacking the β-position of the dipolarophile with respect to the electron-withdrawing group. Electron rich azides were more reactive than electron-poor ones.

Dipolarophiles having two substituents with opposite electronic effects at the same carbon, react with azides to give mixtures of regioisomers. Equation (49) shows some results for phenyl azide[310]. Aziridine **237** is formed from the unstable triazoline (**239**).

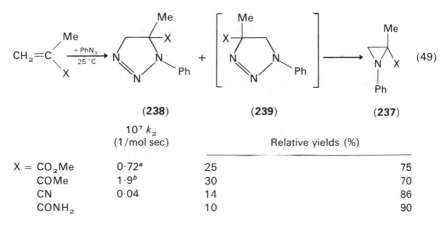

	(238)		
	$10^7\,k_2$		
	(1/mol sec)	Relative yields (%)	
		(239)	(237)
X = CO_2Me	0.72^a	25	75
$COMe$	1.9^b	30	70
CN	0.04	14	86
$CONH_2$		10	90

^{a}CCl$_4$; ^{b}no solvent

Faster reactions and higher regioselectivity were observed when the above mentioned dipolarophiles reacted with methyl and n-butyl azides.

Two regioisomers have also been detected for the reaction of phenyl azide with ω-nitrostyrene[50,245,322].

4. Reactivity and regiochemistry

Table 12 clearly shows that HOMO energies ($-IP_v$) and steric strain relief are not the only factors affecting reaction rates in the cycloadditions of azides. Other elements such as steric factors and stability of the final adducts also play an important role.

The influence of substituents on phenyl residues of aryl azides on the reaction rates is shown in the data of Table 13. p-Methoxyphenyl azide, like phenyl azide (see Section II.G) shows a minimum for cycloadditions with unactivated olefins (cf. cyclopentene). p-Nitrophenyl azide, an electron-poor azide, reacts slowly with electron-poor olefins, while the

TABLE 12. Relative rate constants (k_{rel}) for cycloadditions of phenyl azide with carbon–carbon double bonds at 25 °C in carbon tetrachloride[a]

Dipolarophile	k_{rel}	IP$_v$(eV)	Strain relieved on hydrogenation (kcal/mol)[323]
	580[281]	9.38[281]	26.4[b]
	<0.008[74]	9.43[324]	3.7[b]
	1.00[42]	9.18[324]	−0.39
	0.017[42]	9.12[324]	1.15
	3655[74]	8.69[325]	6.72
	0.349[47]	8.98[324]	−2.58
	101[42]	8.97[326]	9.65[c]

[a] The k_{rel} for *cis*-cyclooctene was determined for reactions in chloroform. However solvent effects have little influence on the value of k_{rel}, e.g. the absolute values k_2 for cyclopentene were 1.86 and 1.83 × 10^{-7} 1/mol sec in CCl_4 and $CHCl_3$ respectively, at 25 °C[42,47].

[b] The values refer to cyclopropene and cyclobutene respectively.

[c] Quite a different value (5.72) was reported by Allinger and Sprague, *J. Amer. Chem. Soc.*, **94**, 5734 (1972).

electron-rich benzyl azide shows very little reactivity with the electron-rich 1-pyrrolidino-cyclohexene[42,47].

These results suggest that for *p*-nitrophenyl azide and generally for azides substituted with electron-withdrawing groups (cyanogen azide, tosyl azide, benzoyl azide, picryl azide) the LUMO azide–HOMO olefin

465

TABLE 13. Rate constants ($10^7 \times k_2$ l/mol sec in benzene at 25 °C) for cycloadditions of azides RN_3 with carbon–carbon double bonds[42]

R =	Maleic anhydride	N-Phenyl maleimide	Cyclopentene	Norbornene	1-Pyrrolidinocyclohexene
$p\text{-}CH_3OC_6H_4$	20·8	66·7	2·15	187	3150
C_6H_5	7·20	27·6	2·40	254	9930
$p\text{-}O_2NC_6H_4$	1·28	10·5	14·9	1530	1,420,000
$C_6H_5CH_2$	53	95	—	221	25·4
ρ (Hammett)	−1·1	−0·8	+0·9	+0·88	+2·54

interaction is dominant (Type III cycloadditions), while for alkyl azides the more important interaction, except with very electron-rich dipolarophiles, is HOMO azide–LUMO olefin (Type I cycloadditions).

The Hammett ρ values found for substituted azides (Table 13) have a rationale in the perturbation approach to 1,3-dipolar cycloadditions[5,184]. The positive (negative) values are considered evidence for the role of electron acceptor (donor) by azides through their LUMO (HOMO). Rate acceleration found for electron-poor (-rich) azides in the reaction with 1-pyrrolidinocyclohexene (maleic anhydride) is due to a decrease in LUMO energies (increase in HOMO), enhancing LUMO azide–HOMO olefin (HOMO azide–LUMO olefin) interaction. A large absolute ρ value is indicative of very small frontier orbital separation of two reacting species.

Figure 15 shows schematically the main interactions between phenyl azide and some dipolarophiles giving rise either to 5- or 4-substituted Δ^2-1,2,3-triazolines in full agreement with experimental data.

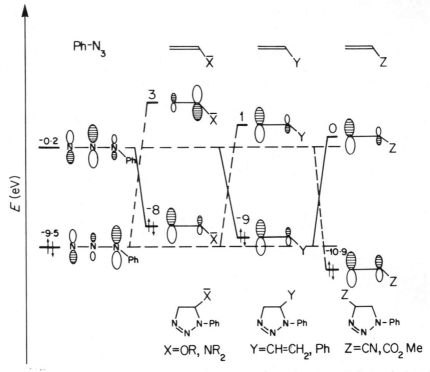

FIGURE 15. Dominant (heavy lines) F.O. interactions for the reactions of phenyl azide with some classes of dipolarophiles and the resulting favoured triazolines.

Reactivity, regioselectivity and Hammett ρ values were rationalized by Huisgen on the basis of a non-synchronous bond formation with partial charge stabilization in the transition states such as **240** and **241**[42].

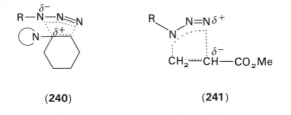

(240)   **(241)**

Bond formation, in the F.O.-perturbation-approach formulation, should be more advanced at the sites of the reactants having the largest coefficients, with an electron flow from one addend to the other. This picture gives theoretical support to the 'charge separation' model.

5. Cycloadditions with α-ester and α-ketophosphorus ylides

Carboalkoxymethylentriphenylphosphoranes are mixtures of equilibrating *cis*-**242** and *trans*-**243** isomers and can be described by the enolate structure **242a** with minor contribution by the canonical forms **242b** and **242c**[327].

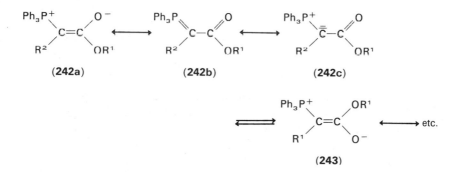

(242a)   **(242b)**   **(242c)**

(243)

Carboalkoxymethylentriphenylphosphoranes **244** were found to react with a variety of azides including tosyl azide[328,329], acyl azides[55], azidoformates[330], *P*-azides[55] and aryl azides[55] to give diazoacetates (**245**) and iminophosphoranes (**246**) as the only isolated products. The second step of equation (50) may be regarded as a retro-1,3-dipolar cycloaddition.

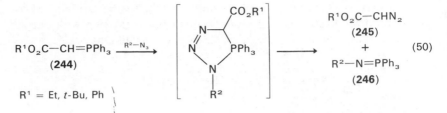

$$R^1 = Et, t\text{-}Bu, Ph$$

In contrast, the first step of the reaction of acyl azides and azidoformates with ylide **247a** resulted in a cycloaddition to the carbon–carbon double bond only (equation 51)[55,330]. The intermediate adducts **249** lose Ph_3PO transforming into the $N_{(1)}$-substituted triazoles (**250**) which in the basic reaction conditions isomerize to $N_{(2)}$-substituted triazoles (**251**)[331].

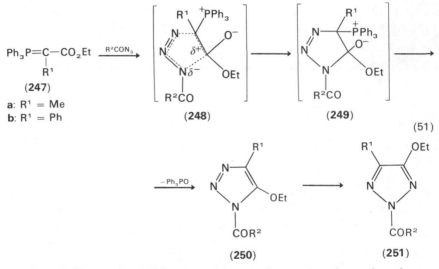

Competitive cycloadditions to carbon–carbon and carbon–phosphorus double bonds were observed in the reaction of aryl azides with the ylides **247a**, **247b** and of acyl azides with **247b**[332].

Acylmethylenetriphenylphosphoranes, which have a fixed *cis*-enolate structure (**252**)[333], react with tosyl azide[328], aryl azides[334], acyl azides[335] and vinyl azides[294] to give, in good yields, the triazoles **253** or **254**, depending on the substituents, according to equation (52). The following order of reactivity was found for azides: $R^2 = $ acyl $> p\text{-}O_2NC_6H_4\text{—} > C_6H_5\text{—} > p\text{-}CH_3OC_6H_4\text{—}$; and for ylides: $R^1 = $ Me $> C_6H_5 > p\text{-}O_2NC_6H_4$. LUMO azide–HOMO ylide interaction is therefore dominant for these cycloadditions.

For all these reactions a concerted cycloaddition mechanism, characterized by a transition state with small charge imbalance, e.g. **248**, is supported

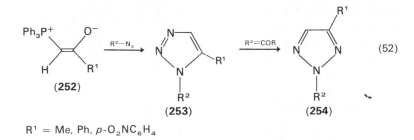

$$R^1 = Me, Ph, p\text{-}O_2NC_6H_4$$

by the small solvent effect and activation parameters found[55,334,335].

6. Cycloadditions with isocyanates and isothiocyanates

Butyl isocyanate was found to be unreactive towards azides. Alkyl
azides (but not aryl azides) were found to react slowly with aryl isocyanates and acyl isocyanates to give 4-aryl and 4-acyl-Δ^2-tetrazolin-
5-ones **255** and **256** respectively, in good yields. Both aryl and alkyl
azides, however, readily react with sulphonyl isocyanates to give 4-
sulphonyl-Δ^2-tetrazolin-5-ones **257** in excellent yields[336].

The trends in reactivity found clearly show that HOMO azide–LUMO
isocyanate is the dominant F.O.interaction.

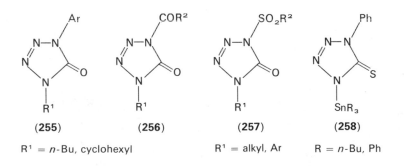

$$R^1 = n\text{-}Bu, \text{cyclohexyl} \qquad R^1 = \text{alkyl, Ar} \qquad R = n\text{-}Bu, Ph$$

The reaction of alkyl azides with sulphonyl isocyanates was found
to be reversible. At temperatures above 100°C an equilibrium is
established between **257** and the addends. A linear Hammett correlation
for substituted sulphonyl isocyanates was found for this retro-1,3-dipolar
cycloaddition with a ρ value of $+1\cdot4$; the rate increased with increasing

electronegativity of the isocyanate moiety as was the case with the forward reaction.

Tri-n-butyltin (at room temperature) and triphenyltin azide (at 140 °C) react with phenyl isothiocyanate only at the carbon–nitrogen double bond to give Δ^2-tetrazoline-5-thiones **258**[336a]. In contrast alkyl azides react at room temperature with arylsulphonyl isothiocyanates only at the carbon–sulphur double bond to give 4-alkyl-5-aryl-sulphonylimino-Δ^2-1,2,3,4-thiatriazolines in good yields[336b].

7. Dimroth reaction

Aryl, vinyl and alkyl azides react with active methylene compounds under basic conditions to give 1,2,3-triazoles. Aryl and vinyl azides are of comparable reactivity while alkyl azides are less reactive. Two reaction pathways have been considered: a two-step mechanism and a concerted 1,3-dipolar cycloaddition to the delocalized carbanion[291]. The former mechanism seems more probable and is supported by recent findings on the reaction of glycosyl azide (**259**) with cyanoacetamide to give a mixture of the two adducts **261** and **262** in a 14·5:1 ratio[337].

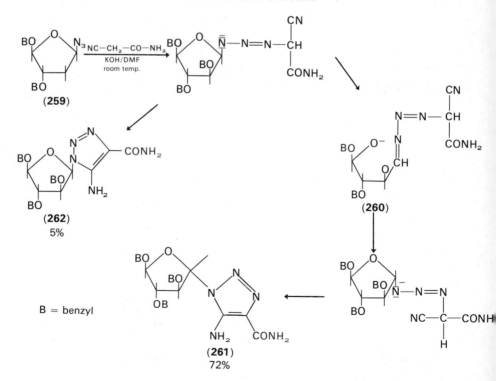

The formation of the two diastereoisomers is undoubtedly due to the inversion of configuration at the carbon of the glycosyl azide bearing the azido group at a stage such as that depicted by the intermediate **260**.

The related reaction of sulphonyl azides with active methylene compounds to give diazo compounds has been reviewed by Regitz[338].

V. CYCLOADDITIONS OF 1,3-DIPOLES WITHOUT AN ORTHOGONAL DOUBLE BOND

A. Azomethine Ylides

1. Preparation and cycloadditions with X=Y

Dehydrohalogenation of immonium salts[1] and thermolysis of the easily accessible aziridines[339] represent the two best routes to open-chain azomethine ylides. Since they are unstable, they are usually prepared *in situ*, in low concentrations, and trapped by the dipolarophile present. N-Benzylisoquinolinium bromide (**263**), for example, reacts with carbon disulphide in a basic medium to give the mesoionic cycloadduct **264** through the intermediate azomethine ylide[340].

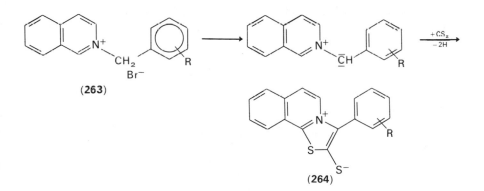

Aziridines, isoelectronic with the cyclopropyl anion, isomerize to azomethine ylides, isoelectronic with the allyl anion, through a conrotatory or disrotatory ring opening upon heating or irradiation respectively[341]. The whole process is described below for the two epimeric dimethyl-1-(4-methoxyphenyl)aziridine-2,3-dicarboxylates, **265** and **266**, to **267** and **268** respectively.

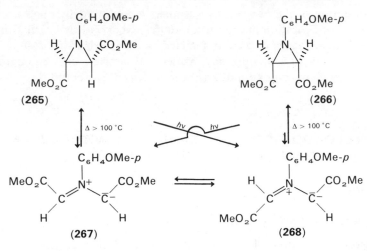

Azomethine ylides **267** and **268** have quite different dipolarophilic reactivities. In fact, **268** reacts with different dipolarophiles to give adducts characterized by complete sterospecificity, while **267** with dipolarophiles which are not particularly active gives rise to mixtures of epimeric cycloadducts.

Assuming that 1,3-dipolar cycloadditions are completely stereospecific, it follows that in the case of **267** isomerization to **268** is competing with the addition, as shown by the reactions with dimethyl azodicarboxylate reported below[342].

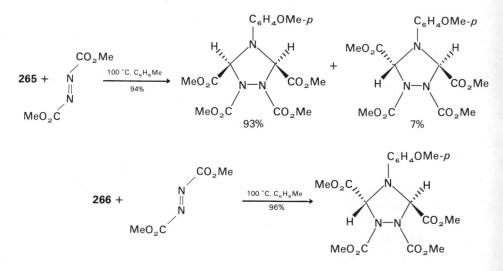

The differing reactivity of **267** and **268** is difficult to explain since they have very similar energy contents. It has been suggested, however, that the different stabilities of the final adducts may be responsible for these results[343].

Unsymmetrically-substituted aziridines react with aldehydes in a regiospecific fashion[344,345]. Structures and configurations of the adducts have been proved with the aid of labelled reagents. Both *cis-* and *trans-*

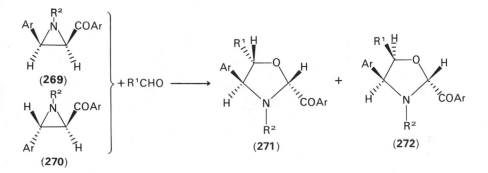

1-alkyl-2-aryl-3-aroylaziridines [(**269**) and (**270**)] react with aliphatic or aromatic aldehydes to give mixtures of the oxazolidines **271** and **272** in which the epimer **271** is dominant. The identical configurations at positions 2 and 4 in the two oxazolidines is rationalized by the isomerization of *cis-* to *trans-*azomethine ylide before the cycloaddition reactions occur.

Several other unsaturated compounds have been reacted with azomethine ylides to give different heterocycles. For azomethine ylides with suitable dipolarophiles, the following heterocycles have been synthesized: imidazolines[346–348], thiazolines[349,350], thiazolidines[348], oxazolines and oxazolidines[344,345,348,351,352] and triazolidines[342,350].

Stable azomethine ylides incorporated into an aromatic system with 5 or 6 atoms are accessible by a variety of syntheses. Some examples are reported below.

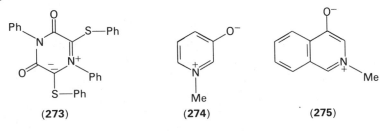

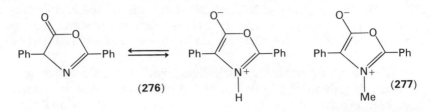

(276)

(277)

The mesoionic piperazine **273** reacts with both C=C and C=O double bonds. The cycloadduct with formaldehyde has been transformed into the corresponding trioxopiperazine on catalytic reduction (equation 53)[353].

An interesting route to the tropone and benzotropone systems is that which goes through the cycloaddition of **274** and **275** with C=C double bonds (equation 54)[354].

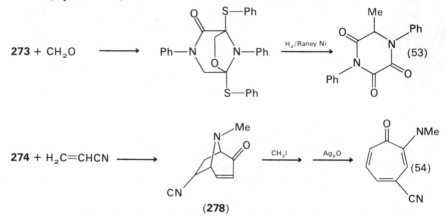

$$\textbf{273} + CH_2O \longrightarrow \qquad \xrightarrow{H_2/Raney\ Ni} \qquad (53)$$

$$\textbf{274} + H_2C{=}CHCN \longrightarrow \qquad \xrightarrow{CH_3I} \xrightarrow{Ag_2O} \qquad (54)$$

(278)

Compound (**274**) reacts with acrylonitrile to give the adduct **278** which in turn is transformed into a tropone derivative on Hoffmann elimination. **274** has also been found to undergo a retro-1,3-dipolar cycloaddition[355]: when dimethyl-7-oxabicyclo[2.2.1]hepta-2,5-diene-2,3-dicarboxylate (**214**) is reacted with **274** in boiling tetrahydrofuran, **279** and **280** are formed in nearly equal yields. Heating the adduct **279** in chlorobenzene to 130 °C gave the thermodynamically favoured isomer **280**.

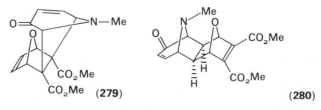

(279)   (280)

As shown in the previous examples **273**, **274** and **275** undergo 1,3-dipolar cycloadditions to give bicyclic compounds containing all the elements of the addends. In contrast, systems **276** and **277**, which are internal anhydrides of *N*-acylaminoacids react with unsaturated compounds to give different types of cycloadducts on extrusion of carbon dioxide[1]. The general reaction of an azalactone with X=Y, shown in

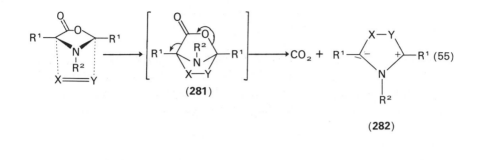

equation (55), yields the non-isolable cycloadduct **281** which decomposes in turn to the new non-aromatic 1,3-dipolar system **282**, which is more reactive than the starting azomethine ylide in cycloadditions.

The Δ^2-oxazolin-5-ones in their tautomeric form of oxazolium 5-oxides[356,357], react smoothly with conjugated alkenes, e.g. 2,4-diphenyl-Δ^2-oxazolin-5-one (**276**) with dimethyl fumarate yields either **283** or **284** depending on the dipolarophile concentration[358].

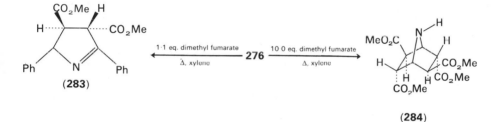

Compound **283** is formed through a prototropic tautomerism process of an intermediate of type **282** ($R^1 = Ph; R^2 = H$) which, in the presence of excess dimethylfumarate, may be trapped to give **284**.

Reaction of azomethine ylides of type **277** with cyclopropenones and with methylenecyclopropenes leads to 4-pyridones and 1,4-dihydro-*N*-methyl-4-methylene-pyridine derivatives respectively, according to the reaction mechanism described below[358a]. Different behaviour is observed

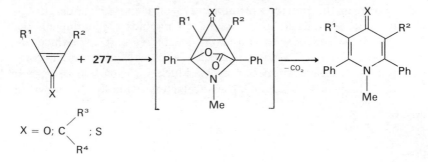

when diphenylcyclopropenone reacts with some substituted azomethine ylides from aziridines. In this case a very close similarity in dipolarophilic reactivity of the carbon–carbon double bond and the carbonyl bond has been found. In fact, as exemplified by the two above reported equations,

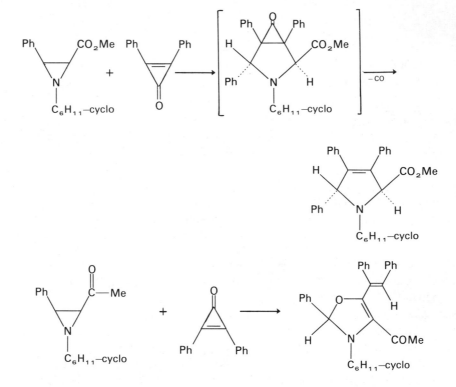

the apparently closely-related 3-carbomethoxy-1-cyclohexyl-2-phenyl-aziridine and 3-acetyl-1-cyclohexyl-2-phenylaziridine react with diphenyl-cyclopropenone selectively to give a pyrroline and oxazoline derivative, respectively[351a].

The mesoionic oxazolones (also called münchnones) are also highly reactive with hetero double bonds. With ketones, aldehydes and keto esters the cycloaddition goes through a bicyclic primary cycloadduct which, after CO_2 elimination, rearranges to an N-acylenamine through electrocyclic ring cleavage[359].

A typical example is the cycloaddition of 3-methyl-2,4-diphenyl-oxazolium-5-olate (277) with aromatic aldehydes.

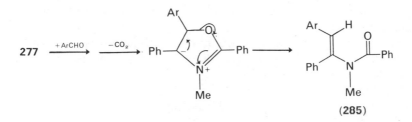

(285)

The relative position of the substituents on the N-acylenamine 285 shows the direction of the cycloaddition. Thiocarbonyl derivatives behave similarly to the carbonyl compounds in their cycloaddition with azo-methine ylides. Münchnone 277 dissolved in carbon disulphide at 20 °C gives the stable 3-methyl-2,4-diphenyl thiazolium-5-thiolate (286) in quantitative yields[360].

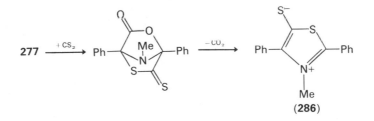

(286)

The driving force of the reaction is attributed to the greater aromaticity of 286 compared to 277. A similar reaction sequence is followed in the formation of 3-methyl-2,4-diphenylthiazolium-5-olate from 277 and carbon oxysulphide.

An unusual dipolarophilic reagent is involved in the reaction of münchnone 277 and p-nitrobenzonitrile in xylene at 120 °C to give compound 287 through several steps[361].

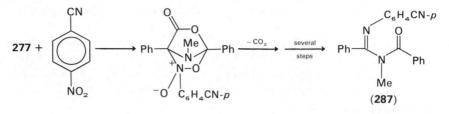

The dipolarophilic reactivity of the nitro group in the 4-nitrobenzo-nitrile is particularly surprising considering the presence in the molecule of the C≡N triple bond which, as is well known, is reactive towards azomethine ylides.

The N=O double bond of the nitrosobenzene also reacts with the azomethine ylide **277** to give the N-methyl-N'-phenyl-N-benzoyl-benzamidine[361].

The reaction of azo compounds with azomethine ylides is yet another example of the reactivity of this 1,3-dipole[361].

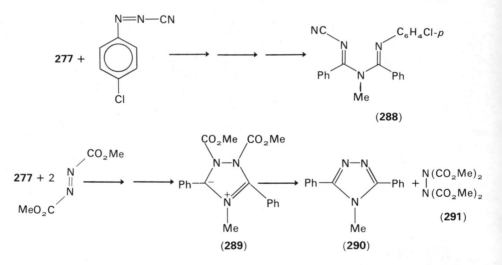

The addition of **277** to 4-chlorobenzene-*anti*-diazocyanide is completely selective, giving only the amine **288** as final product.

Another complex mechanism is involved in the cycloaddition of **277** with dimethyl azodicarboxylate. The reaction proceeds with evolution of CO_2 to give the dipolar triazole derivate **289**, which supposedly transfers two carbomethoxy groups to a second molecule of the azodicarboxylic ester to produce **290** and **291**.

2. Reactivity

A kinetic study[56] of the cycloaddition of 3-methyl-2,4-diphenyl-oxazolium-5-oxide (**277**) (see Table 6 for some examples), with olefins and hetero multiple-bond systems, has shown that the reaction is strictly of the second order. Furthermore, as 1,3-dipolar cycloadditions of azomethine ylides with acetylenes are not faster than those of olefins, e.g. aromatic pyrroles are not formed faster than non-aromatic pyrrolines, it follows that the step from the intermediate cycloadducts to the final products is not rate determinant.

Interestingly enough, the lower reactivity of methyl crotonate $(10^4 \, k_2(\text{l/mol sec}) = 76\cdot7; \, \Delta H^{\neq} = 11\cdot8$ and $\Delta S^{\neq} = -32)$ compared to methyl acrylate, $(10^4 \, k_2 = 26{,}000; \, \Delta H^{\neq} = 12\cdot3 \, \text{kcal/mol}$ and $\Delta S^{\neq} = -19$ e.u.) is mostly due to entropic factors as a consequence of the higher steric requirements of the crotonate. Styrene and non-activated alkenes showed reactivities similar to that of the crotonate.

B. Azomethine Imines

1. Preparation and cycloadditions with X=Y

Sydnones **292**, prepared through cyclization of N-nitroso-α-amino acids, contain the cyclic azomethine imine system in an aromatic ring[1,362].

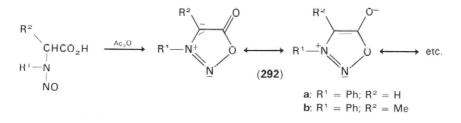

a: $R^1 = Ph; R^2 = H$
b: $R^1 = Ph; R^2 = Me$

Since sydnone LUMO energies are very low, it is reasonable to assume that their cycloadditions are LUMO-dipole controlled. The electron-withdrawing carboxyl group directly attached to the carbon atom of the azomethine imine system of sydnones makes coefficients on the ends of the LUMO similar. Hence, cycloadditions with unsymmetrical dipolarophiles should not result in strict regiochemistry[4,5].

In practice, for the two most studied sydnones, 3-phenylsydnone (**292a**) and 3-phenyl-4-methylsydnone (**292b**), the adducts obtained by cycloaddition with almost all the monosubstituted alkenes[363,364,366], are those resulting from the intermediate adducts **293**, while with propene[364], allylacetate[365] and acrylonitrile[366], mixtures of the product in **294** and its opposite regioisomer **295** were formed.

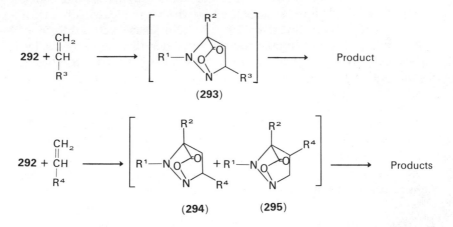

(**293**)

(**294**)   (**295**)

The mechanism of 1,3-dipolar cycloadditions of sydnones with alkenes is illustrated by the reaction of 3-phenylsydnone (**292a**) with ethylene and 1,1-diphenylethylene (equations 56 and 57)[364,365].

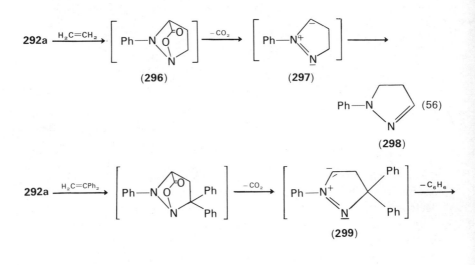

(**296**)   (**297**)

(**298**)

(**299**)

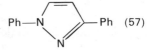

In equation (56) the intermediate **296** readily loses carbon dioxide to give the betaine **297** which in turn gives 1-phenyl-Δ^2-pyrazoline (**298**) through a thermally-allowed 1,4-suprafacial hydrogen shift. In equation (57) the intermediate **299** expels benzene to give the aromatic 1,3-diphenyl-pyrazole.

In agreement with the above proposed reaction mechanism involving intermediate azomethine imines such as **297** and **299**, the 3-phenylsydnone reacts with 1,5-cyclooctadiene to give 10-phenyl-9,10-diazatetracyclo-[6,3,0,0^{4,11},0^{5,9}]undecane(**300**). The Δ^2-pyrazoline **301** was not detected[367].

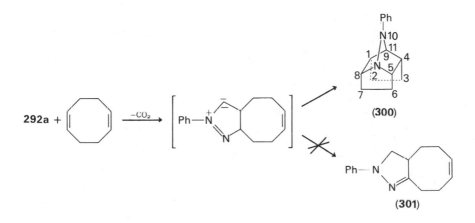

Sydnones react with several classes of hetero double bonds such as phenyl isocyanates[368], aldehydes and aromatic thioketones[369].

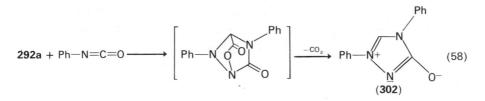

The reactions of 3-phenylsydnone (**292a**) with phenyl isocyanate (equation 58) and 4,4′-dimethoxythiobenzophenone (equation 59) gave the mesoionic 1,2,4-triazolone (**302**) and *N*-thioformylhydrazone (**303**), respectively. Arylidene-*N*-acetylphenylhydrazines (**304**) were obtained from the cycloaddition of 3-phenyl-4-methylsydnone (**292b**) with aromatic

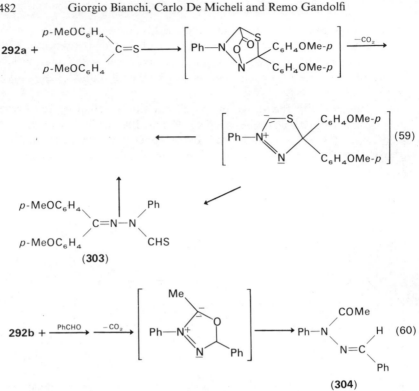

aldehydes (equation 60). All these cycloadditions of sydnones with hetero double bonds invariably showed regiospecificity.

Oxidation of benzylbisphenylhydrazone gives rise to bisphenyl-azostilbene (**305**) which is in equilibrium with the mesoionic anhydro-1-phenylimino-2,4,5-triphenyl-1,2,3-triazoliumhydroxide (**306**)[370]. The latter 1,3-dipole reacts with carbon disulphide to give 2,4,5-triphenyl-1,2,3-triazole, sulphur and phenyl isothiocyanate.

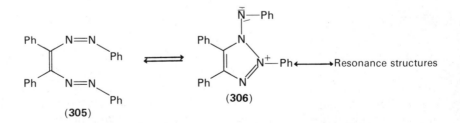

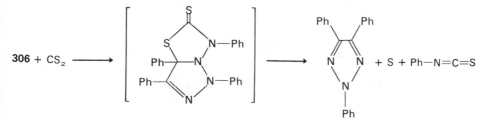

Compound **306** reacts with ethyl acrylate, acrylonitrile, methyl crotonate, phenyl isocyanate and phenyl isothiocyanate to give the related adducts in a regiospecific manner.

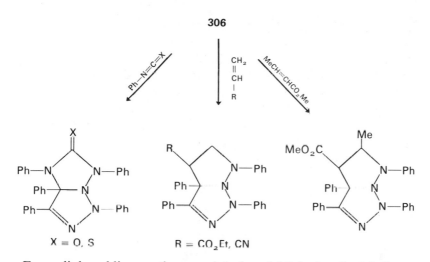

From diphenyldiazomethane and 4-phenyl-1,2,4-triazolin-3,5-dione, a 1,3-dipolar species **307** is produced which may be trapped by a suitable dipolarophile to give adducts such as **308** and **309**[371].

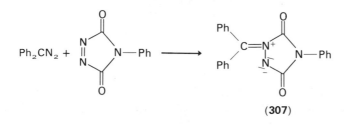

(**307**)

484   Giorgio Bianchi, Carlo De Micheli and Remo Gandolfi

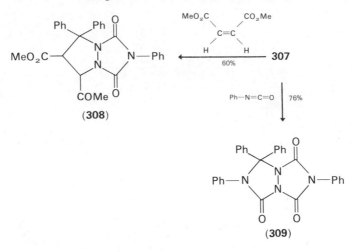

(308)

307

(309)

In the reaction of azodibenzoyl with diphenylketene[372], an azomethine imine intermediate, **310**, has also been postulated. The latter reacts with excess diphenylketene to give **311** and with phenylisocyanate to give the adduct **312**.

Adduct **311** is in equilibrium with **310** and diphenylketene through a retro-1,3-dipolar cycloaddition.

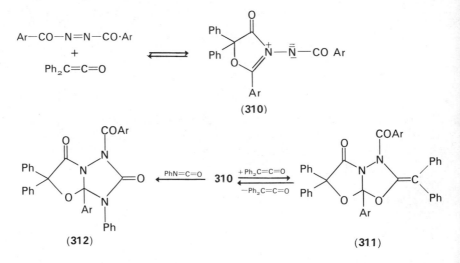

(312)   (311)

N,N'-Disubstituted hydrazines react with carbonyl compounds to give azomethine imines **314** through the intermediate **313** which is seldom isolated[373,374].

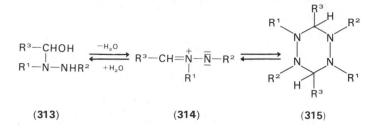

(313)   **(314)**   **(315)**

Azomethine imines **314** tend to dimerize to 1,2,4,5-tetrazines **(315)** which may be used to generate the dipolar species. The tetrazine **316**, for instance, reacts smoothly with the carbon–sulphur double bond of carbon disulphide to give 2-p-methoxyphenyl-3,4-dimethyl-1,3,4-thio-diazolidine-5-thione **(317)** in good yield[375].

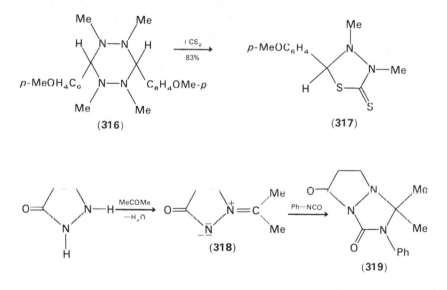

Readily isolable and stable azomethine imines such as **318** are obtained by the reaction of pyrazolin-5-one with aldehydes and ketones. The latter dipoles react with alkynes[374] and isocyanates[376]. Cycloaddition of **318** with phenyl isocyanate is a good method for the synthesis of N'-substi-tuted-bis-carbamoyl-pyrazolidin-3-one **319**.

The intermediate azomethine imine **321** has been postulated in the reaction of aldehyde **320** with N-acyl-N⁻-alkylhydrazines to give **322** and **323**[377].

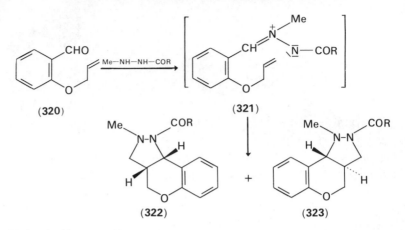

(320)   (321)

(322)   +   (323)

N-Arylamino-3,4-dihydroisoquinolinium salts give the very reactive azomethine imines **324** on treatment with triethylamine or pyridine[1]. The reactions of the latter dipoles with suitable dipolarophiles represent facile routes to pyrazolidines **325**[1], 1,2,4-triazolidin-3-ones, 1,2,3-triazolidin-3-thiones **326**[378], 1,2,4-triazolidines **327**[379], 1,3,4-oxadiazolidines **328**[380] and 3-bora-1,2,4-triazolidines **329**[381].

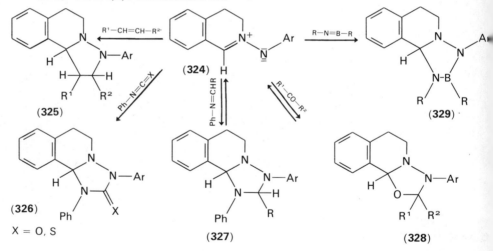

(325)   (324)   (329)

(326)   (327)   (328)
X = O, S

The cycloadditions of azomethine imines to yield **327** and **328** are reversible: adducts give the addends on heating. The cycloadduct of 3,4-dihydroisoquinoline-N-phenylimine with p-nitrobenzaldehyde (**328**, Ar = Ph; $R^1 = p\text{-}O_2NC_6H_5$, R^2 = H), heated in the presence of phenyl isocyanate gives a compound of the type **326** and liberates the aldehyde in good yield[380].

C. Nitrones (Azomethine Oxides)[381a]

1. Structure

Nitrones generally have large dipole moments oriented in the direction of the N—O bond[382–383]. This indicates a high contribution of structure **330** in the ground state of the molecule.

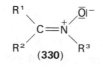

(**330**)

The presence of a double bond in structure **330** may give rise to geometrical isomerism. Below are reported some examples of keto-nitrones[384–386], e.g. **331** and **332**, nitronic esters[387], e.g. **333** and aldonitrones[388], e.g. **334** for which the two isomers have been isolated and characterized by dipole-moment measurements, u.v. or n.m.r. analysis.

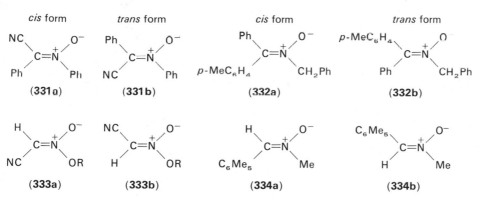

The *trans* isomer is generally more stable than the *cis* isomer and the energy values required to convert one isomer into the other are spread over a large interval. The energy required to convert nitrone **331a** into **331b** ($\Delta E^{\neq} = 24\cdot6\,\text{kcal/mol}$)[389] is quite different from the energy required for the isomerization **332b** → **332a** ($\Delta E^{\neq} = 33\cdot6 \pm 1\cdot4\,\text{kcal/mol}$, $\Delta S^{\neq} = -4 \pm 4\,\text{e.u.}$)[386]. The latter value is similar to the interconversion barrier found for aldonitrone (**334**) for which the energy required to convert the *cis* form to the *trans* at 147 °C is $\Delta G^{\neq} = 33 \cdot 1$ kcal/mol while the reverse process requires $34 \cdot 6$ kcal/mol[388]. The latter interconversions have been found to be catalysed by traces of benzoic acid; the free energies of activation for the catalysed processes were $\Delta G^{\neq} = 23 \cdot 9$ and $25 \cdot 4$ kcal/mol respectively[388].

Aldonitrones are, however, generally constituted by the sole *trans* form at room temperature.

2. Cycloadditions with C=C

a. *Endo–exo and* syn–anti *isomerism. Endo–exo* and *syn–anti* isomerism in 1,3-dipolar cycloadditions can be explored with the aid of both cyclic and open chain nitrones with definite structure. For example 5,5-dimethyl-Δ^1-pyrroline-*N*-oxide reacts with *cis*-3,4-dichlorocyclobutene to give a mixture of the two stereoisomers **337** and **338** whose structures have been elucidated by n.m.r. data and confirmed by X-ray analysis[390]. Adducts **337** and **338** are generated through the *exo*-transition states **335** and **336** respectively, since these, for steric reasons, are more favoured than the corresponding *endo* transition states. The *syn*-chlorine compound is dominant in the reaction mixture as a result of stabilizing interactions

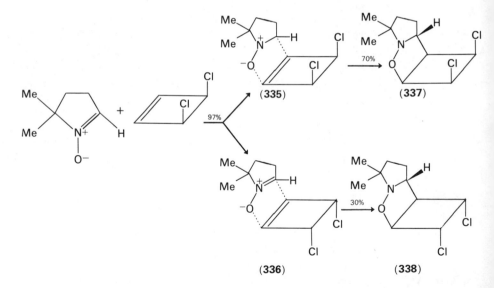

among the two 1,3-dipole ends and the lone pairs of the chlorine atoms. Often, in accordance with the reaction scheme shown in equation (11)

(Section II.L), nitrones react with dipolarophiles to give mainly adducts derived from *endo*-transition states[84,391,387]. The reasons invoked to explain this kind of *endo–exo* isomerism were based on both electronic and steric effects. C-Phenyl-N-phenylnitrone, possessing a *trans* structure, reacts with N-phenylmaleimide to give a mixture of the diastereoisomeric isoxazolidines **341** (63%) and **342** (37%)[391]. Comparable results have been found for some other aryl substituted nitrones and N-arylmaleimides[391].

The prevalence of diastereoisomer **341** over **342** may be rationalized considering the transition states. The *endo* transition state **339** is favoured both by π overlap between the two N-phenyl groups of dipole and dipolarophile and the stabilizing dipole–dipole interaction of the two addends[391]. On the other hand adduct **342**, is formed through the *exo*-transition state **340** in which steric requirements should be less than in **339**. Better stabilization in the *endo* transition state **344** seems to operate in the reactions of nitronic ester **343**, possessing a *cis* structure, with electron-poor dipolarophiles, e.g. dimethyl maleate, to give only the isoxazolidine **345**[392].

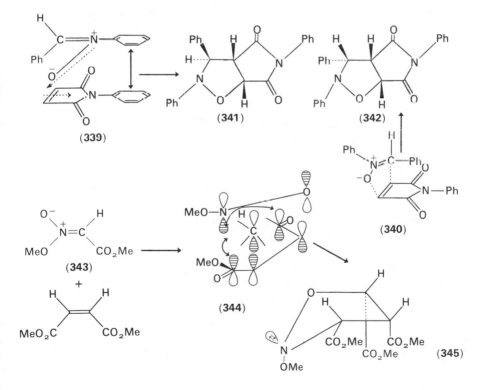

The latter 1,3-dipolar cycloaddition belongs to Type I of Sustmann's classification, in which HOMO dipole–LUMO dipolarophile interaction is dominant[393]. The better stabilization of the *endo* orientation **344** may be attributed to the secondary bonding overlap (a) between the lobes of the nitrogen of the dipole and the carbon atom of the ester which is prevalent over the antibonding interaction (b)[383].

A further aspect in 1,3-dipolar cycloadditions of some nitronic esters with mono- and disubstituted alkenes is the formation of only one invertomer under kinetic control[392,394]. For instance, nitrone **343** reacts with dimethyl maleate to give the invertomer **345** with the configuration at $C_{(3)}$ and conformation shown below. The gradual transformation of the π-orbitals of the two addends involves the lone pair on the nitrogen (see Section II.H), which as depicted in **345**, is only developed on the side opposite to the new σ bonds[392].

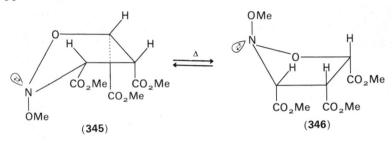

(345)   (346)

As described previously the geometry of the transition state can be deduced from the structure of the final adduct, provided that the configurations of the reagents are known and are not subject to change during the process. On the contrary, a complex situation arises in 1,3-dipolar cycloadditions of nitrones which are in the *trans*-form at room temperature, but which may equilibrate with the *cis* isomer at temperatures of *c*. 80 to 100 °C. The stereochemistry of the final products may then depend on different nitrone–dipolarophile complexes such as **347** and **349** or on transition states such as **347** and **348** with the 1,3-dipole in different configurations, For example the reaction of *C*-phenyl-*N*-methyl nitrone with norbornene gives both diastereoisomers **350** and **351**[74].

The most reasonable transition state leading to **350** is represented by **347**, while compound **351** can be formed *a priori* through two different transition states **348** or **349**. According to the authors[74], transition state

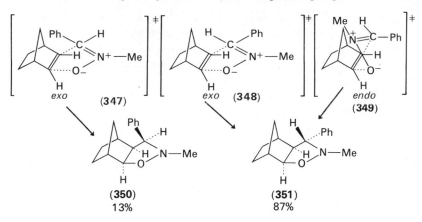

(347)   (348)   endo (349)

exo   exo

(350)
13%

(351)
87%

348 containing the nitrone in its *cis* form is more probable than the sterically more constrained *endo* transition state **349**.

A *cis–trans* isomerization operating at room temperature, may be postulated in the reaction of *C*-benzoyl-*N*-phenyl nitrone with *cis*-3,4-dichlorocyclobutene to give the isoxazolidines **353**, **354** and **355**[395].

Compound **353** is probably formed through the complex **352** with the nitrone in its *cis*-form.

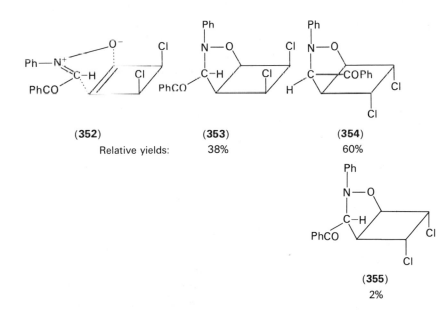

(352)   (353)   (354)

Relative yields: 38%   60%

(355)
2%

b. *Reactivity*. Nitrones are less reactive than the electronically-related azomethine ylides and azomethine imines[5]. The kinetic data available indicate that the reaction rates depend on the structure of the nitrones and also on electronic effects of substituents on the reactants[45]. Hence *cis*-nitrones have been found to be more reactive than the *trans* isomers which in some cases needed relatively drastic reaction conditions in order to enter 1,3-dipolar cycloadditions[387]. A study[45] showed that electron-withdrawing substituents on the nitrone enhance the reaction rates of the 1,3-dipole with ethyl crotonate, while reactions of *C*-phenyl-*N*-methylnitrone are faster with alkenes containing electron-withdrawing substituents (the Hammett ρ value for the cycloadditions of the latter nitrone with *p*-substituted styrenes is $+0.77$).

All of these characteristics of the reactions of nitrones seem to favour their inclusion in Type II of Sustmann's classification. Cycloadditions of nitronic esters, however, as stated previously, belong to Type I. The stability of the reactants strongly affects the reaction rates. When the 1,3-dipolar system is part of an aromatic ring, its reactivity is very low: isoquinoline-*N*-oxide, for instance, was 36,000 times less reactive with ethyl crotonate than the related 3,4-dihydroisoquinoline-*N*-oxide[45].

c. *Regioisomerism*. Nitrones react with electron-rich and moderately electron-poor monosubstituted olefins and 1,1-disubstituted olefins regiospecifically to give a mixture of the two diastereoisomeric 5-substituted isoxazolidines (equation 61)[396–400]. However the reaction is regioselective and orientation reversed when the dipolarophile is substituted by very effective electron-withdrawing groups[401] (equation 62). The regiochemistry found in the cycloaddition of nitrones has been explained by the MO perturbation approach allowing, however, for

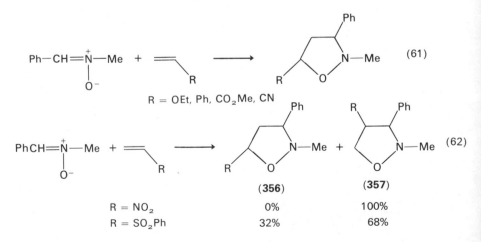

steric and dipole–dipole effects. For nitrone cycloadditions with electron-rich as well as electron-poor substituted alkenes the LUMO dipole, with a large coefficient on carbon, should control regioselectivity generating 5-substituted isoxazolidines[401]. With very electron-poor alkenes, such as nitroethylene or phenylvinylsulphone, the dominant and regiochemistry-determining FO interaction will be HOMO dipole–LUMO dipolarophile, and low regioselectivity is expected due to the nearly identical coefficients on the ends of the 1,3-dipole[401]. The prevalence, and even exclusive formation, of the regioisomer **357** in equation (62) is due to a favoured Coulombic or dipole–dipole interaction of the addends. Cycloadditions involving di- and polysubstituted alkenes generally give rise to a sole adduct whose regiochemistry is supposed to be controlled by steric or electronic effects considered above[381a].

Two cases in which the lack of regiospecificity is to be attributed to the similarity of the coefficients on the double bond of the dipolarophile are represented by the cycloadditions of *C*-phenyl-*N*-methyl nitrone with 1,2-dihydronaphthalene and indene to give four adducts in each case[402]. *C*-Phenyl-*N*-methylnitrone and an excess of 1,2-dihydronaphthalene at 100 °C for 6 days gave the adducts **358**, **359**, **360** and **361** in 97% yield.

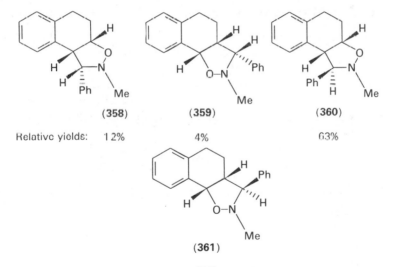

	(358)	(359)	(360)
Relative yields:	12%	4%	63%

(361)

21%

Mixtures of regioisomers were found in the intramolecular kinetically (76 °C) controlled cycloadditions of *N*-alkyl-*C*-6-heptenyl nitrones which are reversible at temperatures ≥200 °C[402a].

d. *Retro*-1,3-*dipolar cycloadditions.* In the case of nitrones, examples of cycloreversions have been found more frequently for reactions with α,β-unsaturated carbonyl compounds as e.g. in the reaction of 5,5-dimethyl-Δ^1-pyrroline-N-oxide with ethyl acrylate. The kinetically-controlled adduct **362** on heating at 100°C is transformed into the thermodynamically-more-stable regioisomer **363**[403]. Further examples are given in the reaction of 3,4-dihydroisoquinoline-N-oxide with methyl methacrylate[400] and of triphenyl nitrone with acrolein[404].

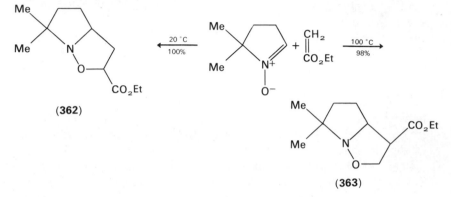

(**362**)

(**363**)

3. Cycloadditions with isocyanates and isothiocyanates

Cycloadditions of nitrones with isocyanates and phenyl isothiocyanate proceed smoothly to give 1,2,4-oxadiazolidin-5-ones and 1,2,4-oxadiazolidin-5-thiones respectively[405,381a], e.g. C-phenyl-N-methylnitrone reacts with phenyl isocyanate and phenyl isothiocyanate to give compound **364**[405]. However, both C=N and C=S double bonds of arylisothiocyanates are reactive while the sole C=S double bond of benzoylisothiocyanate enters the cycloaddition[381a].

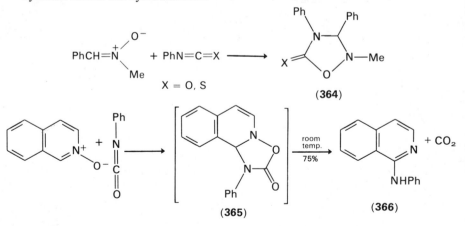

(**364**)

(**365**)

(**366**)

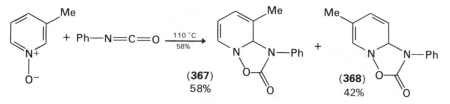

(367)
58%

(368)
42%

In the reaction of aromatic nitrones with phenyl isocyanate, stable cyclo-adducts are not generally isolated. The cycloaddition of isoquinoline-N-oxide with phenyl isocyanate gave compound 366 on expulsion of CO_2 from the intermediate 1,2,4-oxadiazolidine-5-one 365[405]. A unique example where cycloadducts have been isolated from cycloadditions of aromatic nitrones is that of 3-picoline-N-oxide with the C=N double bond of phenyl isocyanate to give compounds 367 and 368[406].

4. Cycloadditions with C=S

Nitrones react with aliphatic thioketones to give 1,2,4-oxathiazolidine 369 in equilibrium with starting compounds at higher temperatures as shown in equation (63)[407].

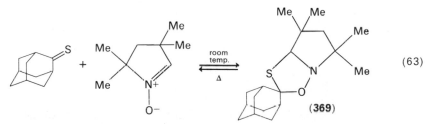

(63)

(369)

The *in situ* prepared sulphenes have been trapped with nitrones to give, through the intermediate cycloadduct 370, the benzoxathiazepine 371[408].

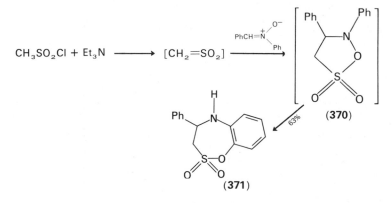

$CH_3SO_2Cl + Et_3N \longrightarrow [CH_2{=}SO_2]$

(370)

(371)

5. Cycloadditions with C=P, N=P, N=S

Several 1,2,5-λ^5-oxazaphospholidines **372** were prepared by the reaction of nitrones with the highly reactive C=P bond of phosphoranes[409].

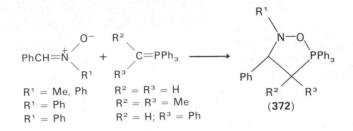

$$R^1 = Me, Ph \quad R^2 = R^3 = H$$
$$R^1 = Ph \quad R^2 = R^3 = Me$$
$$R^1 = Ph \quad R^2 = H; R^3 = Ph$$

Cycloaddition of nitrones to the N=P bond proceeds smoothly to give an intermediate 1,3,5,2-λ^5-oxadiazaphospholidine system (**373**) which fragments to triethylphosphine oxide and compounds **374**[410].

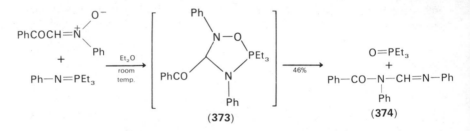

Compounds **376** have been isolated from the reaction of 4,5,5-trimethyl-Δ^1-pyrroline-N-oxide with N-sulphinyl anilines. The proposed mechanism is that of a 1,3-dipolar cycloaddition to give the intermediate **375**[411].

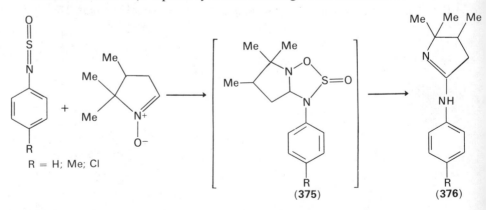

R = H; Me; Cl

An analogous reaction proceeds between *C,N*-diphenylnitrone and *N*-sulphinyl benzene sulphonamide (**377**) to give compound **378**[412].

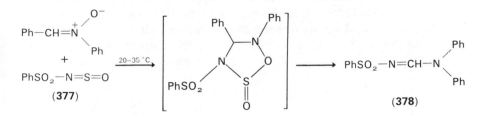

D. Carbonyl Ylides

1. Preparation and cycloadditions with C=C

Carbonyl ylides have never been isolated even though they have been detected by spectroscopy. The reactive 1,3-dipolar species can be prepared in several ways: (i) through valence tautomerism on heating or irradiation of either monocyclic and polycyclic oxiranes[1,413,414]; (ii) by carbene addition to a carbonyl group of an aldehyde or ketone[1,415]; (iii) through elimination of molecular nitrogen from Δ^3-1,3,4-oxadiazolines[416,417], and (iv) through chelotropic extrusion of CO from oxetanes[414].

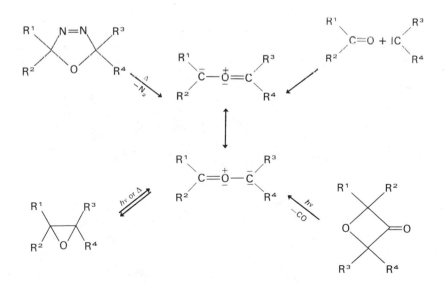

The reactivity of these 1,3-dipoles has been tested on several dipolarophilic systems and studies on kinetic isotopic effects have been carried out for 1,3-dipolar cycloadditions of **20** with labelled olefins (see Section II.D). Carbonyl ylide (**20**) has been found to be reactive towards many alkenes and aromatic and heteroaromatic compounds as well[30,413]. The reaction has been shown to be stereospecific. With benzene, compound **20** gave the mono- and the bis-adducts **379** and **380** respectively. The capacity of

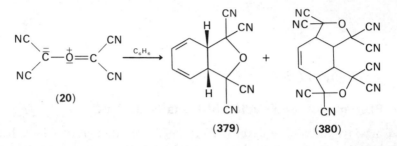

carbonyl ylides to enter 1,3-dipolar cycloadditions with arenes is worth mentioning since only a few 1,3-dipoles (e.g., ozone, azomethine ylides[418]) are known to react with dipolarophiles of such low reactivity.

Isolation of adduct **382** from the pyrylium oxide **381** (in valence–tautomeric equilibrium with 2,3-diphenylindenone oxide) and maleic anhydride, indicates a preferred *exo* (Section II.L) orientation of the two reactants. A similar stereoisomer has been obtained in the reaction of **381** with N-phenylmaleimide whereas with dimethyl maleate the two possible isomers have been isolated. In the case of the reaction of **381** with symmetrically *trans*-disubstituted alkenes the formation of **383** and its diastereoisomer was the rule[419].

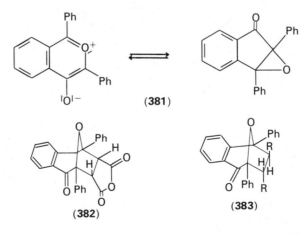

1,3-Dipolar cycloadditions of *cis*- and *trans*-cyanostilbeneoxides gave interesting results[420].

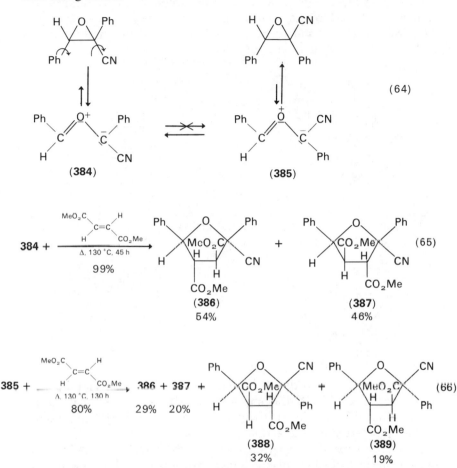

The formation of cycloadducts **386, 387, 388** and **389** is easily understood if it is considered that the C—C bond cleavage of the oxirane is a thermally-allowed conrotatory ring opening (equation 64). Carbonyl ylide **384** adds to dimethyl fumarate to give only **386** and **387**, while **385** gives **388** and **389** as well as **386** and **387**. The formation of the four tetrahydrofuran derivatives in the latter reaction is the proof of a reaction mechanism involving concurrent processes. The less stable carbonyl ylide **385** either reacts as such to give **388** and **389** or undergoes isomerization to the more stable **384** which in turn is trapped by the dipolarophile present.

2. Cycloadditions with C=O

Carbonyl ylides are also reactive towards carbonyl groups. For instance, carbonyl ylide **381** reacts regiospecifically with cyclohexanone to give **390**; **381** was also found to be in equilibrium with its dimer which is tentatively represented as structure **391**[421].

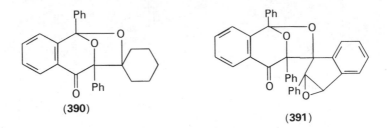

(390)   (391)

The ylides from **392** add to aromatic aldehydes in a regiospecific fashion to give a mixture of diastereoisomeric dioxolanes **393**[422].

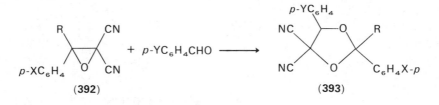

(392)   (393)

The nature of the substituent Y affects the reaction rates and these are higher for electron-withdrawing groups. MO perturbational theory explains this difference in reactivity as due to a stronger HOMO dipole–LUMO dipolarophile interaction.

Competitive reaction experiments have shown that the C=O group of p-nitrobenzaldehyde or of cinnamaldehyde is a more reactive dipolarophile than C=C of methyl fumarate[422].

E. Carbonyl Oxides

1. Photooxidation of diaryl diazoalkanes

When diphenyldiazomethane was photolysed in the presence of oxygen only the product was the diperoxide **394**. However when the photooxidation was carried out in the presence of aldehydes as solvents, no diperoxide was isolated but 1,2,4-trioxolanes (ozonides) **395** were obtained in low yields and the major product was benzophenone[423].

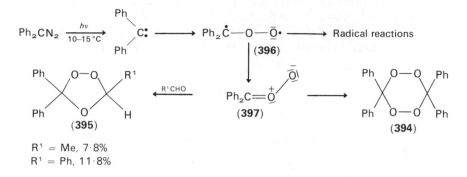

$$R^1 = Me, 7·8\%$$
$$R^1 = Ph, 11·8\%$$

The reaction involved the photolysis of the diazo compound to the triplet carbene which reacts with ground-state oxygen to give the diradical form of the carbonyl oxide (**396**). Most of the diradical form is consumed by radical reactions, however a portion reacts in the dipolar form, **397**, which either dimerizes or, in the presence of aldehydes, gives 1,2,4-trioxolanes. When the photooxidation is carried out using ^{18}O-labelled aldehydes, the labelled oxygen is found only in the ether bridge of the 1,2,4-trioxolanes[424]. This synthesis of carbonyl oxides is limited to diazo compounds containing two aryl substituents. However, it was recently shown that both alkyl and aryl diazoalkanes can be converted to carbonyl oxides by allowing them to react with excited singlet oxygen (equation 67). In the presence of aldehydes, mixtures of *trans*- and *cis*-ozonides **398** were obtained in low yields[425].

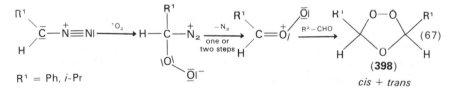

$$R^1 = Ph, i\text{-}Pr$$

These results provide a confirmation of Criegee's theory of final ozonide formation[423–425]; however this interpretation has been more recently criticized[34].

2. From 1,2,3-trioxolanes (primary ozonide)

Criegee's proposed mechanism for ozonolysis[426–429] predicts that the primary ozonide, the 1,2,3-trioxolane **399**, breaks in a concerted manner

to give the carbonyl oxide, **400**, and the carbonyl compound, **401**. These
two remain associated in a solvent cage and subsequently recombine to
give the final ozonide (**402**) unless they are separated because of solvent
polarity, high reaction temperature or association with other species.
Variable amounts of diperoxides **403** and polymeric peroxides **404** may
also be formed. The evidence for the Criegee mechanism includes the

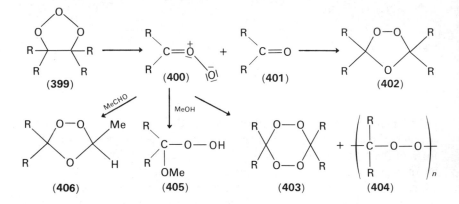

following observations: (i) when ozonolysis is carried out in a participating
solvent, e.g. methanol, the carbonyl oxide is quantitatively trapped to give
a substituted hydroperoxide, e.g. **405**[428,430–433]: (ii) when the ozonolysis
is carried out in the presence of reactive foreign aldehydes, e.g. formaldehyde or acetaldehyde, the corresponding ozonides, e.g. **406** are formed[428];
(iii) ozonolysis of phenylethylenes at temperatures between $-78\,°C$ and
$25\,°C$ in the presence ^{18}O-labelled benzaldehyde[434] as well as that of
ethylene, propylene, cis- and trans-2-butene in the presence of [18]O-
labelled formaldehyde and acetaldehyde at $-95\,°C$ and $-126\,°C$[435],
lead to incorporation of the label only at the ether bridge; (iv) cross
ozonides are formed from unsymmetrical olefins[429] (see equation 69).

The Criegee proposal contains no provision for the dependence of the
cis/trans ratio of the final ozonide on the stereochemistry of the olefinic
precursor. Recently it was shown that the cis/trans ratios of normal
ozonides as well as that of cross ozonides are governed by geometrical
isomerism and substituents on the double bond of the olefin[429,436–441]
(equation 68[438] and 69[439]). The cis/trans ratio of the ozonides formed is
also dependent on temperature, solvent, olefin concentration and fast
or slow warm-up of reaction mixtures[436,438,439,442–445]. In subsequent
discussions, Bailey and coworkers considered Criegee's mechanism valid,
provided that some refinements were introduced[438], while Murray and

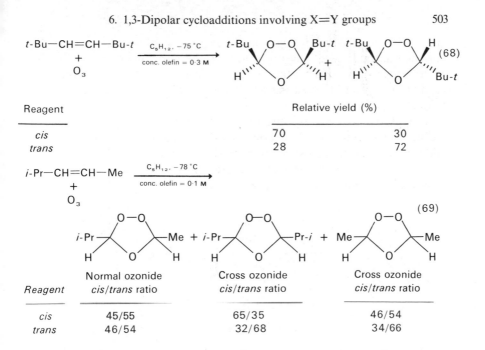

Reagent	Relative yield (%)	
cis	70	30
trans	28	72

Reagent	Normal ozonide cis/trans ratio	Cross ozonide cis/trans ratio	Cross ozonide cis/trans ratio
cis	45/55	65/35	46/54
trans	46/54	32/68	34/66

collaborators postulated that the latter ozonolysis results were not consistent with an exclusive Criegee's mechanism and that an additional pathway was needed, particularly in order to describe low-temperature reactions with high concentrations of foreign added aldehyde. They proposed the reaction scheme described in equation (70)[429,437,441].

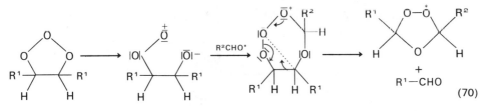

This mechanism requires that with ^{18}O-labelled aldehyde the label only enters into the peroxide bridge of the ozonide. Indeed, in the ozonolysis at $-122\,^{\circ}C$ (pentane, fast warm-up) of trans-diisopropyl ethylene in the presence of a high concentration of ^{18}O-labelled isobutyraldehyde, 60% of the label is found in the ether bridge and 40% in the peroxide bridge[444].

However, more recent reports by Kuczkowski and coworkers[435] dealing with ozonolysis of cis- and trans-diisopropylethylene (in pentane at $-78\,^{\circ}\mathrm{C}$ or at $-111\,^{\circ}\mathrm{C}$ with fast warm-up) in the presence of $^{18}\mathrm{O}$-enriched acetaldehyde, have proved that the enrichment interested the ether bridge only; this finding is in contrast with previous studies according to which the enrichment substantially involved the peroxy bridge[445a].

Recently it has been pointed out[446] that if the first step of Criegee's mechanism (formation of carbonyl oxide) is a retro-1,3-dipolar cycloaddition and the second step (formation of ozonide) a 1,3-dipolar cycloaddition, then a suitable model for stereochemical studies would be one typical of 1,3-dipolar cycloadditions.

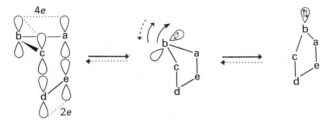

Solid line: cycloaddition; dotted line: retro-cycloaddition

According to the above model the cleavage of 1,2,3-trioxolane should proceed through an initial bending of the molecule in order to obtain an envelope conformation in which the oxygen atom adjacent to the carbon going to zwitterion must begin to move markedly away from the plane of the ring. As a direct consequence of the two possible movements of the oxygen out of the plane of trioxolane, the substituent at the carbon atom going to zwitterion can assume either an axial or an equatorial orientation so as to give rise to a final *syn*-**407** or *anti*-**408** carbonyl oxide respectively.

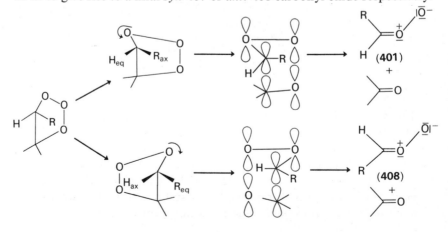

The energies involved in the two transition states will be determined by the sum of interactions between substituents and substituents and ring.

For the cycloaddition giving the final ozonide *endo* or *exo* transition states are possible. The *syn* (or *anti*) carbonyl oxide interacts with an aldehyde to give through the *endo* transition state **A(D)** a *trans* (*cis*), and through the *exo* transition state **B(C)** a *cis* (*trans*), final ozonide as shown below.

The most favoured transition states were considered those in which interaction between the substituent on the aldehyde and end-oxygen lone pairs of the 1,3-dipole is less unfavourable. Molecular models show that **A** and **D** are favoured over **B** and **C**.

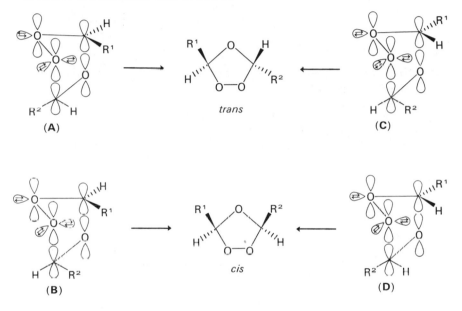

The hypothesis, in addition to that discussed above, that the *syn*-**407** and *anti*-**408** carbonyl oxides do not easily interconvert at low temperatures has allowed the stereochemical results of the ozonolysis to be rationalized[446].

Previous discussion was referred to liquid-phase ozonolysis. For gas-phase ozonolysis a different mechanism has been suggested. The initial ozonide undergoes an homolytic cleavage of an oxygen–oxygen bond to give a diradical which either undergoes a further cleavage to a carbonyl compound and diradical carbonyl oxide or give rise to hydrogen rearrangements[446a].

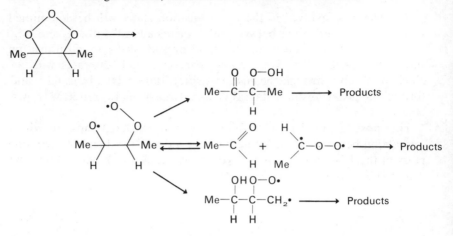

3. Direction of cleavage of unsymmetrical 1,2,3-trioxolanes (primary ozonides)

The study of the factors which determine the cleavage of unsymmetrical ozonides [path (a) or (b) of equation 71] has been made possible by carrying out the ozonolysis in the presence of methanol which quantitatively traps the carbonyl oxides as methoxyhydroperoxides.

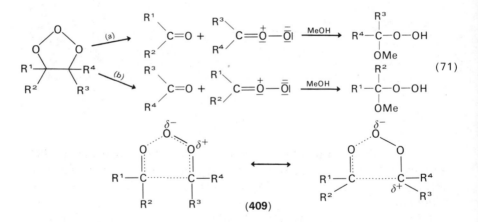

Of the two available routes [paths (a) and (b)] the preferred is that which gives a carbonyl oxide whose substituents are most capable of stabilizing a partial positive charge by inductive, hyperconjugative or mesomeric effects (see transition state **409** for path a)[430–433,447]. The formation of

$RCH=\overset{+}{\underset{}{O}}-\overset{-}{\underset{}{O}}|$ by ozonolysis of monosubstituted ethylenes was shown to be governed by the inductive effect (stabilization in the order t-Bu > i-Pr > Et > Me > H), while the formation of $R^1CH=\overset{+}{\underset{}{O}}\geq\overset{-}{\underset{}{O}}|$ and $R^2CH=\overset{+}{\underset{}{O}}\overset{-}{\underset{}{O}}|$ from *trans*-disubstituted ethylenes is controlled by the hyperconjugative effect (Me > Et > n-Pr > i-Pr > t-Bu, Table 14). For *cis* disubstituted

TABLE 14. Direction of cleavage of the primary ozonides of monosubstituted ethylenes $RCH=CH_2$ and disubstituted *trans*-ethylenes $R^1CH=CHR^2$.

R	$RCH=\overset{+}{\underset{}{O}}-\overset{-}{\underset{}{O}}|$ (%)	R^1	R^2	$R^1-CH=\overset{+}{\underset{}{O}}-\overset{-}{\underset{}{O}}|$ (%)
CH_2Cl	17	Me	Et	56
CH_2Br	20	Me	n-Pr	60
CH_2I	27	Me	i-Pr	71
Ph	60	Me	t-Bu	88
Me	62	Et	i-Pr	65
Et	63	Et	t-Bu	87
n-Pr	60	Me	Ph	82
i-Pr	64	Me	p-MeOC$_6$H$_4$	68
t-Bu	70			
$cyclo$-C$_6$H$_{11}$	86			

ethylenes the order was found to be t-Bu > Me > Et > n-Pr > i-Pr[433]. The ozonolysis of 1,2,3-trisubstituted olefins with alkyl or phenyl groups mainly gives the more substituted carbonyl oxide.

The ratio of paths (a) and (b) has been shown to be independent of both solvent and temperature[432,433]. The temperature independence indicates a nearly identical $\Delta H^{\neq}$ value for the two paths which must, therefore, differ only in $\Delta S^{\neq}$ values.

The values for the activation energy of this retro-cycloaddition in the case of 1-hexene[448] and *trans*-3-hexene[449] were $\simeq 7$ and $\simeq 8$ kcal/mol, respectively.

4. Cycloaddition of carbonyl oxides to C=O and C=C double bonds

The formation of the ozonides is straightforward when the carbonyl compound is an aldehyde. Ketones are generally less reactive towards carbonyl oxide, e.g. the ozonolysis of tetramethylethylene yields only acetone, acetone diperoxide and polymeric peroxides[426]. However, ketones with negative substituents react readily as is shown by the high yield of ozonides obtained from the ozonolysis of 1,4-dibromo-2,3-dimethyl-2-butenes[445]. High reactivity has been also reported for the intramolecular reaction of carbonyl oxides with ketones formed in the ozonolysis of cyclic olefins such as 1,2-dimethylcyclopentene[450], hexamethyl Dewar benzene (equation 72)[451] and octamethylsemibullvalene[452].

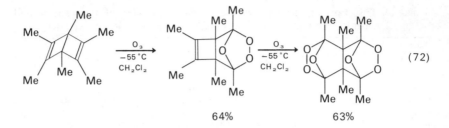

(72)

64%   63%

Aromatic and aliphatic ketones in large excess were found to react with carbonyl oxides: ozonolysis of 2-pentene in acetone gave the ozonides **410** and **411**[453] while ozonolysis of tetraphenylethylene in acetone or benzophenone gave ozonides **412** (74%) and **413** (25%) respectively[454].

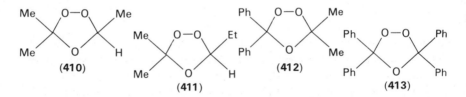

The cycloadditions of carbonyl oxides with carbonyl compounds are always regiospecific.

In general, C=C double bonds are not reactive towards carbonyl oxides, although a few reports have postulated the formation of an intermediate cycloadduct between them[455–457]. This low reactivity is somewhat surprising for 1,3-dipoles whose cycloadditions should be LUMO-dipole controlled.

F. Ozone

Ozone reacts readily both with aliphatic and aromatic double bonds[428]. The reaction has been studied in the gas phase (to 300 °C), and in solutions (to -175 °C) either in neutral solvents, e.g. CCl_4, $CHCl_3$, ether, pentane or in participating solvents such as methanol.

The mechanism of the reaction of ozone with olefins involves a stereo-specific *cis*-addition to the double bond to give the primary ozonides **414**[1,428,429]. These 1,2,3-trioxolanes are unstable and rearrange to form the final ozonides **415** which are sufficiently stable to be handled by gas chromatography. However, the primary adduct ozone-*trans*-di-*t*-butyl-ethylene (**414**, $R^1 = R^2 = t$-Bu) was isolated at $-75\,°C$ as a crystalline compound and could be reduced to a racemic diol (**416**, $R^1 = R^2 = t$-Bu)[458]. The n.m.r. spectrum of the adduct showed a single methine proton absorption, consistent with its symmetrical structure[459]. Formation of the

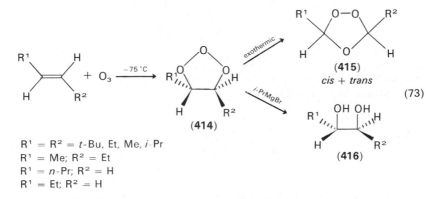

(73)

R¹ = R² = *t*-Bu, Et, Me, *i*-Pr
R¹ = Me; R² = Et
R¹ = *n*-Pr; R² = H
R¹ = Et; R² = H

trioxolane **414** was also detected (at $-110\,°C$), by n.m.r. and by reduction to a diol, for a number of other terminal and *trans*-olefins (equation 73)[449,460–462]. The stability of the primary ozonide obtained from *trans*-olefins is much greater than that from *cis*-olefins. The primary ozonide from ozone and *cis*-3-hexene could be detected by n.m.r. spectroscopy only at $-130\,°C$[462].

Recent n.m.r., u.v. and i.r. studies at very low temperatures ($\leq -150\,°C$) have shown that ozone forms π complexes with the olefins which are in equilibrium with the starting compounds[463,464].

The olefin stereochemistry was retained (except in one case[464a]) in the epoxides which are sometimes the major ozonolysis products, particularly with 1,1-disubstituted olefins having large substituents[423,429,465].

Based on these data, a mechanism may be proposed for the reaction of ozone with olefins (equation 74) in which the first step involves an attraction between O_3 and the olefin to form a π complex, **417**, which either transforms into the primary ozonide through a 1,3-dipolar cycloaddition (path b), or into a σ complex, **418a** or **b** (path a) to give the epoxide[465]. The nature and the role of the π complex is not yet fully clear: in fact it can either be a true intermediate of the epoxides and ozonides, or it can enter into a

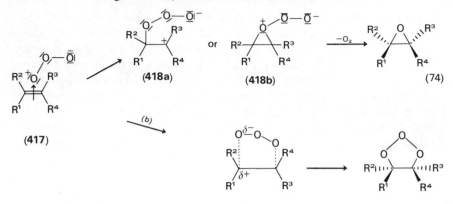

side reaction and dissociation of the π complex must occur prior to further ozone attack (equation 74a).

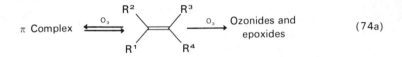

$$\pi \text{ Complex} \rightleftharpoons \xrightarrow{O_3} \quad \xrightarrow{O_3} \text{ Ozonides and epoxides} \qquad (74a)$$

Huisgen first classified ozone–olefin reactions as 1,3-dipolar cyclo-additions and as evidence for a concerted reaction considered the following data for the reactions of ozone with aromatics[1] : (i) reaction rates are little influenced by solvent polarity; (ii) ozone preferentially attacks the aromatic bond possessing the maximum double bond character, (iii) the activation parameters are similar to those of other 1,3-dipolar cyclo-additions (E_a = 13·2 and 10·7 kcal/mol and $\Delta S^{\neq}$ = −23 and −22 e.u. for benzene and mesitylene respectively)[1,466].

Ozone has very low-lying unoccupied orbitals and as a consequence it should show electrophilic characteristics in its 1,3-dipolar cycloaddition (Type III, LUMO controlled). Excellent correlations were found between the electron-donating properties of alkyl-substituted benzenes and their reaction rates with ozone which are relatively small and range (in CCl_4 at 25 °C) from 0·028 to 245 l/mol sec for benzene and hexamethyl-benzene[466].

Less straightforward are the high second-order rate constants found for the reactions of ozone with olefins (Table 15)[467–470].

Table 15 shows a pronounced decrease in reactivity with ethylenes when chlorine is substituted successively for hydrogen and an increase in reactivity for substitution with alkyl groups. These results have been

TABLE 15. Second order rate constants (k, l/mol sec) for the reactions of ozone with olefins in CCl_4 at 25°C[469]

Olefin	k	IP[471]	Olefin	k	IP
$CCl_2=CCl_2$	1.0	9.32	1,3-Butadiene	74,000	9.07
$CHCl=CCl_2$	3.6	9.45	Styrene	103,000	10.47[10]
$CH_2=CCl_2$	22.1	9.79	1-Hexene	76,000	9.46
cis-$CHCl=CHCl$	35.7	9.66	2-Hexene	148,000	9.16
trans-$CHCl=CHCl$	591	9.64	Isobutylene	97,000	9.23
$CH_2=CHCl$	1180	10.0	cis-2-Butene	163,000	9.13
$CH_2=CH_2$	~25,000	10.5	Trimethylethylene	167,000	8.67
$CH_2ClCH=CH_2$	11,000	10.04	Tetramethylethylene	200,000	8.30
$CH_3CH=CH_2$	~80,000	9.73	Cyclopentene	200,000	9.01

considered as evidence for the electrophilic character of ozone in its cycloaddition with olefins.

No correlation, however, can be drawn between reaction rates and IP of the olefins in Table 15; in addition, and contrary to expectation, the increase in reaction rates of halogenated olefins parallels the increase in their IP. As a possible explanation, steric influences have been invoked particularly for cycloadditions with 1,1-disubstituted alkenes[469]. The possibility of a two-step process was also raised[469] and this hypothesis has received further support from studies of the rates of the attack of ozone on 1,2-disubstituted alkyl ethylenes and on phenyl ethylenes. Two different mechanisms were suggested for the two types of alkenes. In the first case the attack by ozone on 1,2-dialkyl ethylenes (equation 75: $R^1 = R^2 =$ Et, n-Pr, i-Pr, t-Bu) was found to follow an overall nucleophilic trend[472]. In CCl_4 solution the kinetics are described adequately by Taft's equation

$$\log k = \log k^\circ + \rho^* \sum \sigma^*$$

where $\sum \sigma^*$ is the sum of Taft's polar constants[473] for the two substituents and $\rho^* = 3{\cdot}75$ (for *trans*) and $\rho^* = 2{\cdot}60$ (for *cis*) are the polar reaction constants. A mechanism implying very fast reversible formation of an intermediate complex (a π or σ complex) by electrophilic attack of ozone on the olefin and a successive slow nucleophilic (rate-determining step, positive ρ) attack by O^- end to give the trioxolane [path (a), equation 75] was considered consistent with the above results. It was therefore argued that steric effects are of minor if any importance for these reactions[472].

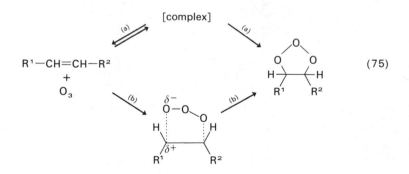

(75)

In the second case however, for reaction of ozone with phenyl ethylenes small negative Hammett ρ values ($-0{\cdot}87$, $-1{\cdot}03$, $-0{\cdot}87$, $-0{\cdot}89$ respectively for **419a, b, c, d**) were found[48,474]. It was proposed that the initial electrophilic attack of ozone on these olefins proceeds rapidly *via* an

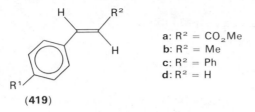

a: $R^2 = CO_2Me$
b: $R^2 = Me$
c: $R^2 = Ph$
d: $R^2 = H$

(419)

irreversible 1,3-dipolar cycloaddition to give the primary ozonide [path (b), equation 75][472,474]†.

The reaction rates of the two sets of ozonolyses were found to be scarcely affected by solvent polarity[48,472].

Interestingly enough the $\Delta H^{\neq}$ (in CCl_4) for the reaction of ozone with *cis-* and *trans-*stilbene and triphenylethylene was found to be $\simeq 0$, whilst the $\Delta S^{\neq}$ was -39, -33 and -39 e.u. respectively[48]. Activation-energy values close to or equal to zero have also been found in the ozonolysis of ethylene and 1-hexene in the gas phase[475], and 1-hexene[470] and 1-dodecene in the liquid phase[476]. These results seem to suggest that the activation entropy term represents the major contribution (the only one in the above cited examples) to the activation free energy in the reaction of ozone with olefins.

Finally the experimental basis for a mechanism of ozonolysis proposed recently by Story and coworkers[477-479] has not been verified in subsequent studies[480-482].

Less is known about the reaction of oxone with hetero double bonds. The reaction with carbon–nitrogen double bonds has been described as involving an electrophilic attack by the ozone either on the carbon or on the nitrogen atom of the dipolarophile[169,483]. An initial 1,3-dipolar cycloadduct has been proposed for the reaction of ozone with unhindered thioketones and sulphines[484].

G. Thiocarbonyl Ylides

1. Structure

Along with classical mesomeric structures common to the other 1,3-dipoles (see Section II.A), thiocarbonyl ylides can be represented by other additional structures, e.g. **420** and **421**, involving the 3d orbitals of sulphur[485].

Probably the planar structure **422** does not represent the correct geometry for the thiocarbonyl ylides, at least for certain examples[486].

---

† Mechanism of path (a) has been also proposed for phenylethylenes by other authors: E. R. Altwicker and J. Basila, *Tetrahedron*, **29**, 1969 (1973).

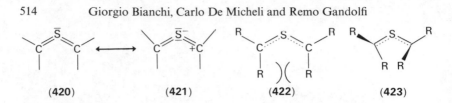

(420)   (421)   (422)   (423)

In fact, tetrasubstituted thiocarbonyl ylides, in a planar structure **422**, would present steric constraints. Therefore the tilted form **423**, in which the strain is relieved by tilting of one p orbital upward and the other downward by the same angle, seem more realistic[486]. The tilted model **423** involves some form of p–d hybridization that the cylindrical symmetry of d orbital might make possible[486].

2. Preparation and reactions

These 1,3-dipoles can be obtained as reactive intermediates in the decomposition of Δ^3-1,3,4-thiadiazolines and then trapped with a suitable dipolarophile, as shown in equation 76[487]. In the absence of a dipolarophile the intermediate thiocarbonyl ylide transforms into the corresponding episulphide, e.g. **427**, through a conrotatory closure[487].

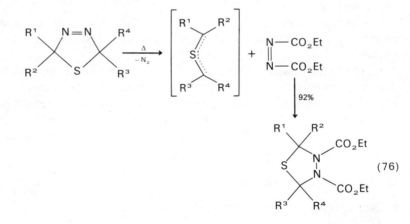

(76)

The thiocarbonyl ylide **424** was found to be reactive with electron-poor dipolarophiles such as N-phenylmaleimide[486], tetracyanoethylene[486] as well as with the C=O bond of diphenylketene[488]. With the latter two dipolarophiles, adducts **425** and **426** were obtained. Norbornadiene was unreactive and the only compound isolated was **427**[486]. These results suggest that the reactions are HOMO-dipole controlled.

Mesoionic 1,3-dithiol-5-ones (**428**) represent a good source of thiocarbonyl ylides. They react with several alkenes at 100 °C to give the stable compounds **429** which are transformed into thiophene derivatives on Pd/C treatment[489]. 1,3-Dipolar cycloadditions of other mesoionic thiocarbonyl ylides such as 1,3-thiazol-4-ones have been thoroughly investigated[489a].

Compound **430**, here represented by three canonical structures, can behave either as thiocarbonyl ylide (**430a**) or azomethine ylide (**430b**).

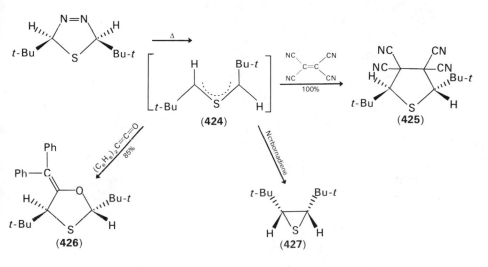

Its 1,3-dipolar cycloadditions with olefinic dipolarophiles are temperature-dependent[490]: e.g. **430** reacts with fumaronitrile in toluene or xylene to give a mixture of adducts **431** and **432** (**431** being readily convertible into **432**

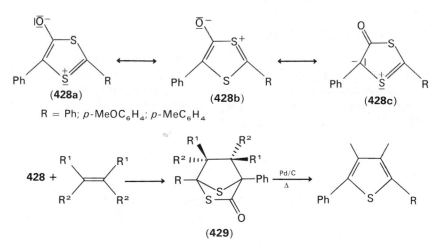

(428a)   (428b)   (428c)

R = Ph; p-MeOC$_6$H$_4$; p-MeC$_6$H$_4$

(429)

on heating), whereas only **433**   is formed in benzene. Experiments on thermal stability and interrelation of adducts **431–433** showed the greater reactivity of the azomethine ylide as compared with the thio-carbonyl ylide system and the greater stability of adduct **431** over **433**[490].

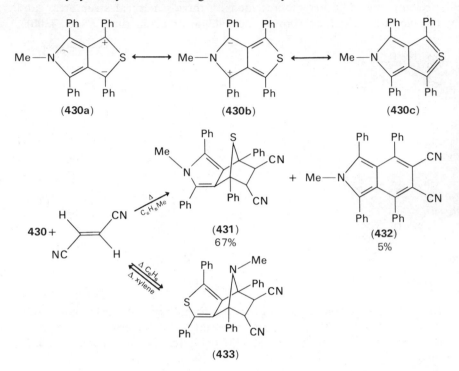

(430a)   (430b)   (430c)

430+

(431)
67%

(432)
5%

(433)

H. Thiocarbonyl Imines

Thiocarbonyl imines have been little studied. This 1,3-dipolar system is present in the compounds **434**[491], **435**[491] and in the mesoionic 1,3,2-oxathiazole derivatives **436**[492]. The latter were reported as reacting with olefins to give the isothiazoles (**438**) through the unstable cycloadducts **437** on CO_2 and hydrogen elimination[492]. Fluorenthione S-benzoylimide

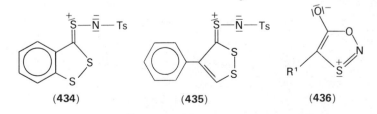

(434)   (435)   (436)

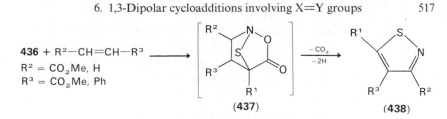

$$436 + R^2—CH=CH—R^3 \longrightarrow$$

$R^2 = CO_2Me, H$
$R^3 = CO_2Me, Ph$

(437)

(438)

(440) containing the thiocarbonyl imine system is readily obtainable from 439 on treatment with Et_3N at $-78\,°C^{493}$. Compound 440 alone, at a

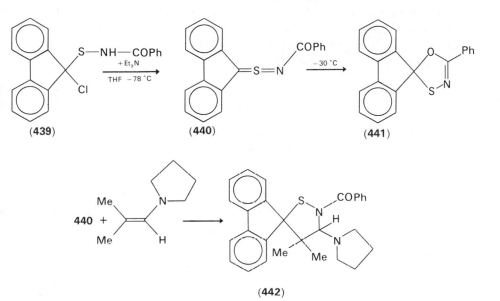

(439)

(440)

(441)

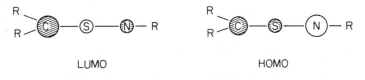

(442)

higher temperature $(-30\,°C)$, undergoes an intramolecular cyclization to give 1,3,4-oxathiazole (441). However when 440 is generated in the presence of N-isobutenylpyrrolidine or N-propenylpiperidine, 1,3-dipolar cycloaddition occurs to give, in the case of the former dipolarophile, cycloadduct 442. Regiochemistry shown by 442 is correct provided that these cycloadditions were $LUMO_{dipole}$ controlled.

LUMO

HOMO

FIGURE 16. Estimated frontier orbital coefficients of R_2CSNR.

Inspection of Figure 16 shows that the greatest coefficient of the $LUMO_{dipole}$ is on the carbon atom; it follows (see Sections II.B and II.J) that the greatest interaction is between the above cited carbon atom and β-carbon of the enamine to give regioisomer **442**[493].

VI. ACKNOWLEDGEMENTS

We wish to thank Professor P. Grünanger and the colleagues P. Caramella, G. De Simoni, P. P. Righetti, G. Tacconi, G. Vidari and P. Vita Finzi of the Institute of Organic Chemistry, University of Pavia, and for helpful suggestions. We also wish to express our gratitude to Professors K. N. Houk, Louisiana State University, and R. Sustmann, Münster University, for preprints of some of their manuscripts. We also thank C.N.R. (Rome) for financial aid.

VII. REFERENCES

1. R. Huisgen, R. Grashey, and J. Sauer, *The Chemistry of Alkenes* (Ed. S. Patai), Interscience, London, 1964, pp. 806–878.
2. R. Huisgen, *Angew. Chem. Int. Ed.*, **7**, 321 (1968).
3. J. W. Linnett, *The Electronic Structure of Molecules*, Methuen, London, 1964.
3a. W. R. Wadt and W. A. Goddard, *J. Amer. Chem. Soc.*, **97**, 3004 (1975); S. P. Walch and W. A. Goddard, *J. Amer. Chem. Soc.*, **97**, 5319 (1975).
4. K. N. Houk, J. Sims, R. E. Duke, Jr., R. W. Strozier, and J. K. George, *J. Amer. Chem. Soc.*, **95**, 7287 (1973).
5. K. N. Houk, J. Sims, C. R. Watts, and L. J. Luskus, *J. Amer. Chem. Soç.*, **95**, 7301 (1973).
6. K. N. Houk, *J. Amer. Chem. Soc.*, **94**, 8953 (1972).
7. J. Bastide, N. El Ghandour, and O. Henri-Rousseau, *Bull. Soc. Chim. France*, 2290 (1973).
8. J. Bastide and O. Henri-Rousseau, *Bull. Soc. Chim. France*, 2294 (1973).
9. R. Sustmann, *Tetrahedron Letters*, 2717 (1971).
10. R. Sustmann and H. Trill, *Angew. Chem. Int. Ed.*, **11**, 838 (1972).
11. J. Bastide, N. El Ghandour, and O. Henri-Rousseau, *Tetrahedron Letters*, 4225 (1972).
12. J. Bastide and O. Henri-Rousseau, *Tetrahedron Letters*, 2979 (1972).
13. D. W. Turner, C. Baker, A. D. Baker, and C. R. Brundle, *Molecular Photoelectron Spectroscopy*, Wiley-Interscience, London, New York, Sydney, Toronto, 1970.
14. K. N. Houk, *J. Amer. Chem. Soc.*, **95**, 4092 (1973).
15. W. C. Herndon and W. B. Giles, *Mol. Photochem.*, **2**, 277 (1970).
16. W. C. Herndon, *Tetrahedron Letters*, 125 (1971).
17. R. Huisgen, M. Seidl, G. Wallbillich, and H. Knupfer, *Tetrahedron*, **17**, 3 (1962).
18. R. Huisgen, H. Knupfer, R. Sustmann, G. Wallbillich, and V. Weberndörfer, *Chem. Ber.*, **100**, 1580 (1967).

19. A. Quilico, G. Stagno d'Alcontres, and P. Grünanger, *Gazz. Chim. Ital.*, **80**, 479 (1950).
20. K. Bast, M. Christl, R. Huisgen, W. Mack, and R. Sustmann, *Chem. Ber.*, **106**, 3258 (1973).
21. M. Christl, R. Huisgen, and R. Sustmann, *Chem. Ber.*, **106**, 3275 (1973).
22. K. v. Auwers and E. Cauer, *Ann. Chem.*, **470**, 284 (1929).
23. P. Eberhard, *Thesis*, University of Munich 1967
24. R. Huisgen and G. Szeimies, *Chem. Ber.*, **98**, 1153 (1965).
25. R. Huisgen, H. Gotthardt, and H. O. Bayer, *Chem. Ber.*, **103**, 2368 (1970).
26. A. Eckell, *Ph.D. Thesis*, University of Munich, 1962.
27. R. Huisgen, R. Grashey, P. Laur, and H. Leitermann, *Angew. Chem.*, **72**, 416 (1960).
28. R. Huisgen, *Angew. Chem. Int. Ed.*, **2**, 565 (1963).
29. R. Huisgen, H. Seidl, R. Grashey, and H. Hauck, *Chem. Ber.*, **102**, 736 (1969).
30. W. J. Linn and R. E. Benson, *J. Amer. Chem. Soc.*, **87**, 3657 (1965).
31. A. Padwa, M. Dharan, J. Smolanoff, and S. I. Wetmore, Jr., *J. Amer. Chem. Soc.*, **95**, 1945 (1973).
32. R. Huisgen, *Angew. Chem. Int. Ed.*, **2**, 633 (1963).
33. R. Huisgen, *J. Org. Chem.*, **33**, 2291 (1968); R. Huisgen, submitted for publication.
34. J. E. Baldwin and R. H. Fleming, *Fortschr. Chem. Forsch.*, **15**, 281 (1970).
35. J. W. McIver, Jr., *J. Amer. Chem. Soc.*, **94**, 4782 (1972); *Accounts Chem. Res.*, **7**, 72 (1974).
35a. C. E. Olsen, *Acta Chem. Scand.* **27**, 2989 (1973).
36. R. A. Firestone, *J. Org. Chem.*, **33**, 2285 (1968).
37. R. A. Firestone, *J. Chem. Soc. (A)*, 1570 (1970).
38. R. A. Firestone, *J. Org. Chem.*, **37**, 2181 (1972).
39. W. F. Baine and E. I. Snyder, *Tetrahedron Letters*, 2263 (1970).
40. W. R. Dolbier and S. Hong Dai, *Tetrahedron Letters*, 4645 (1970).
41. B. M. Benjamin and C. J. Collins, *J. Amer. Chem. Soc.*, **95**, 6145 (1973).
42. R. Huisgen, G. Szeimies, and L. Möbius, *Chem. Ber.*, **100**, 2494 (1967).
43. A. Ledwith and D. Parry, *J. Chem. Soc. (C)*, 1408 (1966).
44. A. Ledwith and Y. Shih-Lin, *J. Chem. Soc. (B)*, 83 (1967).
45. R. Huisgen, H. Seidl, and I. Brünig, *Chem. Ber.*, **102**, 1102 (1969).
46. A. Battaglia and A. Dondoni, *Ric. Scient.*, **38**, 201 (1968).
47. A. S. Bailey and J. E. White, *J. Chem. Soc. (B)*, 819 (1966).
48. H. Henry, M. Zador, and S. Fliszár, *Can. J. Chem.*, **51**, 3398 (1973).
49. C. Reichardt and K. Dimroth, *Fortschr. Chem. Forsch.*, **11**, 1 (1968).
50. P. K. Kadaba, *Synthesis*, 71 (1973).
51. G. Bianchi, C. de Micheli, R. Gandolfi, P. Grünanger, P. Vita Finzi, and O. Vanja de Pava, *J. Chem. Soc. Perkin I*, 1148 (1973).
52. A. P. Krapcho, M. P. Silvon, I. Goldberg, and E. G. E. Jahngen, Jr., *J. Org. Chem.*, **39**, 860 (1974).
53. G. Bianchi, C. de Micheli, A. Gamba, and R. Gandolfi, *J. Chem. Soc. Perkin I*, 137 (1974).
54. P. Scheiner, J. H. Schomaker, S. Deming, W. J. Libbey, and G. P. Nowack, *J. Amer. Chem. Soc.*, **87**, 306 (1965).
55. G. L'abbé, P. Ykman, and G. Smets, *Tetrahedron*, **25**, 5421 (1969).
56. R. Knorr, R. Huisgen, and G. K. Staudinger, *Chem. Ber.*, **103**, 2639 (1970).

57. P. Beltrame and C. Vintani, *J. Chem. Soc.* (*B*), 873 (1970).
58. R. Huisgen and P. Eberhard, *Tetrahedron Letters*, 4343 (1971).
59. J. D. Roberts, *Chem. Ber.*, **94**, 273 (1961).
60. W. C. Agosta and A. B. Smith, III, *J. Org. Chem.*, **35**, 3856 (1970).
61. R. B. Woodward and R. Hoffmann, *Angew. Chem. Int. Ed.*, **8**, 781 (1969).
62. C. De Micheli, R. Gandolfi, and P. Grünanger, *Tetrahedron*, **30**, 3765 (1974).
63. K. N. Houk and C. R. Watts, *Tetrahedron Letters*, 4025 (1970).
64. G. Vinci, *Thesis*, University of Pavia (1973).
65. K. N. Houk and C. J. Luskus, *Tetrahedron Letters*, 4029 (1970).
66. P. Caramella, P. Frattini, and P. Grünanger, *Tetrahedron Letters*, 3817 (1971).
67. M. J. S. Dewar, *Angew. Chem. Int. Ed.*, **10**, 761 (1971).
68. Kei-wei Shen, *J. Chem. Ed.*, **50**, 238 (1973).
69. H. E. Zimmermann, *Accounts Chem. Res.*, **4**, 272 (1971).
70. L. Salem, *J. Amer. Chem. Soc.*, **90**, 543 and 553 (1968).
71. A. Devaquet and L. Salem, *J. Amer. Chem. Soc.*, **91**, 3793 (1969).
72. K. Bast, M. Christl, R. Huisgen, and W. Mack, *Chem. Ber.*, **106**, 3312 (1973).
73. D. Maggi, *Thesis*, University of Pavia, 1973.
74. L. W. Boyle, M. J. Peagram, and G. H. Whithman, *J. Chem. Soc.* (*B*), 1728 (1971).
75. T. Aratani, Y. Nakanisi, and H. Nozaki, *Tetrahedron*, **26**, 4339 (1970).
76. K. R. Henery-Logan and R. A. Clark, *Tetrahedron Letters*, 801 (1968).
77. A. Eckell, R. Huisgen, R. Sustmann, G. Wallbillich, D. Grashey, and E. Spindler, *Chem. Ber.*, **100**, 2192 (1967).
78. W. Fliege and R. Huisgen, *Ann. Chem.*, 2038 (1973).
79. A. Hassner, *J. Org. Chem.*, **33**, 2684 (1968).
80. K. Fukui, *Fortschr. Chem. Forsch.*, **15**, 1 (1970).
81. Ch. Grundmann and P. Grünanger, *The Nitrile Oxides*, Springer-Verlag, Berlin, 1971.
82. K. N. Houk, L. J. Luskus, and N. S. Bhacca, *J. Amer. Chem. Soc.*, **92**, 6392 (1970).
83. K. N. Houk, J. K. George, and R. E. Duke, Jr., *Tetrahedron*, **30**, 523 (1974).
84. M. Joucla, D. Greé, and J. Hamelin, *Tetrahedron*, **29**, 2315 (1973).
85. R. Huisgen, R. Sustmann, and K. Bunge, *Chem. Ber.*, **105**, 1324 (1972).
86. G. Lo Vecchio, G. Grassi, F. Risitano, and F. Foti, *Tetrahedron Letters*, 3777 (1973).
86a. A. Padwa and H. J. Carlsen, *J. Amer. Chem. Soc.*, **97**, 3862 (1975).
86b. Th. Kauffmann, *Angew. Chem. Int. Ed.*, **13**, 627 (1974).
87. G. Boche and D. Martens, *Angew. Chem. Int. Ed.*, **11**, 724 (1972).
88. R. Eidenschink and Th. Kauffmann, *Angew. Chem. Int. Ed.*, **11**, 292 (1972).
88a. J. P. Marino and W. B. Mesbergen, *J. Amer. Chem. Soc.*, **96**, 4050 (1974).
89. Th. Kauffmann, H. Berg, and E. Köppelmann, *Angew. Chem. Int. Ed.*, **9**, 380 (1970).
90. Th. Kauffmann, H. Berg, E. Ludorff, and A. Woltermann, *Angew. Chem. Int. Ed.*, **9**, 960 (1970).
91. Th. Kauffmann and R. Eidenschink, *Angew. Chem. Int. Ed.*, **10**, 739 (1971).
92. Th. Kauffmann and E. Köppelmann. *Angew. Chem. Int. Ed.*, **11**, 290 (1972).
93. Th. Kauffmann, K. Habersaat, and E. Köppelmann, *Angew. Chem. Int. Ed.*, **11**, 291 (1972).
94. Th. Kauffmann and R. Eidenschink, *Angew. Chem. Int. Ed.*, **12**, 568 (1973).

94a. Th. Kauffmann, E. Koppelmann, and H. Berg, *Angew. Chem. Int. Ed.*, **9**, 163 (1970).
95. Th. Kauffmann, D. Berger, B. Scheerer, and A. Woltermann, *Angew. Chem. Int. Ed.*, **9**, 961 (1970).
96. W. Baunwarth, R. Eidenschink, and Th. Kauffmann, *Angew. Chem. Int. Ed.*, **13**, 468 (1974).
97. Th. Kauffmann, A. Busch, K. Habersaat, and B. Scheerer, *Tetrahedron Letters*, 4047 (1973).
98. D. Seebach, *Angew. Chem. Int. Ed.*, **8**, 639 (1969).
99. K. O. Berlin, B. S. Rathore, and M. Peterson, *J. Org. Chem.*, **30**, 226 (1965).
100. R. L. Letsinger and D. F. Pollart, *J. Amer. Chem. Soc.*, **78**, 6079 (1956).
101. J. N. Hines, M. J. Peagram, G. H. Whitham, and M. Wright, *Chem. Comm.*, 1593 (1968).
102. R. B. Bates, L. M. Kroposki, and D. E. Patter, *J. Org. Chem.*, **37**, 560 (1972).
103. P. Eberhard and R. Huisgen, *J. Amer. Chem. Soc.*, **94**, 1345 (1972).
104. K. Bunge, R. Huisgen, R. Raab, and H. Stangl, *Chem. Ber.*, **105**, 1279 (1972).
105. K. Bunge, R. Huisgen, and R. Raab, *Chem. Ber.*, **105**, 1296 (1972).
106. R. Huisgen, H. Stangl, H. J. Sturm, R. Raab, and K. Bunge, *Chem. Ber.*, **105**, 1258 (1972).
107. K. Burger, J. Albanbauer, and F. Manz, *Chem. Ber.*, **107**, 1823 (1974).
108. K. Burger and J. Fehn, *Angew. Chem. Int. Ed.*, **10**, 728 and 729 (1971).
109. K. Burger and J. Fehn, *Chem. Ber.*, **105**, 3814 (1972).
110. K. Burger and J. Fehn, *Tetrahedron Letters*, 1263 (1972).
111. K. Burger and K. Einhellig, *Chem. Ber.*, **106**, 3421 (1973).
112. K. Burger and J. Fehn, *Angew. Chem. Int. Ed.*, **11**, 47 (1972).
113. K. Burger, J. Fehn, and E. Müller, *Chem. Ber.*, **106**, 1 (1973).
114. K. Burger, W. Thenn, and E. Müller, *Angew. Chem. Int. Ed.*, **12**, 155 (1973).
115. A. Padwa, M. Dharan, J. Smolanoff, and S. I. Wetmore, Jr., *Pure Appl. Chem.*, **33**, 269 (1973).
116. P. Claus, T. H. Dopper, N. Gakis, M. Georgarakis, H. Giezendanner, P. Gilgen, H. Heimgartner, B. Jackson, M. Märky, N. S. Narasimhan, H. J. Rosenkranz, A. Wunderli, H. J. Hansen, and H. Schmid, *Pure Appl. Chem.*, **33**, 339 (1973).
117. A. Orahovats, B. Jackson, H. Heimgartner, and H. Schmid, *Helv. Chim. Acta*, **56**, 2007 (1973); A. Orahovats, H. Heimgartner, H. Schmid, and W. Heinzelmann, *Helv. Chim. Acta*, **57**, 2626 (1974).
118. A. Padwa and S. I. Wetmore, Jr., *J. Org. Chem.*, **39**, 1396 (1974).
119. H. Giezendanner, H. Heimgartner, B. Jackson, T. Winkler, H. J. Hansen, and H. Schmid, *Helv. Chim. Acta*, **56**, 2611 (1973).
120. A. Padwa, J. Smolanoff, and S. I. Wetmore, Jr., *J. Org. Chem.*, **38**, 1333 (1973).
121. H. Giezendanner, H. J. Rosenkranz, H. J. Hansen, and H. Schmid, *Helv. Chim. Acta*, **56**, 2588 (1973).
122. B. Singh, A. Zweig, and J. B. Gallivan, *J. Amer. Chem. Soc.*, **94**, 1199 (1972).
123. E. F. Ullman and B. Singh, *J. Amer. Chem. Soc.*, **88**, 1844 (1966).
124. B. Singh and E. F. Ullman, *J. Amer. Chem. Soc.*, **89**, 6911 (1967).
125. A. Padwa, M. Dharan, J. Smolanoff, and S. I. Wetmore, Jr., *J. Amer. Chem. Soc.*, **95**, 1954 (1973).
126. W. Sieber, P. Gilgen, S. Chaloupka, H. J. Hansen, and H. Schmid, *Helv. Chim. Acta*, **56**, 1679 (1973).

522   Giorgio Bianchi, Carlo De Micheli and Remo Gandolfi

127. W. Steglich, P. Grüber, H. V. Heininger, and F. Kneidl, *Chem. Ber.*, **104**, 3816 (1971).
128. B. Jackson, M. Märky, H. J. Hansen, and H. Schmid, *Helv. Chim. Acta*, **55**, 919 (1972); V. Schmid, P. Gilgen, H. Heimgartner, H. J. Hansen, and H. Schmid, *Helv. Chim. Acta*, **57**, 1393 (1974); P. Claus, P. Gilgen, H. J. Hansen, H. Heimgartner, B. Jackson, and H. Schmid, *Helv. Chim. Acta*, **57**, 2173 (1974); P. Gilgen, H. J. Hansen, H. Heimgartner, W. Sieber, P. Uebelhart, and H. Schmid, *Helv. Chim. Acta*, **58**, 1739 (1975).
128a. N. Gakis, H. Heimgartner, and H. Schmid, *Helv. Chim. Acta*, **58**, 748 (1975).
129. A. Padwa and J. Smolanoff, *J. Chem. Soc. Chem. Comm.*, 342 (1973).
130. N. Gakis, M. Märky, H. J. Hansen, and H. Schmid, *Helv. Chim. Acta*, **55**, 748 (1972).
131. A. Padwa, S. Clough, M. Dharan, J. Smolanoff, and S. I. Wetmore, Jr., *J. Amer. Chem. Soc.*, **94**, 1395 (1972).
132. N. S. Narasimhan, H. Heimgartner, H. J. Hansen, and H. Schmid, *Helv. Chim. Acta*, **56**, 1351 (1973).
133. K. Bunge, R. Heisgen, R. Raab, and H. J. Sturm, *Chem. Ber.*, **105**, 1307 (1972).
134. P. Gilgen, H. Heimgartner, and H. Schmid, *Helv. Chim. Acta*, **57**, 1382 (1974).
135. H. Heimgartner, P. Gilgen, U. Schmid, H. J. Hansen, H. Schmid, K. Pfoertner, and K. Bernauer, *Chimia*, **26**, 424 (1972).
136. B. Jackson, N. Gakis, M. Märky, H. J. Hansen, W. von Philipsborn, and H. Schmid, *Helv. Chim. Acta*, **55**, 916 (1972).
137. K. N. Houk, R. W. Strozier, and J. A. Hall, *Tetrahedron Letters*, 897 (1974).
138. H. Giezendanner, M. Märky, B. Jackson, H. J. Hansen, and H. Schmid, *Helv. Chim. Acta*, **55**, 745 (1972).
139. A. Padwa and S. I. Wetmore, Jr., *J. Amer. Chem. Soc.*, **96**, 2414 (1974).
140. N. D. Epiotis, *J. Amer. Chem. Soc.*, **95**, 3087 (1973).
140a. P. Gilgen, B. Jackson, H. J. Hansen, H. Heimgartner, and H. Schmid, *Helv. Chim. Acta*, **57**, 2634 (1974).
141. K. Burger, K. Einhellig, W. D. Roth, and L. Hatzelmann, *Tetrahedron Letters*, 2701 (1974).
142. R. Fusco and C. Musante, *Gazz. Chim. Ital.*, **68**, 147 (1938).
143. S. Y. Hong and J. E. Baldwin, *Tetrahedron*, **24**, 3787 (1968).
144. W. A. F. Gladstone, J. B. Aylward, and K. O. C. Norman, *J. Chem. Soc. (C)*, 2587 (1969).
145. J. S. Clovis, A. Eckell, R. Huisgen, and R. Sustmann, *Chem. Ber.*, **100**, 60 (1967).
146. J. S. Clovis, A. Eckell, R. Huisgen, R. Sustmann, G. Wallbillich, and V. Weberndörfer, *Chem. Ber.*, **100**, 1593 (1967).
147. R. Huisgen, R. Sustmann, and G. Wallbillich, *Chem. Ber.*, **100**, 1786 (1967).
148. R. Paul and S. Tchelitcheff, *Bull. Soc. Chim. France*, 4179 (1967).
149. R. Fusco, G. Bianchetti, and D. Pocar, *Gazz. Chim. Ital.*, **91**, 1233 (1961).
150. V. N. Chistokletov, *Zh. Obshch. Khim.*, **34**, 1190 (1964).
151. T. Sasaki and K. Kanematsu, *J. Chem. Soc. (C)*, 2147 (1971).
152. A. Aspect, P. Battioni, L. Vo-Quang, and Vo-Quang-Yen, *C.R. Acad. Sci. Paris, Ser. C.*, **268**, 1063 (1969).
153. P. Caramella, *Tetrahedron Letters*, 743 (1968).
154. Le Quoc Khanh and B. Laude, *C.R. Acad. Sci. Paris, Ser. C.*, **276**, 109 (1973).
155. M. Ruccia, N. Vivona, F. Piozzi, and M. C. Aversa, *Gazz. Chim. Ital.*, **99**, 588 (1969).

156. W. L. Jorgensen and L. Salem, *The Organic Chemist's Book of Orbitals*, Academic Press, New York, 1973, p. 236.
157. S. Morrocchi, A. Ricca, A. Zanarotti, G. Bianchi, R. Gandolfi, and P. Grünanger, *Tetrahedron Letters*, 3329 (1969).
158. S. Morrocchi, A. Ricca, and A. Zanarotti, *Tetrahedron Letters*, 3215 (1970).
159. V. N. Chistokletov and A. A. Petrov, *Zh. Obshch. Khim.*, **33**, 3558 (1963).
160. S. I. Radchenko, V. N. Chistokletov, and A. A. Petrov, *Zh. Org. Khim.*, **1**, 51 (1965).
161. I. G. Kolokol'Tseva, V. N. Chistokletov, M. D. Stadnichuk, and A. A. Petrov, *Zh. Obshch. Khim.*, **38**, 1820 (1968).
162. M. Noel, Vo-Quang-Yen, and L. Vo-Quang, *C.R. Acad. Sci. Paris, Ser. C.*, **270**, 80 (1970).
163. G. N. Bondarev, V. A. Ryzhov, V. N. Chistokletov, and A. A. Petrov, *J. Org. Chem. U.S.S.R.*, **3**, 789 (1967).
164. J. P. Visser and P. Smael, *Tetrahedron Letters*, 1139 (1973).
165. G. Bianchi, R. Gandolfi, and P. Grünanger, *Tetrahedron*, **29**, 2405 (1973).
166. I. G. Kolokol'Tseva, V. N. Chistokletov, and A. A. Petrov, *Zh. Obshch. Khim.*, **38**, 2819 (1968).
167. V. V. Kosovtsev, V. N. Chistokletov, and A. A. Petrov, *Zh. Obshch. Khim.*, **41**, 2643 (1971).
168. I. G. Kolokol'Tseva, V. N. Chistokletov, B. J. Ionin, and A. A. Petrov, *Zh. Obshch. Khim.*, **38**, 1248 (1968).
169. J. P. Anselme, in *The Chemistry of the Carbon–Nitrogen Double Bond* (Ed. S. Patai), Interscience, London, 1970, p. 314.
170. R. Huisgen, R. Grashey, H. Knupfer, R. Kunz, and M. Seidel, *Chem. Ber.*, **97**, 1085 (1964).
171. R. Huisgen, R. Grashey, E. Aufderhaar, and R. Kunz, *Chem. Ber.*, **98**, 642 (1965).
172. C. Grundmann and K. Flory, *Ann. Chem.*, **721**, 91 (1969).
173. N. Singh, S. Mohan, and J. S. Sandhu, *Chem. Comm.*, 387 (1969).
174. R. Huisgen, R. Grashey, R. Kunz, G. Wallbillich, and E. Aufderhaar, *Chem. Ber.*, **98**, 2174 (1965).
175. R. Huisgen, R. Grashey, M. Seidel, H. Knupfer, and R. Schmidt, *Ann. Chem.*, **658**, 169 (1962).
176. J. Sauer and K. K. Mayer, *Tetrahedron Letters*, 325 (1968).
177. B. F. Bonini, G. Maccagnani, L. Thijs, and B. Zwanenburg, *Tetrahedron Letters*, 3569 (1973).
178. J. P. Snyder, *J. Org. Chem.*, **38**, 3965 (1973).
179. P. I. Paetzold, *Z. Anorg. Allgem. Chem.*, **326**, 64 (1963).
180. S. Hammerum and P. Wolkoff, *J. Org. Chem.*, **37**, 3965 (1972); *Org. Mass. Spectrom.*, **9**, 181 (1974).
181. A. Dondoni, A. Mangini, and S. Ghersetti, *Tetrahedron Letters*, 4789 (1966).
182. G. Barbaro, A. Battaglia, and A. Dondoni, *J. Chem. Soc. (B)*, 588 (1970).
183. A. Dondoni and G. Barbaro, *J. Chem. Soc. Perkin II*, 1769 (1973).
184. R. Sustmann, *Tetrahedron Letters*, 963 (1974).
185. R. Huisgen and M. Christl, *Chem. Ber.*, **106**, 3291 (1973).
186. M. Christl and R. Huisgen, *Chem. Ber.*, **106**, 3345 (1973).
187. M. Arbasino and P. Vita Finzi, *Ric. Scient.*, **36**, 1339 (1966).
188. R. Huisgen and M. Christl, *Angew. Chem. Int. Ed.*, **6**, 456 (1967).

189. G. Bianchi, R. Gandolfi, P. Grünanger, and A. Perotti, *J. Chem. Soc.* (*C*), 1598 (1967).
190. G. Bailo, P. Caramella, G. Cellerino, A. Gamba-Invernizzi, and P. Grünanger, *Gazz. Chim. Ital.*, **103**, 47 (1973).
191. G. Bianchi, C. De Micheli, and R. Gandolfi, *J. Chem. Soc. Perkin I*, 1711 (1972).
192. A. Corsico Coda, P. Grünanger, and P. Veronesi, *Tetrahedron Letters*, 2911 (1966); P. Caramella, private communication.
193. P. Caramella and G. Cellerino, *Tetrahedron Letters*, 229 (1974).
194. N. T. Anh, *Tetrahedron*, **29**, 3227 (1973).
195. K. L. Williamson, Y. F. Li Hsu, R. Lacko, and C. H. Youn, *J. Amer. Chem. Soc.*, **91**, 6129 (1969).
196. R. Lazăr, F. G. Cocu, and N. Barbulescu, *Rev. Roum. Chim.*, **20**, 3 (1969).
197. G. Bianchi, R. Gandolfi, and P. Grünanger, *Tetrahedron*, **26**, 5113 (1970).
198. A. Gamba-Invernizzi, R. Gandolfi, and G. Strigazzi, *Tetrahedron*, **30**, 3717 (1974).
199. G. S. King, P. D. Magnus, and H. S. Rzepa, *J. Chem. Soc. Perkin I*, 437 (1972).
200. G. F. Bettinetti and S. Facchetti, *Org. Mass Spectrom.*, **9**, 753 (1974).
201. R. Huisgen and W. Mack, *Chem. Ber.*, **105**, 2805 (1972).
201a. R. Huisgen and W. Mack, *Chem. Ber.*, **105**, 2815 (1972).
202. S. Morrocchi, A. Ricca, A. Selva, and A. Zanarotti, *Gazz. Chim. Ital.*, **99**, 565 (1969).
203. W. J. Awad and M. Sobhy, *Can. J. Chem.*, **47**, 1473 (1969).
204. S. Morrocchi, A. Ricca, and L. Velo, *Tetrahedron Letters*, 331 (1967).
204a. H. Matsukubo and M. Kato, *J. Chem. Soc. Chem. Commun.*, 412 (1974).
204b. A. Selva, A. Citterio, E. Peila, and R. Torriani, *Org. Mass Spectrom.*, **9**, 1017 (1974).
205. A. Battaglia, A. Dondoni, G. Maccagnani, and G. Mazzanti, *J. Chem. Soc.* (*B*), 2096 (1971); A. Battaglia, A. Dondoni, and G. Mazzanti, *Synthesis*, 378 (1971); A. Dondoni, G. Barbaro, A. Battaglia, and P. Giorgiani, *J. Org. Chem.*, **37**, 3196 (1972).
206. W. O. Foye and J. M. Kauffmann, *J. Org. Chem.*, **31**, 2417 (1966).
207. K. Dickoré and R. Wegler, *Angew. Chem. Int. Ed.*, **5**, 970 (1966).
208. F. Lauria, V. Vecchietti, and G. Tosolini, *Gazz. Chim. Ital.*, **94**, 478 (1964).
209. K. Bast, M. Christl, R. Huisgen, and W. Mack, *Chem. Ber.*, **105**, 2825 (1972).
210. V. Nair, *Tetrahedron Letters*, 4831 (1971).
211. K. H. Magosch and R. Feinauer, *Angew. Chem. Int. Ed.*, **10**, 810 (1971).
212. S. Morrocchi and A. Ricca, *Chim. Ind.* (*Milan*), **49**, 629 (1967).
213. J. P. Gibert, R. Jacquier, C. Petrus, and F. Petrus, *Tetrahedron Letters*, 755 (1974).
214. P. Rajagopalan, *Tetrahedron Letters*, 887 (1964).
215. H. Blaschke, E. Brunn, R. Huisgen, and W. Mack, *Chem. Ber.*, **105**, 2841 (1972).
216. E. H. Burk and D. D. Carlos, *J. Heterocyclic Chem.*, **7**, 177 (1970).
217. P. I. Partzold, *Angew. Chem. Int. Ed.*, **6**, 572 (1967); *Chem. Ber.*, **101**, 2874 (1968).
218. H. J. Bestmann and R. Kunstmann, *Chem. Ber.*, **102**, 1816 (1969).
219. R. Huisgen and J. Wulff, *Chem. Ber.*, **102**, 1833 (1969).
220. J. Wulff and R. Huisgen, *Chem. Ber.*, **102**, 1848 (1969).

221. J. Sheridan, *Advances in Molecular Spectroscopy*, Pergamon Press, New York, 1962, p. 139.
222. J. Bastide, J. Hamelin, F. Texier, and Y. Vo Quang, *Bull. Soc. Chim. France*, 2555 (1973).
223. K. von Auwers and O. Unguemach, *Chem. Ber.*, **66**, 1205 (1933).
224. J. Bastide, N. El-Ghandour, O. Henri-Rousseau, and J. Soulier, *C.R. Acad. Sci. Paris, Ser. C.*, **276**, 113 (1973).
225. J. Bastide, O. Henri-Rousseau, and L. Aspart-Pascot, *Tetrahedron*, **30**, 3355 (1974).
226. C. G. Overberger, N. Weinshenker, and J. P. Anselme, *J. Amer. Chem. Soc.*, **87**, 4119 (1965).
227. E. Stephan, L. Vo Quang, and Y. Vo Quang, *Bull. Soc. Chim. France*, 2795 (1973).
228. H. Kisch, O. E. Polansky, and P. Schuster, *Tetrahedron Letters*, 805 (1969).
229. P. K. Kadaba and T. F. Colturi, *J. Heterocyclic Chem.*, **6**, 829 (1969).
230. E. Heilbronner and H. D. Martin, *Chem. Ber.*, **106**, 3376 (1973).
231. J. H. Atherton and R. Fields, *J. Chem. Soc.* (*C*), 1507 (1968).
232. F. Piozzi, A. Umani-Ronchi, and L. Merlini, *Gazz. Chim. Ital.*, **95**, 814 (1965).
232a. H. de Suray, G. Leroy, and J. Weiler, *Tetrahedron Letters*, 2209 (1974).
233. L. J. Luskus and K. N. Houk, *Tetrahedron Letters*, 1925 (1972).
234. T. L. Jacob in *Heterocyclic Compounds*, Vol. 5, (Ed. R. C. Elderfield), Wiley, New York, 1957, p. 72.
235. M. Frank-Neumann and D. Martina, *Tetrahedron Letters*, 1755, 1759 (1975).
236. S. D. Andrews, A. C. Day, and A. N. McDonald, *J. Chem. Soc.* (*C*), 787 (1969); M. E. Parham, H. G. Braxton, and P. R. O'Connor, *J. Org. Chem.*, **26**, 1805 (1961).
236a. D. C. Dittmer and R. Glassman, *J. Org. Chem.*, **35**, 999 (1970); R. Helder, T. Doornbos, J. Strating, and B. Zwanenburg, *Tetrahedron*, **29**, 1375 (1973); W. R. Parham, F. D. Blake, and D. R. Theissen, *J. Org. Chem.*, **27**, 2415 (1962).
237. M. Franck-Neumann, *Angew. Chem. Int. Ed.*, **8**, 210 (1969).
238. N. El-Ghandour, O. Henri-Rousseau, and J. Soulier, *Bull. Soc. Chim. France*, 2817 (1972).
239. M. Franck-Neumann and M. Sedrati, *Angew. Chem. Int. Ed.*, **13**, 606 (1974).
240. N. S. Zefirov, P. Kadziauskas and Yu. K. Yuriev, *J. Gen. Chem. U.S.S.R.*, **36**, 23 (1966).
241. M. Franck-Neumann and M. Sedrati, *Org. Magn. Resonance*, **5**, 217 (1973).
242. D. I. Davies, P. Mason, and M. J. Parrott, *J. Chem. Soc.* (*C*), 3428 (1971).
243. P. K. Kadaba and J. O. Edwards, *J. Org. Chem.*, **26**, 2331 (1961).
244. P. K. Kadaba and N. F. Fannin, *J. Heterocyclic Chem.*, **4**, 301 (1967).
245. P. K. Kadaba, *Tetrahedron*, **25**, 3053 (1969).
246. K. Burger, J. Fehn, and A. Gieren, *Ann. Chem.*, **757**, 9 (1972).
247. C. D. Gutsche and D. Redmore, *Carbocyclic Ring Expansion Reactions*, Academic Press, New York, 1968, p. 81.
248. F. M. Dean and B. K. Park, *J. Chem. Soc. Chem. Commun.*, 162 (1974).
249. G. W. Cowell and A. Ledwith, *Quart. Rev.*, **24**, 119 (1970); G. F. Bettinetti and A. Donetti, *Gazz. Chim. Ital.*, **97**, 730 (1967).
250. N. P. Gambaryan, L. A. Simonyan, and I. L. Knunyants, *Dokl. Akad. Nauk. S.S.S.R.*, **155**, 833 (1964); *Chem. Abstr.*, **60**, 15725h (1964).
251. B. Eistert, H. Fink and A. Müller, *Chem. Ber.*, **95**, 2403 (1962).

252. B. Eistert, L. Sadek Boulos Gourbran, C. Vamvakaris and T. J. Arackal, *Chem. Ber.*, **108**, 2941 (1975).
252a. W. C. Howell, M. Ktenas and J. M. MacDonald, *Tetrahedron Letters* 1719 (1964).
253. H. Staudinger and T. Reber, *Helv. Chim. Acta*, **4**, 3 (1921).
254. W. Kirmse, *Chem. Ber.*, **93**, 2357 (1960).
255. A. P. Krapcho, D. R. Rao, M. P. Silvon, and B. Abegaz, *J. Org. Chem.*, **36**, 3885 (1971).
256. C. E. Diebert, *J. Org. Chem.*, **35**, 1501 (1970).
257. J. M. Beiner, D. Lecadet, D. Paquer, A. Thuillier, and J. Vialle, *Bull. Soc. Chim. France*, 1979 (1973).
258. S. Holm and A. Senning, *Tetrahedron Letters*, 2389 (1973).
259. J. M. Beiner, D. Lecadet, D. Paquer, and A. Thuillier, *Bull. Soc. Chim. France*, 1983 (1973).
260. B. Zwanenburg, A. Wagenaar, L. Thijs, and J. Strating, *J. Chem. Soc. Perkin I*, 73 (1973).
261. S. Rossi and S. Maiorana, *Tetrahedron Letters*, 263 (1966).
262. W. J. Middleton, *J. Org. Chem.*, **34**, 3201 (1969).
263. H. von Pechmann and A. Nold, *Chem. Ber.*, **29**, 2588 (1896).
264. J. C. Sheehan, and P. I. Izzo, *J. Amer. Chem. Soc.*, **71**, 4059 (1949).
265. D. Martin and W. Mucke, *Ann. Chem.*, **682**, 90 (1965).
266. G. Tomaschewski and D. Zanke, *Zeit. Chem.*, **10**, 145 (1970).
267. *The Chemistry of the Azido Group* (Ed. S. Patai), Interscience, London, 1971.
268. G. L'abbé, *Chem. Rev.*, **69**, 345 (1969).
269. G. L'abbé, *Ind. Chim. Belge*, **34**, 519 (1969).
270. R. Huisgen, L. Möbius, G. Müller, H. Stangl, G. Szeimies, and J. M. Vernon, *Chem. Ber.*, **98**, 3992 (1965).
271. K. Alder, H. J. Ache, and F. H. Flock, *Chem. Ber.*, **93**, 1888 (1960).
272. K. Alder, G. Stein, and W. Schneider, *Ann. Chem.*, **515**, 185 (1935).
273. S. McLean and D. M. Findlay, *Tetrahedron Letters*, 2219 (1969).
274. G. W. Klumpp, A. H. Veefkind, W. L. de Graaf, and F. Bickelhaupt, *Ann. Chem.*, **706**, 47 (1967); B. Halton and A. D. Woodhouse, *Austral. J. Chem.*, **26**, 1619 (1973).
275. M. N. Paddon-Row, *Tetrahedron Letters*, 1409 (1972).
276. K. Mackenzie, *Tetrahedron Letters*, 1203 (1974).
277. D. N. Reinhoudt and C. G. Kouwenhoven, *Tetrahedron Letters*, 2163 (1974).
278. M. G. Barlow, R. N. Haszeldine, W. D. Morton, and D. R. Woodward, *J. Chem. Soc. Perkin I*, 1798 (1973).
279. L. A. Paquette, R. J. Haluska, M. R. Short, L. K. Read, and J. Clardy, *J. Amer. Chem. Soc.*, **94**, 529 (1972).
280. M. Franck-Neumann and C. Bucheker, *Tetrahedron Letters*, 2659 (1969).
281. D. H. Aue and G. S. Helwig, *Tetrahedron Letters*, 721 (1974).
282. P. Scheiner, *Tetrahedron*, **24**, 349 (1968).
283. R. F. Bleiholder and H. Shechter, *J. Amer. Chem. Soc.*, **90**, 2131 (1968).
284. P. Scheiner, *J. Amer. Chem. Soc.*, **88**, 4759 (1966).
285. R. S. McDaniel and A. C. Oehschlager, *Can. J. Chem.*, **48**, 345 (1970).
286. R. Huisgen, K. v. Fraunberg, and H. J. Sturm, *Tetrahedron Letters*, 2589 (1969).
287. J. C. Pezzullo and E. R. Boyko, *J. Org. Chem.*, **38**, 168 (1973).
288. A. L. Logothetis, *J. Amer. Chem. Soc.*, **87**, 749 (1965).

289. G. L'abbé and A. Hassner, *Angew. Chem. Int. Ed.*, **10**, 98 (1971).
290. G. L'abbé and A. Hassner, *J. Heterocyclic Chem.*, **7**, 361 (1970).
291. G. L'abbé, *Ind. Chim. Belge*, **36**, 3 (1971).
292. G. L'abbé, J. E. Galle and A. Hassner, *Tetrahedron Letters*, 303 (1970).
293. G. L'abbé and A. Hassner, *Bull. Soc. Chim. Belge*, **80**, 209 (1971).
294. P. Ykman, G. Mathys, G. L'abbé, and G. Smets, *J. Org. Chem.*, **37**, 3213 (1972).
295. P. Scheiner in *Selective Organic Transformations*, Vol. 1 (Ed. B. S. Thyagarajan), Wiley-Interscience, New York, pp. 327–362.
296. A. S. Bailey and J. J. Wedgwood, *J. Chem. Soc.* (*C*), 682 (1968).
297. M. E. Hermes and F. D. Marsh, *J. Org. Chem.*, **37**, 2969 (1972).
298. J. E. McMurry and A. P. Coppolino, *J. Org. Chem.*, **38**, 2821 (1973).
299. R. L. Hale and L. H. Zalkov, *Tetrahedron*, **25**, 1393 (1969).
300. R. A. Wohl, *J. Org. Chem.*, **38**, 3862 (1973).
301. R. M. Scribner, *Tetrahedron Letters*, 4737 (1967).
302. J. F. Stephen and E. Marcus, *J. Heterocyclic Chem.*, **6**, 969 (1969).
303. M. Regitz and G. Himbert, *Ann. Chem.*, **734**, 70 (1970).
304. G. Bianchetti, R. Stradi, and D. Pocar, *J. Chem. Soc. Perkin I*, 997 (1972).
305. D. Pocar, R. Stradi, and L. M. Rossi, *J. Chem. Soc. Perkin I*, 619 and 769 (1972).
306. R. Huisgen, L. Möbius, and G. Szeimies, *Chem. Ber.*, **98**, 1138 (1965); O. Gerlach, P. L. Reiter, and F. Effenberger, *Ann. Chem.*, 1895 (1974).
307. K. D. Berlin and M. A. R. Khayat, *Tetrahedron*, **22**, 975 (1966).
308. R. A. Wohl, *Tetrahedron Letters*, 3111 (1973).
309. R. Scarpati and M. L. Graziano, *J. Heterocyclic Chem.*, **9**, 1087 (1972) and references cited therein.
310. W. Broeckx, N. Overbergh, C. Samyn, G. Smets, and G. L'abbé, *Tetrahedron*, **27**, 3527 (1971).
311. D. Pocar and R. Stradi, *Ann. Chim.* (*Rome*), **61**, 181 (1971).
312. H. Cardoen, S. Toppet, G. Smets, and G. L'abbé, *J. Heterocyclic Chem.*, **9**, 971 (1972).
313. G. Bianchetti, P. Dalla Croce, D. Pocar, and A. Vigevani, *Gazz. Chim. Ital.*, **97**, 289 (1967).
314. G. Bianchetti, D. Pocar, P. Dalla Croce, and R. Stradi, *Gazz. Chim. Ital.*, **97**, 304 (1967).
315. G. Bianchetti, P. Dalla Croce, and D. Pocar, *Tetrahedron Letters*, 2043 (1965).
316. R. Huisgen, G. Szeimies, and L. Möbius, *Chem. Ber.*, **99**, 475 (1966).
317. M. T. Garcia Lopez, G. Garcia-Munoz, and R. Madronero, *J. Heterocyclic Chem.*, **9**, 717 (1972).
318. B. Stanovnik, *J. Heterocyclic Chem.*, **8**, 1055 (1971).
319. F. Texier and R. Carrié, *Bull. Soc. Chim. France*, 3642 (1971).
320. F. Texier and R. Carrié, *Bull. Soc. Chim. France*, 4119 (1971).
321. F. Texier and R. Carrié, *Bull. Soc. Chim. France*, 258 (1972).
322. G. Rembarz, B. Kirchoff, and G. Dongowski, *J. Prakt. Chem.*, **33**, 199 (1966).
323. P. v. R. Schleyer, J. C. Williams, and K. R. Blanchard, *J. Amer. Chem. Soc.*, **92**, 2377 (1970).
324. P. Bischof and E. Heilbronner, *Helv. Chim. Acta*, 53, 1677 (1970).
325. C. Batich, O. Ermer, E. Heilbronner, and J. R. Wiseman, *Angew. Chem. Int. Ed.*, **12**, 312 (1973).

326. P. Bischof, J. A. Hashmall, E. Heilbronner, and V. Hornung, *Helv. Chim. Acta,* **52**, 1745 (1969).
327. H. I. Zeliger, J. P. Snyder, and H. J. Bestman, *Tetrahedron Letters,* 2199 (1969).
328. G. R. Harwey, *J. Org. Chem.,* **31**, 1587 (1966).
329. V. Schöllkopf and P. Markusch, *Ann. Chem.,* **753**, 143 (1971).
330. G. L'abbé and H. J. Bestman, *Tetrahedron Letters,* 63 (1969).
331. P. Ykman, G. L'abbé, and G. Smets, *Tetrahedron Letters,* 5225 (1970).
332. P. Ykman, G. L'abbé, and G. Smets, *Tetrahedron,* **29**, 195 (1973).
333. H. I. Zeliger, J. P. Snyder, and H. J. Bestman, *Tetrahedron Letters,* 3313 (1970).
334. P. Ykman, G. L'abbé, and G. Smets, *Tetrahedron,* **27**, 845 (1971).
335. P. Ykman, G. L'abbé, and G. Smets, *Tetrahedron,* **27**, 5623 (1971).
336. J. M. Vandensqvel, G. Smets, and G. L'abbé, *J. Org. Chem.,* **38**, 675 (1973).
336a. P. Dunn and D. Oldfield, *Australian J. Chem.,* **24**, 645 (1971).
336b. G. L'abbé, E. Van Look, R. Albert, S. Toppet, G. Verhelst, and G. Smets, *J. Amer. Chem. Soc.,* **96**, 3973 (1974).
337. R. L. Tolman, C. W. Smith, and K. R. Robins, *J. Amer. Chem. Soc.,* **94**, 2530 (1972).
338. M. Regitz, *Angew. Chem. Int. Ed.,* **6**, 733 (1967).
339. H. W. Heine and R. Peavy, *Tetrahedron Letters,* 3123 (1965); H. W. Heine, R. Peavy, and A. J. Durbetaki, *J. Org. Chem.,* **31**, 3924 (1966).
340. J. E. Baldwin and J. A. Duncan, *J. Org. Chem.,* **36**, 627 (1971).
341. R. Huisgen, W. Scheer, G. Szeimies, and H. Huber, *Tetrahedron Letters,* 397 (1966); R. Huisgen, W. Scheer, and H. Huber, *J. Amer. Chem. Soc.,* **89**, 1753 (1967); R. Huisgen, W. Scheer, and H. Maeder, *Angew. Chem. Int. Ed.,* **8**, 602 (1969); H. Hermann, R. Huisgen, and H. Maeder, *J. Amer. Chem. Soc.,* **93**, 1779 (1971).
342. E. Brunn and R. Huisgen, *Tetrahedron Letters,* 473 (1971).
343. R. Huisgen, W. Scheer, H. Maeder, and E. Brunn, *Angew. Chem. Int. Ed.,* **8**, 604 (1969); R. Huisgen, W. Scheer, and H. Maeder, *Angew. Chem. Int. Ed.,* **8**, 602 (1969).
344. G. Dallas, J. W. Lown, and J. P. Moser, *J. Chem. Soc.* (*D*), 278 (1970).
345. G. Dallas, J. W. Lown, and J. P. Moser, *J. Chem. Soc.* (*C*), 2383 (1970).
346. J. W. Lown, J. P. Moser, and R. Westwood, *Can. J. Chem.,* **47**, 4335 (1969).
347. J. W. Lown, R. Westwood, and J. P. Moser, *Can. J. Chem.,* **48**, 1682 (1970).
348. R. Huisgen, V. Martin-Ramos, and W. Scheer, *Tetrahedron Letters,* 477 (1971).
349. J. W. Lown, G. Dallas, and T. W. Maloney, *Can. J. Chem.,* **47**, 3557 (1969).
350. J. W. Lown and K. Matsumoto, *Can. J. Chem.,* **48**, 3399 (1970).
351. J. W. Lown, R. K. Smalley, and G. Dallas, *Chem. Comm.,* 1543 (1968); J. W. Lown, R. K. Smalley, G. Dallas, and T. W. Maloney, *Can. J. Chem.,* 48, 89 (1970).
351a. J. W. Lown, T. W. Maloney, and G. Dallas, *Can. J. Chem.,* **48**, 584 (1970).
352. F. Texier, R. Carrié, and J. Jaz, *J. Chem. Soc. Chem. Commun.,* 199 (1972).
353. J. Honzl and M. Sorm, *Tetrahedron Letters,* 3339 (1969).
354. A. R. Katritzky and Y. Takeuchi, *J. Chem. Soc.* (*C*), 874 (1971); N. Dennis, A. R. Katritzsky, and S. K. Parton, *J. Chem. Soc. Chem. Commun.,* 707 (1972); N. Dennis, A. R. Katritzky, and Y. Takeuchi, *J. Chem. Soc. Perkin I,* 2054 (1972); A. R. Katrizsky and Y. Takeuchi, *J. Chem. Soc.* (*C*), 878 (1971).
355. T. Sasaki, K. Kanematsu, K. Haykawa, and M. Uchide, *J. Chem. Soc. Perkin I,* 2750 (1972).

356. H. Gotthardt, R. Huisgen, and H. O. Bayer, *J. Amer. Chem. Soc.,* **92**, 4340 (1970).
357. G. Kille and J. P. Fleury, *Bull. Soc. Chim. France,* 4636 (1968).
358. R. Huisgen, H. Gotthardt, and H. O. Bayer, *Chem. Ber.,* **103**, 2368 (1970).
358a. Th. Eicher and V. Schäfer, *Tetrahedron,* **30**, 4025 (1974).
359. R. Huisgen, E. Funke, H. Gotthardt, and H. L. Panke, *Chem. Ber.,* **104**, 1532 (1971).
360. E. Funke, R. Huisgen, and F. C. Schoefer, *Chem. Ber.,* **104**, 1550 (1971).
361. E. Brunn, E. Funke, H. Gotthardt, and R. Huisgen, *Chem. Ber.,* **104**, 1562 (1971).
362. R. Huisgen, H. Gotthardt, and R. Grashey, *Chem. Ber.,* **101**, 536 (1968).
363. L. K. Vagina, V. N. Chistokletov, and A. A. Petrov, *J. Org. Chem. U.S.S.R.,* **1**, 1723 (1965).
364. H. Gotthardt and R. Huisgen, *Chem. Ber.,* **101**, 552 (1968).
365. R. Huisgen and H. Gotthardt, *Chem. Ber.,* **101**, 839 (1968).
366. R. Huisgen, R. Grashey, and H. Gotthardt, *Chem. Ber.,* **101**, 829 (1968).
367. P. M. Weintraub, *J. Chem. Soc. (D),* 760 (1970).
368. H. Kato, S. Sato, and M. Ohta, *Tetrahedron Letters,* 4261 (1967).
369. H. Gotthardt, R. Huisgen, and R. Knorr, *Chem. Ber.,* **101**, 1056 (1968).
370. K. B. Sukumaran, C. S. Augadiyavar, and M. V. George, *Tetrahedron,* **28**, 3987 (1972).
371. G. F. Bettinetti and P. Grünanger, *Tetrahedron Letters,* 2553 (1965).
372. J. Markert and E. Fahr, *Tetrahedron Letters,* 769 (1970).
373. W. Oppolzer, *Tetrahedron Letters,* 2199 (1970).
374. H. Dorn and A. Otto, *Chem. Ber.,* **101**, 3287 (1968).
375. R. Grashey, R. Huisgen, K. Kun Sun, and R. M. Moriarty, *J. Org. Chem.,* **30**, 74 (1965).
376. H. Dorn and A. Otto, *Z. Chem.,* **12**, 175 (1972).
377. W. Oppolzer, *Tetrahedron Letters,* 3091 (1970).
378. R. Huisgen, R. Grashey, P. Laur, and H. Leitermann, *Angew. Chem.,* **72**, 416 (1960).
379. R. Grashey, H. Leitermann, R. Schmidt, and K. Adelsberger, *Angew. Chem. Int. Ed.,* **1**, 406 (1962).
380. R. Grashey and K. Adelsberger, *Angew. Chem. Int. Ed.,* **1**, 267 (1962).
381. P. I. Paetzold and H. Maisch, *Chem. Ber.,* **101**, 2870 (1968).
381a. D. St. C. Black, R. F. Crozier, and V. C. Davies, *Synthesis,* 205 (1975).
382. V. I. Minkin, E. A. Medyantseva, I. M. Andreeva, and G. V. Gorshkova, *J. Org. Chem. U.S.S.R.,* **9**, 148 (1973).
383. V. Balich and V. Chandrasekharan, *Ind. J. Chem.,* **10**, 159 (1972).
384. L. Semper and L. Lichtenstadt, *Chem. Ber.,* **51**, 928 (1918).
385. F. Barrow and F. J. Thorneycroft, *J. Chem. Soc.,* 773 (1939).
386. T. S. Dobashi, M. H. Goodrow, and E. J. Grubbs, *J. Org. Chem.,* **38**, 4440 (1973).
387. R. Grée and R. Carrié, *Tetrahedron Letters,* 4117 (1971).
388. J. Bjorgo, D. R. Boyd, and D. C. Neill, *J. Chem. Soc. Chem. Comm.,* 478 (1974).
389. K. Koyano and I. Tanaka, *J. Phys. Chem.,* **69**, 2545 (1965).
390. C. De Micheli and R. Gandolfi, unpublished results.
391. Y. Iwakura, K. Uno, S. J. Hong, and T. Hongu, *Bull. Soc. Chim. Japan,* **45**, 192 (1972).
392. R. Grée and R. Carrié, *Tetrahedron Letters,* 2987 (1972).

393. R. Grée, F. Tonnard, and R. Carrié, *Tetrahedron Letters*, 135 (1974); *Bull. Soc. Chim. France*, 1325 (1975).
394. R. Grée, F. Tonnard, and R. Carrié, *Tetrahedron Letters*, 453 (1973).
395. C. De Micheli and R. Gandolfi, unpublished results.
396. R. Paul and S. Tchelitcheff, *Bull. Soc. Chim. France*, 4179 (1967).
397. R. Huisgen, R. Grashey, H. Hauck, and H. Seidl, *Chem. Ber.*, **101**, 2043 (1968).
398. R. Huisgen, R. Grashey, H. Hauck, and H. Seidl, *Chem. Ber.*, **101**, 2548 (1968).
399. R. Huisgen, R. Grashey, H. Seidl, and H. Hauck, *Chem. Ber.*, **101**, 2559 (1968).
400. R. Huisgen, H. Hauck, R. Grashey, and H. Seidl, *Chem. Ber.*, **101**, 2568 (1968).
401. J. Sims and K. N. Houk, *J. Amer. Chem. Soc.*, **95**, 5798 (1973).
402. A. Cerri, *Thesis*, University of Pavia, 1974.
402a. N. A. LeBel and E. G. Banucci, *J. Org. Chem.*, **36**, 2440 (1971).
403. G. R. Delpierre and M. Lamchen, *J. Chem. Soc.*, 4693 (1963).
404. A. Galbiati, *Thesis*, University of Pavia 1969.
405. H. Seidl, R. Huisgen, and R. Grashey, *Chem. Ber.*, **102**, 926 (1969).
406. T. Hisano, S. Yoshikawa, and K. Muraoka, *Org. Prep. Proceed. Int.*, **5**, 95 (1973).
407. D. St. C. Black and K. G. Watson, *Australian J. Chem.*, **26**, 2491 (1973).
408. W. E. Truce, J. R. Norell, R. W. Campbell, D. G. Brady, and J. W. Fieldhouse, *Chem. Ind.*, 1870 (1965).
409. R. Huisgen and J. Wulff, *Chem. Ber.*, **102**, 746 (1969).
410. R. Huisgen and J. Wulff, *Chem. Ber.*, **102**, 1848 (1969).
411. O. Tsuge, M. Tashiro, and S. Mataka, *Tetrahedron Letters*, 3877 (1968).
412. B. P. Stark and M. H. G. Ratcliffe, *J. Chem. Soc.*, 2640 (1964).
413. P. Brown and R. C. Cookson, *Tetrahedron*, **24**, 2551 (1968).
414. Th. Do-Minh, A. M. Trozzolo, and G. W. Griffin, *J. Amer. Chem. Soc.*, **92**, 1402 (1970).
415. C. W. Martin, J. A. Landgrebe, and E. Rapp, *Chem. Comm.*, 1438 (1971); M. Hamaguchi and T. Ibata, *Tetrahedron Letters*, 4475 (1974).
416. P. Rajagopalan and B. G. Advani, *Tetrahedron Letters*, 2689 (1967).
417. R. W. Hoffmann and H. J. Luthardt, *Chem. Ber.*, **101**, 3861 (1968).
418. R. Huisgen and W. Scheer, *Tetrahedron Letters*, 481 (1971).
419. J. W. Lown and K. Matsumoto, *Can. J. Chem.*, **49**, 3443 (1971).
420. A. Dahmen, H. Hamberger, R. Huisgen, and V. Markowski, *Chem. Commun.*, 1192 (1971).
421. E. F. Ullmann and J. E. Milks, *J. Amer. Chem. Soc.*, **86**, 3814 (1964).
422. A. Robert, J. J. Pommeret, and A. Foucaud, *Tetrahedron Letters*, 231 (1971).
423. R. W. Murray and A. Suzii, *J. Amer. Chem. Soc.*, **95**, 3343 (1973).
424. R. W. Murray and D. P. Higley, *J. Amer. Chem. Soc.*, **95**, 7886 (1973).
425. D. P. Higley and R. W. Murray, *J. Amer. Chem. Soc.*, **96**, 3330 (1974).
426. R. Criegee, *Record Chem. Progr.* **18**, 111 (1957).
427. R. Criegee in *Peroxide Reaction Mechanism* (Ed. J. O. Edwards), Interscience, New York, 1962, p. 29.
428. P. S. Bailey, *Chem. Rev.*, **58**, 925 (1958).
429. R. W. Murray, *Accounts Chem. Res.*, **1**, 313 (1968).

430. W. P. Keaveney and J. J. Pappas, *Tetrahedron Letters,* 841 (1969).
431. S. Fliszár and M. Granger, *J. Amer. Chem. Soc.,* **91**, 3330 (1969).
432. S. Fliszár and M. Granger, *J. Amer. Chem. Soc.,* **92**, 3361 (1970).
433. S. Fliszár and J. Renard, *Can. J. Chem.,* **48**, 3002 (1970).
434. S. Fliszár and J. Carles, *J. Amer. Chem. Soc.,* **91**, 2637 (1969).
435. C. W. Gillies, R. P. Lattimer, and R. L. Kuczkowski, *J. Amer. Chem. Soc.,* **96**, 1536 and 6205 (1974).
436. F. L. Greenwood, *J. Amer. Chem. Soc.,* **88**, 3146 (1966).
437. R. W. Murray, R. D. Youssefyeh, and P. R. Story, *J. Amer. Chem. Soc.,* **89**, 2429 (1967).
438. N. L. Bauld, J. A. Thompson, C. E. Hudson, and P. S. Bailey, *J. Amer. Chem. Soc.,* **90**, 1822 (1968).
439. S. Fliszár and J. Carles, *Can. J. Chem.,* **47**, 3921 (1969).
440. R. P. Lattimer, C. W. Gillies, and R. L. Kuczkowski, *J. Amer. Chem. Soc.,* **95**, 1348 (1973).
441. R. W. Murray and G. J. Williams, *J. Org. Chem.,* **34**, 1891 (1969).
442. R. W. Murray, R. D. Youssefyeh, G. I. Williams, and P. R. Story, *Tetrahedron,* **24**, 4347 (1968).
443. R. W. Murray and R. Hagen, *J. Org. Chem.,* **36**, 1098 (1971).
444. R. W. Murray and R. Hagen, *J. Org. Chem.,* **36**, 1103 (1971).
445. R. Criegee and H. Korber, *Chem. Ber.,* **104**, 1807 (1971).
445a. P. R. Story, C. E. Bishop, J. R. Burgess, R. W. Murray and R. O. Youssefyeh, *J. Amer. Chem. Soc.,* **90**, 1907 (1968).
446. R. P. Lattimer, R. L. Kuczkowski and C. W. Gillies, *J. Amer. Chem. Soc.,* **96**, 348 (1974).
446a. B. J. Finlayson, J. N. Pitts, Jr., and R. Atkinson, *J. Amer. Chem. Soc.,* 96, 5356 (1974).
447. S. Fliszár, J. Renard and O. Z. Simon, *J. Amer. Chem. Soc.,* **93**, 6953 (1971).
448. S. D. Razumovskii and L. V. Berezova, *Izv. Akad. Nauk. S.S.S.R.,* 1, 207 (1968).
449. F. L. Greenwood and L. J. Durham, *J. Org. Chem.,* **34**, 3363 (1969).
450. R. Criegee and G. Lohaus, *Chem. Ber.,* **86**, 1 (1953).
451. H. N. Junker, W. Schäfer, and H. Niedenbrück, *Chem. Ber.,* **100**, 2508 (1967).
452. R. Criegee and H. Korber, *Ann. Chem.,* **756**, 95 (1972).
453. R. W. Murray, P. R. Story, and L. D. Loan, *J. Amer. Chem. Soc.,* **87**, 3025 (1965).
454. R. Criegee and H. Korber, *Chem. Ber.,* **104**, 1812 (1971).
455. R. Criegee and P. Günther, *Chem. Ber.,* **96**, 1564 (1963).
456. H. Kwart and D. M. Hoffman, *J. Org. Chem.,* **31**, 419 (1966).
457. P. R. Story and J. R. Burgess, *J. Amer. Chem. Soc.,* **89**, 5726 (1967).
458. R. Criegee and G. Schröder, *Chem. Ber.,* **93**, 689 (1960).
459. P. S. Bailey, J. A. Thompson, and B. A. Shoulders, *J. Amer. Chem. Soc.,* **88**, 4098 (1966).
460. F. L. Greenwood, *J. Org. Chem.,* **29**, 1321 (1964).
461. F. L. Greenwood, *J. Org. Chem.,* **30**, 3108 (1965).
462. L. J. Durham and F. L. Greenwood, *J. Org. Chem.,* **33**, 1629 (1968).
463. P. S. Bailey, J. W. Ward, and R. E. Hornish, *J. Amer. Chem. Soc.,* **93**, 3552 (1971); P. S. Bailey, J. W. Ward, T. P. Carter, E. Nieh, C. N. Fischer, and A.-J. Khashob, *J. Amer. Chem. Soc.,* **96**, 6136 (1974).

464. L. A. Hull, I. C. Hisatsune, and J. Heicklen, *J. Amer. Chem. Soc.*, **94**, 4856 (1972).
464a. C. W. Gillies, *J. Amer. Chem. Soc.*, **97**, 1276 (1975).
465. P. S. Bailey and A. G. Lane, *J. Amer. Chem. Soc.*, **89**, 4473 (1967).
466. T. W. Nagawa, L. J. Andrews, and R. M. Keefer, *J. Amer. Chem. Soc.*, **82**, 269 (1960).
467. J. J. Bufalini and A. J. Altschuller, *Can. J. Chem.*, **43**, 2243 (1965).
468. D. G. Williamson and R. J. Cvetanovic, *J. Amer. Chem. Soc.*, **90**, 3668 (1968).
469. D. G. Williamson and R. J. Cvetanovic, *J. Amer. Chem. Soc.*, **90**, 4248 (1968).
470. S. D. Razumovskii, *Izv. Akad. Nauk. S.S.S.R.*, **2**, 335 (1970).
471. *Ionization Potentials, Appearance Potentials and Heats of Formation of Gaseous Positive Ions,* U.S. Department of Commerce, National Bureau of Standards, 1969.
472. J. Carles and S. Fliszár, *Adv. Chem. Ser.*, **112**, 35 (1972).
473. R. W. Taft, *J. Amer. Chem. Soc.*, **75**, 4231 (1953); R. W. Taft in *Steric Effects in Organic Chemistry* (Ed. M. S. Newman), Wiley, New York, 1956, pp. 556–675; C. D. Johnson, *The Hammet Equation*, Cambridge University Press, London, 1973.
474. G. Klutsch and S. Fliszár, *Can. J. Chem.*, **50**, 2841 (1972).
475. R. D. Cadle and C. Schadt, *J. Amer. Chem. Soc.*, **74**, 6002 (1952); *J. Chem. Phys.*, **21**, 163 (1953).
476. G. Wagner and A. Greiner, *Z. Phys. Chem.* (*Leipzig*), **215**, 92 (1960).
477. P. R. Story, J. A. Alford, J. R. Burgess, and W. C. Ray, *J. Amer. Chem. Soc.*, **93**, 3042 (1971).
478. P. R. Story, J. A. Alford, W. C. Ray, and J. R. Burgess, *J. Amer. Chem. Soc.*, **93**, 3044 (1971).
479. P. R. Story, E. A. Whited, and J. A. Alford, *J. Amer. Chem. Soc.*, **94**, 2143 (1972).
480. P. S. Bailey, T. P. Carter, C. M. Fischer, and J. A. Thompson, *Can. J. Chem.*, **51**, 1278 (1973).
481. K. R. Kopechy, P. A. Lockwood, J. E. Filby, and R. W. Reid, *Can. J. Chem.*, **51**, 468 (1973).
482. D. R. Kerur and D. G. M. Diaper, *Can. J. Chem.*, **51**, 3110 (1973).
483. R. E. Erickson, P. J. Andrulis, J. C. Collins, M. L. Lungle, and G. D. Mercer, *J. Org. Chem.*, **34**, 2961 (1969); P. Kolsaker and O. Joraandstad, *Acta Chem. Scand.*, B **29**, 7 (1975).
484. B. Zwanenburg and W. A. J. Janssen, *Synthesis,* 617 (1973).
485. R. M. Kellogg and S. Wassenaar, *Tetrahedron Letters*, 1987 (1970).
486. J. Buter, S. Wassenaar, and R. M. Kellogg, *J. Org. Chem.*, **37**, 4045 (1972).
487. R. M. Kellogg, S. Wassenaar, and J. Buter, *Tetrahedron Letters,* 4689 (1970).
488. R. M. Kellogg, *J. Org. Chem.*, **38**, 844 (1973).
489. H. Gotthardt and B. Christl, *Tetrahedron Letters,* 4751 (1968).
489a. K. T. Potts, J. Baum, E. Houghton, D. N. Roy, and U. P. Singh, *J. Org. Chem.*, **39**, 3619 (1974).
490. K. T. Potts and D. McKeough, *J. Amer. Chem. Soc.*, **95**, 2749 (1973).
491. S. Tamagaki and S. Oae, *Tetrahedron Letters,* 1159 (1972).
492. H. Gotthardt, *Chem. Ber.*, **105**, 196 (1972).
493. E. M. Burgess and H. R. Penton, Jr., *J. Org. Chem.*, **39**, 2885 (1974).

CHAPTER **7**

Reactions of carbenes with X=Y groups

ALAN P. MARCHAND

Department of Chemistry, University of Oklahoma,
Norman, Oklahoma 73069, U.S.A.

I. INTRODUCTION

Reactions of alkenes with carbenes were first presented in *The Chemistry of Functional Groups* in a review by Cadogan and Perkins[1]. This article, which was also concerned with reactions of alkenes with free radicals, covered the literature through 1963. The decade which has passed since the appearance of Cadogan and Perkins' review has witnessed an explosive surge of interest in carbene chemistry research, especially as it pertains to the study of reactions of carbenes with double-bonded functional groups. The extent to which this field has developed can be gauged in part by the number of review articles which have appeared since the early 1960's. Comprehensive reviews and treatises which deal at least in part with this subject have proliferated[2–46]. Each of these reviews has contributed in turn to our understanding of carbenes as highly reactive intermediates in organic chemistry and to furthering the development of carbene chemistry as a major area of chemical research.

The purpose of the present article is to survey critically the subject of carbene reactions with double-bonded functional groups, $X=Y$. Our primary concern will be with those developments which have surfaced since the appearance of Cadogan and Perkins' review. In the light of the foregoing discussion, this in itself would appear to be a most ambitious undertaking, and we can only hope to proceed within a relatively limited framework in the space allotted to the present review. Because of the large number and outstanding quality of existing reviews, the present article attempts only to summarize the major results and the conclusions derived therefrom which have accrued over the past decade. To minimize duplication of material and to increase the general utility of the present review, a particular attempt has been exercised to survey the carbene literature from 1971 through April, 1974, and, whenever possible, to draw upon examples thereby secured from the literature current at the time of writing of this review. Omission of some earlier work is inevitable, and most likely reflects space restrictions and the author's avowed desire to minimize duplication with earlier reviews. Such omissions do not necessarily reflect any negative judgment on the author's part.

II. REACTIONS WITH ALKENES (C=C)

A. Stereochemistry of Addition to C=C as a Criterion for Assigning Spin State in the Reacting Carbene

Considerable attention has been devoted in earlier reviews to the question of the electronic configurations of carbene intermediates[13,47,48].

The simplest carbene, methylene, has long been known to exist in two electronic configurations: the bent, singlet ($^1A_1$) (1a) and the more nearly linear, triplet ($^3B_1$) configuration (1b)[49].

(1a)   (1b)

∠ HCH = *c*. 102° (experimental value[50])   ∠ HCH = *c*. 136° (experimental value[51,52])

Although the earliest studies of the $^1A_1$ and $^3B_1$ states of methylene involved spectroscopic investigations, a chemical criterion for demonstrating the spin state of a reacting carbene species in solution or in the gas phase, first developed by Skell[53–56], has been found to be generally useful. Skell's Criterion argues that whereas singlet carbenes undergo concerted, stereospecific addition to olefins, the corresponding triplet carbene *may*

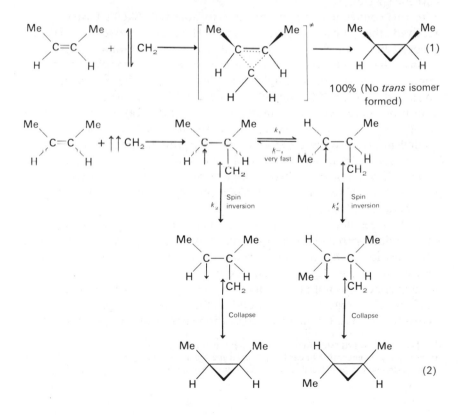

100% (No *trans* isomer formed)

(1)

(2)

add non-stereospecifically. The classical application of Skell's Criterion to the addition of methylene to *cis*-2-butene is illustrated in equations (1) and (2).

According to Skell's treatment, the appearance of *trans*-1,2-dimethyl-cyclopropane (along with the *cis* isomer) among the products of the reactions of the carbene with *cis*-2-butene is taken as *prima facie* evidence that the carbene is reacting in a triplet electronic configuration. In a gas phase reaction, the absence of *trans*-1,2-dimethylcyclopropane in the product mixture implicates a carbene reacting in a singlet electronic configuration. This conclusion is not as clear-cut for reactions carried out in condensed phases where solvent cage effects can be operative in triplet reactions leading to exclusive (or nearly exclusive) formation of products arising via what appears to be a *cis* addition process. There are a number of other potentially complicating factors which are inherent in Skell's Criterion; these have recently been critically assessed[46,53]. Nevertheless, Skell's Criterion has been extensively utilized, and over the years it has proved to be viable for distinguishing between singlet and triplet reacting carbene species.

In this connection, it is worthwhile to note that Skell's Criterion has received theoretical support through Hoffmann's extended Hückel theory calculations[48,57–59]. These calculations substantiate Skell's contention that singlet methylene addition to ethylene should occur in a concerted fashion. Interestingly, the calculations further reveal that the concerted addition of the singlet carbene should prefer the unsymmetrical transition state, **2a**, the symmetrical approach (**2b**) being symmetry forbidden[60]†. The complex **2a** which results from addition of singlet methylene to ethylene correlates with the ground state of the product cyclopropane. However, triplet methylene addition proceeds by way of a complex involving an excited methylene and ground state ethylene; this complex correlates with a stereochemically mobile (excited triplet) configuration of trimethylene[57].

In the foregoing discussion, it is emphasized that the stereochemistry of the addition reactions affords information concerning the electronic configuration of the reacting carbene (in contradistinction to affording information regarding the electronic configuration of the carbene in its ground state). It is not necessarily true that a nascent carbene is produced (and therefore reacts) in its ground state; in fact, the opposite is more generally the case. In practice, information concerning the ground state of

† However, Anastassiou has recently demonstrated that symmetrical addition of a reacting, (highly energetic) linear ($D_{\infty h}$) singlet methylene is both symmetry allowed and, in some cases, energetically feasible; see A. G. Anastassiou, *Chem. Commun.*, 991 (1968).

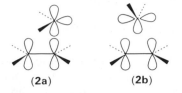

(2a)   (2b)

the carbene can be obtained by first allowing the (often excited) nascent carbene to decay energetically through collisions with an inert medium prior to its undergoing the addition reaction. In the gas phase, this can be accomplished by adding an inert gas (such as helium, argon, xenon, nitrogen, or tetrafluoromethane) in large excess[61-63]. In solution, the high reactivity of singlet carbenes often renders difficult the promotion of intersystem crossing by inert solvents. Nevertheless, a number of investigators have successfully resorted to dilution techniques which employ relatively 'inert' solvents such as perfluoroalkanes[64] or, with less reactive carbenes, hexafluorobenzene[65,66].

A number of examples are presented in Table 1 which serve to illustrate the application of Skell's Criterion toward determining the spin state of the reactive divalent carbon species in a variety of substituted methylene additions to olefins[67-100]. This table is not intended to be exhaustive, but it should serve to illustrate the broad range of applicability of Skell's Criterion for this purpose.

Insofar as possible, an effort has been made to include in Table 1 only those addition reactions which are likely to proceed via 'free' carbene intermediates. Other types of divalent carbon species are known which exist in complexation with metal atoms. These species, termed *carbenoids*, are not 'free' carbenes, but nevertheless they often lead to the formation of the same types of products (e.g., cyclopropanes) as are observed to result from the corresponding reactions of 'free' carbenes. Carbenoids will be dealt with at length in Section II.B.

It is well known that reactions of carbenoids with olefins result in stereospecific *cis* cyclopropanation of the olefinic substrate, and this fact has occasionally been used to distinguish carbenoids from 'free' carbenes[101]. It should be noted, however, that not all divalent carbon species produced from organometallic methylene transfer reagents are carbenoids. As an example, the dihalocarbenes prepared from phenyl(trihalomethyl)-mercury compounds have in many instances been shown to exist as 'free' carbenes[102]. Similarly, dihalocarbenes produced from the reaction of strong bases with haloforms (the Doering–Hoffmann procedure[103]) are likewise generally agreed to exist as 'free' carbene intermediates[103-105]. Both the Doering–Hoffmann[103] and Seyferth[106] procedures afford dihalocarbenes which add in stereospecific *cis* fashion to olefins[107]. Dihalocarbenes produced by these procedures have not been included in Table 1.

Some discussion of the material presented in Table 1 is warranted. It is clear from inspection of the data in Table 1 that a number of carbenes are produced from their respective precursors via non-multiplicity specific processes (i.e., the carbene is formed predominantly, but not exclusively, in one spin state or the other). This situation is often encountered (e.g., in the preparation of mono- and diarylcarbenes[40]). In cases where rapid singlet–triplet equilibration is established in the nascent carbene, it is possible that the carbene in one spin state (singlet or triplet) reacts with the olefinic substrate much more rapidly than does that same carbene in the other spin configuration (triplet or singlet)[24]. In the event that the singlet configuration is the more reactive of the two spin states of a given carbene, stereospecific addition to olefins might be observed. Under such circumstances, the simple, direct application of Skell's Criterion would afford misleading information regarding the spin state of the *nascent* carbene[108].

A further complication arises when the carbene in question is generated from a diazo precursor. When a polar olefin is employed as substrate, it is often the case that pyrazolines are initially formed via 1,3-dipolar cycloaddition of the diazo precursor to the olefin. Cyclopropane formation can then follow upon decomposition of the intermediate pyrazoline[109—112,112a] (equation 3). The ability of pyrazolines to undergo stereospecific *cis* photodecomposition to form cyclopropanes has been noted[113].

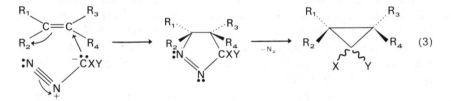

$$(3)$$

The question then arises as to whether these reactions involve carbenes or whether one is simply studying the chemistry of photoexcited diazo compounds. This question can be satisfactorily answered by employing alternative precursors for the carbene; if the species in question is truly a 'free' carbene, then its chemistry should be precursor independent. Often this is indeed found to be the case, e.g. as with arylcarbenes[114,115] and dihalocarbenes[102]. However, the situation is less straightforward for the case of carbalkoxycarbenes and biscarbalkoxycarbenes for which suitable precursors other than diazocompounds are sorely lacking[46]. Although

TABLE 1. Stereochemistry of carbene additions to olefins

Carbene	Precursor	Olefin	Conditions	Additive	Product ratios	Reference
:CH₂	CH₂N₂		hν, direct	—	100% stereospecific *cis* addition	56, 67, 68
			hν	Ph₂CO	66 : 34	69, 70
			hν	:C₃F₈ = 1 : 200	60·4 : 13·3	64
	H₂C=C=O		hν, 366 nm	—	4·0 : 1	71
			hν, 366 nm	CH₃I	2·94 : 1	71
			hν, 253 nm	Hg sens.	3·85 : 1	72
			hν, 366 nm	—	1·9 : 1	73
∴PhCH			hν, direct	—	100% stereospecific *cis* addition	74

TABLE 1. Stereochemistry of carbene additions to olefins (continued)

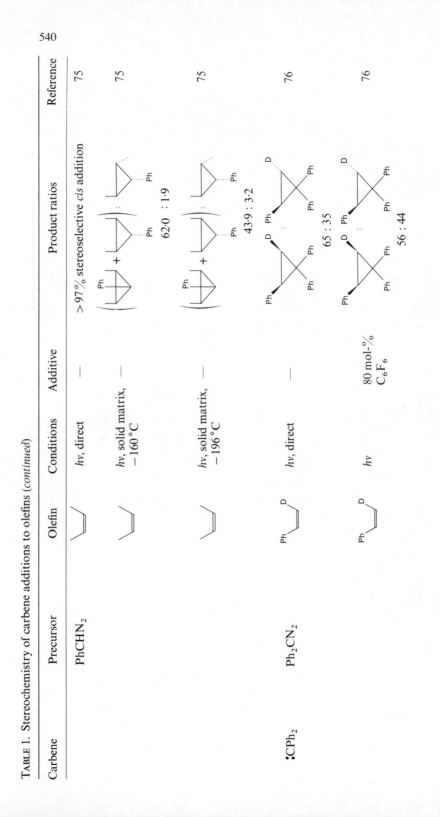

Carbene	Precursor	Olefin	Conditions	Additive	Product ratios	Reference
	PhCHN$_2$		hv, direct	—	>97% stereoselective cis addition	75
			hv, solid matrix, −160 °C	—	62·0 : 1·9	75
			hv, solid matrix, −196 °C	—	43·9 : 3·2	75
:CPh$_2$	Ph$_2$CN$_2$		hv, direct	—	65 : 35	76
			hv	80 mol-% C$_6$F$_6$	56 : 44	76

	Ph–CH=CD (D)	hv	PhC≡CH	60 : 40	76
		hv, direct	—	*cis* : *trans* 40 : 60	77
		hv, direct	—	*cis* : *trans* 83 : 17	78, 79
		hv, direct	—	*cis* : *trans* 53 : 47	77
		hv, direct	—	*cis* : *trans* 65 : 35	77

TABLE 1. Stereochemistry of carbene additions to olefins (*continued*)

Carbene	Precursor	Olefin	Conditions	Additive	Product ratios	Reference
:CH (2-methyl-4-nitrophenyl)	CHN$_2$ (2-Me, 4-NO$_2$ aryl)		*hv*, direct	—	*cis* : *trans* 60 : 40	77
:CF$_2$	F$_2$C (diazirine)		*hv* or Δ	—	100% stereospecific *cis* addition	80
:CHCl	ClCHN$_2$		−30°C	—	100% stereospecific *cis* addition	81
:CHBr	BrCHN$_2$		−30°C	—	100% stereospecific *cis* addition	81
:CHI	CHI$_3$		*hv*, direct	—	100% stereospecific *cis* addition	82
:CClCH$_3$	H$_3$C–C–Cl (diazirine)		*hv*, direct	—	>99% stereoselective *cis* addition	83
:C(CF$_3$)$_2$	(F$_3$C)$_2$C (diazirine)		150°C	—	*cis* : *trans* 39 : 8	84
			150°C	—	100% stereospecific *cis* addition	84

Carbene	Source	Olefin	Conditions	Sensitizer	Ratio	Ref.
:C(CN)$_2$	N$_2$C(CN)$_2$	(olefin)	70°C, neat	—	cis : trans 92 : 8	85
		(olefin)	70°C	Olefin : $\langle$S$\rangle$ = 1:10	cis : trans 60 : 40	85
		(olefins)	70°C	Olefin : $\langle$S$\rangle$ = 1:100	cis : trans 30 : 70	85
:CHOMe	(MeO)$_3$SiCH(OMe)$_2$	(olefin)	125°C	—	100% stereospecific cis addition	86
:CHCO$_2$Et	N$_2$CHCO$_2$Et	Various olefins	140°C	—	100% stereospecific cis addition	87
:CHCO$_2$Me	N$_2$CHCO$_2$Me	(olefin)	hv, direct	—	cis : trans 1·56 : 0·07	88
		(olefin)	hv	Ph$_2$C=O	cis : trans 1·25 : 2·5	88
:C(CO$_2$Me)$_2$	N$_2$C(CO$_2$Me)$_2$	cis-4-Methyl-2-pentene	hv, direct	—	cis : trans 92 : 8	88
		cis-4-Methyl-2-pentene	hv	Ph$_2$C=O	cis : trans 10 : 90	88

TABLE 1. Stereochemistry of carbene additions to olefins (*continued*)

Carbene	Precursor	Olefin	Conditions	Additive	Product ratios	Reference
(2,2-dimethyl-1,3-dioxane-4,6-dione carbene)	(5-diazo-2,2-dimethyl-1,3-dioxane-4,6-dione)	(olefin structure)	$h\nu$, direct	—	Very little adduct formed, (1–2%)	88
		(olefin structure)	$h\nu$	$Ph_2C{=}O$	*cis : trans* 14 : 86	88
:ClCO₂Et	N₂ClCO₂Et	(olefin structure)	$h\nu$, direct	—	100% stereospecific *cis* addition	89
:CBrCO₂Et	N₂CBrCO₂Et	(olefin structure)	$h\nu$, direct	—	100% stereospecific *cis* addition	89
:CClCO₂Et	N₂CClCO₂Et	(olefin structure)	$h\nu$, direct	—	100% stereospecific *cis* addition	89
Me₃Ge—C:—CO₂Et	Me₃Ge—C=N₂, CO₂Et	(olefin structure)	$h\nu$, direct	—	c. 97% stereoselective *cis* addition	90
Me₃Sn—C:—CO₂Et	Me₃Sn—C=N₂, CO₂Et	(olefin structure)	$h\nu$, direct	—	100% stereospecific *cis* addition	90
Me₃Pb—C:—CO₂Et	Me₃Pb—C=N₂, CO₂Et	(olefin structure)	$h\nu$, direct	—	100% stereospecific *cis* addition	90

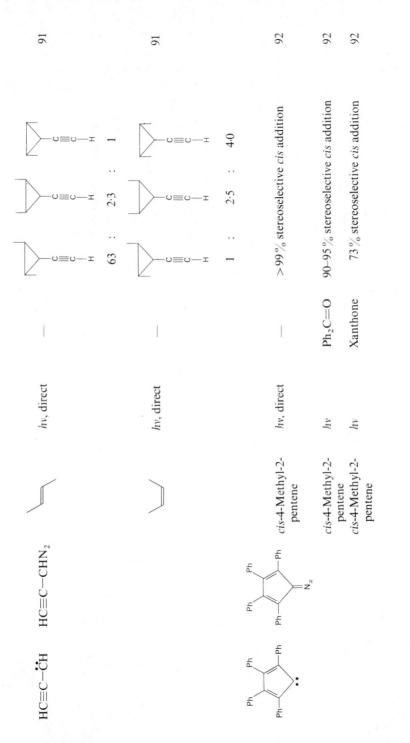

HC≡C—ĊH		trans-butene	hv, direct	—	63 : 2·3 : 1	91
HC≡C—CHN₂		cis-butene	hv, direct	—	1 : 2·5 : 4·0	91
		cis-4-Methyl-2-pentene	hv, direct	—	>99% stereoselective cis addition	92
		cis-4-Methyl-2-pentene	hv	Ph₂C=O	90–95% stereoselective cis addition	92
		cis-4-Methyl-2-pentene	hv	Xanthone	73% stereoselective cis addition	92

TABLE 1. Stereochemistry of carbene additions to olefins (*continued*)

Carbene	Precursor	Olefin	Conditions	Additive	Product ratios	Reference
		cis-4-Methyl-2-pentene	*hv*, direct	—	>99% stereoselective *cis* addition	92
		cis-4-Methyl-2-pentene	*hv*	$Ph_2C{=}O$	90–95% stereoselective *cis* addition	92
		cis-4-Methyl-2-pentene	*hv*, direct	—	*cis* : *trans* 61·5 : 1	93
		cis-4-Methyl-2-pentene	*hv*	Mol fraction $C_6F_6 = 0{\cdot}94$	*cis* : *trans* 32·3 : 1	93
		cis-4-Methyl-2-pentene	*hv*, direct	—	*cis* : *trans* 55 : 45	94
			hv, direct	—	100% stereospecific *cis* addition	95, 96
			hv, direct	—	100% stereospecific *cis* addition	95, 96

Carbene	Diazo precursor	Alkene	Conditions	Additive	Products (ratio)	Ref.
(fluorenylidene)	(fluorenyl–N_2)	(2-butene)	hv, direct	—	*cis* : *trans* 1.95 : 1	65, 66
(fluorenylidene)	(fluorenyl–N_2)	(2-butene)	hv	90–100 mol-% C_6F_6	*cis* : *trans* 0.25 : 1	65, 66
(2-oxocyclohexylidene)	(2-diazocyclohexanone)	(2-butene)	hv	$Ph_2C=O$	*cis* : *trans* 1 : 6 1 : 5 : 3	97
			hv	$Ph_2C=O$	5 : 3	97
			hv	$Ph_2C=O$	8 : 1	97
$H_3C-\overset{..}{C}-\overset{..}{C}H$, $\overset{\|}{O}$	$H_3C-C-CHN_2$, $\overset{\|}{O}$		hv	$Ph_2C=O$		97
$Ph-\overset{..}{C}-\overset{..}{C}H$, $\overset{\|}{O}$	$Ph-C-CHN_2$, $\overset{\|}{O}$	(2-butene)	hv, direct	—	50 : 40 : 10	98

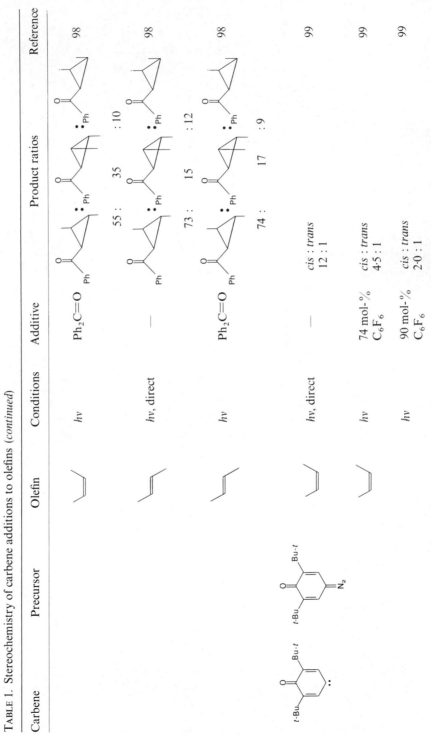

TABLE 1. Stereochemistry of carbene additions to olefins (*continued*)

Carbene	Precursor	Olefin	Conditions	Additive	Product ratios	Reference
			hv	Ph$_2$C=O	55 : 35 : 10	98
			hv, direct	—		98
			hv	Ph$_2$C=O	73 : 15 : 12	98
					74 : 17 : 9	
			hv, direct	—	*cis* : *trans* 12 : 1	99
			hv	74 mol-% C$_6$F$_6$	*cis* : *trans* 4·5 : 1	99
			hv	90 mol-% C$_6$F$_6$	*cis* : *trans* 2·0 : 1	99

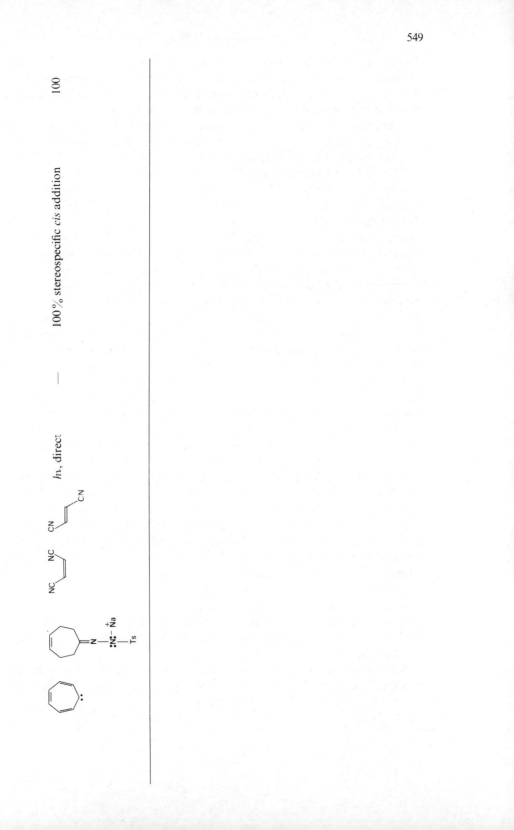

100% stereospecific *cis* addition

hv, direct

—

there is an impressive body of evidence which argues against the intermediacy of pyrazolines in reactions of diazoacetic or diazomalonic esters
with simple olefins[31], the exact nature of the reacting species in these
diazoester decomposition reactions awaits clarification.

It has been argued that it is necessary to observe the reactions of both
the singlet and triplet states of a given carbene before the spin state of that
carbene in a particular reaction may be regarded as having been firmly
established[47]. However, in only a very few instances have systematic
studies of both singlet and triplet states of a given carbene been studied.
Among these are studies of singlet and triplet methylene[61,64,73,116],
fluorenylidene[65,66], dicyanocarbene[85], diphenylcarbene[76], and some
carbalkoxycarbenes[88]. For photochemical reactions in solution which
produce a given carbene in an excited singlet electronic configuration,
decay to a lower-lying triplet state (i.e., intersystem crossing) can be
promoted by the addition of a photosensitizing agent such as benzophenone. In this way, both singlet and triplet states of a given carbene can be
generated and their respective chemistries can be compared. This approach
has been applied with some success to the study of reactions of olefins
with carbenes produced by photodecomposition of diazoesters[88,117,118];
this subject has recently been reviewed[46].

Recently, the promotion of intersystem crossing in nascent carbenes by
'internal heavy atom' effects has been studied. The examples reported
involved the photodecomposition of diazo compounds possessing heavy
metal atoms adjacent to the developing carbene carbon atom. The
diazoesters studied possessed the general structures $Me_3M-C(N_2)-$
CO_2Et (where M = Si, Ge, Sn, and Pb[90]) and $MeHg-C(N_2)-X$ (where
X = CN and CO_2Me[119-121]). The diazo esters having the structure
$Me_3M-C(N_2)-CO_2Et$ as well as $MeHgC(N_2)CO_2Me$ all afforded
carbenes which added stereospecifically to cis-2-butene. Since the heavy
atom, M, should promote very rapid singlet–triplet equilibration, it was
concluded that the reactions observed were those of the ground state of
the reacting carbene in question[90,119-121]. In contrast to this result,
$MeHg-\ddot{C}-CN$ was found to add in non-stereospecific fashion to
cis-2-butene, a result which is suggestive of a triplet ground state for
$MeHg-\ddot{C}-CN$[119-121]. It was concluded that such internal heavy atom
effects might provide useful criteria when studying carbene additions to
olefins for determining the ground state spin multiplicities of the carbenes
in question.

An alternative to Skell's Criterion for determining the spin state of a
reacting carbene has recently been proposed by Shimizu and

Nishida[122,122a]. These investigators have suggested utilizing 1,1-dicyclo-propylethylene (**3**) as substrate. They have argued that concerted addition of a singlet carbene ($\downarrow\!\!\uparrow$CXY) should afford adduct **4**, whereas a triplet carbene ($\uparrow\uparrow$CXY) might afford an intermediate diradical which could subsequently rearrange to products **5** and **6** (equations 4 and 5). 'Shimizu's

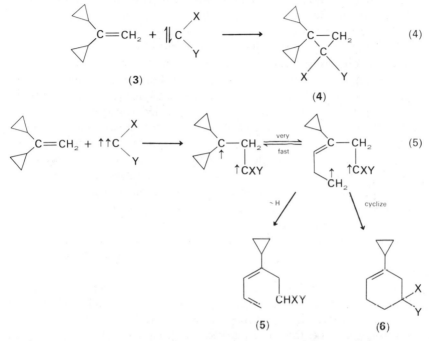

$$(4)$$

$$(5)$$

(3)   (4)

(5)   (6)

Criterion' has been successfully applied to the study of reactions of singlet and triplet fluorenylidene with **3**[122]. This approach has similarly been used to examine the reaction of **3** with ethyl diazoacetate in the presence of copper sulphate; this reaction afforded only the addition product, (**4**, X = H, Y = CO_2Et)[123]. The generality of Shimizu's approach remains yet to be fully demonstrated and exploited. However, a clear limitation of this method is that it is only useful in instances where the relative rate of rearrangement (to **5** or **6**) in the intermediate diradical can successfully compete with that of ring closure to the cyclopropane, **4**.

In passing, it is worth noting that chemically-induced, dynamic nuclear polarization (CIDNP) studies have been found useful in examining reactions which proceed through radical pair intermediates[124–128]. Since triplet carbenes are themselves diradical species, it might be expected that CIDNP techniques might find application in the study of reactions of triplet carbenes. Indeed, this approach has been found useful for studying triplet carbene 'insertion' reactions, (actually, abstraction–recombination processes)[129–134]. However, CIDNP studies have not yet been routinely applied to the investigation of triplet carbene additions to olefins. Such an approach would appear to be capable of yielding useful information in this regard, and further pursuit of this application of CIDNP techniques would appear to be a worthwhile effort.

The foregoing discussion has been concerned primarily with utilizing carbene additions to carbon–carbon double bonds in order to distinguish between singlet and triplet reacting carbenes. Once it was determined that a given carbene could be produced in different spin states, the chemistries of these species could then be individually studied. Along with the contrast in the degree of stereospecificity of singlet vs. triplet carbene additions to olefins, it has become apparent that the two spin states of a given carbene display other, strikingly-different, features of their chemical behaviour toward olefinic substrates. Although there are relatively few instances where both spin states of a given carbene have been studied, it is nevertheless possible to note some generalizations regarding the contrasting chemistries of singlet vs. triplet states of carbenes:

(i) In general, singlet carbenes behave as electrophilic species[41,81,135–137], although steric factors can influence the outcome of reactions of bulky carbenes with olefins[117,118].

(ii) The relative rates of addition of singlet vs. triplet carbenes to mono-olefins are often comparable although a rate preference for addition by the singlet species is sometimes observed. However, when the olefin substrate is a diene, there is a marked rate preference for addition by the triplet carbene over addition by the corresponding singlet species[88,117,138–140]. This difference in relative rates of addition probably reflects the fact that addition of the triplet carbene to one double bond in a diene affords a diradical intermediate which is stabilized by (allylic) resonance. No such stabilization can intervene in the (concerted) addition of the singlet carbene[55]. In their behaviour toward dienes, triplet carbenes, then, display 'radical-like' character, as one might have intuitively expected[117].

(iii) One misconception has recently received considerable attention. Carbon–hydrogen 'insertion' by triplet carbenes into allylic C—H bonds was once widely regarded as being an important competing pathway in reactions of triplet carbenes with olefins. This process is now known to involve an abstraction–recombination process, and does not appear to be as important a process as was once believed[65,73,117,140a].

(iv) Although a number of claims of 1,4-addition of carbenes to olefins have been made, there appears to be only one *bona fide* example, namely the formation of 9,9-dicyanobicyclo[4.2.1]nona-2,4,7-triene (**7**) in the thermal reaction of dicyanocarbene with cyclooctatetraene (equation 6)[141]. One might naively expect a triplet carbene to be more likely to afford 1,4-

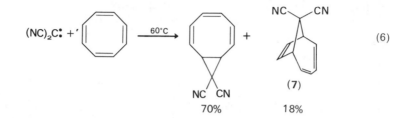

addition products in its reaction with dienes than is the corresponding singlet species (since in the former case an intermediate diradical is formed which might have options available other than direct 1,2-closure). However, this expectation is not borne out by experiment; triplet carbenes form only cyclopropanes (1,2-adducts) with acyclic dienes[88]. The subject of 1,4-additions of carbenes to X=Y will be discussed in some detail in Section V.B.

B. Carbenoid Additions to Olefins

The term 'carbenoid' was coined by Closs and Moss[142] in 1964 to describe the divalent carbon species which they generated via α-elimination resulting from the reaction of organolithium reagents with alkyl halides[39]. This term is now employed in a more general sense to denote a metal-complexed carbene as differentiated from a non-complexed ('free') carbene. Although carbenoids undergo some reactions characteristic of free carbenes (such as addition to olefins), it is clear from studies of their chemistry that they are not, in fact, simple divalent carbon species (*vide infra*). What is not clear is exactly how in, e.g. carbenoid additions to olefins, the metal element and its ligands (if any), the divalent carbon atom, and the olefinic substrate might be interrelated in the transition state of the addition process.

Although metal catalysts have for decades been employed in the generation of carbenoids from their various precursors, there have been relatively few critical studies aimed toward delineating the detailed mechanism of carbenoid addition reactions. The first important evidence to suggest that transition states of carbenoid reactions might indeed involve both the divalent carbon species and the metal catalyst was obtained through studies of the catalytic decomposition of diazo compounds, $RR'C=N_2$, in the presence of olefins. In a number of instances when optically active, homo-geneous metal catalysts were employed, decomposition of the diazo compounds and subsequent addition of the resulting carbenoid to the olefinic substrate led to the formation of optically active cyclopropanes in low optical yields[143–149]. Some examples illustrating this phenomenon are shown in equations 7–9.

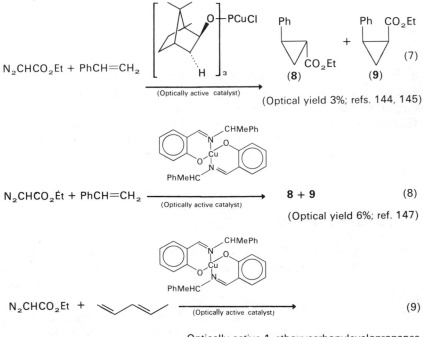

(Optical yield 6%; ref. 147)

Optically active 1-ethoxycarbonylcyclopropanes
(ref. 146)

These results have generally been interpreted as evidence for the inter-mediacy of a metal-bound divalent carbon atom in carbenoid additions to olefins. The evidence generally considered to support this contention is twofold. Firstly, it has been shown that in the partial asymmetric synthesis of 1-methyl-2-phenylcyclopropane, the optical yield of the product remained constant when the [N-(R)-α-phenethylsalicylaldiminato]copper-(II)-catalysed reaction of diazomethane with trans-propenylbenzene was repeatedly carried out under various degrees of dilution with n-hexane[148].

This result effectively rules out the possibility that free carbenes are involved: the observed optical induction being due to asymmetric solvation by the soluble, optically active catalyst.

Secondly, when 1:1 mixtures of cyclohexene and 1-methylcyclohexene are allowed to react with dimethyl diazomalonate in the presence of various soluble copper(II) catalysts, the relative yields of the two norcarane adducts (10 and 11, respectively) are found to vary markedly with the type of catalyst employed[150]. Results obtained for this reaction carried out both in the absence of catalyst and in the presence of various copper(II) catalysts are summarized in Table 2[150]. Note that the relative yields of norcaranes 10 and 11 are actually reversed in going from either the direct or the benzophenone-sensitized photochemical reactions to the corresponding copper(II)-catalysed cases. These results suggest that quite different chemical species are involved in the uncatalysed and copper(II)-catalysed reactions. Furthermore, there is considerable variation in the product ratio 10:11 among the competition reactions when the ligand on the copper(II) catalyst is varied. This result strongly suggests that the transition state of the addition reaction involves some sort of metal-complexed species. However, it is appropriate to introduce a caveat at this point: nearly all of the results which have thus far been rationalized in terms of carbenoid reactions involving a metal-complexed carbene can be equally accounted for by assuming a metal-complexed olefin which suffers

TABLE 2. Competitive reactions of dimethyl diazomalonate with cyclohexene and 1-methylcyclohexene (1:1) in the presence of various copper(II) catalysts[150]

Catalyst	Solvent	Temperature (°C)	Product ratio, 10:11
$(MeO)_3P$–CuI	None	Reflux	4·90:1
$Cu(BF_4)_2$	None	Reflux	13·82:1
$Cu(BF_4)_2$	90 mol-% C_6F_6	Reflux	2·29:1
$Cu(acac)_2$	None	Reflux	2·73:1
$Cu(acac-f_3)^a$	None	Reflux	2·96:1
$Cu(acac-f_6)^b$	None	Reflux	2·58:1
Direct photolysis	None	35	1·00:3
Photolysis, Ph_2CO sens.	None	35	1·00:2

^{a} Copper(II) α,α,α-trifluoroacetylacetonate.
^{b} Copper(II) hexafluoroacetylacetonate.

rate-determining attack by a free (non-complexed) diazo compound. The resulting intermediate metal-complexed pyrazoline might then rapidly collapse to afford the observed reaction products[150]. This possibility has yet to be rigorously excluded (*vide infra*).

In general, carbenoids are less reactive species than are free carbenes; hence, they are generally more selective in their reactions with olefins than are their free carbene counterparts. This renders carbenoid reactions correspondingly more suitable for preparative purposes (e.g., for methylenation) since in this case side-reaction products (such as those arising via carbon–hydrogen bond insertion by the divalent carbon species) are minimized†. Indeed, methylenation with methylene carbenoid, the familiar Simmons–Smith reaction[44], has become a major standard method for methylenation of olefins. However, despite the wide-ranging synthetic utility of carbenoid additions to olefins, surprisingly little is known of their detailed mechanisms. Although Cu(0) and Cu(I) have often been utilized as catalysts in diazoester decompositions[144,145], it is only relatively recently that evidence has been garnered which suggests that copper(II) may in fact be the active catalytic species in these reactions[153–155]. However, there remains some controversy regarding the question of the oxidation state of copper in copper–carbenoid complexes[156,157], and this issue must at present be regarded as being unsettled.

Various models have been postulated for the transition state of carbenoid–olefin addition reactions. Most investigators appear to agree that the carbene carbon atom, the metal atom, and the olefin are all involved in the final transition state leading to products, but even this most fundamental point has not gone unchallenged[158]. A model involving second-order reaction of a complexed carbene with a complexed olefin has been ruled out by kinetic studies[159,160]. Moser[144,145] has pictured the transition

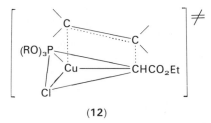

(12)

---

† It has long been accepted as fact that copper carbenoids are incapable of inserting into aliphatic carbon–hydrogen bonds (see Kirmse, ref. 36, pp. 85–88). However, intermolecular insertion of the copper carbenoids derived from ethyl diazoacetate and diazoacetophenone into a carbon–hydrogen bond of cyclohexane has recently been achieved [see L. T. Scott and G. J. DeCicco, *J. Amer. Chem. Soc.*, **96**, 332 (1974); for some earlier examples, see B. W. Peace and D. S. Wulfman, *Synthesis*, 137 (1973)].

state for the trialkylphosphitecopper(I) chloride-catalysed addition of ethyl diazoacetate to olefins as **12**. By contrast, Wulfman and coworkers view these reactions as being bimolecular carbene-transfer reactions with attack occurring exclusively at the carbene carbon atom and not involving direct interaction between the copper atom [in this case, copper(II)] and the olefin[150]. Wulfman's suggested transition state for the copper(II)-catalysed addition of dimethyl diazomalonate to olefins is depicted by structure **13**[150].

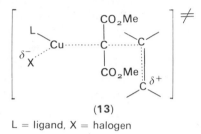

(**13**)

L = ligand, X = halogen

It is clear from the foregoing discussion that carbenoid additions to olefins are enormously complicated processes whose complexity has only recently been fully appreciated. The elucidation of the mechanism of these reactions offers a meaningful challenge to future generations of research chemists.

Before leaving the subject of carbenoid additions to olefins, it is worthwhile to examine some of the methods which have been utilized for the generation of carbenoids. The earliest studies involved generation of carbenoids via metal-promoted α-elimination in alkyl halides and poly-halomethanes[142,161–163]. In addition to copper[157], a number of other metals have been found to be capable of promoting diazoester decompositions: among these are Ag$^+$ [164], di-μ-chloro-di-π-allyldipalladium[165,166], and organotin[167], organonickel[168–170], organoaluminium[171–173], organorhodium[174,175], and organosodium compounds[176], among others. Unlike phenyl(trihalomethyl)mercury compounds, which generally appear to afford free carbenes[102], the corresponding methylene transfer reactions of bis(halomethyl)mercury compounds proceed via carbenoid inter-mediates[177,178]. Seyferth has suggested the transition state depicted by **14** for the addition of bis(bromomethyl)mercury to olefins[177].

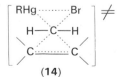

(**14**)

Studies of the relative reactivities of carbenoids towards olefins and the question of stereoselectivity of carbenoid additions to olefins will be considered in Sections II.C and II.D respectively.

C. Reactivity

It was Doering who some years ago referred to the methylene intermediate which was generated via direct photolysis of diazomethane as being 'the most indiscriminate reagent in organic chemistry'[179]. However, later investigations revealed that the unbridled reactivity of methylene can be modulated somewhat by successive substitution on the carbene carbon atom of one or both of the methylene hydrogen atoms. Functional groups such as halogen, aryl, OR, COR, or CO_2R can stabilize the carbene by some combination of electronic (polar and resonance) substituent effects, producing a species which is somewhat more tractable than is methylene for synthetic purposes and for use in mechanistic studies.

The application of relative reactivity studies to carbene and carbenoid additions to olefins has recently been the subject of a lengthy review by Moss[41]. Some of the important conclusions drawn by Moss and by other investigators are summarized in the paragraphs which follow.

As might be expected, steric effects and electronic effects operate, often in delicate balance, in both the carbene (or carbenoid) and in the olefinic substrate. It is often difficult to separate the relative contribution of each of these factors in a given situation. To aid us in assessing the relative contributions of each of these factors, it is worthwhile examining them in turn insofar as cases can be found where it is possible to isolate any one factor from the others. This is normally accomplished by attempting to vary one factor while holding the others as constant as possible through a series of olefins or of carbenes. However, since variations in steric effects in the transition state very often manifest themselves as the result of the operation of subtle electronic effects, it is clear that these two effects are generally interrelated to the point of redundancy. Attempts to separate them artificially are ultimately, therefore, doomed to failure. This is a most serious limitation, and it should be borne firmly in mind by the reader as he evaluates the discussion which follows.

1. Electronic effects in the carbene

The resonance effects alluded to in the beginning of this section can operate in one of two ways, both of which have a net stabilizing effect on the carbene relative to methylene. In the case of electron-withdrawing (+ E) substituents (e.g., carbethoxyl), stabilization results from conjugative interaction with the lone electron pair on the carbene carbon atom (**15a** and **15b**). The contribution of structure **15b** might be expected to render

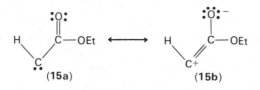

(15a)   (15b)

carbethoxycarbene somewhat more electrophilic than methylene itself; this has indeed been found to be the case[46].

Alternatively, lone-pair substituents, —$\ddot{\text{X}}$, can act as electron-donating (—E) groups. These stabilize the divalent carbon atom through conjugation with the vacant orgital (**16a** and **16b**). By donating electronic charge

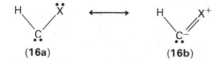

(16a)   (16b)

in this fashion, the substituent —X might be expected to increase the nucleophilicity of the carbene relative to methylene. Indeed, carbenes displaying nucleophilic properties have been extensively studied; discussion of these species is deferred to Section II.F.

The effect of *meta* and *para* substituents in phenylcarbenes generated from substituted phenyldiazomethanes has been examined by Closs and Moss[142,161]. The relative reactivities of these carbenes toward isobutene and *trans*-2-butene were found to vary in the order m-Cl > p-Cl > H > p-CH$_3$ > p-OCH$_3$. This order parallels the expected order of increasing carbene stability (cf. the reduced electrophilicity of p-methoxyphenyl-carbene compared with phenylcarbene due to the resonance interaction in the former which is indicated by structures **17a** and **17b**).

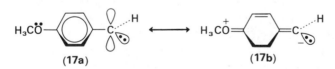

(17a)   (17b)

Relative rate studies of the type used above have been conventionally utilized to distinguish between carbenes and carbenoids. Free carbenes should show a pattern of relative reactivities which is precursor independent, whereas carbenoids will vary in their reactivities depending upon the conditions under which they are generated. This criterion works best when one is dealing with fairly selective carbenes or carbenoids which show a wide range of relative rates of addition to the series of olefins studied. Indeed, it was in this way that dihalocarbenes produced from a variety of precursors were shown to have been generated as free carbenes rather than as carbenoids[104] (cf. discussion in Section II.A).

It is often difficult to assess quantitatively the electronic effects of substituents in a carbene upon the relative reactivity of that carbene. This is partly due to the fact that there are two spin states for every carbene. It is therefore necessary to have at hand the carbene in both singlet and triplet spin states (or to have some idea to what extent these two forms are present in cases where a carbene is produced via a non-multiplicity specific process)†. Another complicating factor is that many carbenes show so little discriminating ability when reacted competitively with different olefins that it is difficult to assess quantitatively their selectivities (and hence their relative reactivities).

Nevertheless, some unifying concepts emerge upon examination of the literature. An order of increasing electrophilicities of carbenes has been developed by Harrison[180]: $:CF_2 < :CHF < :CH_2$ and $:CF_2 < :CCl_2 < :CBr_2 < :CI_2 < :CH_2$. Here, the concept of electrophilicity is seen to correlate with the 'freeness of a vacant p-orbital' in the carbene; those carbenes having the 'freer' p-orbital are classed as being the more electrophilic[104]. More strongly electron-donating groups, X and Y, in a carbene :CXY are seen to increase the negative (nucleophilic) character of the carbene carbon atom and thus raise the activation energy for electrophilic addition to olefins[41]. Since fluorine is the strongest π-donor and iodine the weakest π-donor in the halogen series, the order of increasing electrophilicities of the dihalocarbenes emerges as suggested by Harrison[180].

Increasing carbene selectivity is generally encountered with increasing substitution of one (or, better, both) of the methylene hydrogen atoms by substituents such as halogen, carbalkoxyl, aryl, and alkyl groups. Generally, the greatest enhancement of selectivity is arrived at when the methylene is disubstituted[41]. An order of reagent selectivity toward addition to olefins has been developed by Skell and Cholod[104]: $:CH_2 < :CBr_2 < Me_2C{=}C{=}C: < :CCl_2 < :CFCl < :CF_2$. Skell also interpreted this as the order of decreasing electrophilicities of these carbenes. The reader is referred to Moss' excellent review[41] for an extended discussion of the electronic effects of substituents, X and Y, in carbenes (:CXY) upon their relative reactivities.

---

† The unusual reactivity of triplet carbenes toward dienes has been noted earlier (Section II.A). However, it is interesting to note in this connection that singlet and triplet biscarbomethoxycarbene show strikingly similar relative reactivities toward a large number of monoolefinic substrates[117,118].

Very little systematic work has been aimed at delineating the effects of substituents on the divalent carbon atom on the relative reactivities of *carbenoid* additions to olefins. Moss[41] has noted that selectivities of carbenoid additions, unlike those of the corresponding addition reactions of the free arylcarbenes, do not correlate with Hammett σ-values for *para* substituents in the carbenoid.

Moser has examined the effects of *ortho, meta* and *para* substituents in the triarylphosphite ligand upon the stereoselectivity of (triarylphosphite)-copper(I) chloride-catalysed additions of ethyl diazoacetate to cyclohexene[144,145]. A small negative value of Hammett $\rho(-0.18)$ was found for this reaction. The primary effect of substituents in the ligand was considered to be their ability to induce electrically changes in carbenoid-to-substrate bond lengths in the transition state. These effects could then, in turn, induce slight differences in steric crowding in the transition state, the extent of the crowding being slightly different for different substituents in the catalyst ligands. The observed differences in isomer ratios, then, were considered to result from these subtle differences in steric effects in the transition states for variously ligand-substituted carbenoid additions to olefins[144,145].

There are some known cases which allow comparison of the relative reactivity of a carbenoid with that of the corresponding free carbene. In general, the transition state for addition of a carbenoid to olefins appears to involve a greater degree of bond formation to the divalent carbon atom than is encountered in the corresponding addition reaction involving the free carbene. This conclusion is supported by relative reactivity data as well as by observations regarding the stereoselectivity of the addition process[30,41,142,161].

Interestingly, there are a large number of cases known where *cis* addition of monosubstituted carbenes and carbenoids to, e.g. *cis*-2-butene, affords predominantly the thermodynamically less-stable *syn* isomer (equation 10)[24,30,181]. An explanation based upon presumed steric effects

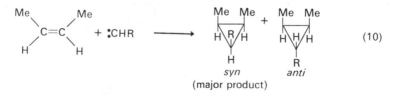

(10)

encountered in the transition state for the addition process would lead to the opposite conclusion (i.e. that the *anti* isomer should be the major addition product). The observed effect might be electronic in origin, involving the carbenic substituent, R, and the alkyl groups present in the olefin, and it clearly must be an attractive interaction. The phenomenon in its more generalized sense has recently been termed 'steric attraction', and explanations for its origin have been forthcoming[182–187]. One picture recently put forward envisages the primary interaction taking place in the transition state of highly exothermic carbene and carbenoid additions to olefins as being that which involves the lowest unoccupied molecular orbital (LUMO) of the acceptor (divalent carbon atom) and the highest occupied molecular orbital (HOMO) of the donor (olefin π-orbital). The contrathermodynamic stereoselectivity observed in many of these additions is then thought to result from a secondary, attractive effect operating between the (non-bonded) substituents on the divalent carbon atom and those on the olefin[183].

2. Electronic effects in the olefin

Since electrophilic attack is experienced by the carbene or carbenoid on the olefinic substrate during the addition process, it might be expected that the reaction would be facilitated by the presence of electron-donating substituents in the olefin. Such substituents could assist the reaction by increasing the nucleophilicity of the olefinic substrate. This expectation is borne out quantitatively by linear free energy correlations. In all cases, negative Hammett ρ-values are obtained for additions of carbenes or carbenoids to substituted olefins (Table 3[188–192]).

The selectivity which carbenes and carbenoids display toward highly alkylated olefins is well documented[41,53]. Less is known, however, about the electronic effects of other types of substituents in the olefinic substrate. In general, it appears that phenyl has a slight accelerating influence; carbene and carbenoid additions to styrene and α-methylstyrene generally proceed more rapidly than do the corresponding additions to cyclo-hexene[148,188,193,194]. In contrast, fluorine substitution on the olefinic substrate has been shown to have a decelerating influence on the carbene addition reaction[195–198]. Similar deactivating effects have been noted for 1-carbalkoxy[199] and 1-acetyl[200] substitution on the olefin.

Moss[199,199a] has emphasized a transition-state analysis approach in accounting for electronic substituent effects in the olefin upon reactivity. In this analysis, it is the inductive effect of the substituent which is of paramount importance. This contrasts with a ground-state analysis of relative reactivities for which resonance effects would be expected to be more important. The former analysis was found to be applicable in rationalizing the observed trend in relative reactivities of 1-substituted olefins toward addition by dichlorocarbene[199].

563

TABLE 3. Linear free energy correlations of carbene and carbenoid additions to olefins.

Carbene or carbenoid	Substrates	Conditions	ρ-value	Reference
:CCl$_2$	styrene, C=CH$_2$, CH$_3$, X-substituted	CHCl$_3$, KOBu-t, t-BuOH, 0°C	-0.38; (correlates with σ^+)	188
:CCl$_2$	C=CH, CH$_3$, X-substituted (X = H or OMe)	CHCl$_3$, KOBu-t, t-BuOH, 0°C	-0.53; (correlates with σ^+, two points only)	188
:CCl$_2$	RCH$_2$CH=CH$_2$, (R = i-Pr, MeO, Cl, PhCH$_2$, PhO)	Cl$_3$CCO$_2$Et + NaOMe, pentane, 0–5°C	-0.74; (correlates with σ^*)	189
:CCl$_2$	X–CH=CH$_2$ (X = Me, H, F, Cl, CF$_3$)	PhHgCCl$_2$Br, benzene, 80·3°C	-0.619 (correlates with σ^+)	191
$:C=C=C\!\begin{smallmatrix}CH_3\\CH_3\end{smallmatrix}$	X–CH=CH$_2$ (X = H, Me, Cl)	Me$_2$C=CBr$_2$ + MeLi, ether, -40°C	-4.3; (correlates with σ^+)	190
:C=C=CMe$_2$	X–CH=CH$_2$ (X = Me, H, Cl)	Me$_2$C=C=CHBr + KOBu-t,t-BuOH -10°C	-0.95; (correlates with σ^+)	190
EtZnCH$_2$I (methylenecarbenoid: 'Simmons–Smith Reagent')	X–CH=CH$_2$ (X = p-Me, H, p-F, p-Cl, m-CF$_3$)	Et$_2$Zn + CH$_2$I$_2$, benzene, 78·6°C	-1.61; (correlates with σ)	192

In a number of instances, reactions of carbenes with functional groups which contain one or more unshared electron pairs have been shown to proceed via ylide intermediates (see Section II.G)[46]. A number of investigators have devised systems to test the coordinating (ylide-forming) ability of carbenes with, e.g. oxygen-containing functionalities in appropriately-substituted olefinic substrates†. This was done in an effort to determine whether the lone-pair substituent might be capable of exerting a directive effect on the overall reaction of the carbene with the olefinic substrate. Such synergistic activity by a ketal substituent has been cited to account for the observed 3:1 preference for attack by dichlorocarbene on the central double bond of olefin **18** (equation 11[204—205]). There are a number

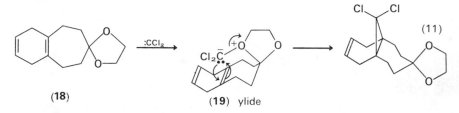

(11)

(**18**)   (**19**) ylide

of other examples where synergism of this type has been suggested to account for *cis* (or *syn*) stereoselectivity in carbene and carbenoid additions to cyclic olefins bearing oxygen-containing substituents[206—213].

This question has recently been reinvestigated by Moss[199,214], who found no support for the concept of synergism in a series of dichlorocarbene additions to olefins bearing oxygen-containing substituents

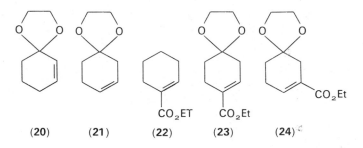

(**20**)   (**21**)   (**22**)   (**23**)   (**24**)

(**20–24**). Instead, Moss rationalized his results in terms of electronic-substituent effects which operate on a highly-unsymmetrical transition state leading to the :CCl₂–olefin adduct. Again, a transition state analysis was preferred to ground-state analysis of the addition process[199,199a].

---

† Reactions of dichlorocarbene with benzaldehyde and with benzophenone have been postulated as proceeding via intermediate carbonyl ylides[201–203].

3. Steric effects in the carbene and in the olefin

Relatively few cases have been studied where steric hindrance of olefin addition, due to substituents present in the attacking carbene, could be unequivocally demonstrated. A better case can be made for those steric effects which manifest themselves in carbenoid additions to olefins. Since carbenoids are less reactive toward olefins than are carbenes, they proceed through 'tighter' transition states (i.e., transition states which are product-like[215,216]). Steric hindrance would be expected to be more important for carbenoid additions to olefins, for which there exists a greater degree of olefin bonding to the divalent carbon atom in the transition state, than for the corresponding addition process involving free carbenes[30]. Additionally, the effective steric bulk of carbenoids can be enhanced through solvation effects in the presence of polar solvents[117] and/or by the proximity of the metal atom and ligands surrounding the metal to the three-centre addition site in the transition state[217].

The task of demonstrating the importance of steric effects of substituents in the olefin has proved to be much more straightforward, since addition reactions of a given carbene (or carbenoid) with a series of olefins of varying degrees of alkylation can be performed readily under identical environmental conditions. Steric hindrance to olefin addition can be demonstrated even for the case of a relatively small and reactive free carbene such as dichlorocarbene[218-221]. The increased steric demand presented by the carbon–bromine bonds in dibromocarbene relative to the carbon–chlorine bonds in dichlorocarbene has been demonstrated for additions of these carbenes in turn to styrene and to vinylmesitylene (competitively)[193].

The reactions of diphenylcarbene with olefins have recently been reinvestigated[76]. Diphenylcarbene is known to react with olefins to give the normal addition products and also by radical abstraction of hydrogen followed by radical recombination to afford carbon–hydrogen 'insertion' products. Jones and coworkers have recently demonstrated that the partitioning of the reaction between these two paths is determined largely by steric factors presented by substituents in the olefin. The competing abstraction–recombination process, observed some years ago for reactions of diphenylcarbene with 2-butenes[24,222,223], is unusual in that other triplet carbenes do not generally undergo this reaction with olefins. Jones' study demonstrates that the abstraction–recombination reaction is not an inherent feature of the reactivity of triplet diphenylcarbene, but instead it depends as well upon the nature of the olefin with which the triplet diphenylcarbene reacts[76].

Steric effects of substituents in the carbene and in olefinic substrates upon the course of the carbene–olefin addition reaction have been discussed extensively in recent reviews by Moss[30,41]. The reader is referred to these articles for additional examples and more extensive discussion.

D. Stereoselectivity: Orienting Effects of Substituents in the Carbene and in the Olefinic Substrate

In the preceding section, we considered some of the factors in the divalent carbon species and in the substrate which affect the ability of a carbene or carbenoid to undergo addition to an olefin. It was concluded that a subtle blend of steric and electronic factors determined the reactivity of a given divalent carbon species toward addition to an olefin. It was also pointed out that quantitative evaluation of any one of these factors was generally not possible owing to their simultaneous action and resulting mutual inseparability. Similar considerations (and frustrations) apply to the question of the stereoselectivity of addition of carbenes and carbenoids to olefins.

A divalent carbon species is considered as displaying stereoselectivity in its addition to an olefin when it is able to discriminate to some degree

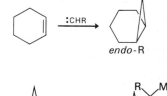

(12)

(13)

between the various possible modes of *cis* addition which are available to it. We have seen that *cis* addition to a substituted ethylene (acyclic olefin) can proceed in two different ways: the cyclopropane product having the substituent, R, in :CHR *cis* to the largest number of alkyl groups in the olefin is termed *syn*, and the other product is termed *anti* (equation 10). In addition to these *syn* and *anti* modes, two additional possibilities, termed *exo* and *endo*, can result from carbene or carbenoid additions to monocyclic and bicyclic olefins (equations 12 and 13, respectively).

In discussing matters related to reactivity in carbene–olefin addition reactions (preceding Section), it was occasionally relevant to discuss some points which actually had more direct bearing upon the question of stereoselectivity of carbene and carbenoid additions. However, there have been a number of important studies which have been concerned with this question, the results of which have been of significant synthetic and mechanistic import. Accordingly, it would appear worthwhile to examine in greater depth the question of stereoselectivity in carbene–olefin addition reactions. Again, this subject has recently been reviewed[30], and the paragraphs which follow only attempt to summarize some of the more salient and general conclusions extant.

(i) Interesting and synthetically useful results have been obtained from studies of methylenation of bicyclic systems related to norbornene. Reaction of norbornene with methylene carbenoid (generated via the Simmons–Smith procedure[223a] or via cuprous chloride catalysed decomposition of diazomethane[224]) affords the *exo*-tricyclooctane as the exclusive product. Both mono- and dihalocarbenes and carbenoids similarly exclusively add *exo* to the norbornene double bond[35,225,226]. Interestingly, in the *exo*-tricyclooctane formed by addition of mono- or dihalocarbenoids to norbornene, the *syn* halogen has been found to be thermally labile.

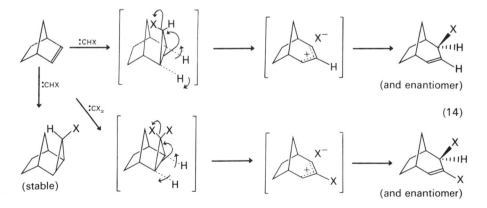

(14)

Simultaneous loss of this halogen and stereospecific disrotatory (electro-cyclic) opening of the cyclopropane ring[60] provides a synthetically-useful entry into the bicyclo[3.2.1]oct-2-ene system (equation 14)[35]. Both the addition and the rearrangement steps occur with *exo* stereospeci-ficity[227-231].

In contrast to the behaviour of norbornene, carbene and carbenoid additions to norbornadiene also afford some *endo*- as well as *exo*-monomethylenated products[35,223]. This establishes that steric hindrance presented by the *endo*-ethano(5,6-) bridge protons in norbornene toward attack by the divalent carbon species is more severe than is the corres-ponding steric effect presented by the *syn*-7- proton in norbornene and norbornadiene.

(ii) Perhaps the most important single concept to be developed in recent years which relates to the question of stereoselectivity in carbene–olefin addition reactions is the suggestion that the frequently-observed contra-thermodynamic addition of carbenes and carbenoids to alkylated olefins might have its origin in the 'steric attraction' effects which were discussed in the preceding section[183]. This theory offers a unifying concept, based on molecular orbital considerations, which at present offers at least a qualitative explanation for those cases where addition proceeds with predominant formation of the *syn* isomer (equation 10). The theory of steric attraction is a general one, and its applications are not restricted to carbene–olefin addition reactions. It is therefore anticipated that the future development and refinement of this theory will have a substantial impact upon the general theory of organic reaction mechanisms: 'a consummation devoutly to be wish'd'[231a].

(iii) The stereoselectivities observed in a number of individual carbene and carbenoid additions to olefins have been rationalized. It is instructive to consider a few of the better understood examples. Free halocarbenes show almost no stereoselectivity in addition reactions with olefins[81,232]. However, monohalocarbenoids generated from methylene halides and alkyllithium reagents generally afford mainly the *syn*-cyclopropanated product with alkylated acyclic[81,233,234] and cyclic[235,236] olefins. Fluoro-chlorocarbene[237] and fluorobromocarbene[194] both add with slight *anti*-fluoro stereoselectivity to cyclohexene. Phenylhalocarbenes and car-benoids generally show *syn*-halo (*anti*-phenyl) stereoselectivity in their additions to olefins[30].

Additions of carbalkoxycarbenes (produced photolytically from the corresponding diazoesters) to acyclic olefins afford predominantly *anti* products. Similarly, *exo* products predominate from the corresponding carbalkoxycarbene additions to cyclic olefins. The preference for *anti* and *exo* products is even more pronounced when carbalkoxycarbenoids are employed[30,144,145,238].

In contrast to the stereoselectivity shown by carbalkoxycarbenes and their corresponding carbenoids, arylcarbenes and arylcarbenoids display a marked preference for *syn* stereoselectivity in their additions to acyclic olefins. This preference was found to be stronger in all cases studied for the arylcarbenoid than for the corresponding free carbene[142,239,240]. However, repulsive steric effects operating in additions of arylcarbenoids to olefins have been noted. The *syn/anti* ratios for additions of p-tolyl-carbenoids to a series of 1-alkenes, $RCH=CH_2$, were seen to decrease with increasing size of the substituent R[241]. Such differences in stereo-selectivity between carbene and carbenoid additions to alkylated olefins are now well documented for a number of different divalent carbon species. These observations suggest that stereoselectivity in olefin addition reactions may provide a criterion for distinguishing between free carbenes and carbenoids, much in the same way as observations concerning the stereo-specificity of addition and relative reactivity studies in carbene–olefin addition reactions have been utilized in this context (see Sections II.A and II.C).

The stereoselectivities shown by oxycarbenoids in their additions to olefins appear to be quite sensitive to the (repulsive) steric effects of substituents in the olefin. Thus, the stereoselectivity of alkoxycarbenoid and aryloxycarbenoid additions depends very strongly upon the nature of the olefin substrate and, particularly, upon the degree (and type) of substitution in the olefin[217,242–244].

Phenoxycarbenoids produced from $PhOCH_2X$ (where X = F, Cl, and Br) with salt-free *n*-butyllithium were found to afford predominantly *exo*-7-phenoxynorcarane upon addition to cyclohexene. However, the *exo/endo* ratio varied with the nature of the leaving group, X, being greatest for X = F and smallest for X = Br. Interestingly, when lithium bromide was added to the *n*-butyllithium base, the epimer ratio was found to be nearly independent of the nature of the leaving group, X. These results were considered as suggesting the intermediacy of a true carbenoid (lithium-bound phenoxycarbene) in these reactions[217]. In contrast to the behaviour of phenoxycarbenoid, its corresponding thio[245,246] and seleno[247] analogues (phenylthiocarbene and phenylselenocarbene, respectively) display predominant *endo* stereoselectivity in their additions to cyclohexene.

E. 'Foiled' Carbene Additions to Olefins

There exists a plethora of examples of intramolecular additions of carbenes and carbenoids to olefins[248]. Many of these are of synthetic utility, offering the best available route to highly-strained, polycyclic systems. Some examples illustrating the synthetic applications of intramolecular carbene and carbenoid additions to carbon–carbon double bonds appear in equations 15–19.

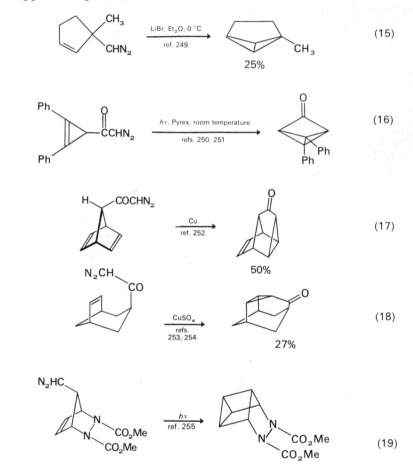

It is clear from these examples that such intramolecular carbene additions to carbon–carbon double bonds can be synthetically useful despite the high degree of internal bond strain which is often present in the adducts. The question thus arises as to the limitations which steric strain in the products might impose upon this reaction. It is also of interest to consider what alternative reaction paths might be available to the carbene if the culmination of its (normally preferred) intramolecular addition process is rendered sterically impossible. These questions have been considered in detail by Gleiter and Hoffmann[256]. Their conclusions have important bearing on the question of the relative energetics of addition vs. carbon–hydrogen insertion processes and on the nature of stabilization of divalent carbon species by neighbouring carbon–carbon double bonds.

Gleiter and Hoffmann introduced the term 'foiled-reaction' methylene to describe the situation which develops when a carbene species attempts an intramolecular addition to a carbon–carbon double bond but finds the culmination of the methylenation process to be sterically impossible[256]. Extended Hückel calculations were performed for carbenes 25–27. It was concluded that non-classical stabilization (as in 25b) could occur pre-

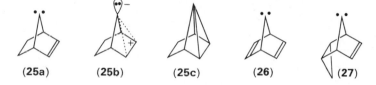

(25a)　　(25b)　　(25c)　　(26)　　(27)

ferentially in the ground states but not in the first excited states of these carbenes. The geometry of 25b (and of analogous structures in the case of carbenes 26 and 27) was visualized as being somewhere between that of the initial, unreacted carbene (25a) and that of the final adduct (25c). Importantly, this type of stabilization requires that these carbenes exist as ground state singlets[256].

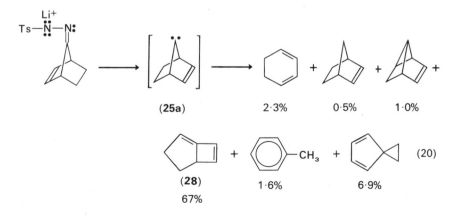

(25a)　　　　　2·3%　　　0·5%　　　1·0%

(28)　　　1·6%　　　　6·9%　　　(20)
67%

572   Alan P. Marchand

Carbene **25** has been prepared, and its intramolecular reactions have been studied by Moss and coworkers[257]. The reaction products obtained upon pyrolysis of the lithium salt of 7-norbornenone tosylhydrazone are shown in equation (20). It was suggested that the 'foiled methylene' adduct (**25c**) might be a transient intermediate leading to the formation of product **28** (equation 21).

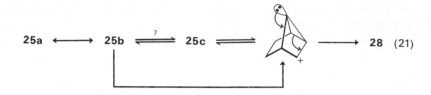

$$25a \longleftrightarrow 25b \underset{?}{\rightleftharpoons} 25c \rightleftharpoons \quad \longrightarrow 28 \quad (21)$$

A corresponding study of carbene **26** would be of interest. However, this carbene has not yet been prepared although an attempt at its synthesis has recently been reported[258]. The corresponding saturated carbene, norbornan-7-ylidene, has been studied by Moss and Whittle[259].

The intramolecular chemistry of bicyclo[3.3.1]non-2-en-9-ylidene (**29**) has also been explained in terms of a 'foiled methylene' intermediate[260], (equation 22). The directive ('non-classical') influence of the double bond

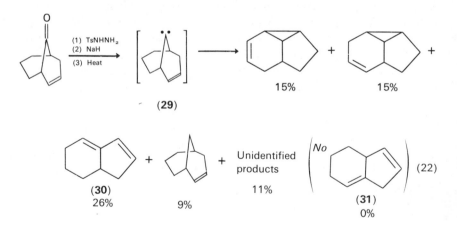

was evidenced by the fact that the major rearranged product was **30** and not the isomeric compound, **31**. Rearrangement of **29** to **30** must have involved participation of the double bond, whereas the corresponding rearrangement to **31** would not similarly involve the double bond (see equation 23).

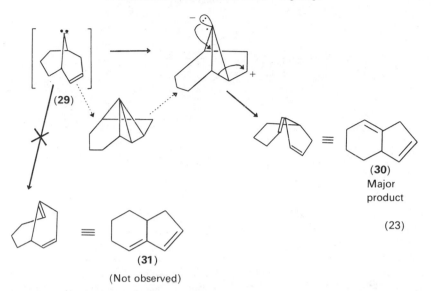

(30)
Major
product

(23)

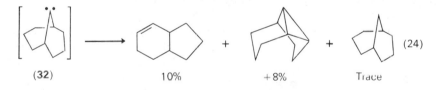

(31)

(Not observed)

Further evidence in this regard is provided by the intramolecular reactions of the saturated carbene, **32**, closely related to carbene **29** (equation 24). Whereas **29** affords mainly rearranged product, its saturated

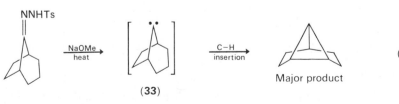

(24)

(32)   10%   +8%   Trace

analogue (**32**) affords mainly carbon–hydrogen insertion product. Thus, the rearrangement **29 → 30** is seen to require the presence of the double bond in **29**, and such a tendency toward rearrangement is not an inherent feature of the bicyclo[3.3.1]nonyl ring system[260]. Similar intramolecular behaviour is shown by the carbene pair **33** and **34** (equations 25 and 26, respectively)[261].

NNHTs

$\xrightarrow[\text{heat}]{\text{NaOMe}}$

$\xrightarrow[\text{insertion}]{\text{C–H}}$

(25)

Major product

(33)

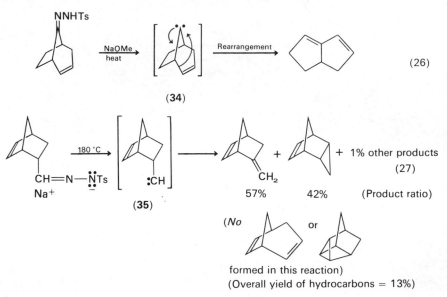

(26)

(34)

(27)

+ 1% other products

CH₂

57%   42%   (Product ratio)

(35)

(*No*   or

formed in this reaction)
(Overall yield of hydrocarbons = 13%)

Another interesting example is provided by Freeman's study of the epimeric 5-norbornenyl- and 2-norbornylcarbenes (equations 27–30)[262]. The ratio of products arising via hydrogen migration to those arising via carbon–hydrogen insertion was found to be similar for norbornyl- and norbornenylcarbene intermediates possessing the same geometry (*exo* or *endo*). However, the fact that carbene **35** did not suffer ring expansion

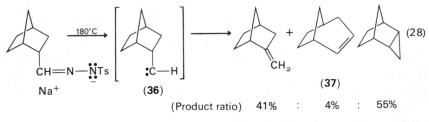

(28)

CH₂

(36)   (37)

(Product ratio)   41%   :   4%   :   55%

(Overall yield of hydrocarbons = 43%)

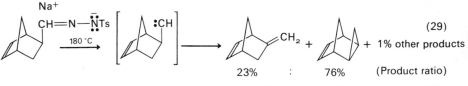

(29)

CH₂ +   + 1% other products

23%   :   76%   (Product ratio)

(Overall yield of hydrocarbons = 9%)

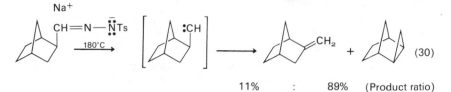

11%   :   89%   (Product ratio)

(Overall yield of hydrocarbons = 14%)

whereas the corresponding saturated carbene **36** did give some ring-expanded product (**37**) was taken as evidence that **35** might indeed possess some non-classical carbene character[262].

Other potentially non-classical 'foiled methylenes' have been studied. Among these are bicyclo[4.2.1]nona-2,4,7-trienylidene (**38**)[263–265], the interesting bishomoconjugated carbene, **39**[266], and the cyclopropyl-carbenes **40** and **41**[267]. In addition, the results of an elegant deuterium-

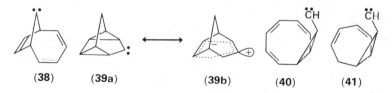

labelling study have suggested that a tetrahedrally-symmetric intermediate, perhaps tetrahedrane itself, is a transient species in the intramolecular rearrangement of the potentially non-classical carbene **43** derived from bistosylhydrazone **42**[268] (equation 31).

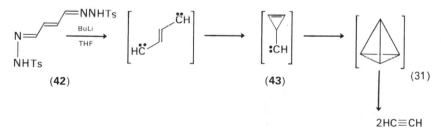

$$2HC\equiv CH$$

It is worthwhile to assess critically the applicability of the types of evidence which have been put forward to support claims of non-classical stabilization of 'foiled methylenes' in the foregoing examples. A consistent picture emerges when the chemistry of a potentially non-classical carbene is compared with that of the corresponding fully saturated species (*vide supra*). However, it has not been clearly established that the patterns of behaviour shown by such potentially non-classical carbenes are indeed viable operational criteria for establishing the presence or absence of homoconjugative stabilization in these carbenes. Along these lines, it should be noted that the type of stabilization depicted in **25b** would be expected to confer nucleophilic properties upon carbene **25** (see Section II.F), whereas resonance stabilization of carbene **35** (e.g., **35b**) should render that species electrophilic. Experimental evidence garnered from appropriate intermolecular carbene–olefin addition reactions which would have bearing on this point is sorely needed, and would be most welcome.

Goldstein and Hoffmann have recently extended the concept of Hückel aromaticity to π-electron interaction topologies which involve up to four 'ribbons'[269]. Here, a ribbon is defined as 'an intact conjugated polyene segment' subject to the restrictions that 'interactions between ribbons occur only at their termini', and that 'the two termini of any ribbon must remain indistinguishable both in the number of their connections and in the sense (σ or π) that such connections are made'[269]. A further restriction is that Möbius ribbons are excluded from consideration.

The simplest type of interaction, that which occurs between the two termini of a single ribbon, was termed *pericyclic*. Pericyclic interactions are possible for two- and three-ribbon interactions as well (**44** and **45**). Another type of interaction is possible between two ribbons; this was termed *spirocyclic* (**46**). Three types of interactions are possible for three interacting ribbons: these are *pericyclic* (**45**), *longicyclic* (**47**), and *laticyclic* (**48**).

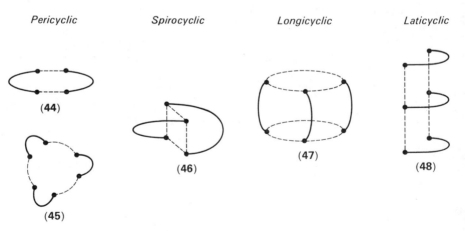

behaviour

| Pericyclic | Spirocyclic | Longicyclic | Laticyclic |

(44)

(46)

(47)

(48)

(45)

Goldstein and Hoffmann did not explicitly consider application of their treatment to the stabilization of carbene intermediates, although such an extension would appear to be both possible and highly desirable. A detailed discussion of their treatment is beyond the scope of this review, and the reader is accordingly referred to their original paper for details[269]. However, it is interesting to note some potential 'foiled methylenes' which are predicted to receive homoconjugative stabilization via spirocyclic, longicyclic, and laticyclic interactions; these are depicted in structures **49–51**, respectively. As yet, there does not appear to be any experimental evidence which might have bearing on these predictions.

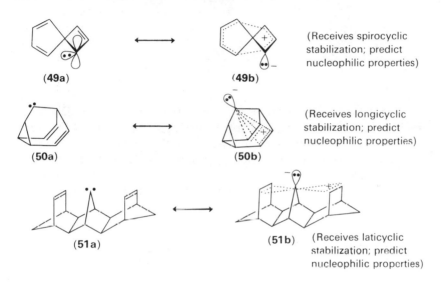

(Receives spirocyclic stabilization; predict nucleophilic properties)

(49a) **(49b)**

(Receives longicyclic stabilization; predict nucleophilic properties)

(50a) **(50b)**

(Receives laticyclic stabilization; predict nucleophilic properties)

(51a) **(51b)**

F. Reactions of Nucleophilic Carbenes with Olefins

Consider the monocyclic, conjugated carbene, **52**: for the case of

(52)

$(n + m)$ = even integer (including zero), the monocyclic carbene **52** will contain an odd number of carbon atoms. In such cases, a planar, conjugated, monocyclic system will be Hückel-aromatic (i.e. it will contain $(4n + 2)$ delocalizable electrons) if it bears a positive charge. This requires that the carbene carbon atom be left with a negative charge occupying an sp^2

hybrid orbital which is orthogonal to the π-system in the ring. One would therefore predict that the carbene in question should show nucleophilic character[270]. Two simple examples are shown (structures **53** and **54**).

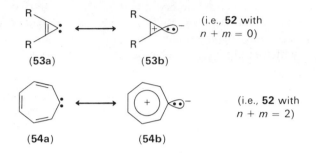

Evidence exists in support of the prediction that carbenes **53** and **54** should indeed be nucleophilic. Diphenylcyclopropenylidene (**53**, R = Ph) has been generated by α-elimination from the diphenylcyclopropenyl carbanate, **55**. This carbene could be trapped successfully by dimethyl fumarate (an electron-poor olefin), a result consistent with the anticipated nucleophilic character of the carbene (equation 32)[271–274]. Interestingly,

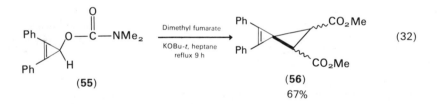

$$(32)$$

addition of cyclohexene to this reaction produced no norcarane-type adducts, nor was the yield of the fumarate adduct (**56**) affected by the presence of cyclohexene[273]. Diphenylcyclopropenylidene was found to react with fumaronitrile, diethyl fumarate, and N,N,N′,N′-tetramethyl-fumaramide, but no addition took place with *trans*-stilbene, tetrachloro-ethylene, diphenylacetylene, or dimethyl acetylenedicarboxylate[273].

To gain evidence regarding the spin state of the reacting carbene, an attempt was made to determine the stereochemistry of the addition of diphenylcyclopropenylidene to dimethyl maleate and fumarate. Both substrates gave the same adduct in this reaction; only one of the two possible spiropentenes was isolated, no trace of the other isomer being detectable. No conclusion regarding the stereochemistry of the addition process could be reached since the absence of the second spiropentene isomer leaves unanswered the question of the geometrical stability of the products under the reaction conditions[274].

Cycloheptatrienylidene (**54**) has been prepared directly via tropone tosylhydrazone (equation 33)[275,276] and indirectly via rearrangement of phenylcarbene (equation 34)[277–287]. An analogous rearrangement has been observed for diphenylcarbenes, affording aryl-substituted cyclo-heptatrienylidenes (equation 35)[284,288].

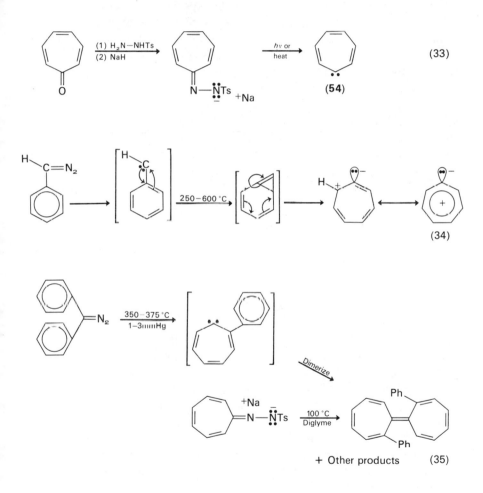

The nucleophilic character of **54** has been investigated through a Hammett study of its additions to *meta*- and *para*-substituted styrenes[289].

Correlation of the relative rates of addition of **54** (generated photolytically from the sodium salt of tropone tosylhydrazone) to these substituted styrenes with Hammett σ afforded a straight line having $\rho = +1 \cdot 05 \pm 0 \cdot 05$ (correlation coefficient $0 \cdot 982$). The observed positive value of ρ demonstrates conclusively the nucleophilic character of cycloheptatrienylidene in its additions to olefins. Also consistent with this conclusion is the demonstrated reluctance of **54** to react with electron-rich olefins[290]. Furthermore, its ability to add stereospecifically to maleinitrile suggests that **54** preferentially adds as a singlet species to electron-deficient double bonds[100]. Jones has recently reported that **54** adds to *cis-* and *trans*-1,3-pentadiene and to styrene; the spirocyclic adducts are thermally labile, and they readily rearrange via stepwise, diradical processes (equations 36 and 37, respectively)[291].

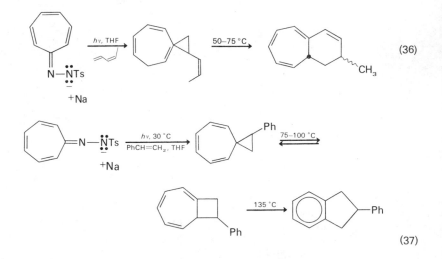

$$(36)$$

$$(37)$$

The reactions of **54** with 1,3-dicyanocyclooctatetraene[292] and with 2,3-dicyanobicyclo[2.2.2]octatriene[293] have recently been studied (equations 38 and 39). The former reaction was considered as proceeding with initial electron transfer from **54** to the substrate (path A, equation 38) although the possibility that the reaction might involve direct nucleophilic attack of carbene **54** upon 1,2-dicyanocyclooctatetraene (path B, equation 38) could not be rigorously excluded[292]. The latter process was believed to proceed via initial nucleophilic attack of **54** upon the electron-poor 2,3-double bond in 2,3-dicyanobicyclo[2.2.2]octatriene, although a direct, homo-1,4-addition of **54** to this substrate could not be excluded (equation 39).

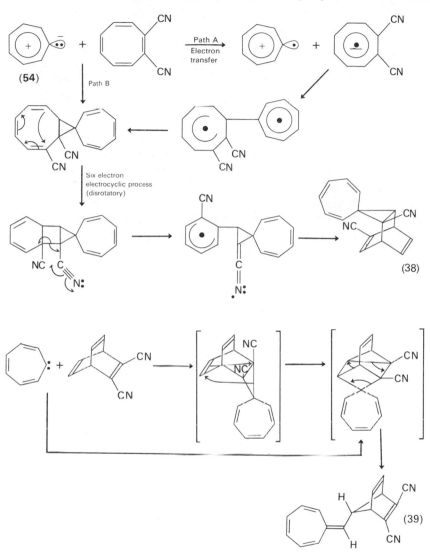

(38)

(39)

Benzo analogues of **54** have also been studied; these include **57**[281], **58**[95,294,295], and **59**[95,294]. E.s.r. studies of carbenes **58** and **59** have revealed that these species have triplet ground states[95]. However, when produced via photodecomposition of the corresponding diazo precursor, carbenes **58** and **59** have been found to add stereospecifically to olefins[96]. Such behaviour suggests that **58** and **59** are not nucleophilic carbenes. However,

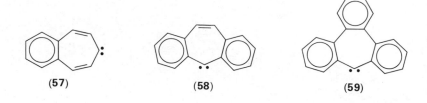

(57)   (58)   (59)

58 does appear to act as a nucleophile in its addition reaction with tetra-cyanoethylene[295]. Formation of the methylenated adduct is believed to arise via a charge transfer complex in which electron transfer occurs from the carbene to the electron-poor olefin[295].

Another carbocyclic carbene which is expected to show nucleophilic properties is 4,9-methano[11]annulenylidene (**60**). Generation of **60** from the corresponding ketone **61** via its tosylhydrazone in the presence of

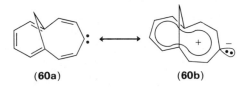

(60a)   (60b)

dimethyl fumarate or styrene afforded the spirocyclopropane adducts, **62** and **63**, respectively (equation 40)[296,297]. It will be of interest once the

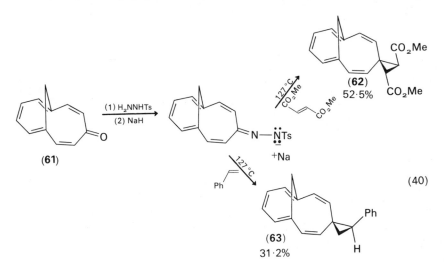

(61)

(1) H$_2$NNHTs
(2) NaH

127 °C
CO$_2$Me
CO$_2$Me

(**62**)
52·5%   CO$_2$Me   CO$_2$Me

N—NTs
+Na

127 °C
Ph

(40)

(**63**)
31·2%   Ph   H

Hammett ρ-value for the addition of **60** to substituted styrenes has been determined to compare quantitatively the nucleophilicity of **60** with that of cycloheptatrienylidene (**54**).

Carbenes which bear lone-pair (heteroatom) substituents adjacent to the divalent carbon atom might also be expected to display nucleophilic properties. Here, stabilization of the carbene might result from resonance interaction of the electron-donating lone-pair substituent with the empty orbital of the (singlet) divalent carbon atom (**64a, b** and **c**). Although

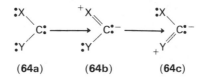

(64a)   (64b)   (64c)

dialkoxy- and dialkylthiocarbenes have been prepared, they do not appear to display marked nucleophilicity toward electron-poor olefins (e.g., attempts to trap dimethoxycarbene have met with repeated failure[298-300], and bis(methylthiocarbene adds to ketene dimethylacetal, an electron-rich olefin[301,302]). However, the siloxycarbene **65** can be trapped by reaction with diethyl fumarate (equation 41)[303]. Carbene **65** cannot be

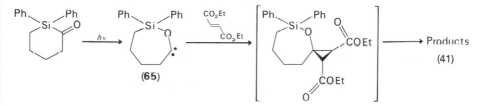

(65)   (41)

trapped by more electron-rich olefins such as cyclohexene or tetramethylethylene, nor does it react with ketene dimethylacetal or tetracyanoethylene[303].

Additional stabilization of heteroatom-substituted carbenes can be obtained when the delocalization extends over a conjugated, cyclic system. Stabilization by virtue of Hückel aromaticity is then possible (e.g., equation 42). This situation (**66**) is analogous to the type of resonance

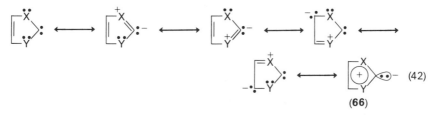

(42)

(66)

584   Alan P. Marchand

stabilization depicted by structures **53b** and **54b**. Carbenes **67**[304,305] and **68**[306] serve as examples of carbenes of the type **66** which have been prepared and whose additions to olefinic double bonds have been studied. Carbene **67** (R = Et) has been found to be capable of undergoing addition to

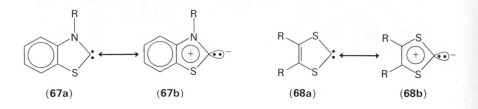

(67a)   (67b)   (68a)   (68b)

cyanoethylene[304]. However, carbene **68** does not simply add to electron-deficient olefins to produce methylenated products. Instead, **68** adds in Michael fashion to dimethyl maleate, thereby permitting isomerization of the olefin to dimethyl fumarate, with concomitant regeneration of the original carbene[306].

Another example is provided by carbene **69** which was generated by decarboxylation of the *N*-methyl betaine, **70**, in aprotic solvent (equation

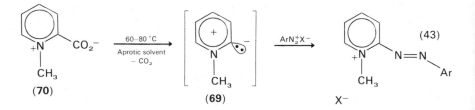

43)[307]. The carbene can be trapped by reaction with electrophiles such as aryl diazonium salts.

Quast[308] has considered the relative nucleophilicities of oxy-, thio-, and aminocarbenes. Among these various types of heteroatom-containing carbenes, the order of increasing nucleophilicities emerges as indicated in equation (44).

G. Addition vs. Ylide Formation in Reactions of Carbenes and Carbenoids with Olefins Bearing Heteroatomic Substituents

It was noted earlier (Section II.C) that carbenes are capable of co-ordinating to first- and second-row electronegative (lone-pair containing) elements, resulting in ylide formation[41,201–203,309–311]. There are a number of examples known where carbenoids have similarly reacted with lone-pair substituents to form ylides[166,312–316]. An interesting situation arises when a molecule contains both a carbon–carbon double bond and a heteroatomic (lone-pair) substituent elsewhere in the molecule. In this case, a competition is established between carbene (or carbenoid) addition to the olefin and ylide formation with the heteroatomic substituent, $-\ddot{X}$. The ultimate product which results from intermediate ylide formation is usually the C—X 'insertion' product, which probably arises via an abstraction–recombination mechanism (equation 45)[46]. Some examples

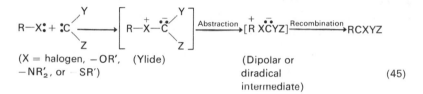

(X = halogen, $-OR'$,   (Ylide)   (Dipolar or
$-NR'_2$, or — SR')   diradical   (45)
                                            intermediate)

which illustrate the outcome of the competition between addition and ylide formation for some carbene and carbenoid reactions with allylic systems, $H_2C=CH-CH_2\ddot{X}$ (where $-\ddot{X}$ is a lone-pair heteroatomic substituent) will be presented in the ensuing discussion.

Reaction of methylene carbenoid (generated via cuprous halide-catalysed decomposition of diazomethane) with allylic ethers results in both cyclopropane formation and apparent allyl-oxygen bond 'insertion'[166]. However, the 'insertion' reaction was found to proceed with allylic inversion; the mechanism proposed for this process is depicted in equation (46). Cyclopropane formation and C—X 'insertion' accompanied

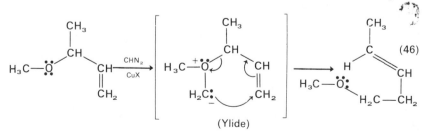

(Ylide)

by allylic rearrangement have also been observed for reactions of methylene carbenoid with allylic sulphides[166] and halides[317]. However, the corresponding reaction with allylamines does not afford the allylic rearrangement product[316].

Dichlorocarbene reacts with allylic sulphides to afford products derived via allylic rearrangement of the corresponding intermediate sulphur ylides, unaccompanied by the formation of methylenated products[318]. However, the corresponding reaction of dichlorocarbene with allylamines proceeds to afford mainly the carbene–olefin addition product. No trace of products arising via rearrangement of the (expected) intermediate nitrogen ylide could be detected[319]. Dichlorocarbene also reacts with allyl sulphides to afford some product which arises via the allylic inversion route[318,320,321].

Cyclopentadienylidene has been found to add to the double bond of allylic ethers, but to afford both addition and carbon–sulphur 'insertion' products with allyl sulphides[322]. Interestingly, the 'insertion' product, **72**, formed in the reaction of cyclopentadienylidene with allyl ethyl sulphide was considered as resulting via [3,3]-sigmatropic rearrangement of the intermediate sulphonium ylide, **71** (equation 47). The ratios of insertion to

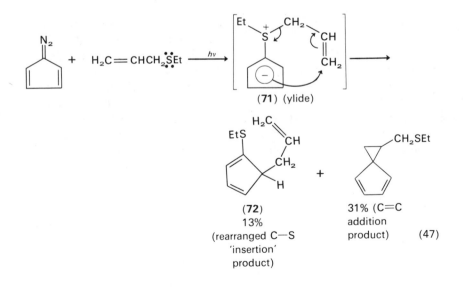

(**71**) (ylide)

(**72**)   31% (C=C
13%   addition
(rearranged C—S   product)   (47)
'insertion'
product)

addition products in reactions of this type were found to vary with the structure of the allyl sulphide substrate (from a high of 4·3 for γ-methylallyl ethyl sulphide to 1·5 for β-methylallyl methyl sulphide)[322].

Recently, a number of excellent studies have examined in detail the reactions of carbalkoxycarbenes, biscarbalkoxycarbenes, and their corresponding carbenoids with systems of the general type $H_2C=CH-CH_2X$[323]. Singlet, electrophilic biscarbomethoxycarbene, produced photolytically from dimethyl diazomalonate, interacts with the lone electron pair on a sulphur or oxygen atom much more rapidly than it attacks an olefinic π-bond[324,325]. By comparison, the corresponding triplet carbene, produced via benzophenone-sensitized photodecomposition of dimethyl diazomalonate, prefers addition to the olefinic double bond in allyl sulphides and ethers rather than electrophilic attack on sulphur or oxygen to form the corresponding ylide[326]. Biscarbomethoxycarbenoid reacts with allyl sulphides and ethers to afford products arising solely via initial attack of the carbenoid on sulphur or oxygen, respectively. An exception to this behaviour occurred with allyl t-butyl sulphide and allyl t-butyl ether, where the bulky t-butyl substituent effectively hindered attack by the carbenoid on the heteroatom[312,313,326].

Similar findings have resulted from the study of reactions of diazomethane[315,317], ethyl diazoacetate[312,313,327], and dimethyl diazomalonate[312,313,327] with allyl halides. Again, the relative yields of addition products and C—X 'insertion' products were found to depend upon the mode of decomposition of the diazo compound. The results for some reactions of diazo compounds with allyl halides are summarized in Table 4. The general trends previously observed for reactions of singlet and triplet carbenes and for reactions of carbenoids with allyl ethers and sulphides, also appear to hold true for the corresponding reactions with allyl halides.

The effect of added methylene halide solvents upon the reaction of biscarbomethoxycarbene with allyl chloride has been studied by Ando and coworkers[328]. The added diluents had the effect of increasing the relative amounts of methylenated products at the expense of the carbon–chlorine 'insertion' product. It was suggested that the diluents served to catalyse intersystem crossing (singlet to triplet) in the carbene. However, it should be recognized that the methylene halide diluents employed in this study are not inert to biscarbomethoxycarbene[329], and the increased preference for carbene–olefin addition in their presence might consequently result simply via preferential entrapment of the singlet carbene by the solvent[328].

TABLE 4. Reactions of diazo compounds with allyl chloride (73), β-methylallyl chloride (74), γ-methylallyl chloride (75) and allyl bromide (76)

Diazo compound[a]	Substrate	Method of decomposition of diazo compound	C—X insertion (%)	C=C addition (%)	Reference
A	73	hv (direct)	53	23	327
A	73	hv (Ph$_2$C=O sens.)	5	88	327
A	73	110°C, CuCl	32	3	313
B	73	hv (direct)	21	18	327
B	73	hv (Ph$_2$C=O sens.)	7	69	327
A	74	hv (direct)	25	22	327
A	74	hv (Ph$_2$C=O sens.)	trace	86	327
A	74	110°C C, CuCl	35	4	313
B	74	hv (direct)	57	30	312
B	74	hv (Ph$_2$C=O sens.)	6	85	328
A	75	hv (direct)	38	15	328
A	75	hv (Ph$_2$C=O sens.)	trace	49	328
A	75	110°C, CuCl	35	trace	313
A	73	hv, 10 mol-% C$_6$H$_{12}$ solvent	49	15	328
A	73	hv, 50 mol-% C$_6$H$_{12}$ solvent	18	7	328
A	73	hv, 90 mol-% C$_6$H$_{12}$ solvent	3·3	1·7	328

589

		Conditions			Ref.
A	73	$h\nu$, 10 mol-% CH_2Cl_2 solvent	42	18	328
A	73	$h\nu$, 50 mol-% CH_2Cl_2 solvent	24	11	328
A	73	$h\nu$, 90 mol-% CH_2Cl_2 solvent	5·4	4·6	328
A	73	$h\nu$, 10 mol-% CH_2Br_2 solvent	25	11	328
A	73	$h\nu$, 50 mol-% CH_2Br_2 solvent	8·6	5·4	328
A	73	$h\nu$, 90 mol-% CH_2Br_2 solvent	trace	4	328
A	73	$h\nu$, 10 mol-% CH_2I_2 solvent	18	9	328
B	75	$h\nu$ (direct)	16	8	312
B	75	80°C, 60 min, CuCl	9	2	312
B	75	80°C, 9 min, CuCl	12	3	312
B	75	80°C, 60 min, Cu	6	3	312
A	76	$h\nu$ (direct)	38	6	327
A	76	$h\nu$ ($Ph_2C{=}O$ sens.)	4	30	327
C	73	$CuCl_2$, pentane solvent, 10–15°C	3·9	25	317
C	73	$CuCl_2$, CH_2Cl_2 solvent, 10–15°C	17	11·5	317
C	76	$CuCl_2$, pentane solvent, 10–15°C	13	3·6	317
C	76	$CuCl_2$, CH_2Cl_2 solvent, 10–15°C	26	1·7	317
C	74	$CuCl_2$, pentane solvent, 10–15°C	4·5	13	317
C	74	$CuCl_2$, CH_2Cl_2 solvent, 10–15°C	18	10	317

[a] A = $N_2C(CO_2Me)_2$; B = N_2CHCO_2Et; C = N_2CH_2.

H. Reactions of Unsaturated Carbenes with Olefins

Unsaturated carbenes are species of the type X=C: in which the divalent carbon atom itself is directly incorporated into an unsaturated (double-bonded) system. Two types of unsaturated carbenes will be discussed in this section: alkylidenecarbenes (R_2C=C:) and alkenylidenecarbenes ('vinylidenecarbenes', R_2C=C=C:)†. Alkenylidenecarbenes have been generated by way of base-promoted α- or γ-elimination from haloallenes and propargyl halides, respectively[190,334–337,337a]. They have also been prepared via reaction of base with derivatives of ethynylcarbinols[338–340].

The results of kinetic studies indicate that base promoted reactions of 3-chloro-3-methyl-1-butyne (77) and of 1-chloro-3-methyl-1,2-butadiene (78) proceed via the same intermediate[334,335,337]. Isobutenylidenecarbene

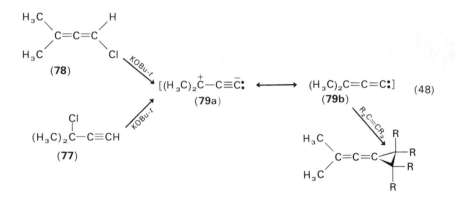

(79) was suggested as an intermediate in these reactions on the basis of trapping experiments with added olefins (equation 48)[334,335]. Identical rates of addition of 79 to various olefins were observed when the carbene was generated by either α- or γ-elimination from 77 or 78, respectively. Thus, either free carbenes are involved in these reactions, or, alternatively, complexed carbenes (carbenoids) may be present which are complexed in

† There has been considerable confusion in the literature regarding the question of nomenclature of unsaturated carbenes. Hence $(H_3C)_2C$=C: has been referred to variously as dimethylethylidenecarbene (one word)[330], dimethylethylidene carbene (two words)[331], dimethylmethylidene[190], and as a 'substituted vinylidene'[332]. The correct name[333] is based on the alkylidenecarbene (R_2C=C:) root; thus, $(H_3C)_2C$=C: should correctly be termed *isopropylidenecarbene*. An extension of this method of nomenclature to alkenylidenecarbenes would require that $(H_3C)_2C$=C=C: be named *isobutenylidenecarbene*. Throughout this Section, nomenclature based on the above system will, whenever possible, be consistently employed.

an identical manner whether generated from **77** or from **78**. Of these two possibilities, the latter appears to be more probable on the basis of Skell's observations with the related species, free C_3 and cyclopropenylidenecarbene (**80**; see Section II.I)[341]. Both of the free carbenes studied by

$\triangleright\!=\!C\!=\!C\text{:}$

(**80**)

Skell are non-selective reagents at 77 K, whereas **79** is a selective, electrophilic carbene at c. 273 K.

Skell's group has succeeded in generating isopropylidenecarbene (**81**) by reaction of diatomic carbon with propylene (see Section II.I)[332]. Isopropylidenecarbene has also been generated via photofragmentation of 2,2-diphenyl-1-isopropylidenecyclopropane (**82**). The intermediate carbene produced by photofragmentation of **82** can be trapped by cyclohexene to afford the addition product **83**, or by cyclohexane to afford the carbon–hydrogen insertion product, **84** (equation 49)[342].

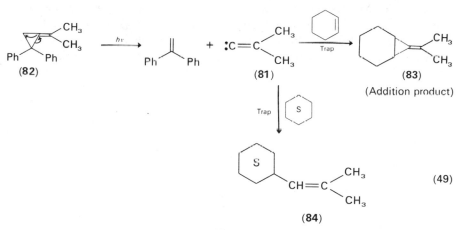

(49)

(**84**)

(C—H insertion product)

Substituted alkylidenecarbenes have also been synthesized by base-promoted decomposition of N-nitrosooxazolidones (**85**, equation

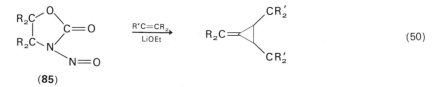

(50)

(**85**)

$50)^{330,343-345}$. A recent refinement of this procedure employs N-nitroso-acetylamino alcohols (**86**) for the production of alkylidenecarbenes

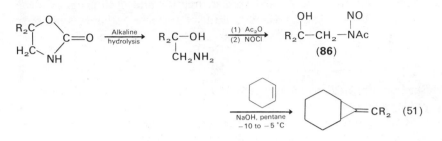

(equation 51)^{346,346a}. Isopropylidenecarbene generated by the method of Newman and Patrick^{330,345} does not insert into cyclohexane carbon–hydrogen bonds, and it is thus a different species from the one which was generated photolytically by Gilbert and Butler^{342}. It has been suggested that the former species may be a carbenoid whereas the latter species is probably the free isopropylidenecarbene^{342}.

Dichloromethylenecarbene ($Cl_2C{=}C{:}$) has been generated via photolytic decomposition of phenylperchlorovinylmercury (**87**). The divalent carbon species thus produced has been found capable of insertion into a carbon–hydrogen bond of cyclohexane and of addition to the double bond of cyclohexene (equation 52)^{347}. An unusual solvent effect was observed

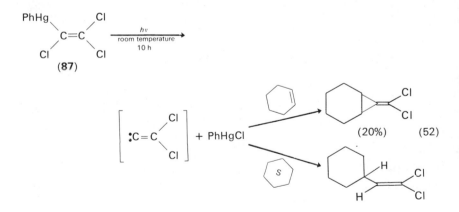

when the decomposition of **87** was carried out in ether (THF or dioxane) solvents. In such cases, the principal reaction products were benzene, trichloroethylene, hexachlorobutadiene and metallic mercury. These products were considered as resulting from a competing free radical process.

Isopropylidenecarbene (generated via base-promoted decomposition of 5,5-dimethyl-N-nitrosooxazolidone) has been reacted competitively with a variety of olefins[330]. This carbene has been shown to be subject to a marked steric effect in its additions to highly-substituted olefins. The origin of this steric effect becomes evident upon examination of the transition state for the addition process (**88**). In this diagram, the 2p orbitals of

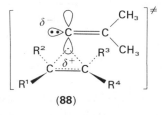

(**88**)

the double-bonded carbons in the carbene have been omitted for clarity. Note the unavoidable, unfavourable non-bonded interactions between one of the carbene methyl groups and the substituents R^3 and R^4 in the substrate. No such difficulty is experienced for addition of carbenes of the type $X_2C{=}C{=}C{:}$ to olefins because in this case, the vacant 2p orbital on the divalent carbon atom is orthogonal to the plane occupied by the C—R bonds (**89**). This explanation has been offered to account for the observation that isopropylidenecarbene shows unusually low reactivity toward

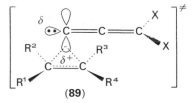

(**89**)

tetramethylethylene relative to cyclohexene[330] whereas $Ph_2C{=}C{=}C{:}$[348], $(t\text{-}Bu)_2C{=}C{=}C{:}$[349], and $(H_3C)_2C{=}C{=}C{:}$[334,335] all react considerably more rapidly with tetramethylethylene than with cyclohexene.

A quantitative comparison of the reactivities of isopropylidenecarbene and isobutenylidenecarbene toward olefins has been made possible through linear free energy correlations of their respective additions to *para*-substituted styrenes[190]. The relative rates of addition of isobutenylidene-carbene (generated via base-promoted α-elimination on 1-bromo-3-

methyl-1,2-butadiene) to styrene, p-methylstyrene, and p-chlorostyrene correlated with σ^+ to afford a Hammett ρ-value of $-0\cdot95$. Isopropylidene-carbene (generated via reaction of base with either 1,1-dibromo-2-methylpropene or 1-bromo-2-methylpropene) showed much greater selectivity toward addition to these styrenes than did isobutenylidene-carbene. For the isopropylidenecarbene additions, correlation with σ^+ afforded an unusually large, negative Hammett ρ-value of $-4\cdot3$. In this way, the relative degree of electrophilicity of these carbenes in their additions to substituted styrenes could be established quantitatively. Furthermore, the high degree of electrophilicity observed for isopropyli-denecarbene additions to these olefins offers strong support for the earlier contention that the observed low reactivity of this carbene toward highly-substituted olefins must be the manifestation of a specific steric effect (as in **88**)[330].

In much of the work which has been aimed toward studying the chemistry of alkylidenecarbenes and alkenylidenecarbenes, there remains unsettled the question of the nature of the divalent carbon intermediates, i.e. whether they are free carbenes or carbenoids. This is an especially troublesome point when considering the chemistry of those divalent carbon species which are produced via base-promoted α- or γ-eliminations, or via decomposition of heterocyclic precursors. Although the term 'carbene' has been used throughout this section in referring to the divalent carbon species thus produced, it should be noted that this has been done mainly for convenience and for economy of style. Thus, the uncertainty remains unresolved at present, and this question accordingly merits further, detailed consideration.

A new method for generating unsaturated carbenes from primary vinyl triflates has recently been developed by Stang and coworkers[350,350a]. Treatment of β,β-dialkylvinyl triflates (**90**) with KOBu-t in a variety of

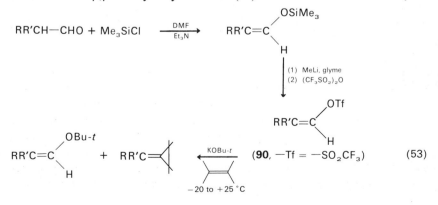

olefins afforded the corresponding methylenecyclopropanes along with *t*-butyl vinyl ethers (equation 53). The potential synthetic utility of this new method is indicated by the high yields of methylenecyclopropanes which result from the reaction of olefins with primary vinyl triflates (Table 5).

The origin of the *t*-butyl vinyl ethers in this reaction has been sought through a study of the reaction of $(H_3C)_2C=CH-OTf$ (91) with KOBu-*t* in the presence of excess *t*-BuOD. Substantial deuterium incorporation was observed for the product ether, whereas the recovered unreacted triflate contained no incorporated deuterium. These results establish the carbene pathway shown in equation (54) for the reaction of 91 with KOBu-*t*.

$$Me_2C=CHOTf \xrightarrow[\alpha\text{-elimination}]{KOBu\text{-}t} [Me_2C=C\ddot{:}] \xrightarrow{t\text{-BuOD}} Me_2C=C\begin{smallmatrix}OBu\text{-}t\\ \\ D\end{smallmatrix} \qquad (54)$$

(91)

To gain further insight into the nature of the intermediate which is produced in the reaction of primary vinyl triflates with KOBu-*t*, a study of the base-promoted reactions of *E*- and *Z*-β,β-ethylmethylvinyl triflates (92 and 93, respectively) with isobutylene was undertaken. Both starting materials afforded identical mixtures of methylenecyclopropanes and *t*-butyl vinyl ethers (94a–d, equations 55 and 56). This observation is a

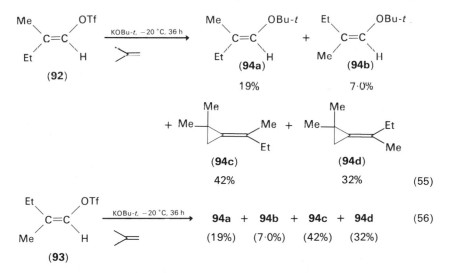

necessary (but not sufficient) condition for establishing the intermediacy of a free carbene in this reaction. In this connection, it is also of interest to note that the addition of isopropylidenecarbene (generated from the

TABLE 5. Reaction of primary vinyl triflates with olefins and KOBu-t at 0 °C[350,350a].

Triflate	Olefin	Reaction time (h)	Products (relative % yield)[a]	
			tert-Butyl vinyl ether	Methylenecyclopropane
H₃C, OTf / C=C / H₃C H (structure)	EtO—CH=CH₂	1	Me₂C=CHOBu-t (7%)	CH₃ CH₃ EtO (structure) (91%)
(same)	cyclohexene	24	Me₂C=CHOBu-t (32%)	CH₃ CH₃ (structure) (68%)
(same)	Ph—C(CH₃)=CH₂	1·5	Me₂C=CHOBu-t 22%	CH₃ CH₃ Ph (structure) (78%)
Et, OTf / C=C / Et H (structure)	EtO—CH=CH₂	4	Et₂C=CHOBu-t (7%)	C(Et)₂ EtO (structure) (93%)
(same)	cyclohexene	4	Et₂C=CHOBu-t (17%)	C(Et)₂ (structure) (83%)
(same)	Ph—C(CH₃)=CH₂	4	Et₂C=CHOBu-t (13%)	C(Et)₂ CH₂ Ph (structure) (87%)
cyclohexylidene C OTf / H (structure)	EtO—CH=CH₂	4	=CHOBu-t (7%)	EtO (structure) (93%)
cyclohexylidene C OTf / H (structure)	cyclohexene	20	=CHOBu-t (23%)	(structure) (77%)

[a] Uncalibrated relative areas by v.p.c.; estimated uncertainty ± 2–5%[350].

corresponding primary vinyl triflate) to *cis*- and *trans*-2-butene and to *cis*-and *trans*-2-methoxy-2-butene proceeds in 100% stereospecific *cis* fashion[350,350a,350b]. The foregoing results are suggestive of a free carbene intermediate which reacts with olefins in a singlet electronic configuration in these base-promoted reactions of β,β-dialkylated primary vinyl triflates with olefins.

Contrasting behaviour was observed for base-promoted reactions of olefins with primary vinyl triflates which bear one or two aryl substituents in the β position. In these cases, intramolecular rearrangements occur yielding substituted acetylenes to the extent of completely excluding methylenecyclopropanes or *t*-butyl vinyl ethers (equation 57). It was not

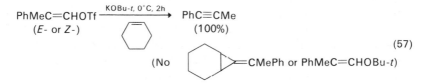

$$\begin{array}{ll} \text{PhMeC=CHOTf} \xrightarrow{\text{KOBu-}t,\ 0^\circ\text{C},\ 2\text{h}} & \text{PhC}\equiv\text{CMe} \\ (E\text{- or } Z\text{-}) & (100\%) \end{array}$$

(57)

(No ⟨⟩=CMePh or PhMeC=CHOBu-*t*)

established whether carbenoid reactions are involved in such cases[23] or whether free carbenes are formed which suffer intramolecular rearrangement more rapidly than they can be trapped by olefins.

The last of the unsaturated carbenes to be considered in this section is carbonylcarbene, :C=C=O, which may be regarded here as being structurally related to the alkenylidenecarbenes. Carbonylcarbene (C_2O) can be generated by photodecomposition of carbon suboxide (O=C=C=C=O) and the carbene thus produced reacts with simple olefins to afford allenes[351-353]. The mechanism of allene formation in the reaction of C_2O with ethylene was clarified by Mullen and Wolf[352] who observed that photodecomposition of $O=C={}^{14}C=C=O$ in the presence of ethylene afforded allene-^{14}C as the major product (*c.* 38% yield). The ^{14}C activity distribution in the allene thus produced was 92% in the central position

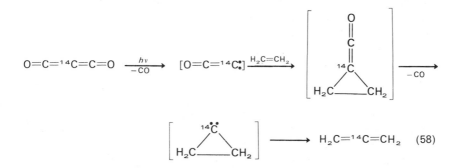

$$O=C={}^{14}C=C=O \xrightarrow[-\text{CO}]{h\nu} [O=C={}^{14}C:] \xrightarrow{H_2C=CH_2} \quad \longrightarrow$$

(58)

($C_{(2)}$) and 4% at each of the terminal positions ($C_{(1)}$ and $C_{(3)}$). The mechanism proposed by Mullen and Wolf[354] to account for the major reaction product is indicated in equation (58). This mechanism finds support in the observation that cyclopropylidenecarbenes are known to collapse to form allenes[355]. However, an alternative mechanism involving insertion of C_2O into an ethylene carbon–hydrogen bond followed by hot-molecule (i.e., excited allene) isomerization was not rigorously excluded, although it was considered to be unlikely.

The exact nature of the reactive intermediate could not be directly inferred from the foregoing results. Formation of the intermediate cyclo-propylidenecarbene in equation (58) via direct addition of atomic carbon ($C_{(1)}$) to ethylene could be ruled out on thermochemical grounds[353,356,357] and on the basis of other considerations[358–362]. In a reinvestigation of the photolytic reaction of carbon suboxide (C_3O_2) with ethylene, Bayes[353] found that two different reactive intermediates could be generated depending upon the wavelength of light employed in the photodecom-position of C_3O_2. The reactive intermediate which was generated when C_3O_2 was irradiated at shorter wavelengths (240–280 nm) was found to be inert toward oxygen and nitric oxide; this behaviour is suggestive of a carbene in a singlet electronic configuration. The corresponding species produced upon photodecomposition of C_3O_2 utilizing radiation of longer wavelength (>290 nm) reacted preferentially with oxygen (rather than with ethylene). Thus, allene formation could be quenched effectively by added oxygen when the reactive intermediate was generated from C_3O_2 using longer wavelength radiation. Bayes concluded that the 'short wavelength intermediate' was excited, singlet $C_2O(\tilde{a}^1\Delta)$ whereas the 'long wavelength intermediate' was a lower energy triplet C_2O, [probably the ground state, $C_2O(\tilde{X}^3\Sigma)$][353]. However, Bayes was careful to point out that his results could also be rationalized in terms of molecular C_3O_2 inter-mediates possessing different degrees of electronic excitation.

The chemistry of the 'long wavelength intermediate', $C_2O(\tilde{X}^3\Sigma)$, has been the subject of additional investigations. Willis and Bayes[363] have found that C_2O produced via photodecomposition of C_3O_2 at 300 nm always adds to olefins to afford allenes in preference to forming the cor-responding acetylenes (which are thermodynamically more stable). The percentage of acetylenes among the products was found to be a function of the nature of the olefinic reactant and of the olefin pressure. The $C_2O(\tilde{X}^3\Sigma)$ thus produced was found to be capable of isomerizing cis-2-butene to the trans isomer, a result in keeping with its expected triplet (diradical) character. Spin conservation arguments along with observations of specific effects which they noted when the C_2O–ethylene reaction was carried out in the presence of other gases, led Willis and Bayes to conclude that the allenes were being formed initially in triplet states in these C_2O–olefin reactions[363]. Additional evidence for the triplet nature of 'long wavelength' C_2O emerged from a detailed examination of its reactions with oxygen and with nitric oxide[364].

Further evidence regarding the nature of the reactive intermediates in C_3O_2 photolyses and regarding the mechanism of the photolytic reaction of C_3O_2 with olefins has been sought through relative reactivity studies[362,365–367]. Several methods have been employed to measure the relative reactivities of C_2O with olefins. The results obtained by Bayes and his collaborators[365,366] are self-consistent, but they stand in complete disagreement with the results obtained by Baker, Kerr, and Trotman-Dickenson[362,367]. The British group found that C_2O behaves as a nucleophilic carbene toward substituted olefins and that both $C_2O(\tilde{X}^3\Sigma)$ and $C_2O(\tilde{a}^1\Delta)$ show approximately the same reactivity toward a series of olefins[362,367]. However, Bayes concluded that $C_2O(\tilde{X}^3\Sigma)$ produced via photolytic decomposition (300 nm) of C_3O_2 is an electrophile, whereas $C_2O(\tilde{a}^1\Delta)$ similarly generated using radiation of wavelength 250 nm reacts indiscriminately with olefins[365]. More recent theoretical (extended Hückel and INDO) calculations support the suggestion that $C_2O(\tilde{X}^3\Sigma)$ is indeed an electrophilic carbene[368]. Nevertheless, the discrepancy noted above cannot at present be regarded as having been satisfactorily resolved.

Before leaving the subject of carbonylcarbene reactions with olefins, it is worthwhile to note an amusing application of this reaction to the synthesis of highly strained ring systems (equation 59)[369]. The results of

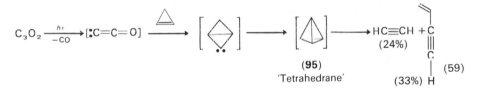

analysis of the deuterium distribution in the products obtained from the reaction of C_2O with 3,3-dideuteriocyclopropene required that an intermediate (such as **95**) possessing tetrahedral symmetry be invoked[369].

I. Reactions of Elemental Carbon and Carbynes with Olefins

It is worthwhile to consider a number of other reactive intermediates which bear formal resemblance to carbenes in that they possess a reactive carbon atom having non-bonded electrons and one (or more) low-lying, empty orbitals. Species of interest to us which fall into this classification are the various forms of elemental carbon $(C_1–C_5)$[369a] and carbynes (RĊ:). Some reactions of these species with olefins will be explored in the discussion which follows.

The simplest of the carbene analogues is atomic carbon, C_1. A number of early studies of the reactions of C_1 with hydrocarbons involved generation of the atomic carbon species by nuclear recoil processes[370–374]. The C_1 species thus generated inserts into carbon–hydrogen bonds and adds to olefinic pi-bonds[374,375]. However, these species are highly energetic, and the intermediates derived from them often suffer subsequent skeletal rearrangements and fragmentations, thereby obfuscating attempts to unravel the basic mechanism of the C_1–olefin reaction. Thus, the reaction of C_1 (produced by nuclear recoil) with ethylene affords several products, among which are allene, methylacetylene, acetylene, vinylacetylene, pent-1-yne, and ethylallene[376].

A refinement on this approach has been to moderate the reactivity of C_1 produced by nuclear techniques through collision of the 'hot' carbon with inert gases. In this way, carbon atoms can be generated which possess thermal kinetic energies, and a number of 'hot' carbon processes may be suppressed†. The reaction of C_1 with ethylene has been studied in the solid phase (rare gas–ethylene matrices)[379]. Under these conditions, fragmentation of the resulting (lower energy) intermediates is minimized, and the yield of fragmentation products (such as acetylene) is diminished. At higher moderation, the C_3H_4 adduct appears preferentially to afford allene which can be accounted for in terms of relatively low energy processes. The yield of this product is significantly higher than is the

---

† Recently, a chemical method of formation of C_1 which involves decomposition of 'quadricylanilidene' has been reported[377]:

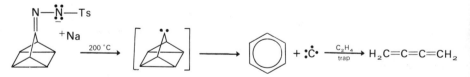

Atomic carbon has also been formed chemically via thermal decomposition of 5-tetrazoyl-diazonium chloride[358,378]:

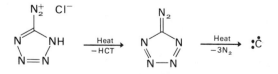

corresponding yield of allene produced in a non-moderated C_1–ethylene reaction[379]. Similar findings were reported for reactions of 'hot' vs. moderated C_1 with cyclopentadiene[380].

Skell and his collaborators have succeeded in producing mixtures of ground and excited state C_1–C_5 species utilizing a low intensity carbon arc under high vacuum. The carbon vapour thereby produced was deposited on a surface at 77 K, together with other substances with which the various carbon species were allowed to react. Through an analysis of the product mixture, the behaviour of each individual component of the carbon vapour could be studied[341].

Monatomic carbon was among the first of the components of carbon vapour to be studied extensively by Skell[381,382]. Three spin states of C_1 had been recognized earlier: the (triplet) ground state (^3P) and the first two (singlet) excited states (^1D and ^1S, respectively)[383]. The electronic configurations and excitation energies of these three electronic states of C_1 are depicted in Table 6. The reactivity of C_1 in different electronic

TABLE 6. Electronic configurations and excitation energies of the ^3P, ^1D, and ^1S states of C_1[383]

Electronic state	Electronic configuration				Excitation energy (eV)
	2s	$2p_x$	$2p_y$	$2p_z$	
^3P (ground state)	⥮	↑	↑	—	0
^1D	⥮	⥮	—	—	1·3
^1S	⥮	↑	↓	—	2·7

configurations with olefins could be conveniently studied at 77 K on an inert matrix. Reactions of C_1 in its ^1S state could be studied by depositing carbon vapour and olefin substrates simultaneously on the inert matrix at 77 K[384,385]. If instead the olefin was introduced after the carbon vapour had been allowed to age for 2 min or more on the inert matrix, the reaction was found to involve the ^3P (ground) state of C_1, (sufficient time having elapsed prior to introduction of the olefin to permit decay of C_1 from its metastable ^1S state to its ground state)[138,139,381].

Interestingly, the different electronic states of C_1 afford different products in their reactions with olefins[386]. Ground state (^3P) C_1 reacts with two molecules of olefin to form spiropentanes[138,381], whereas excited (^1S) C_1 reacts with olefins to afford allenes as the only significant products[384]. By way of contrast, the reaction of $C_1(^1D)$ with olefins affords spiropentanes and other, unidentified, products, but no allenes[384].

The addition of $C_1(^3P)$ to olefins to form spiropentanes is a two-step process, the first step of which is a stereospecific, singlet addition to afford a triplet cyclopropylidene which subsequently adds non-stereospecifically to a second molecule of olefin. This process is shown in equation $(60)^{138,381}$. Further evidence for this mechanism is provided by the results

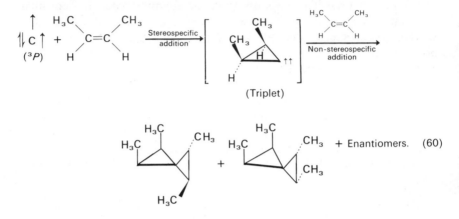

of competition experiments whereby mixtures of olefins were allowed to react with C_1 on a paraffin surface at 123–173 K. Analysis of the product mixture permitted calculation of the relative rate constants for both steps of the $C_1(^3P)$–olefin addition process. Butadiene displayed low reactivity towards $C_1(^3P)$ in the first step of the addition process. However, in the second step which involves (presumably triplet) cyclopropylidene as the reactive carbene species, butadiene displayed higher reactivity than did any of the monoolefins studied. It was concluded that $C_1(^3P)$ reacts with olefins preferentially through its filled and unfilled (rather than its half-filled) orbitals with conservation of spin angular momentum to afford a triplet cyclopropylidene. This new species then reacts as a typical triplet carbene, i.e. it adds in non-stereospecific fashion to the 2-butenes and it displays higher reactivity toward dienes than toward monoolefins[138].

A corresponding study of the stereochemistry of addition of C_1 meta-stable states (^1D and ^1S) to cis-2-butene has been performed[382]. Whereas spiropentane formation with $C_1(^3P)$ was 100% non-stereospecific[381], the corresponding additions of the C_1 metastable species to cis-2-butenes were completely stereospecific, indicating that the reacting metastable species is indeed in a singlet electronic configuration[382]. Subsequent investigations involving time-delay studies indicated that the metastable species which adds stereospecifically to cis-2-butene to afford the corresponding spiro-pentane is $C_1(^1D)$ (equation 61)[384]. The higher energy $C_1(^1S)$ reacts with

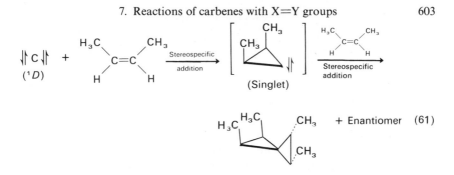

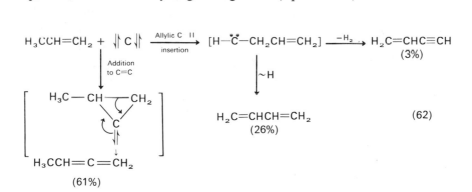

either *cis-* or *trans-*2-butene to afford exclusively the corresponding allene (2,3-pentadiene) in yields of 42 and 43 mol-%, respectively[384]. Under the experimental conditions, i.e. liquid nitrogen cooling and high olefin concentrations, Skell and Engel found that none of the intermediate singlet cyclopropylidene (formed from reaction of $C_1(^1D)$ with olefins) rearranged to allene. Hence they concluded that $C_1(^1S)$ is the only C_1 species which is capable of reacting with olefins to afford allenes[384].

A more recent study of the reaction of metastable C_1 species with olefins has been reported by Skell and coworkers[385]. In this study, metastable singlet C_1 atoms were found to react with olefins to afford dienes in addition to the allenes previously reported[384]. Although it is likely that $C_1(^1S)$ is solely responsible for allene formation[386a], $C_1(^1D)$ and/or $C_1(^1S)$ may participate in diene formation. The diene probably arises via insertion of singlet, metastable C_1 into allylic carbon–hydrogen bonds, with subsequent occurrence of hydrogen migration (equation 62). Reactions of

metastable, singlet C_1 with *cis-* and *trans-*2-butenes afforded dienes in each case with predominant retention of stereochemistry (*cis* or *trans* respectively) about the original double bond[385].

Interestingly, spiropentanes were not found to be significant products of the reaction which takes place when carbon vapour (produced in a low-intensity carbon arc under high vacuum) is co-condensed with olefin on a surface maintained at 77 K. This failure to observe the expected products of the $C_1(^3P)$-olefin reaction is not presently understood, and awaits further clarification.

Diatomic carbon (C_2) is also a constituent of carbon vapour, and its reactions with olefins have also been the subject of recent investigations[332,387a]. The reaction of C_2 with two molecules of propylene at 77 K afforded the isomeric dienes **96** and **97** in $c.\ 5\%$ combined yield (equation

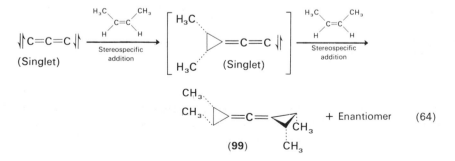

63); here, C_2 is behaving as a dicarbene. Reaction of C_2 with one molecule of propylene afforded 1-penten-4-yne (4% yield), 2-methylbutenyne (1% yield), and 1-pentyne (2% yield)[332]. A vinylidene intermediate (**98**) was suggested by the formation of **96** and **97** from C_2 (equation 63). This intermediate could not be trapped by added 1,3-butadiene; accordingly, it was suggested that **98** was reacting in a singlet electronic configuration with a second molecule of propylene to afford **96** and **97**.

Triatomic carbon is another major constituent of carbon vapour which, like C_2, can behave as a dicarbene in its reactions with olefins[341,387]. One molecule of C_3 can add to two olefin molecules to afford bisethano-allenes[341,387]. When C_3 (generated via carbon arc) is deposited on a solid hydrocarbon surface at 77 K and aged for 9 sec to 2 h, decay to the ground state occurs. Subsequent introduction of *cis*-2-butene results in the formation of only one bisethanoallene, **99** (i.e., that which results exclusively from two successive stereospecific addition processes, as indicated in equation 64). Simultaneous deposition of C_3 and either *cis*- or *trans*-2-

butene on a hydrocarbon surface at 77 K results in the formation of two bisethanoallenes (**99** and **100**) via the mechanism indicated in equation

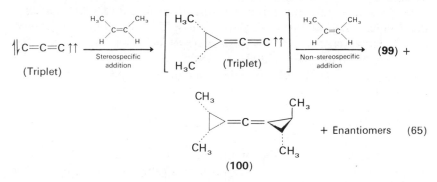

+ Enantiomers   (65)

$(65)^{341}$. The fraction of metastable (triplet) C_3 in carbon vapour produced via a carbon arc was found to increase with increasing arc voltage[369a].

Studies of reactions of singlet C_3 with a variety of olefins reveal that this species is a non-selective reagent. Relative reactivities of $C_1(^1S)$, $C_1(^3P)$, and singlet C_3 toward additions to olefins are compared in Table 7.

TABLE 7. Relative rates for $C_1(^1S)$, $C_1(^3P)$, and singlet C_3 additions to olefins

Olefin	$C_1(^1S)^{384}$	Relative rates $C_1(^3P)^{341}$	Singlet $C_3{}^{139}$
	(1·00)	(1 00)	(1·00)
	1·81	0·6	2·2
	—	1·25	0·6
	0·94	0·05	0·36
	4·92	—	1·2
	1·36	0·04	0·94

Another dicarbene, C_4, has been identified as a minor constituent of carbon vapour[388]. Arc-generated C_4 reacts with a variety of hydrogen-containing materials, (including *cis*-2-pentene) to afford ethylacetylene, methylallene, 1,3-butadiene, vinylacetylene, and trace amounts of diacetylene[388]. However, no report has yet appeared which might indicate whether C_4 is capable of forming methylenated adducts with olefins.

Carbynes are the last of the carbene analogues to be discussed in this section. The simplest example, methine (:ĊH), can exist in two spin configurations: a doublet [(↓↑)↑C] and a quartet [↑↑↑C], analogous to singlet and triplet spin states of carbenes, respectively. Methine has been examined spectroscopically by Braun and his collaborators[389]. Occasionally, methine has been postulated as an intermediate in reactions of 'hot' carbon atoms with organic substrates. However, it is generally not clear whether the products of these reactions were necessarily formed via an intermediate methine or whether their chemistry can be accounted for by direct reaction of carbon atoms[370,387a,390-392]. A possible exception to this statement may be the reaction of 'hot' $^{11}C_1$ with hydrogen–ethylene mixtures. Of the many products of this reaction, one of them, 1-pentene, is difficult to rationalize in terms of C_1, $:CH_2$, or $\cdot CH_3$ as precursors. A mechanism involving methine as a precursor for 1-pentene in this reaction has been suggested (equation 66)[393].

$$:C: \xrightarrow{H_2} [:\overset{\bullet\bullet}{C}H + \cdot H] \xrightarrow{H_2C=CH_2} [H_2C \cdots CH \cdots CH_2{}^\bullet] \xrightarrow{H_2C=CH_2} [H_2C=CHCH_2CH_2\overset{\bullet}{C}H_2]$$

H-abstraction
or radical
disproportionation

↓

1-Pentene

(66)

Carbethoxycarbyne ($:\dot{C}\!-\!CO_2Et$) has been generated conveniently by photolysis of diethyl mercurybisdiazoacetate [**101**, $Hg(N_2CCO_2$-$Et)_2$][394,395]. Like methine[396,397], carbethoxycarbyne has been shown by low-temperature e.s.r. studies to possess a doublet ground state[394]. Carbethoxycarbyne (generated photolytically from **101**) has been reported as reacting with cyclohexene to afford both carbon–hydrogen insertion and methylenated products (equation 67). That the addition products (**102** and **103**) did not result via hydrogen abstraction by the carbyne followed by addition of the resulting carbethoxycarbene to the substrate was revealed by the product ratio **102**/**103** = 31/17[394] (this result should be compared with the corresponding value of **102**/**103** = 1/1·6 which was

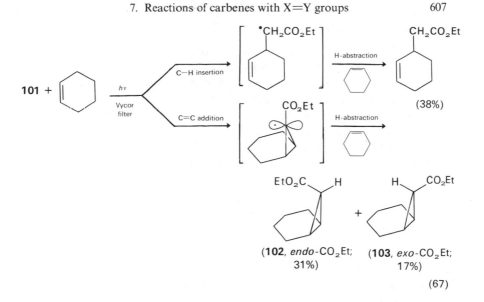

$$(102, endo\text{-}CO_2Et; \quad (103, exo\text{-}CO_2Et; \\ 31\%) \qquad\qquad 17\%)$$

$$(67)$$

observed some years ago for the addition of carbethoxycarbene to cyclo-hexene[398]). In the carbyne addition to cyclohexene, preferential formation of the thermodynamically less stable *endo* isomer (102) could readily be accounted for in terms of the addition–abstraction mechanism shown in equation (67)[394].

The stereospecificities of carbethoxycarbyne additions to *cis*- and *trans*-2-butenes have been studied[395]. Carbethoxycarbyne produced via photo-decomposition of the organomercury precursor 101 was found to add in stereospecific *cis* fashion to 2-butenes. This result suggests that carbethoxy-carbyne adds to olefins through its ground (doublet) state.

III. REACTIONS WITH C=X

In Sections III and IV, reactions of carbenes and carbenoids with X=Y (where either or both X and Y are heteroatoms) will be discussed. Reactions of this type have been reviewed by Kirmse[36], whose coverage of the carbene literature extends through mid-1970. In the remaining sections, a brief survey of the major features of carbene additions to X=Y functionalities will be presented, with particular emphasis on those contributions which have appeared since the publication of the second edition[36] of Kirmse's book.

A. Reactions with C=O and C=S

Simple addition of carbenes or carbenoids to C=O and C=S would be expected to result in the formation of oxiranes (epoxides) and thiiranes, respectively. In practice, very few examples are known which involve direct addition of free carbenes to carbonyl and thiocarbonyl compounds, and these are rarely of synthetic value. A possible exception to this statement is the addition of dihalocarbenes (generated via thermolysis of phenyl(trihalomethyl)mercury precursors) to carbonyl groups which bear highly electronegative substituents (equation 68)[399-401,401a]. A mechanism

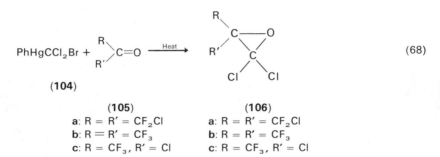

$$PhHgCCl_2Br + \begin{array}{c} R \\ R' \end{array}\!\!C{=}O \xrightarrow{\text{Heat}}$$

(68)

(**104**)

	(**105**)	(**106**)
	a: R = R′ = CF$_2$Cl	**a**: R = R′ = CF$_2$Cl
	b: R = R′ = CF$_3$	**b**: R = R′ = CF$_3$
	c: R = CF$_3$, R′ = Cl	**c**: R = CF$_3$, R′ = Cl

for this reaction which involves the formation of an intermediate carbonyl ylide has been suggested (equation 69)[400]. However, it is wise to be warned

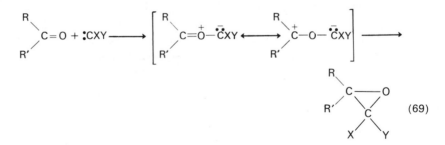

(69)

that reactions of PhHgCX$_2$Y with substrates containing lone-pair substituents might proceed via direct nucleophilic attack of the heteroatom upon the phenyl(trihalomethyl)mercury reagent. Such reactions might involve direct trnasfer of :CX$_2$ from the organomercury reagent to the substrate; free carbenes would not necessarily be involved in such cases[400,402-405].

Other carbene additions to carbonyl groups have been reported. Photolysis of perfluoropropylene oxide afforded :CF$_2$ which was found to be capable of addition to the carbon–oxygen double bond of $(F_3C)_2C=O$ and of $(F_3CCF_2)_2C=O^{406}$. Similarly, $(F_3C)_2C:$ (generated via pyrolysis of the corresponding diazirene) was found to add to $F_2C=O^{407,408}$. It is not clear why this reaction should be facilitated by the presence of strongly electron-withdrawing substituents in both the carbene (or carbenoid) and the substrate.

Diphenyldiazomethane reacts with $(MeO_2)P(O)—C(O)—R$ ($R = CH_3$ or Ph) to afford oxiranes. The carbonyl group appears to be activated in this reaction by the presence of the electron-withdrawing $(MeO)_2P(O)—$ group[409].

Diazoesters undergo a number of interesting reactions with ketones[410] and with thioketones[411,412]. However, many of these reactions are catalysed by copper or by copper salts, and they generally do not result in either oxirane or thiirane formation. Photolytic decomposition of diazomethane ($\lambda > 320$ nm) in the presence of acetone affords 2,2-dimethyloxirane, along with other products[413].

A rare example of addition of an unsaturated carbene to the carbonyl group of ketones and aldehydes has recently been reported[414]. Addition of fluorenylidenecarbene (107) to $RR'C=O$ occurs to afford a 1:2 adduct (in low yield) which is believed to result via formation of an allene oxide intermediate (108, equation 70).

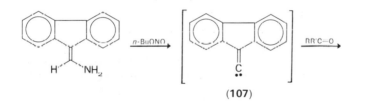

(107)

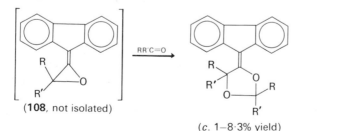

(108, not isolated)

(c. 1–8·3% yield)

(70)

The addition of $:CX_2$ (produced via carbene transfer from $PhHgCX_3$) to carbon–sulphur double bonds has received attention[405,415,416]. The reaction of $PhHgCCl_2Br$ with thiophosgene affords perchlorothiirane in 36% yield (equation 71)[405]. The corresponding reaction with CS_2 also

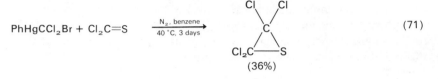

$$PhHgCCl_2Br + Cl_2C{=}S \xrightarrow[\text{40 °C, 3 days}]{\text{N}_2\text{, benzene}} \qquad (71)$$

$$(36\%)$$

affords perchlorothiirane in low yield (equation 72)[415]. Also, the addition of $:CClF$ (generated from $PhHgCCl_2F$) to thiobenzophenone has been

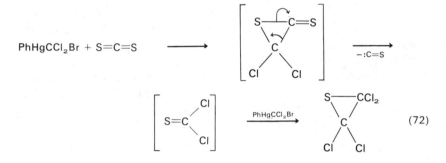

$$PhHgCCl_2Br + S{=}C{=}S \longrightarrow \qquad \xrightarrow{-:C=S}$$

$$\xrightarrow{PhHgCCl_2Br} \qquad (72)$$

observed recently[416]. As in the case of reactions of phenyl(trihalomethyl)-mercury compounds with aldehydes and ketones (discussed previously), there is a distinct possibility that the reactions discussed above do not involve free carbene additions to carbon–sulphur double bonds.

Diazo compounds have also been found to react with carbon–sulphur double bonds. Recent examples are shown in equations 73–76[412,417–419].

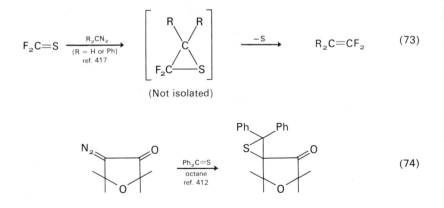

$$F_2C{=}S \xrightarrow[\substack{(R = H\ or\ Ph) \\ \text{ref. 417}}]{R_2CN_2} \qquad \xrightarrow{-S} \qquad R_2C{=}CF_2 \qquad (73)$$

$$(\text{Not isolated})$$

$$\xrightarrow[\substack{\text{octane} \\ \text{ref. 412}}]{Ph_2C=S} \qquad (74)$$

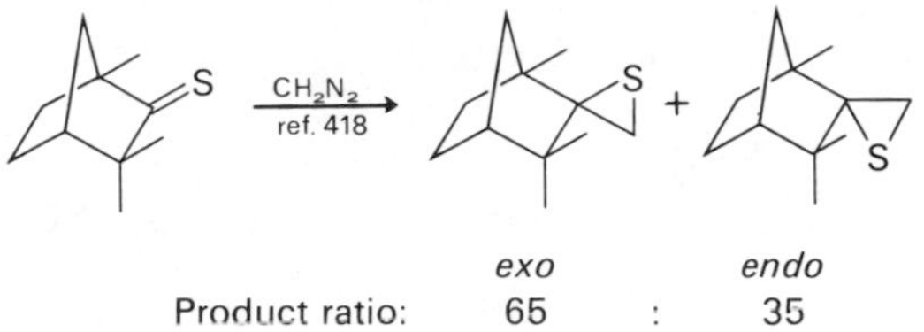

(75)

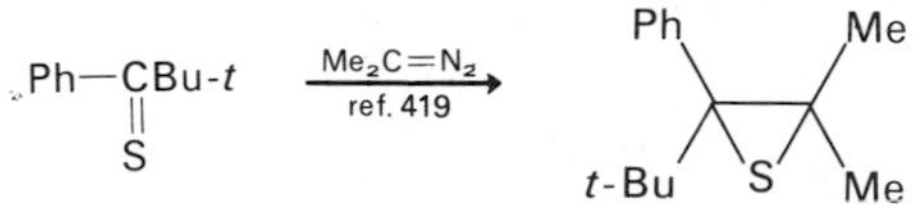

(76)

Earlier examples of the formation of thiiranes through reactions of diazo compounds with diarylthioketones have been reviewed by Sander[411].

It has been known for many years that diazo compounds can react with thioketones to afford a variety of products. Prominent among these are thiiranes and 1,3-dithiolanes. The course of such reactions has been found to be strongly dependent upon the nature of the diazo compound employed and of the thioketone substrate. Schönberg and coworkers established that thioketones react with diaryldiazomethanes to afford thiiranes whereas the corresponding reaction with either diazomethane or ethyl diazoacetate produces mainly 1,3-dithiolanes[420]. A later investigation revealed that diazomethane also reacts with diaryl thioketones to afford the corresponding 1,3-dithiolanes. However, the corresponding reaction with primary diazoalkanes affords thiiranes or 1,3-dithiolanes, whereas secondary diazoalkanes react with diaryl thioketones mainly to produce the corresponding thiiranes[421].

Recently, some reactions of diazoalkanes with α,β-unsaturated thioketones have been investigated[422] and both thiiranes and 1,3-dithiolanes are produced in these reactions. Interestingly, the relative proportion of these two products was found to be dependent upon three factors: the nature of the diazoalkane[421], the temperature of the reaction (higher temperatures favouring dithiolane formation), and the order of addition of the reactants. Interestingly, addition of the diazoalkane to the thioketone was found to favour dithiolane formation, whereas the inverse order of addition of these reagents favoured thiirane formation[422].

In all of these diazoalkane–thioketone reactions, it is likely that the initial reaction which occurs involves the undissociated diazo compound, affording thiadiazoline intermediates (equation 77)[411,418,419,422]. It is

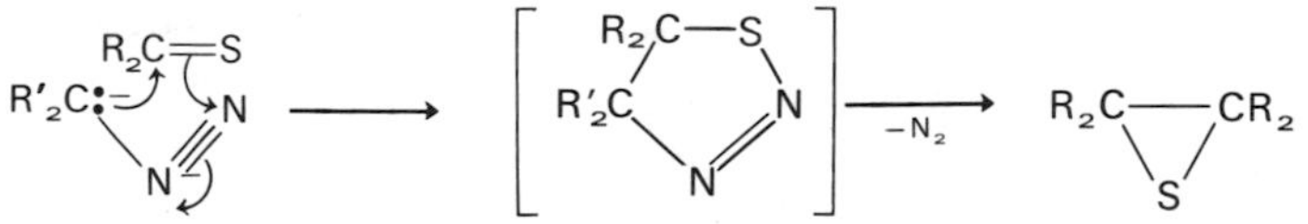

(77)

therefore possible that these reactions do not involve carbenes at all. Additional studies which might serve to clarify this point further would be most welcome.

The reaction of diazoalkanes with elemental sulphur has proved to be a useful synthetic method for the preparation of thiiranes (equation

$$R_2CN_2 + 1/8S_8 \xrightarrow[-N_2]{} [R_2C{=}S] \xrightarrow[-N_2]{R_2CN_2} \underset{S}{R_2C{-}CR_2} \qquad (78)$$

78)[423,424]. The reaction of PhHgCCl$_2$Br with elemental sulphur proceeds similarly to afford a thiirane (equation 79)[405].

$$PhHgCCl_2Br + 1/8S_8 \longrightarrow [Cl_2C{=}S] \xrightarrow{PhHgCCl_2Br} PhHgBr + \underset{S}{Cl_2C{-}CCl_2} \quad (79)$$

B. Reactions with C=N and C=P

Several carbene and carbenoid additions to the carbon–nitrogen double bond of imines have been reported[425]. The earliest report of reactions of this type was in 1959 by Fields and Sandri, who isolated 2,2-dichloro-1,3-diphenylaziridine from the reaction of dichlorocarbene with N-benzylideneaniline (equation 80)[426]. This same reaction received further attention in the early 1960's[427,428].

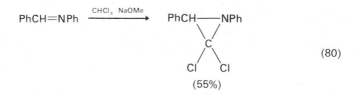

$$(80)$$

(55%)

A later investigation by Deyrup and Greenwald involved the reaction of N-benzylideneaniline with LiCHCl$_2$[429]. Interestingly, only one of the two possible aziridine products resulted from this reaction (equation 81).

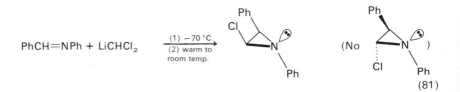

$$(81)$$

A reasonable mechanism might involve stereoselective *cis* addition of lithium monochlorocarbenoid to the imine carbon–nitrogen double bond.

Additions of phenyl(trihalomethyl)mercury-derived dihalocarbenes to carbon–nitrogen double bonds have been extensively studied by Seyferth and coworkers. The transfer of $:CX_2$ from $PhHgCX_3$ to C=N was seen to be facilitated by the presence of electron-withdrawing groups on the imine carbon atom (which served to depress the nucleophilic character of the imine nitrogen)[430]. Thus, reaction of $RN=CCl_2$ (where R = phenyl, cyclohexyl or *iso*-propyl) with $PhHgCCl_2Br$ afforded the corresponding aziridines in yields of 53%, 29%, and 43%, respectively[431]. Attempts to add dichlorocarbene (generated via $PhHgCCl_2Br$) to $PhCH=NPh$, $PhCH=NMe$, and $Me_2C=NPh$ were all unsuccessful[432]. Attempted addition of dibromocarbene (generated via $PhHgCBr_3$) across the carbon–nitrogen double bond of *N*-neopentylidene-*t*-butylamine failed[431], but the corresponding addition of dichlorocarbene (produced via reaction of chloroform with potassium *t*-butoxide) to this same substrate afforded the corresponding aziridine in low yield[433].

A recent study of the reaction of $PhHgCFCl_2$ with $PhN=CCl_2$ has produced the first example of $:CFCl$ addition to a carbon–nitrogen double bond[416]. Fluorocarbalkoxycarbene ($F-\ddot{C}-CO_2R$) generated via $PhHg-CFXCO_2R$ (where X = Cl or Br, and R = CH_3 or CH_2CH_3) has likewise been found capable of undergoing addition to the carbon–nitrogen double bond of $PhN=CCl_2$ to afford the corresponding aziridine (equation 82)[434].

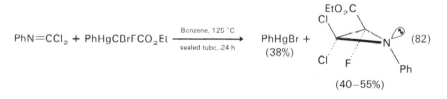

$$PhN=CCl_2 + PhHgCBrFCO_2Et \xrightarrow[\text{sealed tube, 24 h}]{\text{Benzene, 125 °C}} \quad PhHgBr + \quad (82)$$
$$(38\%)$$
$$(40-55\%)$$

Reactions of isocyanates with phenyl(trihalomethyl)mercury compounds have also been studied[415]. Isocyanates are very much less reactive toward $PhHgCCl_2Br$ than are carbon–carbon double bonds (e.g., equations 83[106] and 84[415]). However, addition of dichlorocarbene

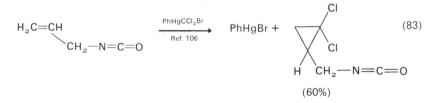

$$H_2C=CH \xrightarrow[\text{Ref. 106}]{PhHgCCl_2Br} \quad PhHgBr + \quad (83)$$
$$(60\%)$$

(84)

(84%)

(generated via PhHgCCl₂Br) to the carbon–nitrogen double bond of phenyl isocyanate does occur at elevated temperatures (80°C, benzene solvent) and in the presence of a five-fold excess of phenyl isocyanate, (equation 85)[415].

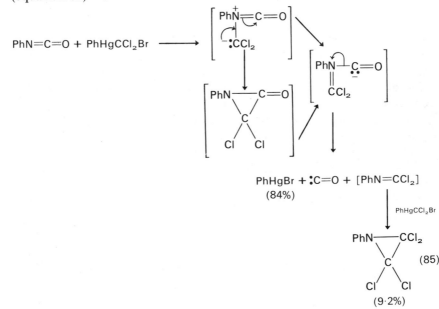

Other carbene additions to the carbon–nitrogen double bond of isocyanates have also been reported. The first example of a carbene addition to an isocyanate was due to Sheehan (equation 86)[435,436].

An unusual nucleophilic carbene addition to the carbon–nitrogen double bond of aryl isocyanates and of aryl isothiocyanates has recently been reported by Hoffmann and coworkers (equation 87)[437]. In this reaction, hydantoin formation (which results from the reaction of dimethoxycarbene with p-tolylisocyanate) occurs 11-times faster than does dithiohydantoin formation from the corresponding reaction with p-tolylisothiocyanate. This is true regardless of the mode of generation of the carbene, i.e. the relative rates of addition to p-tolylisocyanate and p-

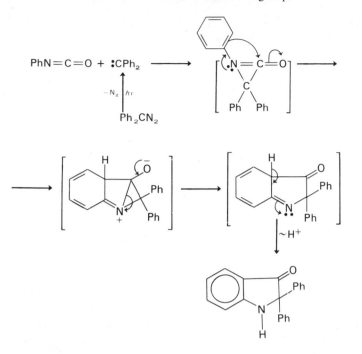

(86)

tolylisothiocyanate are the same whether dimethoxycarbene is generated via **109** (equation 87) or from trimethyl orthoformate.

The reaction of phenyl(trihalomethyl)mercury compounds with carbodiimides results in fragmentation in a manner analogous to that observed

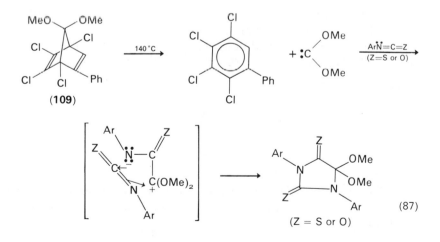

(87)

for the corresponding reaction with isocyanates (equation 88)[404,415]. Other products which were observed in this reaction probably arose via

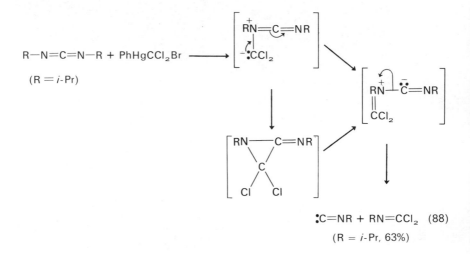

$$:C{=}NR + RN{=}CCl_2 \quad (88)$$

(R = *i*-Pr, 63%)

a complex reaction of PhHgCCl₂Br with the product isonitrile. Carbodiimides were found to be more reactive than RN=CCl₂ but less reactive than isonitriles toward PhHgCCl₂Br.

The reaction of azirines with diazomethane at room temperature leads ultimately to the formation of allyl azides[438,439] and diphenyldiazomethane[440]. These reactions probably involve 1,3-dipolar cycloaddition of the diazoalkane to the carbon–nitrogen double bond of the azirine, resulting in initial 1,2,3-triazoline formation. Subsequent decomposition of the triazoline leads to allyl azide formation. Accordingly, it is unlikely that carbenes are directly involved in these reactions.

Carbene and carbenoid additions of :CCl₂ to azirines have recently been observed by Hassner and coworkers (equation 89)[441]. In these reactions,

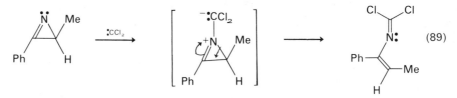

none of the azabicyclobutane which would be expected to arise via simple addition of :CCl₂ to the azirine carbon–nitrogen double bond was observed.

The reaction of carbethoxycarbenoid with *N*-benzylideneaniline has been reported as giving ethyl 3-anilinocinnamate, presumably via the intermediate aziridine[442]. However, a later study has revealed that the formation of ethyl 3-anilinocinnamate in this reaction probably does not occur via the intermediate aziridine[443]. The question of the reaction mechanism of ethyl diazoacetate with imines has recently received attention[444]. Ethyl diazoacetate reacts with enolizable imines in the presence of transition-metal catalysts to form small amounts of the corresponding aziridine along with a number of other products. The corresponding reaction with non-enolizable imines generally affords both the corresponding aziridine and an *N*-substituted ethyl 3-aminocinnamate (e.g., equation 90)[444]. From a study of the reactions of ethyl diazoacetate

$$PhCH=NPh + N_2CHCO_2Et \xrightarrow[\text{S}]{\text{Cu, 80 °C, 24 h}}$$

Ph

CO$_2$Et   Ph

(15%)

+

PhNH   CO$_2$Et

C=C

Ph   H

(40%)   (90)

with a number of non-enolizable imines, it was concluded that the aziridine-forming addition step is both stereospecific and subject to kinetic control.

Additions to carbon–phosphorus double bonds, although relatively rare, have nevertheless occasionally been reported[445]. In one example, electrophilic attack by dihalocarbenes upon phosphorus ylides has been found to afford haloolefins (equation 91)[446—448]. Interestingly, the

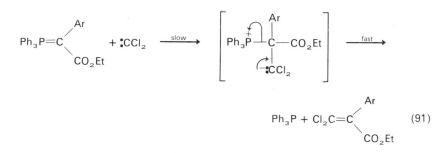

$$Ph_3P + Cl_2C=C\begin{array}{c}Ar\\ \\CO_2Et\end{array}$$ (91)

corresponding reaction of phenylchlorocarbene with $Ph_3P^+ - \overset{\cdot\cdot}{C}HCO_2Et$ afforded only ethyl *trans*-β-chlorocinnamate under conditions where *cis*–*trans* isomerization of the cinnamate ester does not occur[448]. The

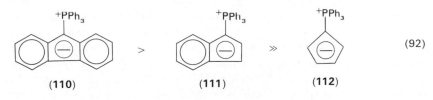

$$> \qquad (92)$$

(110) (111) (112)

relative reactivity of ylides **110**, **111**, and **112** toward reaction with dichlorocarbene followed the order shown in equation (92) (in fact, **112** was found to be unreactive toward dichlorocarbene). This observed reactivity sequence was rationalized in terms of calculated pi-electron localization energy differences within **110**, **111**, and **112**[446].

Finally, a novel, base-promoted reaction of dihaloalkanes with phospholes has recently been reported (equation 93)[449]. A mechanism for this

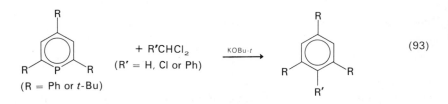

$$(93)$$

reaction has been suggested which involves attack of $R'\overset{..}{\overset{..}{C}}Cl_2$ on the phosphorus atom, followed by norcaradiene formation (**113**, equation 94).

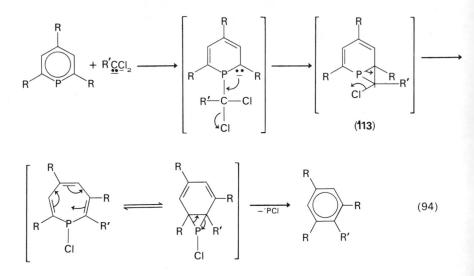

$$(94)$$

Alternatively, the norcaradiene intermediate (113) could be generated directly via carbene (or carbenoid) addition of R′—C̈—Cl to the phosphole carbon–phosphorus (aromatic) double bond.

IV. REACTIONS WITH X=Y

A. Reactions with N=N

Although relatively rare, a few examples of carbene and carbenoid additions to nitrogen–nitrogen double bonds have been reported[450]. One of the earliest examples of this type of process is the thermal reaction of ethyl diazoacetate with diethyl azodicarboxylate, which was reported as giving the corresponding diaziridine (equation 95)[451]. Recently, Seyferth

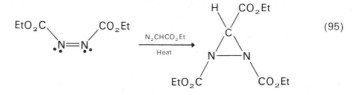

(95)

and Shih have found that dichlorocarbene (generated either via decarboxylation of sodium trichloroacetate or via thermal decomposition of PhHgCCl₂Br) enters into reaction with diethyl azodicarboxylate[452]. However, instead of the expected diaziridine, an isomeric product, $(EtO_2C)_2—N—N=CCl_2$, was actually isolated. A tentative mechanism which may involve an intermediate 1,3,4-oxadiazoline[453] (114) could account for the observed course of this reaction (equation 96)[452]. Similar

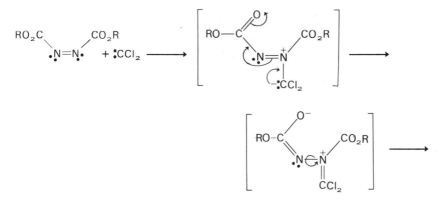

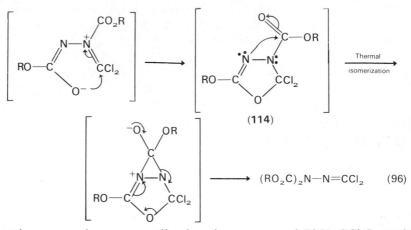

(114)

$$(RO_2C)_2N-N=CCl_2 \quad (96)$$

reactions occur between azodicarboxylate esters and $PhHgCCl_2Br$ and $PhHgCCBr_3$, leading to the formation of 2-halo-1,3,4-oxadiazolin-5-ones.

In a recent extension of the foregoing study, Seyferth and coworkers have investigated the reaction of azoarenes and azoxyarenes with $PhHgCCl_2Br$[430]. With the former substrate, cleavage of the nitrogen–nitrogen double bond occurs, affording $ArN=CCl_2$ and an arylnitrene (equation 97). The reaction of $PhHgCCl_2Br$ with azoxybenzene (1:1)

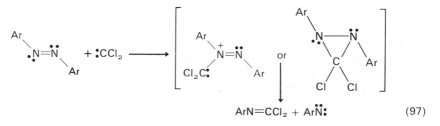

$$ArN=CCl_2 + Ar\ddot{N}: \quad (97)$$

resulted in initial deoxygenation by dichlorocarbene[454] to afford the corresponding azoarene (equation 98). The use of excess $PhHgCCl_2Br$ in this reaction increased the yield of the aziridine to 35%, and no

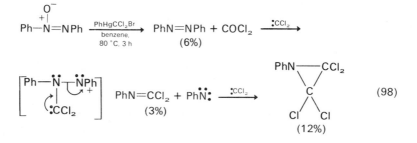

$PhN=CCl_2$ was obtained when an excess of the mercurial was so employed.

The reaction of unsaturated carbenes with the nitrogen–nitrogen double bond in azobenzene has recently been studied[413]. A 1:1 adduct was isolated, but its spectral properties did not correspond with those expected for the diazirine (**115**). It was suggested that the expected diazirine might have been formed, but that it may have subsequently suffered intramolecular rearrangement (equation 99).

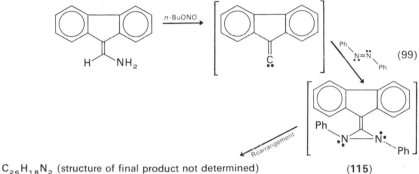

$C_{26}H_{18}N_2$ (structure of final product not determined)   (**115**)

Fluorenylidene has been found to add to the nitrogen–nitrogen double bond of N-phenylazodicarboximide to afford a 1,3-dipole[445]. This species dimerizes when treated with acetic acid, and it reacts readily with a variety of dipolarophiles to afford [2 + 3] cycloadducts (equation 100).

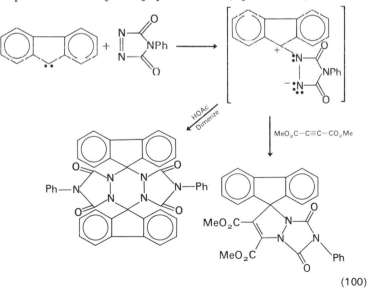

(100)

B. Reactions with N=O and N=S

There do not appear to be any examples of additions of carbenes to nitrogen–oxygen double bonds. The first example of carbene addition to a non-carbon-containing multiple bond was reported by Stoffer and Musser, who observed products which were probably formed by addition of diphenylcarbene to the nitrogen–sulphur double bond of N-sulphinyl-aniline (equation 101)[456].

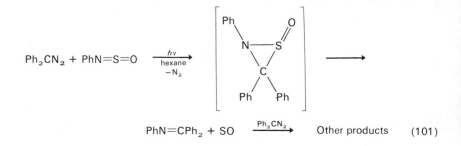

$$PhN{=}CPh_2 + SO \xrightarrow{Ph_2CN_2} \text{Other products} \qquad (101)$$

V. MISCELLANEOUS REACTIONS WITH X=Y

A. 1,3-Dipolar Cycloadditions of Carbenes to X=Y

Under appropriate structural circumstances, carbenes (or carbenoids) can react as dipolar species with unsaturated substrates ('dipolarophiles') to afford heterocyclic compounds via 1,3-dipolar cycloaddition. An example is provided by carbalkoxycarbenes which can function in the manner indicated in equation (102). However, there are relatively few *bona*

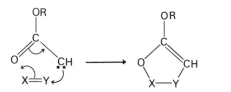

$$(102)$$

fide examples of 1,3-dipolar cycloadditions of carbenes or carbenoids to unsaturated (double-bonded) systems. Some reasons for the observed reluctance of, e.g. carbalkoxycarbenes, to enter into 1,3-dipolar cyclo-addition reactions with unsaturated substrates have been discussed recently[457].

One example of this type of reaction is provided by the photolytic decomposition of ethyl trifluoroacetyldiazoacetate, which provides a carbene (116) which is capable of reacting with acetone to afford a 1,3-dioxole (equation 103)[458–460]. Carbene 116 also undergoes 1,3-dipolar cycloaddition with ketene[461].

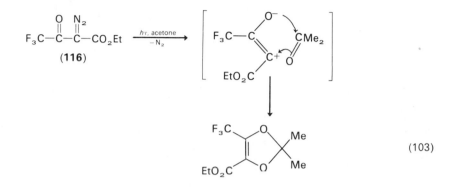

(103)

A similar reaction has been observed for reactions of the carbene (117) derived from tetrachlorobenzen-1,2-diazooxide with ketones and with carbon disulphide (equation 104)[462]. Carbene 117 can also be trapped as

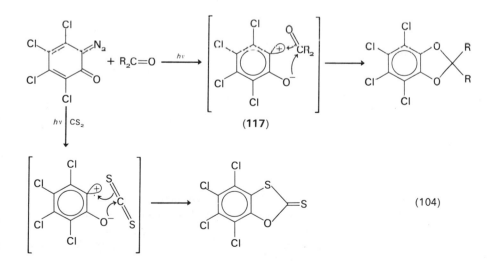

(104)

a 1,3-dipole by olefins[463,464] and by phenyl isocyanate[465] (equation 105).

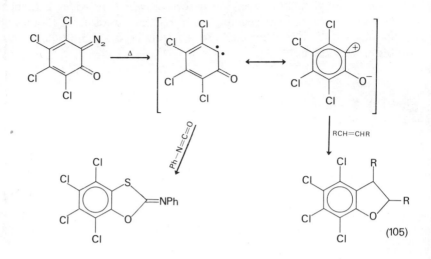

(105)

1,3-Dipolar cycloaddition of carbene **118** to carbon disulphide has also been observed (equation 106)[466].

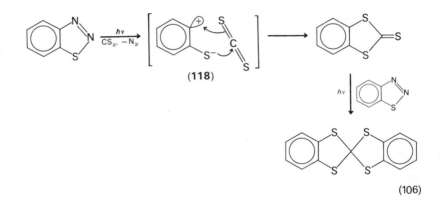

(106)

The thermal decomposition of diazoacetophenone in the presence of benzonitrile has been studied by Huisgen and coworkers[467]. Interestingly, the major reaction at 145–155°C is a Wolff rearrangement of the keto-carbene (**119**) leading to the formation of phenylketene. The ketene is then trapped by additional ketocarbene to afford the 1,3-cycloaddition product (**120**) and a dimeric product (**121**, equation 107).

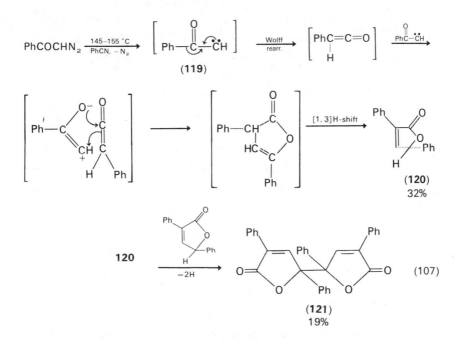

B. 1,4-Additions of Carbenes to Conjugated Systems

Jones and his collaborators have recently assessed the known examples of 1,4-carbene additions to conjugated systems[88]. The number of these is exceedingly small; indeed, until very recently, only one *bona fide* example of a 1,4-carbene addition to a diene system had been reported, namely that which arises via reaction of dicyanocarbene with cyclooctatetraene (**7**, equation 6)[141]. The results of dilution studies indicated that the 1,2-addition product (equation 6) was formed via singlet dicyanocarbene addition to cyclooctatetraene, whereas the 1,4-adduct (**7**) arose chiefly via reaction of the corresponding triplet carbene with this substrate[141].

Mitsuhashi and Jones have recently reported a second such example[468]. 2,3-Diphenylcyclopropenylidene has been found to react with tetracyclone to afford a product (**123**) which was probably formed via an intermediate 1,4-cycloadduct (**122**, equation 108)[468].

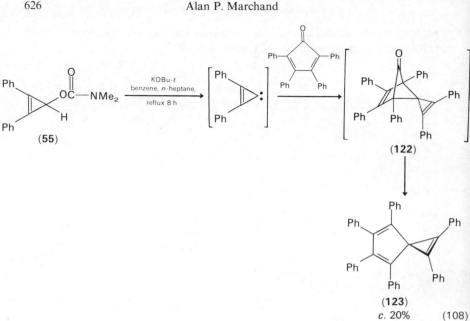

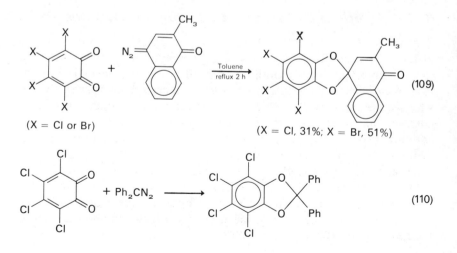

There exist a few reports of 1,4-additions of :CXY to heteroatom-containing conjugated systems. In most of these cases, however, it is not clear whether a divalent carbon intermediate is necessarily involved in the addition process. Two examples of potential 1,4-carbene additions to conjugated diketones are shown in equations (109)[469] and (110)[470].

Net 1,4-addition of carbethoxycarbenoid to α,β-unsaturated ketones has been reported (equation 111)[471,472]. A similar 1,4-addition of difluoro-

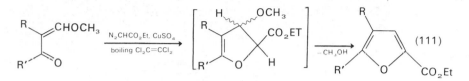

(111)

carbene to steroidal β-methoxyenones has also been reported[473].

Ketene has been found to react with diazoacetophenone to afford a 1,4-addition product (equation 112)[474].

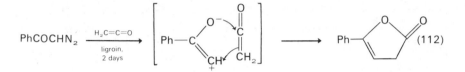

(112)

Diazoalkanes react with thioacyl isocyanates to afford 1,4-cycloadducts (equation 113)[475]. Recently, 1,4-addition of methylene to a 'sulphur

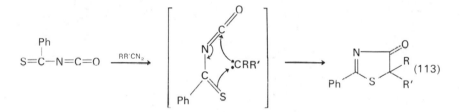

(113)

diimide' has been reported (equation 114)[476].

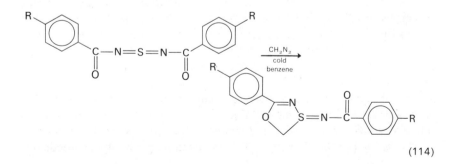

(114)

Evidence bearing on the 'carbenoid activity'[477] of benzonitrile oxide has recently been reported by LoVecchio and coworkers[478]. Net 1,4-addition of benzonitrile oxide to the arylidene double bond of 3-phenyl-4-aryliden-isoxazol-5-ones has been observed (equation 115)[478].

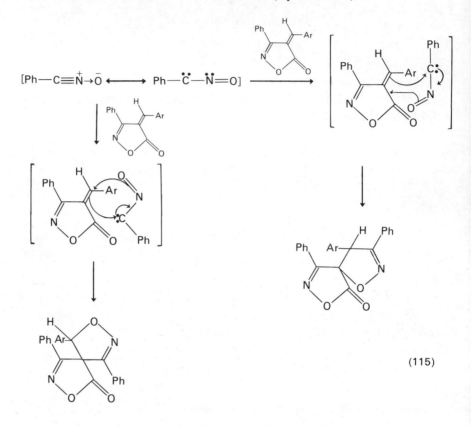

(115)

Seyferth and Shih have observed 1,4-addition of dichlorocarbene (generated via $PhHgCCl_2Br$ or $PhHgCCl_3$) to $PhC(O)—N=N—C(O)Ph$ and to $EtO_2C—N=C(CO_2Et)_2$[479]. However, only simple carbon–carbon double bond addition by dichlorocarbene occurred when the corresponding reaction was attempted utilizing methyl acrylate or mesityl oxide as substrates.

Two other novel reactions merit consideration in this section. The first of these is the unusual cycloaddition of methylene to thiocarbonyl compounds which leads to the formation of cyclopenta-1,3-dithiolane derivatives (equation 116)[421,480]. This reaction probably proceeds via

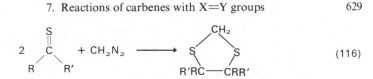

$$(116)$$

initial attack of diazomethane on the thiocarbonyl compound; accordingly, it is unlikely that carbene intermediates are directly involved in this reaction. The second reaction is the novel cycloaddition of dihalocarbenes to norbornadiene (which leads to the corresponding [2 + 2 + 2] cyclo-adducts as indicated in equation 117)[481,481a].

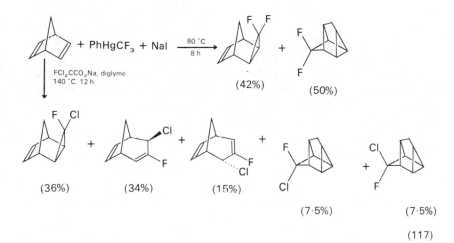

$$(117)$$

VI. ADDENDUM

The material presented in this section deals mainly with highlights of pertinent literature published between April 1974 and November 1974. Although coverage is necessarily not exhaustive, the references cited in this section should serve to indicate the directions taken by research current at the time of publication of this review.

Linear free energy correlations for dihalocarbene additions to carbon–carbon double bonds[482,483] and carbon–nitrogen double bonds[483] have been reported. Utilizing a dual substituent parameter equation[484], Moss has determined the relative selectivities of free carbenes, :CXY, toward addition to simple alkenes. The following order of increasing relative electrophilic selectivities was thereby obtained: H_3C—$\ddot{C}$—Cl > Ph$\ddot{C}$Br >

PhČCl > :CCl$_2$ > :CF$_2$[482]. Hammett analysis of difluorocarbene additions to substituted styrenes in benzene at 80°C afforded $\rho = -0.57$, suggesting that this carbene species is indeed an electrophile. Interestingly, the electrophilicity of :CF$_2$ relative to :CCl$_2$ was found to be less when styrenes were employed as substrates than was found for the case of the corresponding additions to alkylethylenes[482].

Recently performed kinetic studies reveal that electrophilic carbenes such as dichlorocarbene react more readily with carbon–nitrogen double bonds than with carbon–carbon double bonds[482]. The results of competition reactions suggest that dichlorocarbene adds approximately 1·5-times more rapidly to benzylideneaniline than to 1,1-diphenylethylene[482].

A number of carbenoid additions to olefins have recently been studied. The first successful methylenation involving a diphenylcarbenoid, generated via treatment of diphenyldiazomethane with zinc chloride, has been reported[485]. Reaction of this carbenoid with excess cyclopentadiene at 22°C affords 6,6-diphenylbicyclo[3.1.0]hex-2-ene in moderate yield (35%) along with benzophenone azine (24%) and benzophenone (29%). The latter two products were thought to be formed via reaction of the carbenoid (an α-chlorozinc chloride intermediate, Ph$_2$C(Cl)ZnCl) with diphenyldiazomethane and with water respectively[485]. In another study, it was reported that decomposition of diphenyldiazomethane by copper(II) bromide in acetonitrile solution at 30°C likewise leads to the formation of benzophenone and benzophenone azine along with a small amount of tetraphenylethylene[486].

The acid-induced reaction of aryldiazomethanes with olefins affords cyclopropanes along with other products[487]. The methylenated products are formed stereospecifically in this reaction. A mechanism was formulated which involves slow proton transfer from the acid to the diazo compound. Marked variations in *syn* addition stereoselectivity with change in the nature of the acid or solvent employed suggested the intermediacy of a carbenoid (rather than a free carbene) in this reaction[487].

The first example of asymmetric induction in methylenation of olefins via carbene transfer from an optically-active iron complex has been reported[487a]. Also, asymmetric induction in a methylenation reaction (70% enantioselective) has been observed for the reaction of ethyl diazoacetate with styrene when carried out in the presence of bis-[(+)-camphorquinonedioximato]cobalt(II) catalyst[488]. By way of contrast, Wulfman and coworkers[489] have observed that the reaction of ethyl diazoacetate with styrene in the presence of a chiral catalyst. 2-phenoxy-5, 10, 10-trimethyl-1,3-dioxa-2-phosphatricyclo [5.2.1^{5,8} 0^{4,9}]-decanecopper(I) iodide (**124**), afforded optically-inactive cyclopropanes as products (equation 118). Wulfman's results also stand in conflict with Moser's earlier findings[144,145] (see equation 7 and discussion in Section II.B).

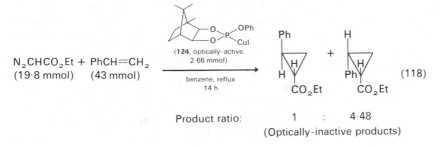

$$N_2CHCO_2Et + PhCH=CH_2$$

(19·8 mmol)   (43 mmol)

(124, optically-active, 2·66 mmol)

benzene, reflux 14 h

(118)

Product ratio: 1   :   4·48

(Optically-inactive products)

Wulfman[489] has suggested that Moser's results might be explained in terms of an asymmetric solvation effect as a result of Moser's having originally employed large quantities of chiral ligands in his study. Destruction of complex **124** by the diazo compound is believed to occur, and the resulting products of this decomposition process are thought to be specific solvents for the transition state in the carbene transfer process[489].

A number of other carbenoid addition studies have been reported. These include reactions of copper vinylcarbenoids[490], zinc iodocarbenoid (formed via the reaction of triethylzinc with iodoform)[491], and α-fluorocarbenoids[492].

A particularly important paper by Moss and Pilkiewicz has appeared which suggests a new method for generating free carbenes (as distinguished from carbenoids) via base-induced α-elimination reactions[493]. In this method, the reaction of a phenyldihalomethane or of a trihalomethane with potassium t-butoxide (in homogeneous solution) is carried out in the presence of a crown ether, 1,4,7,10,13,16-hexaoxacyclooctadecane ('18-crown-6'). The crown ether complexes potassium ion, effectively preventing it from associating with the divalent carbon species produced in this reaction. Thus, identical olefin selectivities were shown by phenylbromocarbenes generated either via photolysis of phenylbromodiazirine or via the reaction of phenyldibromomethane with potassium t-butoxide/18-crown-6. The same α-elimination reactions when carried out in the absence of crown ether afforded phenylbromocarbenoid. This latter species demonstrated a substantially different selectivity toward olefin addition than was observed for the identical reaction when carried out in the presence of 18-crown-6. Thus, it should now be possible to determine whether the divalent carbon species produced via base-induced α-elimination is a carbene or a carbenoid simply by observing the olefin selectivity of that species in the presence and in the absence of crown ether[493]. It is anticipated that this new method will find wide application in years to come.

New evidence has recently been presented which lends support to the suggestion of a bicyclo[4.1.0]heptatriene intermediate in the phenyl-carbene-cycloheptatrienylidene rearrangement[277-288] (see equations 34 and 35 and discussion in Section II.F)[494-496]. Semi-empirical INDO calculations have been performed on phenylcarbene and on cyclohepta-trienylidene[497]. The previously-observed effect of annelation upon the phenylcarbene–cycloheptatrienylidene equilibrium[284] could be explained in terms of the calculated stabilizing effects of annelation (which preferentially stabilize phenylcarbene, thereby shifting the phenylcarbene–cycloheptatrienylidene equilibrium to the left)[497].

Seyferth and coworkers have reported a number of new reactions of phenyl(trihalomethyl)mercury compounds with double bonded functional groups. Thermolysis of phenyl(1-bromo-1,2,2,2-tetrafluoroethyl)mercury at 155°C (24 h) affords the corresponding tetrafluoroethylidene which, when generated in the presence of olefins, affords *gem*-fluoro(trifluoromethyl)cyclopropanes[498]. Phenyl(chlorobromocarbomethoxymethyl)-mercury and phenyl(dichlorocarbomethoxymethyl)mercury have been found to be useful reagents for carbene transfer of chlorocarbomethoxy-carbene[499].

The reaction of phenyl(trihalomethyl)mercury compounds with azodicarboxylate esters in benzene at 80°C affords compounds of the type $(RO_2C)_2N-N=CX_2$, where X = Cl and/or Br[500]. The mechanism shown in equation (119) was postulated for the reaction of dimethylazodicarboxy-

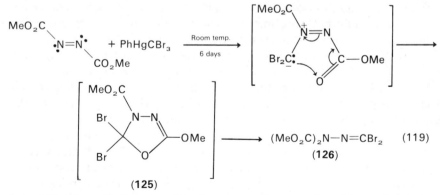

(119)

late with phenyl(tribromomethyl)mercury. Support for this mechanism was obtained via isolation of the 1,3,4-oxadiazoline intermediate **125** and through a kinetic study of the first order, thermal rearrangement of **125** to the final carbonyl halide hydrazone (**126**)[500]. Additionally, a process for adding difluorocarbene (generated via reaction of sodium iodide with alkyl- or aryl(trifluoromethyl)mercury compounds) to olefins has been patented[501].

Nikiforov and coworkers have studied a number of olefin addition reactions involving the carbene (**128**) derived from 3,5-di-*t*-butylbenzene-1,4-diazooxide, (**127**). Pyrolysis of diazooxide **127** in the presence of either *trans*-2-butene or *trans*-stilbene affords the corresponding *trans* adduct (*c.* 98% stereoselective *cis* addition)[502]. Interestingly, dilution with hexafluorobenzene was reported to have no effect on the degree of *cis* stereoselectivity of the addition reaction. This observation stands in apparent conflict with the results of an earlier study by Pirkle[99], who found that dilution with hexafluorobenzene markedly diminished the stereospecificity of photolytically-generated **128** in its addition to *cis*-2-butene (cf. Table I).

The reaction of carbene **128** derived via pyrolysis of **127** with 1,3-dienes was found to afford 1,4-adducts, whereas photolytically-generated **128** afforded 1,2-adducts at the less highly substituted double bond[503]. It is likely that the 1,4-adducts resulted via a [1,3]-sigmatropic rearrangement of initially-formed 1,2-adducts (**129**, equation 120). The results of a CIDNP

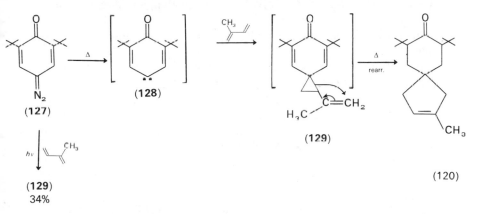

(127)

(128)

(129)

(129)
34%

(120)

study of reactions of carbene **128** with halogenated solvents suggest that this carbene reacts in a singlet electronic configuration[504].

Other recent studies by Russian investigators include the addition of dihalocarbenes to simple olefins[505–509], to acyclic dienes[510–512], to cyclopentadiene[513,514], and to the carbon–nitrogen double bond of imines[483,515]. Reactions of carbalkoxycarbenes with olefins have also received attention[516–519]. A brief carbene review by Nefedov and Ioffe has recently appeared[520].

An important study of reactions of carbethoxycarbyne with olefins has recently been reported[521]. The carbyne, produced in its doublet ground state via short wavelength photolysis of $(N_2CCO_2Et)_2Hg$, displays chemical behaviour which bears a close resemblance to that of singlet carbenes[521].

Some additional carbene addition studies merit brief mention. Additions of dihalocarbenes to α,β-unsaturated ketones have been reported[522]. The first example of epoxide formation via reaction of a diazoalkane with an ester carbonyl group has appeared[523]. Additions of halocarbenes to bicyclo[4.2.2]deca-2,4,7,9-triene have been studied[524].

Sulphinylcarbenes have been produced via low temperature decomposition of α-diazolsulphoxides. Benzenesulphinylcarbene thus generated has been found to add to olefins in a singlet electronic configuration, displaying a high degree of *anti* stereoselectivity in the addition process (the ratio of *anti* to *syn* addition is 34:1 for addition to cyclohexene and $> 99:1$ for addition to *cis*-2-butene)[525].

The electrophilic character of cyclopentadienylidene has recently been demonstrated via a study of its additions to *meta*- and *para*-substituted styrenes. Correlation of the relative rates of addition of singlet cyclopentadienylidene (generated photolytically from diazocyclopentadiene) to these styrenes with Hammett σ afforded $\rho = -0.76 \pm 0.10$; this result confirms the expectation that cyclopentadienylidene is indeed an electrophile[526].

A versatile procedure for the specific 'insertion' of carbenes into carbon–carbon bonds has recently been reported[527]. The product of this carbene 'insertion' process actually arises via a two-step process, namely addition of the carbene to an enamine ester, followed by hydrolysis of the resulting methylenated adduct. The net result is conversion of a β-keto ester to a γ-keto ester in which the β-carbon atom is derived from the carbene (equation 121)[527].

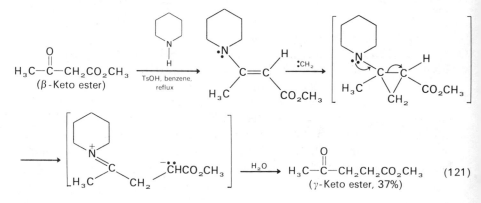

Some years ago, it was reported that a benzocyclopropene resulted from the reaction of 1,4-benzoquinone N,N'-dibenzenesulphonylimine with diphenyldiazomethane[528]. This reaction has recently been reinvestigated by two groups[529,530], and the major reaction product has been identified as a substituted bicyclo[4.1.0]-3-heptene rather than a benzocyclopropene (equation 122).

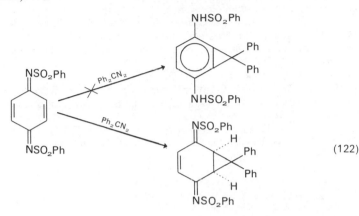

(122)

VII. ACKNOWLEDGEMENTS

I thank Professors D. S. Wulfman, P. J. Stang, J. A. Langrebe, and Dr. H. D. Roth for providing information and results in advance of publication. Financial support in the form of a summer fellowship, which I hereby gratefully acknowledge, was provided by the University of Oklahoma Graduate College. Finally, I thank my wife, Dr. Nancy Wu Marchand, for her invaluable assistance with the preparation of the manuscript and for other contributions.

VIII REFERENCES

1. J. I. G. Cadogan and M. J. Perkins in *The Chemistry of Alkenes* (Ed. S. Patai), Interscience, New York, 1964, Chapter 9, pp. 585–680.
2. H. W. Wanzlick, *Angew. Chem.*, **74**, 129 (1962).
3. C. C. Kuo, *Hua Hsueh Tung Pao*, 27 (1962); *Chem. Abstr.*, **58**, 12377 (1963).
4. P. Miginiac, *Bull. Soc. Chim. France*, 2000 (1962).
5. E. Chinoporos, *Chem. Rev.*, **63**, 235 (1963).

6. E. Chinoporos, *Chim. Chron. A,* **28**, 168 (1963).
7. W. E. Parham and E. E. Schweizer, *Org. React.,* **13**, 55 (1963).
8. H. Kloosterziel, *Chem. Weekbl.,* **59**, 77 (1963).
9. W. Kirmse, *Carbene Chemistry,* Academic Press, New York, 1964.
10. J. Hine, *Divalent Carbon,* Ronald Press, New York, 1964.
11. H. M. Frey, *Progr. React. Kinet.,* **2**, 131 (1964).
12. W. Kirmse, *Progr. Org. Chem.,* **6**, 164 (1964).
13. J. A. Bell, *Progr. Phys. Org. Chem.,* **2**, 1 (1964).
14. W. B. DeMore and S. W. Benson, *Advan. Photochem.,* **2**, 219 (1964).
15. C. W. Rees and C. E. Smithen, *Advan. Heterocycl. Chem.,* **3**, 57 (1964).
16. A. Ledwith, *The Chemistry of Carbenes,* Royal Inst. Chem. Lecture Series, No. 4, 1964.
17. W. Kirmse, *Angew. Chem., Int. Ed.,* **4**, 1 (1965).
18. B. J. Herold and P. P. Gaspar, *Fortschr. Chem. Forsch.,* **5**, 89 (1965).
19. G. G. Rozantsev, A. A. Fainzil'berg, and S. S. Novikov, *Russ. Chem. Rev.,* **34**, 69 (1965).
20. W. Reid and H. Mengler, *Fortschr. Chem. Forsch.,* **5**, 1 (1965).
21. E. Müller, H. Kessler, and B. Zeeh, *Fortschr. Chem. Forsch.,* **7**, 128 (1966).
22. U. Schöllkopf, *Angew. Chem., Int. Ed.,* **7**, 588 (1968).
23. G. Köbrich, *Angew. Chem., Int. Ed.,* **6**, 41 (1967).
24. G. L. Closs, *Top. Stereochem.,* **3**, 193 (1968).
25. T. L. Gilchrist and C. W. Rees, *Carbenes, Nitrenes, and Arynes,* Appleton-Century-Crofts, New York, 1969.
26. W. Kirmse, *Chem. Unserer Zeit,* **3**, 184 (1969).
27. W. Kirmse, *Carbene, Carbenoide, und Carbenanaloge,* Verlag-Chemie, Weinheim, 1969.
28. R. A. Moss, *Chem. Eng. News,* **47**, 60 (June 16, 1969); *Chem. Eng. News,* **47**, 50 (June 30, 1969).
29. D. Bethell, *Advan. Phys. Org. Chem.,* **7**, 153 (1969).
30. R. A. Moss in *Selective Organic Transformations,* Vol. 1 (Ed. B. S. Thyagarajan), Wiley, New York, 1970, pp. 35–88.
31. V. Dave and E. W. Warnhoff, *Org. React.,* **18**, 217 (1970).
32. L. P. Danilkina, M. I. Komendantov, R. R. Kostikov, T. V. Mandel'shtam, V. V. Razin, and E. M. Kharicheva. *Vestn. Leningrad Univ., Fiz. Khim.,* 123 (1970); *Chem. Abstr.,* **73**, 34367x (1970).
33. G. W. Cowell and A. Ledwith, *Quart. Rev.,* **24**, 119 (1970).
34. A. M. Van Leusen and J. Strating, *Quart. Rep. Sulfur Chem.,* **5**, 67 (1970).
35. C. W. Jefford, *Chimia,* **24**, 357 (1970).
36. W. Kirmse, *Carbene Chemistry,* 2nd ed., Academic Press, New York, 1971.
37. C. G. Krespan and W. J. Middleton, *Fluorine Chem. Rev.,* **5**, 57 (1971).
38. C. A. Buehler, *J. Chem. Education,* **49**, 239 (1972).
39. G. Köbrich, *Agnew. Chem., Int. Ed.,* **11**, 473 (1972).
40. W. J. Baron, M. R. DeCamp, M. E. Hendrick, M. Jones, Jr., R. H. Levin, and M. B. Sohn in *Carbenes,* Vol. 1 (Eds. M. Jones, Jr. and R. A. Moss), Wiley, New York, 1973, Chapter 1, pp. 1–151.
41. R. A. Moss in *Carbenes,* Vol. 1 (Eds. M. Jones, Jr. and R. A. Moss), Wiley, New York, 1973, Chapter 2, pp. 153–304.
42. G. W. Griffin and N. R. Bertoniere in *Carbenes,* Vol. 1 (Eds. M. Jones, Jr. and R. A. Moss), Wiley, New York, 1973, Chapter 3, pp. 305–349.

42a. N. R. Bertoniere and G. W. Griffin in *Organic Photochemistry*, Vol. 3 (Ed. O. L. Chapman), Marcel Dekker, New York, 1973, Chapter 2, pp. 115–195.
43. T. L. Gilchrist, *Chem. Ind. (London)*, 881 (1973).
43a. D. Bethell in *Organic Reactive Intermediates* (Ed. S. P. McManus), Academic Press, New York, 1973, Chapter 2, pp. 61–126.
44. H. E. Simmons, T. L. Cairns, S. A. Vladuchick, and C. M. Hoiness, *Org. React.*, **20**, 1 (1973).
45. B. M. Trost, *Fortschr. Chem. Forsch.*, **41**, 1 (1973).
46. A. P. Marchand and N. M. Brockway, *Chem. Rev.*, **74**, 431 (1974).
47. P. P. Gaspar and G. S. Hammond in W. Kirmse, ref. 9, Chapter 12, pp. 235–274.
48. J. F. Harrison in W. Kirmse, ref. 36, Chapter 5, pp. 159–193.
49. M. J. S. Dewar, R. C. Haddon, and P. K. Weiner, *J. Amer. Chem. Soc.*, **96**, 253 (1974).
50. G. Herzberg and J. W. C. Johns, *Proc. Roy. Soc., Sec. A*, **295**, 107 (1967).
51. E. Wasserman, W. A. Yager, and V. J. Kuck, *Chem. Phys. Lett.*, **7**, 409 (1970).
52. E. Wasserman, V. J. Kuck, R. S. Hutton, and W. A. Yager, *J. Amer. Chem. Soc.*, **92**, 7491 (1970).
53. For a review, see W. Kirmse, ref. 36, Chapter 8, pp. 267–362.
54. P. S. Skell and A. Y. Garner, *J. Amer. Chem. Soc.*, **78**, 3409 (1956).
55. P. S. Skell and A. Y. Garner, *J. Amer. Chem. Soc.*, **78**, 5430 (1956).
56. R. C. Woodworth and P. S. Skell, *J. Amer. Chem. Soc.*, **81**, 3383 (1959), and references cited therein.
57. R. Hoffmann, *J. Amer. Chem. Soc.*, **90**, 1475 (1968).
58. R. Hoffmann, D. M. Hayes, and P. S. Skell, *J. Phys. Chem.*, **76**, 664 (1972).
59. H. Fujimoto, S. Yamabe, and K. Fukui, *Bull. Chem. Soc. Japan*, **45**, 2424 (1972). 2424 (1972).
60. R. B. Woodward and R. Hoffmann, *Angew. Chem., Int. Ed.*, **8**, 781 (1969).
61. R. F. W. Bader and J. I. Generosa, *Can. J. Chem.*, **43**, 1631 (1965).
62. F. A. L. Anet, R. F. W. Bader, and A. M. van der Auwera, *J. Amer. Chem. Soc.*, **82**, 3217 (1960).
63. H. M. Frey, *J. Amer. Chem. Soc.*, **82**, 5947 (1960).
64. D. F. Ring and B. S. Rabinovitch, *J. Phys. Chem.*, **72**, 191 (1968).
65. M. Jones, Jr. and K. R. Rettig, *J. Amer. Chem. Soc.*, **87**, 4013 (1965).
66. M. Jones, Jr. and K. R. Rettig, *J. Amer. Chem. Soc.*, **87**, 4015 (1965).
67. P. S. Skell and R. C. Woodworth, *J. Amer. Chem. Soc.*, **78**, 4496 (1956).
68. P. S. Skell and R. C. Woodworth, *J. Amer. Chem. Soc.*, **78**, 6427 (1956).
69. K. R. Kopecky, G. S. Hammond, and P. A. Leermakers, *J. Amer. Chem. Soc.*, **83**, 2397 (1961).
70. K. R. Kopecky, G. S. Hammond, and P. A. Leermakers, *J. Amer. Chem. Soc.*, **84**, 1015 (1962).
71. C. McKnight, P. S. T. Lee, and F. S. Rowland, *J. Amer. Chem. Soc.*, **89**, 6802 (1967).
72. F. J. Duncan and R. J. Cvetanović, *J. Amer. Chem. Soc.*, **84**, 3593 (1962).
73. T. W. Eder and R. W. Carr, Jr., *J. Phys. Chem.*, **73**, 2074 (1969).
74. H. Kristinsson and G. W. Griffin, *J. Amer. Chem. Soc.*, **88**, 1579 (1966).
75. R. A. Moss and U.-H. Dolling, *J. Amer. Chem. Soc.*, **93**, 954 (1971).
76. W. J. Baron, M. E. Hendrick, and M. Jones, Jr., *J. Amer. Chem. Soc.*, **95**, 6286 (1973).

77. S. H. Goh, *Chem. Commun.*, 512 (1972).
78. S. H. Goh, *J. Chem. Soc., C,* 2275 (1971).
79. G. L. Closs and S. H. Goh, *J. Chem. Soc., Perkin I,* 2103 (1972).
80. R. A. Mitsch, *J. Amer. Chem. Soc.,* **87**, 758 (1965).
81. G. L. Closs and J. J. Coyle, *J. Amer. Chem. Soc.,* **87**, 4270 (1965).
82. N. C. Yang and T. A. Marolewski, *J. Amer. Chem. Soc.,* **90**, 5644 (1968).
83. R. A. Moss and A. Mamantov, *J. Amer. Chem. Soc.,* **92**, 6951 (1970).
84. D. M. Gale, W. J. Middleton, and C. G. Krespan, *J. Amer. Chem. Soc.,* **88**, 3617 (1966).
85. E. Ciganek, *J. Amer. Chem. Soc.,* **88**, 1979 (1966).
86. W. H. Atwell, D. R. Weyenberg, and J. G. Uhlmann, *J. Amer. Chem. Soc.,* **91**, 2025 (1969).
87. See footnote 3 in T. DoMinh and O. P. Strausz, *J. Amer. Chem. Soc.,* **92**, 1766 (1970).
88. M. Jones, Jr., W. Ando, M. E. Hendrick, A. Kulczycki, Jr., P. M. Howley, K. F. Hummel, and D. S. Malament, *J. Amer. Chem. Soc.,* **94**, 7469 (1972).
89. M. Reetz, U. Schöllkopf, and B. Bánhidai, *Ann. Chem.,* 599 (1973).
90. U. Schöllkopf, B. Bánhidai, and H.-U. Scholz, *Ann. Chem.,* **761**, 137 (1972).
91. P. S. Skell and J. Klebe, *J. Amer. Chem. Soc.,* **82**, 247 (1960).
92. H. Dürr and W. Bujnoch, *Tetrahedron Letters,* 1433 (1973).
93. R. A. Moss and J. R. Przybyla, *J. Org. Chem.,* **33**, 3816 (1968).
94. E. T. McBee and K. J. Sienkowski, *J. Org. Chem.,* **38**, 1340 (1973).
95. S.-I. Murahashi, I. Moritani, and M. Nishino, *J. Amer. Chem. Soc.,* **89**, 1257 (1967).
96. I. Moritani, S.-I. Murahashi, M. Nishino, Y. Yamamoto, K. Itoh, and N. Mataga, *J. Amer. Chem. Soc.,* **89**, 1259 (1967).
97. M. Jones, Jr. and W. Ando, *J. Amer. Chem. Soc.,* **90**, 2200 (1968).
98. D. O. Cowan, M. M. Couch, K. R. Kopecky, and G. S. Hammond, *J. Org. Chem.,* **29**, 1922 (1964).
99. W. H. Pirkle and G. F. Koser, *Tetrahedron Letters,* 3959 (1968).
100. W. M. Jones, B. N. Hamon, R. C. Joines, and C. L. Ennis, *Tetrahedron Letters,* 3909 (1969).
101. W. Kirmse, ref. 36, Chapter 3, pp. 85–128.
102. D. Seyferth, *Accounts Chem. Research,* **5**, 65 (1972).
103. W. von E. Doering and A. K. Hoffmann, *J. Amer. Chem. Soc.,* **76**, 6162 (1954).
104. P. S. Skell and M. S. Cholod, *J. Amer. Chem. Soc.,* **91**, 6035, 7131 (1969).
105. G. Köbrich, H. Büttner, and E. Wagner, *Angew. Chem., Int. Ed.,* **9**, 169 (1970).
106. D. Seyferth, J. M. Burlitch, R. J. Minasz, J. Y.-P. Mui, H. D. Simmons, Jr., A. J. H. Treiber, and S. R. Dowd, *J. Amer. Chem. Soc.,* **87**, 4259 (1965).
107. W. Kirmse, ref. 36, Chapter 4, pp. 129–157.
108. W. J. Baron, M. R. DeCamp, M. E. Hendrick, M. Jones, Jr., R. H. Levin, and M. B. Sohn, ref. 40, pp. 64–73.
109. W. M. Jones, *J. Amer. Chem. Soc.,* **81**, 3776 (1959).
110. A. Pudovik, R. D. Gareev, L. A. Stavrovskaya, G. I. Evstaf'ev, and A. B. Remozov, *J. Gen. Chem. U.S.S.R.,* **42**, 77 (1972), and references cited therein.
111. I. S. Lishanskii, V. I. Pomerantsev, and L. D. Turkova, *J. Org. Chem. U.S.S.R.,* **8**, 263 (1972).

112. D. H. White, P. B. Condit, and R. G. Bergman, *J. Amer. Chem. Soc.*, **95**, 1348 (1972), and references cited therein.
112a. See D. H. Aue and G. S. Helwig, *Tetrahedron Letters*, 721 (1974), and references cited therein.
113. T. V. Van Auken and K. L. Rinehart, Jr., *J. Amer. Chem. Soc.*, **84**, 3736 (1962).
114. H. Dietrich, G. W. Griffin, and R. C. Petterson, *Tetrahedron Letters*, 153 (1968).
115. R. L. Smith, A. Manmade, and G. W. Griffin, *Tetrahedron Letters*, 553 (1970).
116. R. J. Cvetanović, H. E. Avery, and R. S. Irwin, *J. Chem. Phys.*, **46**, 1993 (1967).
117. M. Jones, Jr., W. Ando, and A. Kulczycki, Jr., *Tetrahedron Letters*, 1391 (1967).
118. M. Jones, Jr., A. Kulczycki, Jr., and K. F. Hummel, *Tetrahedron Letters*, 183 (1967).
119. P. S. Skell, S. J. Valenty, and P. W. Hunter, *J. Amer. Chem. Soc.*, **95**, 5041 (1973).
120. P. S. Skell and S. J. Valenty, *J. Amer. Chem. Soc.*, **95**, 5042 (1973).
121. S. J. Valenty and P. S. Skell, *J. Org. Chem.*, **38**, 3937 (1973).
122. N. Shimizu and S. Nishida, *Chem. Commun.*, 389 (1972).
122a. N. Shimizu and S. Nishida, *J. Amer. Chem. Soc.*, **96**, 6451 (1974).
123. O. M. Nefedov, I. E. Dolgii, I. B. Shevedova, and R. N. Shafran, *Izv. Akad. Nauk S.S.S.R., Ser. Khim.*, 1885 (1972); *Chem. Abstr.*, **77**, 164071g (1972).
124. H. D. Roth, *Mol. Photochem.*, **5**, 91 (1973).
125. S. H. Pine, *J. Chem. Education*, **49**, 664 (1973).
126. H. Fischer, *Accounts Chem. Research*, **2**, 110 (1969).
127. H. R. Ward, *Accounts Chem. Research*, **5**, 18 (1972).
128. R. G. Lawler, *Accounts Chem. Research*, **5**, 25 (1972).
129. H. D. Roth, *J. Amer. Chem. Soc.*, **94**, 1761 (1972).
130. H. D. Roth, *J. Amer. Chem. Soc.*, **93**, 1527 (1971).
131. H. D. Roth, *J. Amer. Chem. Soc.*, **93**, 4935 (1971).
132. H. D. Roth, *J. Amer. Chem. Soc.*, **94**, 1400 (1972).
133. H. D. Roth, *Ind. Chim. Belg.*, **36**, 1068 (1971).
134. M. Cocivera and H. D. Roth, *J. Amer. Chem. Soc.*, **92**, 2573 (1970).
135. R. A. Moss, *J. Org. Chem.*, **31**, 3296 (1966).
136. W. M. Jones, M. H. Grasley, and W. S. Brey, *J. Amer. Chem. Soc.*, **85**, 2754 (1963).
137. M. Jones, Jr., A. M. Harrison, and K. R. Rettig, *J. Amer. Chem. Soc.*, **91**, 7462 (1969).
138. P. S. Skell and R. R. Engel, *J. Amer. Chem. Soc.*, **87**, 2493 (1965).
139. P. S. Skell and R. R. Engel, *J. Amer. Chem. Soc.*, **88**, 3749 (1966).
140. S. Kryzanowski and R. J. Cvetanović, *Can. J. Chem.*, **45**, 665 (1967).
140a. M. Jones, Jr., W. J. Baron, and Y. H. Shen, *J. Amer. Chem. Soc.*, **92**, 4745 (1970).
141. A. G. Anastassiou, R. P. Cellura, and E. Ciganek, *Tetrahedron Letters*, 5267 (1970).
142. G. L. Closs and R. A. Moss, *J. Amer. Chem. Soc.*, **86**, 4042 (1964).
143. H. Nozaki, S. Moriuti, H. Takaya, and R. Noyori, *Tetrahedron Letters*, 5239 (1966).
144. W. E. Moser, *J. Amer. Chem. Soc.*, **91**, 1135 (1969).
145. W. E. Moser, *J. Amer. Chem. Soc.*, **91**, 1141 (1969).

146. I. S. Lishanskii, V. I. Pomerantsev, N. G. Illarionova, A. S. Khatchaturov, and T. I. Vakorina, *J. Org. Chem. U.S.S.R.,* **7**, 1870 (1971).
147. H. Nozaki, S. Moriuti, M. Yamabe, and R. Noyori, *Tetrahedron Letters,* 59 (1966).
148. H. Nozaki, S. Moriuti, H. Takaya, and R. Noyori, *Tetrahedron,* **24**, 3655 (1968).
149. M. D. Cooke and E. O. Fischer, *J. Organometal. Chem.,* **56**, 279 (1973).
150. D. S. Wulfman, unpublished results; personal communication, with permission to cite.
151. B. W. Peace and D. S. Wulfman, *Chem. Commun.,* 1179 (1971).
152. B. W. Peace, F. Carman, and D. S. Wulfman, *Synthesis,* 658 (1971).
153. B. W. Peace and D. S. Wulfman, *Tetrahedron Letters,* 3799 (1971).
154. D. S. Wulfman and B. W. Peace, *Tetrahedron Letters,* 3903 (1972).
155. M. I. Komendantov, I. A. D'yakonov, and T. S. Smirnova, *J. Org. Chem. U.S.S.R.,* **2**, 561 (1966).
156. R. G. Salomon and J. K. Kochi, *J. Amer. Chem. Soc.,* **95**, 3300 (1973).
157. T. Saegusa, K. Yonezawa, I. Murase, T. Konoike, S. Tomita, and Y. Ito, *J. Org. Chem.,* **38**, 2319 (1973).
158. D. S. Wulfman, B. W. Peace, and E. K. Steffen, *Chem. Commun.,* 1360 (1971).
159. A. G. Vitenberg, I. A. D'yakonov, and A. Zindel, *J. Org. Chem. U.S.S.R.,* **2**, 1516 (1966).
160. A. G. Vitenberg, I. A. D'yakonov, and A. Zindel, *J. Org. Chem. U.S.S.R.,* **2**, 1532 (1966).
161. R. A. Moss, *Ph.D. Dissertation,* University of Chicago, 1963.
162. I. Fleming and E. J. Thomas, *Tetrahedron,* **28**, 5003 (1972).
163. G. Kobrich and R. H. Fischer, *Tetrahedron,* **24**, 4343 (1968).
164. W. Kirmse and H. Arold, *Angew. Chem., Int. Ed.,* **7**, 539 (1968).
165. R. K. Armstrong, *J. Org. Chem.,* **31**, 619 (1966).
166. W. Kirmse and M. Kapps, *Chem. Ber.,* **101**, 994 (1968).
167. W. von E. Doering and J. F. Coburn, Jr., *Tetrahedron Letters,* 991 (1965).
168. B. Bogdanović, M. Körner, and G. Wilke, *Ann. Chem.,* **699**, 1 (1968).
169. D. J. Yarrow, J. A. Ibers, Y. Tatsuno, and S. Otsuka, *J. Amer. Chem. Soc.,* **95**, 8590 (1973).
170. H. Werner and J. H. Richards, *J. Amer. Chem. Soc.,* **90**, 4976 (1968).
171. U. Burger and R. Huisgen, *Tetrahedron Letters,* 3049 (1970).
172. U. Burger and R. Huisgen, *Tetrahedron Letters,* 3057 (1970).
173. R. Huisgen and U. Burger, *Tetrahedron Letters,* 3053 (1970).
174. R. Paulissen, A. J. Hubert, and P. Teyssie, *Tetrahedron Letters,* 1465 (1972).
175. R. Paulissen, E. Hayez, A. J. Hubert, and P. Teyssie, *Tetrahedron Letters,* 607 (1974).
176. L. Friedman, R. S. Honour, and J. G. Berger, *J. Amer. Chem. Soc.,* **92**, 4640 (1970).
177. D. Seyferth, R. M. Turkel, M. A. Eisert, and L. J. Todd, *J. Amer. Chem. Soc.,* **91**, 5027 (1969).
178. R. Scheffold and U. Michel, *Angew. Chem., Int. Ed.,* **11**, 231 (1972).
179. W. von E. Doering, R. G. Buttery, R. G. Laughlin, and H. Chaudhuri, *J. Amer. Chem. Soc.,* **78**, 3224 (1956).
180. J. F. Harrison, *J. Amer. Chem. Soc.,* **93**, 4112 (1971).
181. W. Kirmse, ref. 36, pp. 288–291.

182. N. D. Epiotis, *J. Amer. Chem. Soc.*, **95**, 3087 (1973).
183. R. Hoffmann, C. C. Levin, and R. A. Moss, *J. Amer. Chem. Soc.*, **95**, 629 (1973).
184. T. Fueno, S. Nagase, K. Tatsumi, and K. Yamaguchi, *Theor. Chim. Acta*, **26**, 43 (1972).
185. N. D. Epiotis and W. Cherry, *Chem. Commun.*, 278 (1973).
186. C. W. Jefford and W. Wojnarowski, *Tetrahedron Letters*, 193 (1968).
187. C. W. Jefford and W. Wojnarowski, *Tetrahedron*, **25**, 2089 (1969).
188. I. H. Sadler, *J. Chem. Soc., B*, 1024 (1969).
189. E. V. Couch and J. A. Landgrebe, *J. Org. Chem.*, **37**, 1251 (1972).
190. T. B. Patrick, E. C. Haynie, and W. J. Probst, *J. Org. Chem.*, **37**, 1553 (1972).
191. D. Seyferth, J. Y.-P. Mui, and R. Damrauer, *J. Amer. Chem. Soc.*, **90**, 6182 (1968).
192. J. Nishimura, J. Furukawa, N. Kawabata, and M. Kitayama, *Tetrahedron*, **27**, 1799 (1971).
193. O. M. Nefedov and R. N. Shafran, *J. Gen. Chem. U.S.S.R.*, **37**, 1482 (1967).
194. W. Funasaka, T. Ando, H. Yamanaka, H. Kanehira, and Y. Shimokawa, *Symposium on Organic Halogen Compounds*, Tokyo, Nov. 29, 1967, *Abstracts*, pp. 25ff.
195. F. Casas, J. A. Kerr, and A. F. Trotman-Dickenson, *J. Chem. Soc.*, 3655 (1964).
196. F. Casas, J. A. Kerr, and A. F. Trotman-Dickenson, *J. Chem. Soc.*, 1141 (1965).
197. N. C. Craig, T. Hu, and P. H. Martyn, *J. Phys. Chem.*, **72**, 2234 (1968).
198. R. T. K. Baker, J. A. Kerr, and A. F. Trotman-Dickenson, *J. Chem. Soc., A*, 1641 (1967).
199. R. A. Moss, *J. Amer. Chem. Soc.*, **94**, 6004 (1972).
199a. R. A. Moss and D. J. Smudin, *Tetrahedron Letters*, 1829 (1974).
200. O. M. Nefedov and R. N. Shafran, *Izv. Akad. Nauk S.S.S.R., Ser. Khim.*, 538 (1965); *Chem. Abstr.*, **63**, 475h (1965).
201. C. W. Martin and J. A. Landgrebe, *Chem. Commun.*, 15 (1971).
202. C. W. Martin, J. A. Landgrebe, and E. Rapp, *Chem. Commun.*, 1438 (1971).
203. J. A. Landgrebe, C. W. Martin, and E. Rapp, *Angew. Chem., Int. Ed.*, **11**, 326 (1972).
204. W. Grimme, J. Reisdorf, W. Junemann, and E. Vogel, *J. Amer. Chem. Soc.*, **92**, 6335 (1970).
205. Private communication from E. Vogel to P. S. Skell, cited in P. S. Skell and M. S. Cholod, *J. Amer. Chem. Soc.*, **91**, 6035 (1969), footnote 22.
206. W. G. Dauben and G. H. Berezin, *J. Amer. Chem. Soc.*, **85**, 468 (1963).
207. S. Winstein and J. Sonnenberg, *J. Amer. Chem. Soc.*, **83**, 3235 (1961).
208. A. C. Cope, S. Moon, and C. H. Park, *J. Amer. Chem. Soc.*, **84**, 4843 (1962).
209. C. D. Poulter, E. C. Friedrich, and S. Winstein, *J. Amer. Chem. Soc.*, **91**, 6892 (1969).
210. J. J. Sims, *J. Amer. Chem. Soc.*, **87**, 3511 (1965).
211. J. J. Sims and L. H. Selman, *Tetrahedron Letters*, 561 (1969).
212. T. Hanafusa, L. Birladeanu, and S. Winstein, *J. Amer. Chem. Soc.*, **87**, 3510 (1965).
213. T. Hanafusa, L. Birladeanu, and S. Winstein, *J. Amer. Chem. Soc.*, **88**, 2315 (1966).

214. R. A. Moss and C. B. Mallon, *Tetrahedron Letters,* 4481 (1973).
215. G. S. Hammond, *J. Amer. Chem. Soc.,* **77**, 334 (1955).
216. M. G. Evans and M. Polanyi, *Trans. Faraday Soc.,* **32**, 1340 (1936).
217. U. Schöllkopf and H. Görth, *Ann. Chem.,* **709**, 97 (1967).
218. O. M. Nefedov, M. N. Manakov, and A. A. Ivashenko, *Izv. Akad. Nauk S.S.S.R., Otd. Khim. Nauk,* 1242 (1962); *Chem. Abstr.,* **58**, 5528 (1963).
219. R. A. Moss and A. Mamantov, *Tetrahedron Letters,* 3425 (1968).
220. A. Bezaguet and M. Bertrand, *C. R. Acad. Sci., Paris, Ser. C,* **262**, 428 (1966).
221. D. Seyferth, M. E. Gordon, J. Y.-P. Mui, and J. M. Burlitch, *J. Amer. Chem. Soc.,* **89**, 959 (1967).
222. G. L. Closs and L. E. Closs, *Angew. Chem., Int. Ed.,* **1**, 334 (1962).
223. G. L. Closs, unpublished results, reported in R. A. Moss and U.-H. Dolling, *J. Amer. Chem. Soc.,* **93**, 954 (1971).
223a. H. E. Simmons, E. P. Blanchard, and R. D. Smith, *J. Amer. Chem. Soc.,* **86**, 1347 (1964).
224. R. E. Pincock and J. I. Wells, *J. Org. Chem.,* **29**, 965 (1964).
225. C. W. Jefford, E. H. Yen, and R. Medary, *Tetrahedron Letters,* 6317 (1966).
226. C. W. Jefford and R. Medary, *Tetrahedron,* **23**, 4123 (1967).
227. C. W. Jefford, S. Mahajan, J. Waslyn, and B. Waegell, *J. Amer. Chem. Soc.,* **87**, 2183 (1965).
228. C. W. Jefford, *Proc. Chem. Soc.,* 64 (1963).
229. L. Ghosez and P. Laroche, *Proc. Chem. Soc.,* 90 (1963).
230. R. C. De Selms and C. M. Combs, *J. Org. Chem.,* **28**, 2206 (1963).
231. E. Bergmann, *J. Org. Chem.,* **28**, 2210 (1963).
231a. W. Shakespeare, *Hamlet,* Act III, Scene 1.
232. G. L. Closs and G. M. Schwartz, *J. Amer. Chem. Soc.,* **82**, 5729 (1960).
233. M. Schlosser and G. Heinz, *Angew. Chem., Int. Ed.,* **7**, 820 (1968).
234. W. L. Dilling, *J. Org. Chem.,* **29**, 960 (1964).
235. E. A. LaLancette and R. E. Benson, *J. Amer. Chem. Soc.,* **85**, 2853 (1963).
236. T. J. Katz and P. J. Garratt, *J. Amer. Chem. Soc.,* **86**, 4876 (1964).
237. T. Ando, H. Yamanaka, S. Terabe, A. Horike, and W. Funasaka, *Tetrahedron Letters,* 1123 (1967).
238. P. H. Mazzochi and H. J. Tamburin, *J. Org. Chem.,* **38**, 2221 (1973).
239. G. L. Closs, R. A. Moss, and J. J. Coyle, *J. Amer. Chem. Soc.,* **84**, 4985 (1962).
240. S. H. Goh, L. E. Closs, and G. L. Closs, *J. Org. Chem.,* **34**, 25 (1969).
241. R. A. Moss, *J. Org. Chem.,* **30**, 3261 (1965).
242. U. Schöllkopf, A. Lerch, and W. Pitterhoff, *Tetrahedron Letters,* 241 (1962).
243. U. Schöllkopf and W. Pitterhoff, *Chem. Ber.,* **97**, 636 (1964).
244. U. Schöllkopf, A. Lerch, and J. Paust, *Chem. Ber.,* **96**, 2266 (1963).
245. U. Schöllkopf, G. J. Lehmann, J. Paust, and H.-D. Härtl, *Chem. Ber.,* **97**, 1527 (1964).
246. U. Schöllkopf and G. J. Lehmann, *Tetrahedron Letters,* 165 (1962).
247. U. Schöllkopf and H. Küppers, *Tetrahedron Letters,* 105 (1963).
248. W. Kirmse, ref. 36, pp. 328–342.
249. G. L. Closs and R. B. Larrabee, *Tetrahedron Letters,* 287 (1965).
250. J. Trotter, C. S. Gibbons, N. Nakatsuka, and S. Masamune, *J. Amer. Chem. Soc.,* **89**, 2792 (1967).
251. S. Masamune, *J. Amer. Chem. Soc.,* **86**, 735 (1964).
252. R. M. Coates and J. L. Kirkpatrick, *J. Amer. Chem. Soc.,* **90**, 4126 (1968).

253. J. E. Baldwin and W. D. Foglesong, *Tetrahedron Letters*, 4089 (1966).
254. J. E. Baldwin and W. D. Foglesong, *J, Amer. Chem. Soc.*, **90**, 4305 (1968).
255. B. M. Trost, R. M. Cory, P. H. Scudder, and H. B. Neubold, *J. Amer. Chem. Soc.*, **95**, 7813 (1973).
256. R. Gleiter and R. Hoffmann, *J. Amer. Chem. Soc.*, **90**, 5457 (1968).
257. R. A. Moss, U.-H. Dolling, and J. R. Whittle, *Tetrahedron Letters*, 931 (1971).
258. S.-I. Murahashi, K.-I. Hino, Y. Maeda, and I. Moritani, *Tetrahedron Letters*, 3005 (1973).
259. R. A. Moss and J. R. Whittle, *Chem. Commun.*, 341 (1969).
260. M. H. Fisch and H. D. Pierce, Jr., *Chem. Commun.*, 503 (1970).
261. G. N. Fickes and C. B. Rose, *J. Org. Chem.*, **37**, 2898 (1972).
262. P. K. Freeman and K. B. Desai, *J. Org. Chem.*, **36**, 1554 (1971).
263. T. A. Antkowiak, D. C. Sanders, G. B. Trimitsis, J. B. Press, and H. Shechter, *J. Amer. Chem. Soc.*, **94**, 5366 (1972).
264. M. J. Goldstein, *J. Amer. Chem. Soc.*, **89**, 6359 (1967).
265. R. E. Leone and P. von R. Schleyer, *Angew. Chem., Int. Ed.*, **9**, 860 (1970).
266. R. G. Bergman and V. J. Rajadhyaksha, *J. Amer. Chem. Soc.*, **92**, 2163 (1970).
267. M. Jones, Jr., S. D. Reich, and L. T. Scott, *J. Amer. Chem. Soc.*, **92**, 3118 (1970).
268. L. B. Rodewald and H.-K. Lee, *J. Amer. Chem. Soc.*, **95**, 623 (1973).
269. M. Goldstein and R. Hoffmann, *J. Amer. Chem. Soc.*, **93**, 6193 (1971).
270. H. Dürr, *Fortschr. Chem. Forsch.*, **40**, 103 (1973).
271. W. M. Jones and J. M. Denham, *J. Amer. Chem. Soc.*, **86**, 944 (1964).
272. W. M. Jones and M. E. Stowe, *Tetrahedron Letters*, 3459 (1964).
273. W. M. Jones, M. E. Stowe, E. E. Wells, Jr., and E. W. Lester, *J. Amer. Chem. Soc.*, **90**, 1849 (1968).
274. S. D. McGregor and W. M. Jones, *J. Amer. Chem. Soc.*, **90**, 123 (1968).
275. W. M. Jones and C. L. Ennis, *J. Amer. Chem. Soc.*, **89**, 3069 (1967).
276. T. Mukai, T. Nakazawa, and T. Isobe, *Tetrahedron Letters*, 565 (1968).
277. W. D. Crow and M. N. Paddon-Row, *J. Amer. Chem. Soc.*, **94**, 4745 (1972).
278. R. C. Joines, A. B. Turner, and W. M. Jones, *J. Amer. Chem. Soc.*, **91**, 7754 (1969).
279. P. O. Schissel, M. E. Kent, D. J. McAdoo, and E. Hedaya, *J. Amer. Chem. Soc.*, **92**, 2147 (1970).
280. T. Mitsuhashi and W. M. Jones, *J. Amer. Chem. Soc.*, **94**, 677 (1972).
281. K. E. Krajca, T. Mitsuhashi, and W. M. Jones, *J. Amer. Chem. Soc.*, **94**, 3661 (1972).
282. G. G. Vander Stouw, A. R. Kraska, and H. Shechter, *J. Amer. Chem. Soc.*, **94**, 1655 (1972).
283. W. M. Jones, *Angew. Chem., Int. Ed.*, **11**, 325 (1972).
284. W. M. Jones, R. C. Joines, J. A. Myers, T. Mitsuhashi, K. E. Krajca, E. E. Waali, T. L. Davis, and A. B. Turner, *J. Amer. Chem. Soc.*, **95**, 826 (1973).
285. W. J. Baron, M. Jones, Jr., and P. P. Gaspar, *J. Amer. Chem. Soc.*, **92**, 4739 (1970).
286. E. Hedaya and M. E. Kent, *J. Amer. Chem. Soc.*, **93**, 3283 (1971).
287. C. Wentrup and K. Wilczek, *Helv. Chim. Acta*, **53**, 1459 (1970).
288. J. A. Myers, R. C. Joines, and W. M. Jones, *J. Amer. Chem. Soc.*, **92**, 4740 (1970).

644   Alan P. Marchand

289. L. W. Christensen, E. E. Waali, and W. M. Jones, *J. Amer. Chem. Soc.*, **94**, 2118 (1972).
290. W. M. Jones and C. L. Ennis, *J. Amer. Chem. Soc.*, **91**, 6391 (1969).
291. E. E. Waali and W. M. Jones, *J. Amer. Chem. Soc.*, **95**, 8114 (1973).
292. K. Saito and T. Mukai, *Tetrahedron Letters*, 4885 (1973).
293. K. Saito, Y. Yamashita, and T. Mukai, *Chem. Commun.*, 58 (1974).
294. S.-I. Murahashi, I. Moritani, and M. Nishino, *Tetrahedron*, **27**, 5131 (1971).
295. Y. Yamamoto, S.-I. Murahashi, and I. Moritani, *Tetrahedron Letters*, 589 (1973).
296. R. A. LaBar and W. M. Jones, *J. Amer. Chem. Soc.*, **95**, 2359 (1973).
297. P. R. Gebert, R. W. King, R. A. LaBar, and W. M. Jones, *J. Amer. Chem. Soc.*, **95**, 2357 (1973).
298. D. M. Lemal, E. P. Gosselink, and S. D. McGregor, *J. Amer. Chem. Soc.*, **88**, 582 (1966).
299. R. W. Hoffmann, *Angew. Chem., Int. Ed.*, **10**, 529 (1971).
300. R. W. Hoffmann and H. Häuser, *Tetrahedron*, **21**, 891 (1965).
301. D. Seebach, M. Braun, and N. Du Preez, *Tetrahedron Letters*, 3509 (1973).
302. U. Schöllkopf and E. Wiskott, *Angew. Chem.*, **75**, 725 (1963).
303. A. G. Brook, H. W. Kucera, and R. Pearce, *Can. J. Chem.*, **49**, 1618 (1971).
304. H. Balli, *Angew. Chem., Int. Ed.*, **3**, 809 (1964).
305. H. W. Wanzlick and H. J. Kleiner, *Angew. Chem.*, **75**, 1024 (1963).
306. H. D. Hartzler, *J. Amer. Chem. Soc.*, **95**, 4379 (1973), and references cited therein.
307. H. Quast and E. Frankenfeld, *Angew. Chem., Int. Ed.*, **4**, 691 (1965).
308. H. Quast and S. Hünig, *Chem. Ber.*, **99**, 2017 (1966).
309. W. H. Pirkle and G. F. Koser, *J. Amer. Chem. Soc.*, **90**, 3598 (1968).
310. W. Kirmse, ref. 36, Chapter 11, pp. 407–456.
311. A. P. Marchand and N. M. Brockway, ref. 46, Section IV.
312. W. Ando, T. Yagihara, S. Kondo, K. Nakayama, H. Yamoto, S. Nakaido, and T. Migita, *J. Org. Chem.*, **36**, 1732 (1971).
313. W. Ando, S. Kondo, K. Nakayama, K. Ichibori, H. Kohoda, H. Yamato, I. Imai, S. Nakaido, and T. Migita, *J. Amer. Chem. Soc.*, **94**, 3870 (1972).
314. M. Kapps and W. Kirmse, *Angew. Chem., Int. Ed.*, **8**, 75 (1969).
315. W. Kirmse and M. Kapps, *Angew. Chem., Int. Ed.*, **4**, 691 (1965).
316. W. Kirmse and H. Arold, *Chem. Ber.*, **101**, 1008 (1968).
317. W. Kirmse, M. Kapps, and R. B. Hager, *Chem. Ber.*, **99**, 2855 (1966).
318. W. E. Parham and S. H. Groen, *J. Org. Chem.*, **31**, 1694 (1966).
319. W. E. Parham and J. R. Potoski, *J. Org. Chem.*, **32**, 275, 278 (1967).
320. W. E. Parham and S. H. Groen, *J. Org. Chem.*, **29**, 2214 (1964).
321. W. E. Parham and S. H. Groen, *J. Org. Chem.*, **30**, 728 (1965).
322. W. Ando, Y. Saiki, and T. Migita, *Tetrahedron*, **29**, 3511 (1973).
323. A. P. Marchand and N. M. Brockway, ref. 46, Section IV.C.3.
324. W. Ando, T. Yagihara, S. Tozune, and T. Migita, *J. Amer. Chem. Soc.*, **91**, 2786 (1969).
325. W. Ando, T. Yagihara, S. Tozune, I. Imai, J. Suzuki, T. Yoyama, S. Nakaido, and T. Migita, *J. Org. Chem.*, **37**, 1721 (1972).
326. W. Ando, K. Nakayama, K. Ichibori, and T. Migita, *J. Amer. Chem. Soc.*, **91**, 5164 (1969).

327. W. Ando, S. Kondo, and T. Migita, *J. Amer. Chem. Soc.,* **91**, 6516 (1969).
328. W. Ando, S. Kondo, and T. Migita, *Bull. Chem. Soc. Japan,* **44**, 571 (1971).
329. T. Migita, W. Ando, S. Kondo, H. Matsuyama, and M. Kosugi, *Nippon Kagaku Zasshi,* **91**, 374 (1970); *Chem. Abstr.,* **73**, 44856v (1970).
330. M. S. Newman and T. B. Patrick, *J. Amer. Chem. Soc.,* **91**, 6461 (1969).
331. R. A. Moss, ref. 41, p. 257.
332. P. S. Skell, J. E. Villaume, and F. A. Fagone, *J. Amer. Chem. Soc.,* **94**, 7866 (1972).
333. See footnote 2 in M. S. Newman and T. B. Patrick, *J. Amer. Chem. Soc.,* **92**, 4312 (1970).
334. H. D. Hartzler, *J. Amer. Chem. Soc.,* **83**, 4990, 4997 (1961).
335. H. D. Hartzler, *J. Org. Chem.,* **29**, 1311 (1964).
336. V. J. Shiner, Jr. and J. S. Humphrey, Jr., *J. Amer. Chem. Soc.,* **89**, 622 (1967).
337. W. J. LeNoble, Y. Tatsukami, and H. F. Morris, *J. Amer. Chem. Soc.,* **92**, 5681 (1970).
337a. R. Bloch, F. Leyendecker, and N. Toshima, *Tetrahedron Letters,* 1025 (1973).
338. H. Reimlinger and R. Paulissen, *Tetrahedron Letters,* 3143 (1970).
339. H. D. Hartzler, *J. Amer. Chem. Soc.,* **93**, 4527 (1971).
340. H. D. Hartzler, *J. Amer. Chem. Soc.,* **88**, 3155 (1966).
341. P. S. Skell, L. D. Wescott, Jr., J.-P. Golstein, and R. R. Engel, *J. Amer. Chem. Soc.,* **87**, 2829 (1965).
342. J. C. Gilbert and J. R. Butler, *J. Amer. Chem. Soc.,* **92**, 7493 (1970).
343. M. S. Newman and A. O. M. Okorodudu, *J. Amer. Chem. Soc.,* **90**, 4189 (1968).
344. M. S. Newman and A. O. M. Okorodudu, *J. Org. Chem.,* **34**, 1220 (1969).
345. M. S. Newman and T. B. Patrick, *J. Amer. Chem. Soc.,* **92**, 4312 (1970).
346. M. S. Newman and Z. ud Din, *J. Org. Chem.,* **38**, 547 (1973).
346a. See also M. S. Newman and M. C. Vander Zwan, *J. Org. Chem.,* **39**, 761 (1974).
347. T. Sakakibara, Y. Odaira, and S. Tsutsumi, *Tetrahedron Letters,* 503 (1968).
348. D. J. Northington and W. M. Jones, *Tetrahedron Letters,* 317 (1971).
349. H. D. Hartzler, *J. Amer. Chem. Soc.,* **93**, 4527 (1971).
350. P. J. Stang, personal communication with permission to cite.
350a. P. J. Stang, M. G. Mangum, D. P. Fox, and P. Haak, *J. Amer. Chem. Soc.,* **96**, 4562 (1974).
350b. P. J. Stang and M. G. Mangum, *J. Amer. Chem. Soc.,* **97**, 1459 (1975).
351. K. D. Bayes, *J. Amer. Chem. Soc.,* **83**, 3712 (1961).
352. K. D. Bayes, *J. Amer. Chem. Soc.,* **84**, 4077 (1962).
353. K. D. Bayes, *J. Amer. Chem. Soc.,* **85**, 1730 (1963).
354. R. T. Mullen and A. P. Wolf, *J. Amer. Chem. Soc.,* **84**, 3215 (1962).
355. W. M. Jones, M. H. Grasley, and W. S. Brey, Jr., *J. Amer. Chem. Soc.,* **85**, 2754 (1963).
356. H. B. Palmer and T. J. Hirt, *J. Amer. Chem. Soc.,* **84**, 113 (1962).
357. C. MacKay, P. Polak, H. E. Rosenberg, and R. Wolfgang, *J. Amer. Chem. Soc.,* **84**, 308 (1962).
358. S. Kammula and P. B. Shevlin, *J. Amer. Chem. Soc.,* **95**, 4441 (1973).
359. T. J. Hirt and H. B. Palmer, *Carbon,* **1**, 65 (1963).
360. B. D. Kybett, G. K. Johnson, C. K. Barker, and J. L. Margrave, *J. Phys. Chem.,* **69**, 3603 (1965).

361. C. Devillers, *C. R. Acad. Sci., Paris, Ser. C,* **262**, 1485 (1966).
362. R. T. K. Baker, J. A. Kerr, and A. F. Trotman-Dickenson, *J. Chem. Soc., A,* 975 (1966).
363. C. Willis and K. D. Bayes, *J. Phys. Chem.,* **71**, 3367 (1967).
364. D. G. Williamson and K. D. Bayes, *J. Amer. Chem. Soc.,* **89**, 3390 (1967).
365. D. G. Williamson and K. D. Bayes, *J. Amer. Chem. Soc.,* **90**, 1957 (1968).
366. C. Willis and K. D. Bayes, *J. Amer. Chem. Soc.,* **88**, 3203 (1966).
367. R. T. K. Baker, J. A. Kerr, and A. F. Trotman-Dickenson, *Chem. Commun.,* 358 (1965).
368. J. F. Olsen and L. Burnelle, *Tetrahedron,* **25**, 5451 (1969).
369. P. B. Shevlin and A. P. Wolf, *J. Amer. Chem. Soc.,* **92**, 406 (1970).
369a. For a current review, see P. S. Skell, J. J. Havel, and M. J. McGlinchey, *Accounts Chem. Research,* **6**, 97 (1973).
370. A. P. Wolf, *Advan. Phys. Org. Chem.,* **2**, 201 (1964).
371. R. L. Wolfgang, *Progr. Reaction Kinetics,* **3**, 99 (1965).
372. C. MacKay and R. Wolfgang, *Science,* **148**, 899 (1965).
373. J. Dubrin, C. MacKay, and R. Wolfgang, *J. Amer. Chem. Soc.,* **86**, 959 (1964).
374. M. Marshall, C. MacKay, and R. Wolfgang, *J. Amer. Chem. Soc.,* **86**, 4741 (1964).
375. J. Dubrin, C. MacKay, and R. Wolfgang, *J. Amer. Chem. Soc.,* **86**, 4747 (1964).
376. C. MacKay, J. Nicholas, and R. Wolfgang, *J. Amer. Chem. Soc.,* **89**, 5758 (1967).
377. P. B. Shevlin and A. P. Wolf, *Tetrahedron Letters,* 3987 (1970).
378. P. B. Shevlin, *J. Amer. Chem. Soc.,* **94**, 1379 (1972).
379. J. Nicholas, C. MacKay, and R. Wolfgang, *J. Amer. Chem. Soc.,* **88**, 1610 (1966).
380. T. Rose, C. MacKay, and R. Wolfgang, *J. Amer. Chem. Soc.,* **88**, 1064 (1966).
381. P. S. Skell and R. R. Engel, *J. Amer. Chem. Soc.,* **87**, 1135a (1965).
382. P. S. Skell and R. R. Engel, *J. Amer. Chem. Soc.,* **87**, 1135b (1965).
383. G. Herzberg, *Atomic Spectra and Atomic Structure,* Dover Publications, New York, 1944, p. 142.
384. P. S. Skell and R. R. Engel, *J. Amer. Chem. Soc.,* **89**, 2912 (1967).
385. P. S. Skell, J. E. Villaume, J. H. Plonka, and F. A. Fagone, *J. Amer. Chem. Soc.,* **93**, 2699 (1971).
386. For a theoretical discussion of C_1-olefin reactions, see: (a) M. J. S. Dewar, E. Haselbach, and M. Shanshal, *J. Amer. Chem. Soc.,* **92**, 3505 (1970); (b) N. Bodor, M. J. S. Dewar, and Z. B. Maksic, *J. Amer. Chem. Soc.,* **95**, 5245 (1973).
387. P. S. Skell and L. D. Wescott, Jr., *J. Amer. Chem. Soc.,* **85**, 1023 (1963).
387a. P. S. Skell and R. F. Harris, *J. Amer. Chem. Soc.,* **88**, 5933 (1966).
388. R. F. Harris and P. S. Skell, *J. Amer. Chem. Soc.,* **90**, 4172 (1968), and references cited therein.
389. W. Braun, J. R. McNesby, and A. M. Bass, *J. Chem. Phys.,* **46**, 2071 (1967).
390. A. P. Wolf and G. Stöcklin, Abstracts, *146th National Meeting of the American Chemical Society,* Denver, Colorado, January, 1964, p. 32C.
391. G. Stöcklin and A. P. Wolf, *Proceedings of the Conference on Methods of Preparing and Storing Labeled Molecules,* Brussels, Euratom, 1963.

392. D. E. Clark and A. F. Voigt, *J. Amer. Chem. Soc.* **87**, 5558 (1965).
393. J. Nicholas, C. MacKay, and R. Wolfgang, *J. Amer. Chem. Soc.,* **88**, 1065 (1966).
394. T. DoMinh, H. E. Gunning, and O. P. Strausz, *J. Amer. Chem. Soc.,* **89**, 6785 (1967).
395. O. P. Strausz, T. DoMinh, and J. Font, *J. Amer. Chem. Soc.,* **90**, 1930 (1968).
396. G. Herzberg, *Spectra of Diatomic Molecules,* 2nd ed., Van Nostrand, Princeton, New Jersey, 1961, p. 341.
397. P. W. B. Pearse and A. G. Gaydon, *The Identification of Molecular Spectra,* 3rd ed., Wiley, New York, 1963, pp. 100–101.
398. P. S. Skell and R. M. Etter, *Proc. Chem. Soc.,* 443 (1961).
399. D. Seyferth and W. Tronich, *J. Organometal. Chem.,* **18**, P8 (1969).
400. D. Seyferth and W. E. Smith, *J. Organometal. Chem.,* **26**, C55 (1971).
401. D. Seyferth and S. P. Hopper, *J. Organometal. Chem.,* **51**, 77 (1973).
401a. D. Seyferth, W. Tronich, W. E. Smith, and S. P. Hopper, *J. Organometal. Chem.,* **67**, 341 (1974).
402. D. Seyferth, J. K. Heeren, G. Singh, S. O. Grim, and W. B. Hughes, *J. Organometal. Chem.,* **5**, 267 (1966).
403. D. Seyferth, M. E. Gordon, and R. Damrauer, *J. Org. Chem.,* **32**, 469 (1967).
404. D. Seyferth and R. Damrauer, *Tetrahedron Letters,* 189 (1966).
405. D. Seyferth and W. Tronich, *J. Amer. Chem. Soc.,* **91**, 2138 (1969).
406. E. P. Moore, Jr., *U.S. Patent* 3,338,978 (August 29, 1967); *Chem. Abstr.,* **68**, 114045c (1968).
407. W. Mahler, *J. Amer. Chem. Soc.,* **90**, 523 (1968).
408. W. Mahler and P. R. Resnick, *J. Fluorine Chem.,* **3**, 451 (1973).
409. A. N. Pudovik, R. D. Gareev, and L. A. Stabrovskaya, *J. Gen. Chem. U.S.S.R.,* **40**, 668 (1970).
410. I. K. Korobitsyna, O. P. Studzinskii, and Z. I. Ugorets, *J. Org. Chem. U.S.S.R.,* **5**, 1090 (1969).
411. M. Sander, *Chem. Rev.,* **66**, 297 (1966).
412. I. K. Korobitsyna and O. P. Studzinskii, *J. Org. Chem. U.S.S.R.,* **5**, 1454 (1969).
413. J. N. Bradley and A. Ledwith, *J. Chem. Soc.,* 3480 (1963).
414. Y.-N. Kuo and M. J. Nyc, *Can. J. Chem.,* **51**, 1995 (1973).
415. D. Seyferth, R. Damrauer, H. Shih, W. Tronich, W. E. Smith, and J. Y.-P. Mui, *J. Org. Chem.,* **36**, 1786 (1971).
416. D. Seyferth and G. J. Murphy, *J. Organometal. Chem.,* **49**, 117 (1973).
417. W. J. Middleton, E. G. Howard, and W. H. Sharkey, *J. Org. Chem.,* **30**, 1375 (1965).
418. J. M. Beiner, D. Lecadet, D. Paquer, A. Thuillier, and J. Vialle, *Bull. Soc. Chim. France,* 1979 (1973).
419. J. M. Beiner, D. Lecadet, D. Paquer, and A. Thuillier, *Bull. Soc. Chim. France,* 1983 (1973).
420. A. Schönberg and E. Frese, *Chem. Ber.,* **96**, 2420 (1963), and references cited therein.
421. A. Schönberg, B. König, and E. Singer, *Chem. Ber.,* **100**, 767 (1967).
422. P. Metzner, *Bull. Soc. Chim. France,* 2297 (1973).
423. A. Schönberg and E. Frese, *Chem. Ber.,* **95**, 2810 (1962).
424. N. Latif and I. Fathy, *J. Org. Chem.,* **27**, 1633 (1962).

425. W. Kirmse, ref. 36. pp. 412–414.
426. E. K. Fields and J. M. Sandri, *Chem. Ind. (London)*, 1216 (1959).
427. P. K. Kadaba and J. O. Edwards, *J. Org. Chem.*, **25**, 1431 (1960).
428. A. G. Cook and E. K. Fields, *J. Org. Chem.*, **27**, 3686 (1962).
429. J. A. Deyrup and R. B. Greenwald, *Tetrahedron Letters*, 321 (1965).
430. See D. Seyferth, W. Tronich, and H.-M. Shih, *J. Org. Chem.*, **39**, 158 (1974), and references cited therein.
431. D. Seyferth and W. Tronich, *J. Organometal. Chem.*, **21**, P3 (1970).
432. R. Damrauer, *Ph.D. Dissertation*, Massachusetts Institute of Technology, 1967.
433. J. C. Sheehan and J. H. Beeson, *J. Amer. Chem. Soc.*, **89**, 362 (1967).
434. D. Seyferth and R. A. Woodruff, *J. Org. Chem.*, **38**, 4031 (1973).
435. J. C. Sheehan and J. W. Frankenfeld, *J. Amer. Chem. Soc.*, **83**, 4792 (1961).
436. J. C. Sheehan and I. Lengyel, *J. Org. Chem.*, **28**, 3252 (1963).
437. R. W. Hoffmann, K. Steinbach, and B. Dittrich, *Chem. Ber.*, **106**, 2174 (1973).
438. A. L. Logothetis, *J. Org. Chem.*, **29**, 3049 (1964).
439. V. Nair, *J. Org. Chem.*, **33**, 2121 (1968).
440. J. H. Bowie, B. Nussey, and A. D. Ward, *Aust. J. Chem.*, **26**, 2547 (1973).
441. A. Hassner, J. O. Currie, Jr., A. S. Steinfeld, and R. F. Atkinson, *J. Amer. Chem. Soc.*, **95**, 2982 (1973).
442. A. J. Speziale, C. C. Tung, K. W. Ratts, and A. Yao, *J. Amer. Chem. Soc.*, **87**, 3460 (1965).
443. J. A. Deyrup, *J. Org. Chem.*, **34**, 2724 (1969).
444. P. Baret, H. Buffet, and J.-L. Pierre, *Bull. Soc. Chim. France*, 2493 (1972).
445. K. Dimroth, *Fortschr. Chem. Forsch.*, **38**, 1 (1973).
446. Y. Ito, M. Okano, and R. Oda, *Tetrahedron*, **22**, 2615 (1966).
447. R. Oda, Y. Ito, and M. Okano, *Tetrahedron Letters*, 7 (1964).
448. M. Okano, Y. Ito, and R. Oda, *Bull. Inst. Chem. Research, Kyoto Univ.*, **42**, 217 (1964); *Chem. Abstr.*, **63**, 6929a (1965).
449. G. Märkl and A. Merz, *Tetrahedron Letters*, 1269 (1971).
450. W. Kirmse, ref. 36, pp. 419–420.
451. E. Müller, *Chem. Ber.*, **47**, 3001 (1914).
452. D. Seyferth and H.-M. Shih, *J. Amer. Chem. Soc.*, **94**, 2509 (1972).
453. E. Fahr, K. Königsdorfer, and F. Scheckenbach, *Ann. Chem.*, **690**, 138 (1965).
454. E. E. Schweizer and G. J. O'Neill, *J. Org. Chem.*, **28**, 2460 (1963).
455. W. Ried and S.-H. Lim, *Ann. Chem.*, 1141 (1973).
456. J. O. Stoffer and H. R. Musser, *Chem. Commun.*, 481 (1970).
457. A. P. Marchand and N. M. Brockway, ref. 46, Section VI.A.
458. F. Weygand, *Angew. Chem.*, **73**, 70 (1961).
459. F. Weygand, H. Dworschak, K. Koch, and S. Konstas, *Angew. Chem.*, **73**, 409 (1961).
460. H. Dworschak and F. Weygand, *Chem. Ber.*, **101**, 289 (1968).
461. W. Ried and H. Mengler, *Ann. Chem.*, **678**, 113 (1964).
462. R. Huisgen, G. Binsch, and H. König, *Chem. Ber.*, **97**, 2868 (1964).
463. R. Huisgen, H. König, G. Binsch, and H. J. Sturm, *Angew. Chem.*, **73**, 368 (1961).
464. G. Binsch, R. Huisgen, and H. König, *Chem. Ber.*, **97**, 2893 (1964).
465. R. Huisgen, G. Binsch, and H. König, *Chem. Ber.*, **97**, 2864 (1964).
466. R. Huisgen and V. Weberndörfer, *Experientia*, **17**, 566 (1961).

467. R. Huisgen, G. Binsch, and L. Ghosez, *Chem. Ber.*, **97**, 2628 (1964).
468. T. Mitsuhashi and W. M. Jones, *Chem. Commun.*, 103 (1974).
469. W. Ried and W. Radt, *Ann. Chem.*, **688**, 170 (1965).
470. A. Schönberg and B. König, *Chem. Ber.*, **101**, 725 (1968).
471. S. T. Murayama and T. A. Spencer, *Tetrahedron Letters*, 4479 (1969).
472. D. L. Storm and T. A. Spencer, *Tetrahedron Letters*, 1865 (1969).
473. P. Hodge, J. A. Edwards, and J. H. Fried, *Tetrahedron Letters*, 5175 (1966).
474. W. Ried and H. Mengler, *Ann. Chem.*, **678**, 113 (1964).
475. J. Goerdler and R. Schimpf, *Chem. Ber.*, **106**, 1496 (1973).
476. E. S. Levchenko and E. M. Dorokhova, *J. Org. Chem. U.S.S.R.*, **10**, 39 (1974).
477. R. Huisgen, *Angew. Chem.*, **75**, 742 (1963).
478. G. LoVecchio, G. Grassi, F. Risitano, and F. Foti, *Tetrahedron Letters*, 3777 (1973).
479. D. Seyferth and H. Shih, *J. Amer. Chem. Soc.*, **95**, 8464 (1973).
480. T. Machiguchi, Y. Yamamoto, M. Hoshino, and Y. Kitahara, *Tetrahedron Letters*, 2627 (1973).
481. C. W. Jefford, nT. Kabengele, J. Kovace, and U. Burger, *Tetrahedron Letters*, 257 (1974).
481a. C. W. Jefford, nT. Kabengele, J. Kovacs, and U. Burger, *Helv. Chim. Acta*, **57**, 104 (1974).
482. R. A. Moss and C. B. Mallon, *J. Org. Chem.*, **40**, 1368 (1975).
483. R. R. Kostikov, A. F. Khlebnikov, and K. A. Ogloblin, *J. Org. Chem. U.S.S.R.*, **9**, 2360 (1973).
484. S. Ehrenson, R. T. C. Brownlee, and R. W. Taft, *Progr. Phys. Org. Chem.*, **10**, 1 (1973).
485. (a) D. S. Crumrine, T. J. Haberkamp, and D. J. Suther, Abstracts of Papers, *168th National Meeting of the American Chemical Society*, Atlantic City, N.J., Sept. 9–13, 1974, Paper No. ORGN-32; (b) D. S. Crumrine, T. J. Haberkamp, and D. J. Suther, in press.
486. D. Bethell and M. F. Eeles, *J. Chem. Soc., Perkin II*, 704 (1974).
487. G. L. Closs and S. H. Goh, *J. Org. Chem.*, **39**, 1717 (1974).
487a. A. Davison, W. C. Krusell, and R. C. Michaelson, *J. Organometal. Chem.*, **72**, C7 (1974).
488. Y. Tatsuno, A. Konishi, A. Nakamura, and S. Otsuka, *Chem. Commun.*, 588 (1974).
489. D. S. Wulfman, N. v. Thinh, R. S. McDaniel, B. W. Peace, C. W. Heitsch, and M. T. Jones, Jr., *J. Chem. Soc., Dalton*, 522 (1975).
490. Y. Ito, K. Yonezawa, and T. Saegusa, *J. Org. Chem.*, **39**, 1763 (1974).
491. S. Miyano and H. Hashimoto, *Bull. Chem. Soc. Japan*, **47**, 1500 (1974).
492. J. Hahnfeld and D. J. Burton, Abstracts of Papers, *10th Midwest Regional Meeting of the American Chemical Society*, Iowa City, Iowa, Nov. 7–8, 1974, Paper No. ORGN-526.
493. R. A. Moss and F. G. Pilkiewicz, *J. Amer. Chem. Soc.*, **96**, 5632 (1974).
494. W. E. Billups, L. P. Lin, and W. Y. Chow, *J. Amer. Chem. Soc.*, **96**, 4026 (1974).
495. T. T. Coburn and W. M. Jones, *J. Amer. Chem. Soc.*, **96**, 5218 (1974).
496. T. T. Coburn, *Ph.D. Dissertation*, University of Florida, Gainesville, 1973.
497. R. L. Tyner, W. M. Jones, Y. Öhrn, and J. R. Sabin, *J. Amer. Chem. Soc.*, **96**, 3765 (1974).

498. D. Seyferth and G. J. Murphy, *J. Organometal. Chem.*, **52**, C1 (1973).

499. D. Seyferth and R. A. Woodruff, *J. Organometall. Chem.*, **71**, 335 (1974).

500. D. Seyferth and H.-M. Shih, *J. Org. Chem.*, **39**, 2329 (1974).

501. D. Seyferth and S. P. Hopper, *U.S. Patent* 3,803,251 (April 9, 1974); *Chem. Abstr.*, **80**, 145549x (1974).

502. B. D. Sviridov, G. A. Nikiforov, A. U. Stepanyants, V. P. Lezina, and V. V. Ershov, *Izv. Akad. Nauk S.S.S.R., Ser. Khim.*, 2052 (1973); *Chem. Abstr.*, **80**, 59552m (1974).

503. G. A. Nikiforov, B. D. Sviridov, and V. V. Ershov, *Izv. Akad. Nauk S.S.S.R., Ser. Khim.*, 373 (1974); *Chem. Abstr.*, **81**, 37311a (1974).

504. G. A. Nikiforov, Sh. A. Markaryan, L. G. Plekhanova, B. D. Sviridov, T. Pehk, E. Lippmaa, and V. V. Ershov, *Izv. Akad. S.S.S.R., Ser. Khim.*, 93 (1974); *Chem. Abstr.*, **80**, 119900k (1974).

505. O. M. Nefedov and A. A. Ivashenko, *Izv. Akad. Nauk S.S.S.R., Ser. Khim.*, 1651 (1974); *Chem. Abstr.*, **81**, 104944z (1974).

506. A. A. Retinskii, L. I. Borodin, V. V. Keiko, and S. M. Shostakovskii, *Izv. Akad. Nauk S.S.S.R., Ser. Khim.*, 1613 (1974); *Chem. Abstr.*, **81**, 135493f (1974).

507. O. M. Nefedov and E. S. Agavelyan, *Izv. Akad. Nauk S.S.S.R., Ser. Khim.*, 2045 (1973); *Chem. Abstr.*, **80**, 59120u (1974).

508. Y. V. Savinykh, V. S. Aksenov, and L. E. Tskhai, *Izv. Sib. Otd. Akad. Nauk S.S.S.R., Ser. Khim. Nauk*, 112 (1974); *Chem. Abstr.*, **80**, 132862u (1974).

509. S. M. Shostakovskii, V. S. Aksenov, and V. A. Filimoshkina, *U.S.S.R. Patent* 425,892 (April 30, 1974); *Chem. Abstr.*, **81**, 63231u (1974).

510. R. R. Kostikov, A. P. Molchanov, and A. Ya. Bespalov, *J. Org. Chem. U.S.S.R.*, **10**, 8 (1974).

511. Y. M. Slobodin, I. Z. Egenburg, and A. S. Khachaturov, *J. Org. Chem. U.S.S.R.*, **10**, 18 (1974).

512. R. R. Kostikov, A. P. Molchanov, and K. A. Ogloblin, *J. Org. Chem. U.S.S.R.*, **9**, 2473 (1973).

513. I. D. Kushina, S. F. Politanskii, V. U. Shevchuk, I. M. Gutov, A. A. Ivashenko, and O. M. Nefedov, *Izv. Akad. Nauk S.S.S.R., Ser. Khim.*, 946 (1974); *Chem. Abstr.*, **81**, 24623d (1974).

514. I. D. Kushina, O. M. Nefedov, A. A. Ivashenko, S. F. Politanskii, V. U. Shevchuk, I. M. Gutov, and A. Ya. Steinshneider, *Izv. Acad. Nauk S.S.S.R., Ser. Khim.*, 728 (1974); *Chem. Abstr.*, **81**, 12743e (1974).

515. N. S. Kozlov, V. D. Pak, and V. V. Mashevskii, *Khim. Geterotskikl. Soedin.*, 84 (1974); *Chem. Abstr.*, **80**, 120656s (1974).

516. L. P. Danilkina and R. N. Gmyzina, *J. Org. Chem. U.S.S.R.*, **10**, 129 (1974).

517. E. M. Kharicheva, T. V. Mandel'shtam, and L. M. Emel'yanova, *J. Org. Chem. U.S.S.R.*, **10**, 762 (1874).

518. I. E. Dolgii, E. A. Shapiro, and O. M. Nefedov, *Izv. Akad. Nauk S.S.S.R., Ser. Khim.*, 957 (1974); *Chem. Abstr.*, **81**, 37280q (1974).

519. I. E. Dolgii, I. B. Shevedova, and O. M. Nefedov, *U.S.S.R. Patent* 425,899 (April 30, 1974); *Chem. Abstr.*, **81**, 49326x (1974).

520. O. M. Nefedov and A. I. Ioffe, *Zh. Vses. Khim. Obshchest.*, **19**, 305 (1974); *Chem. Abstr.*, **81**, 62542j (1974).

521. O. P. Strausz, G. J. A. Kennepohl, F. X. Garneau, T. Do Minh, B. Kim, S. Valenty, and P. S. Skell, *J. Amer. Chem. Soc.*, **96**, 5723 (1974).

522. R. Barlet, *C. R. Acad. Sci. Paris, Ser. C,* **278**, 621 (1974).
523. F. M. Dean and B. K. Park, *Chem. Commun.,* 162 (1974).
524. G. Schröder, J. Thio, and J. F. M. Oth, *Tetrahedron Letters,* 3649 (1974).
525. C. G. Venier, H. J. Barager, III, and M. A. Ward, *J. Amer. Chem. Soc.,* **97**, 3238 (1975).
526. H. Dürr and F. Werndorff, *Angew. Chem., Int. Ed.,* **13**, 483 (1974).
527. H. Bieräugel, J. M. Akkerman, J. C. Lapierre Armande, and U. K. Pandit, *Tetrahedron Letters,* 2817 (1974).
528. A. Mustafa and M. Kamel, *J. Amer. Chem. Soc.,* **75**, 2934 (1953).
529. G. W. Jones, D. R. Kerur, T. Yamazaki, H. Schechter, A. D. Woolhouse, and B. Halton, *J. Org. Chem.,* **39**, 492 (1974).
530. A. G. Pinkus and J. Tsuji, *J. Org. Chem.,* **39**, 497 (1974).